Underwater Structures

Underwater Structures provides the necessary background theory of shell structures for the geometries encountered in submarine pressure hulls and deep-diving submersibles. Both analysis and design issues are considered, using analytical as well as numerical approaches. Classical thin-walled structures theory is implemented, and both elastic stability and collapse are considered for all types of structural geometries presented. These include ring-stiffened cylindrical shells, spheres, hemispherical shells, truncated cones, and torispherical end closures. This book also discusses structural design issues, including the selection of materials and the use of design codes. This book is suitable for university students at both the undergraduate and postgraduate level, and it also serves as a reference text for academic teachers, engineering professionals, and researchers.

Key Features:

- Covers various types of submarines, both naval and civilian, and deep-diving submersibles.
- Presents the latest in submersible structural design and includes both classical structures theory and finite element procedures.
- Includes coverage of corrosion and fatigue in submarine structures.

Underwater Structures
The Strength of Submarines and Deep-Diving Submersibles

Piero Caridis

CRC Press
Taylor & Francis Group
Boca Raton London New York

CRC Press is an imprint of the
Taylor & Francis Group, an **informa** business

Designed cover image: The Hellenic Navy S/M *Papanikolis*

First edition published 2025
by CRC Press
2385 NW Executive Center Drive, Suite 320, Boca Raton FL 33431

and by CRC Press
4 Park Square, Milton Park, Abingdon, Oxon, OX14 4RN

CRC Press is an imprint of Taylor & Francis Group, LLC

ISBN: 978-1-032-78306-2 (hbk)
ISBN: 978-1-032-78716-9 (pbk)
ISBN: 978-1-003-48915-3 (ebk)

DOI: 10.1201/9781003489153

Typeset in Times
by codeMantra

In memory of my parents

Contents

Preface

The purpose of this book is to provide an introduction to the theory of submarine pressure hull analysis and design as well as that of deep-diving submersibles.

Submarine pressure hull design has until recently been the responsibility of navies in a number of advanced industrial countries. In recent years, this has changed since classification societies have published rules for submarine and submersible structural design and construction so that the private sector has also become involved in these activities. As a result, the interest in pressure hull analysis and design has expanded significantly.

The emphasis of this book is on the pressure hull strength of underwater structures. These include submarines, both naval and civilian, and deep-sea submersibles. Classical thin-walled structures theory is used, and both elastic stability and collapse are considered for the structural geometries used. These include ring-stiffened cylindrical shells, spheres, hemispherical shells, truncated cones, and torispherical end closures. Structural design, including the selection of materials is discussed, as also is the use of design codes. In discussing analysis and design issues, both analytical and numerical approaches are presented.

The material can be read with either submarines or submersibles in mind. In Chapter 1, a description of modern underwater vehicles is provided, and this is supplemented with a brief historical overview of the development of both submarines and submersibles with an emphasis on structural issues. Chapter 2 deals with the different types of materials used in pressurised underwater structures, both submarines and submersibles.

Submarine pressure hull theory is discussed in the next four chapters. Chapter 3 provides an introduction to the theory of shells of revolution and also to the strength of unstiffened cylindrical shells. Chapters 4 and 5 that follow deal with the elastic stability of ring-stiffened cylindrical shells, which are the most important component of submarine pressure hulls. One chapter deals with interframe behaviour, whereas the following chapter deals with the problem of general instability. In discussing interframe strength, the theory is developed in stages, proceeding from the infinitely long ring-stiffened cylindrical shell subjected to radial pressure to that in which axial compression also acts, and in the last part of the chapter, the strength of shells with initial geometrical distortions is considered. Chapter 4 deals with the general instability (overall collapse strength) of a ring-stiffened cylindrical shell between fixed ends (transverse bulkheads). The two classical solutions to this problem by Bryant and Kendrick are presented and compared. In the last section, the effect of out of circularity on general instability is considered. Chapter 6 deals with the various collapse modes of ring-stiffened cylindrical shells. These include interframe collapse (yielding and lobar buckling) and also the effect that ring frame eccentricity (internal or external ring frames) has on both interframe and overall strength. The various collapse modes of internal ring frames and the influence of intermediate heavy frames on general instability are described. The role of initial distortions and that of residual stresses are discussed. In the last section of this chapter, a brief account of submarine accidents is provided using statistical data.

Chapter 7 deals with the strength of one other important structural type that forms part of submarine pressure hulls, that of truncated cones. These are used to connect cylindrical shells whose diameter differs and are thus encountered in most submarine pressure hulls. The mathematical background to the solution of the equations of equilibrium is provided, and this is followed by a description of several series of laboratory tests used to provide an understanding of cone behaviour, to validate the mathematical models proposed, and to provide guidance in design. The design of truncated cones is considered, and reference is made to rules and design equations. The last part of the chapter deals with structural issues pertaining to the major internal structural components, which are transverse bulkheads and decks. The function and form of these elements are described, and design issues are considered.

Chapter 8 deals with the application of the finite element method to the analysis and design of underwater pressure hulls. The use of the modified Newton–Raphson method and the arc-length method is described in order to overcome the difficulties that arise in modelling nonlinear response and inflection points. The use of finite element procedures to determine the collapse strength of ring-stiffened cylindrical shells and also to assess the effect of major internal structures is described. The case study of a small tourist submarine, whose cylindrical shell is fitted with a large number of viewing ports, is presented. In the last part of the chapter, procedures that can be used to ensure the accuracy and reliability of numerical results are described.

Chapter 9 deals with the structural design of submarine pressure hulls, with a focus on ring-stiffened cylindrical shells. The pressure hull form is discussed, and single-skin pressure hulls are compared with double-skin pressure hulls. The design philosophy followed in ring-stiffened cylindrical shells is presented following which a preliminary design procedure that is based on the theory described in Chapters 3–6 is outlined, followed by an example to illustrate its use. The use of the finite element method in pressure hull design is discussed in subsequent sections, and the results obtained from analytical design formulas are compared with those from numerical analysis. Strength predictions using classification society rules, analytical approaches, and finite element software are compared in the last section of the chapter.

In Chapter 10, issues related to the effect that the marine environment has over time on underwater pressure hulls are presented. These include corrosion and fatigue, especially corrosion–fatigue. The nature of corrosion and the use of numerical tools to predict corrosion rates in submarine pressure hulls are described and compared with analytical design formulas. The factors that affect the fatigue behaviour of submarine pressure are discussed, and the use of linear elastic fracture mechanics and the test results are presented.

Chapters 11 and 12 deal with the analysis and design of the pressure hulls of deep-diving submersibles. The pressure hull of deep-diving submersibles is generally spherical, and the classical, elastic buckling pressure of perfect spheres is obtained. Inelastic buckling is considered next, and this is followed by a description of several series of tests carried out on the buckling of spherical shells. In the next part of the chapter, barrelled shells are considered as these have potential use in underwater structures, especially in the case of multi-hulled structures. Their failure modes are considered, and in the last part of the chapter, multi-segment hulls are described, both multi-sphere and multi-barrel types.

Chapter 12 deals with the design of the pressure hulls of deep-diving submersibles. The weight-to-displacement ratio is used as a measure of merit, and a number of different pressure hull forms for use in submersibles are compared. The design of spherical shells is presented, and a procedure based on test results mentioned in Chapter 11 is described, followed by a numerical example to illustrate its use. The role of geometrical imperfections and alternative definitions for design purposes are discussed. The structural design of spherical acrylic pressure hulls is considered. In the sections that follow, the design of barrelled shells and multi-segment shells is discussed. The last part of the chapter deals with the optimum design of underwater pressure hulls, including ring-stiffened cylinders, multi-sphere pressure hulls, barrelled shells, and torispheres.

The appendices that follow provide supplementary information on submersible designs, on mathematical aspects of analytical solutions presented in the text, and on extracts of design rules relating to the design of spherical shells used in underwater structures.

Acknowledgements

The author first became acquainted with submarine structural analysis and design as an undergraduate student at Glasgow University. This led to a lifelong interest in pressure hull theory, which presents some of the most serious challenges to the structural naval architect, both at a theoretical level and at a practical level. At an early stage during the writing of this book a number of chapters were reviewed by Dr. David Chalmers, RCNC, and his assistance is gratefully acknowledged. A large part of the text has been written following the author's retirement from university teaching and as a result access to library facilities has been limited during this period. The author wishes to express his gratitude to the Library of the National Technical University of Athens that permits access to journal and other material to retired faculty staff through the use of the internet. In a number of cases librarian Mr. P. Lembesis assisted in providing material that was difficult to access through normal channels, and his assistance is gratefully acknowledged.

Lastly I wish to express my gratitude to my wife Roma for her patience in putting up with my monologues on underwater structures.

Author

Dr. Piero Caridis received his education in Greece and attended the universities of Glasgow and London (UCL) where he studied naval architecture at an undergraduate level and a postgraduate level. After his studies, he was trained and subsequently assumed the responsibilities of a marine superintendent in a shipping company in Piraeus, Greece. During this period, he was involved in repairs and maintenance of oil tankers and bulk carriers. He subsequently carried out research at Glasgow University, which led to his Ph.D. degree in nonlinear structural mechanics. He then joined the School of Naval Architecture and Marine Engineering of the National Technical University of Athens where he taught courses in ship strength and inspection, repair, and maintenance of ship structures for 30 years.

His research interests included the nonlinear behaviour of ship structures, marine accident analysis, ship repair and maintenance, and the history of ship structural design. He was a member of the International Ship and Offshore Structures Congress (ISSC) committees during the period 1991–2021. He retired from NTUA in 2016 following which he taught a number of courses at the University of Strathclyde in Glasgow, Scotland, during the period 2017–2019. From 2022 onwards, he has been contributing to online teaching courses on ship and submarine structures.

1 Evolution and Present-Day Types of Underwater Vehicles

1.1 INTRODUCTION

In this chapter, we shall discuss the evolution of the structure of vessels that operate below the surface of the sea, and following that, we shall describe the types of submersibles and submarines that operate at the present time. A large variety of such vessels exists, and it is therefore considered necessary, before focussing on structural matters, to provide a brief description of their layout and development. This has been affected by function and operational requirements; these in turn influence form, structural capacity, and materials used.

Before discussing the different types of vessels, it is useful to turn to the two terms that are currently used in this context: *submarine* and *submersible*. It is helpful to understand the origin of their current use by discussing their background. To the uninitiated, the term submersible ought to include the term submarine. Submarines have evolved from surface vessels that originally were designed so that they could operate underwater as well, whereas submersibles were and are conceived of as vessels that are designed to operate submerged, without any expectation that they remain on the surface for any significant length of time, other than for entry and exit. They are built to perform a variety of specific activities (scientific, exploratory, search, mechanical) and are appropriately designed and equipped for these.[1]

The term "submarine" was initially used as an adjective ("submarine boat"), and for this reason, it has always denoted a large, crewed vessel capable of autonomous operation underwater. The activities of submarines parallel those of surface ships but within the constraints of the underwater environment. At the same time, designers strive to take advantage of the environment within which they operate, which for naval vessels is the difficulty in detecting them and in operating for long periods submerged, without the need for recourse to the surface. An overwhelming proportion of submarines are naval vessels, although in recent years both passenger (tourist) submarines and private submarines are being built in increasing numbers. For the latter, visual contact with the underwater environment is an important attribute.

Before discussing the evolution and development of the structural features of submersibles and submarines through time, we shall refer to the features of the environment within which these vessels operate and the challenges facing designers.

1.2 CHALLENGES FACING THE STRUCTURE OF UNDERWATER VEHICLES

1.2.1 Hydrostatic Pressure

For depths that do not exceed a certain level, the effects of compressibility of seawater are negligible, and hence, the relationship between pressure and depth is linear. The total pressure at a depth h is thus given by

$$p = p_a + \rho g h \tag{1.1}$$

where p_a is the atmospheric pressure (1 atm, or approximately 0.1 MPa). At great depths, the compressibility of water does become important (Figure 1.1a), and to obtain an accurate value of pressure, an additional quadratic term has to be introduced in (1.1), so that

DOI: 10.1201/9781003489153-1

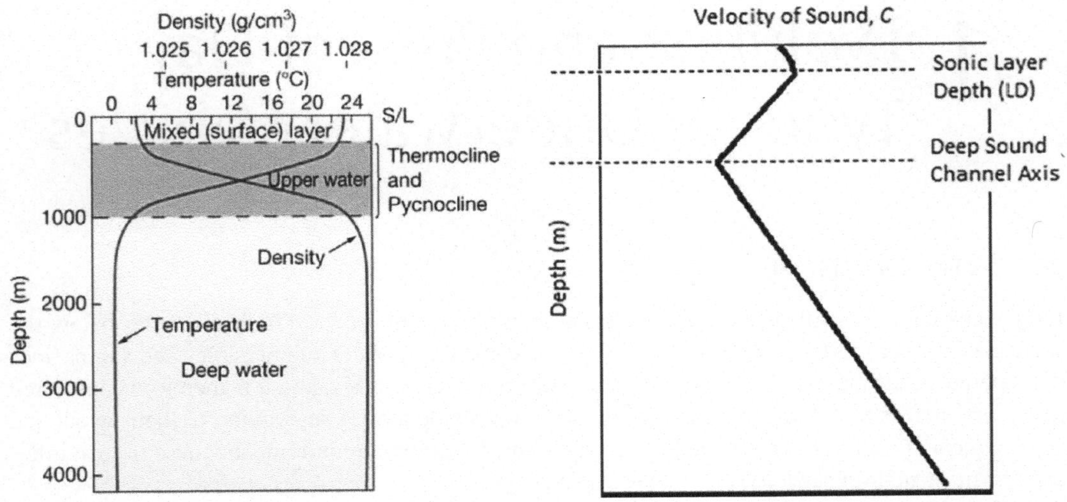

FIGURE 1.1 Variation of density, temperature, and speed of sound with depth.

$$p = p_a + \rho g h + k h^2 \tag{1.2}$$

where k=0.017 m/s per increase in metre depth. The *Leroy formula* gives the hydrostatic pressure p as a function of depth, h [14]:

$$p = 10.196 + k_1(1 + k_2 \sin\phi)h + k_3 h^2 \tag{1.3}$$

where

φ=latitude in degrees
$k_1 = 1.0052405$
$k_2 = 5.28 \times 10^{-3}$
$k_3 = 2.36 \times 10^{-6}$

The pressure obtained from (1.3) is in MPa × 10^4.

Density is also affected by temperature and salinity, and this affects vessel buoyancy.

1.2.2 Temperature

The most important effect that temperature has on metallic structures that operate below the sea surface is the possible reduction in ductility. Materials need to retain their fracture toughness throughout the operating temperature range. Particular attention is required in the case of structural details that involve complex welding arrangements in the way of openings and viewports. We note from Figure 1.1a that below a depth of around 3,500 m, the temperature remains approximately constant at 4°C. The other effect that the drop in temperature with depth has is the loss of heat from within the vessel, and this affects the crew.

1.2.3 Acoustics

The transmission of sound waves within the ocean affects the operation of submerged vehicles in different ways. In seawater, light and radio waves attenuate rapidly, whereas sound waves travel over large distances. The speed of sound varies from 1,450 to 1,570 m/s, and it increases with increasing temperature, salinity, and pressure. The increase in pressure is due to the increase in bulk modulus. Because the speed is not constant, sound does not travel along straight lines within the ocean.

Refraction occurs when a sound wave travels through a medium of varying density. Thus, as a sound wave travels vertically, its speed will decrease as the temperature decreases. As a result, the upper part of the sound beam travels faster than the lower part. Refraction effects impose limits to the range for which a submersible can be tracked from the ocean surface.

In the case of submersibles, it is necessary to maintain communication with a ship on the surface and good transmission is important. In contrast, in the case of submarines, it is necessary for them to remain undetected. In order to succeed in this, the vessel has to operate at depths at which it cannot be detected. At the same time, any noise signal (audible or otherwise) that the vessel emits has to be eliminated as effectively as possible.

Acoustic quieting is employed in the case of submarines in order to reduce the transmission of sound that results from machinery operation and the resulting vibration. Sources of sound waves are various types of machinery in operation as well as collapsing bubbles of steam that are formed by cavitating propellers. From the structural point of view, it is necessary to dampen vibration, reduce the intensity of emitted sound by use of suitable design and layout of interior spaces, and also use of appropriate materials such as *anechoic tiles* (Figure 1.2). These consist of rubber or synthetic sections that contain a large number of regularly spaced small voids. The tiles are attached to the outer hull surface, and their role is twofold. On the one hand, they absorb sound waves emitted by active sonar, and on the other, they attenuate the sound waves transmitted by machinery on board, thereby reducing the range of possible detection. Modern materials consist of a number of layers and different-sized voids, each targeted at a specific sound frequency range at different depths.

FIGURE 1.2 Anechoic tiles. (a) Anechoid (acoustic) tiles fitted to *Akula* class submarine (b) Action of anechoid tile.[2]

1.2.4 Survivability and Vulnerability

In addition to hydrostatic pressure and the effects of the marine environment, naval submarines have to be able to withstand a whole range of loads and conditions that are related to wartime action. It is therefore necessary to take these into account at the design stage. The question of *survivability* arises, and it is necessary to consider the types of threats that may arise, their range of magnitude, and their corresponding probability of occurrence. Such threats include torpedoes, mines, depth charges, and possibly, nuclear explosions. Additionally, it is possible that underwater collisions with other vessels as well as impact with the sea bed may occur.

Having defined the potential threats, the *vulnerability* of a vessel needs to be considered. This concerns the entire system, and this includes the items in direct contact with the sea environment as well as the pressure hull itself and its contents. In order to assess the consequences of such threats, specific scenarios need to be studied. Damage assessment needs to consider the effects of shock on the pressure hull and its contents, whipping that may result from a nearby underwater explosion as well as the growth and propagation of flooding, fire, and smoke within the pressure hull. The outcome of such studies has repercussions in submarine design throughout, as well as that of the pressure hull. With regard to the pressure hull, major issues that arise are:

1. Whether a single, double, or mixed single/double hull is the best solution. Advantages and disadvantages for each of these from a structural perspective are discussed in Section 9.2.2.
2. How the pressure hull is to be subdivided into compartments. Such an arrangement permits safeguarding vital operations onboard such as command and control, crew survivability, and the continued operation of specific items of equipment. From a structural perspective, the internal subdivision of the pressure hull determines the position and type of bulkheads that are required. This concept is implemented in surface warships as well.

In addition to the above, a number of other questions arise in relation to detailed structural design as well as broader issues such as vibration isolation and the acoustic signature of individual items. A systematic procedure is required, and we shall briefly list the various stages of a particular approach [2].

1. Identification of threats and definition of shock levels
2. Acceptance criteria for structure and equipment
3. Threat basket identification (definition of probable load combinations and scenarios)
4. Lessons from past experience
5. Concept option comparisons (for example, single versus double hull, equipment redundancy, compartment roles)
6. Shock protection strategy
7. Subdivision strategy
8. Vulnerability management through design
9. Trade-off assessments
10. Safety case support
11. Magazine safety case support
12. Design acceptance
13. Through-life change management

Clearly, the question of survivability, in addition to structural issues, touches upon other fields such as stability, subdivision, and equipment operation, and thus, an interdisciplinary approach is necessary.

1.2.5 Effects of the Environment on Material Performance: Corrosion and Fatigue

In submersibles, the problem of corrosion has particular aspects:

- Galvanically incompatible materials
- Geometrical complexity of components rendering cathodic protection impractical and difficult
- Galvanic corrosion
- Crevice corrosion

Fatigue also affects underwater vehicles, primarily in relation to residual welding stresses. These topics are discussed in Chapter 10 in relation to naval submarines.

1.3 THE EVOLUTION OF UNDERWATER VEHICLES

1.3.1 Early Forms of Subsea Equipment: From the Diving Bell to the Modern Submersible

From early on, the seabed was scoured by divers in search of valued items such as seafood, sponges (Crete 3,000 BCE), and pearls (China 2,200 BCE). These activities were generally performed in an unaided manner although simple aids such as ropes and weights were used. Reeds were used to assist submergence and hinder detection and divers participated in military operations (Trojan War, c. 1200 BCE). By the 3rd century BCE, diving was a recognised activity in ancient Greece with rules and terms of engagement [9].

The first recorded mechanically assisted underwater operations were performed with the use of the *diving bell*, which encloses air used for breathing. A written record is provided by Aristotle during the 4th century BCE:

> … ἔοικε δέ· ὁμοίως γὰρ ἀναπνοὴν ποιοῦσι τοῖς κολυμβηταῖς λέβητα καταφέντες. οὐ πίμπλαται γὰρ οὗτος τοῦ ὕδατος, ἀλλὰ τηρεῖ τὸν ἀέρα μετὰ βίας γὰρ ἡ κάθεσις. Ορθὸν γὰρ ὁτιοῦν παρεγκλιθὲν εἰσρεῖ
>
> Αριστοτέλους Προβλήματα Φυσικά, Βιβλίον λβ΄ Οσα Περί ώτα, Κεφ ε΄.

> "… for they enable the divers to respire equally well by letting down a cauldron; for this does not fill with water, but retains the air, for it is forced down straight into the water; since, if it inclines at all from an upright position, the water flows in." [1]

From this quotation, we infer that the bell was made of some kind of metal, most probably an alloy of iron.

A different type of vessel is alluded to in a medieval chronicle originating from France, entitled *Le Roman d' Alexandre* (*Alexandriade*). This mentions that Alexander the Great entered a glass chamber that was lowered into the water, enabling him to observe the marine environment. Figure 1.3 is an illustration that accompanies the text, and it depicts a vessel that is just below or floats on the sea surface. It appears to be made of glass, thus permitting observation of the marine environment. We can surmise that such a vessel would have operated just below or close to the surface.

The diving bell continued to be used essentially in its original form until the Renaissance.[4]

Towards the end of the 17th century, Edmund Halley designed a sizable and sophisticated bell that received a patent in 1691 (Figure 1.4a). Dives were made to depths of 20 m and lasted up to 90 minutes, during which divers used the bell as a source of supply of fresh air. In 1707, an improved design was launched with glass ports and an air replenishment system.

FIGURE 1.3 Alexander observing marine life from a glass chamber (*Alexandriade*).[3]

During the 19th and early 20th centuries, a number of patents were taken out on diving bells and diving suits (Figure 1.4b and c). The first true modern submersible may have been built in 1929 by Nishimura and was intended for fisheries research [5].

1.3.2 Deep Diving Records

A particularly important stage in the development of submersibles was reached when attempts to reach ever greater depths were made during the 20th century. The deepest part of the ocean lies at approximately 11,000 m, and this poses formidable challenges.

During the 1930s, Beebe, a zoologist, made several excursions to hitherto unreached depths in a suspended spherical steel submersible, which he termed a "bathysphere" (Figure 1.5). He reached a record depth of 923 m during a dive performed in 1934 in the Bermudas.

The pressure hull of the submersible had a diameter of 1,450 mm and a thickness of 31.75 mm. It was fitted with three viewing ports made of fused quartz of 75 mm thickness. From the operational point of view, the bathysphere had an important and potentially serious disadvantage. Its weight to displacement ratio exceeded unity ($W/\Delta = 1.49$), so that if the cable to which it was attached to broke, it would sink to the bottom of the sea. The next step was achieved by the submersibles designed over a 20-year period by the Swiss Auguste Piccard, culminating in the *Trieste*. Piccard made a number of important advances, including the use of ballast tanks that were used to compensate for the pressure sphere's weight that exceeded displacement ($W/\Delta > 1$) (Figure 1.6a, p.10). In this way, the submersible was rendered autonomous. In 1939, Piccard designed the *FNRS II*.[5] Its construction was interrupted by the Second World War, and following completion, it made its first dive in 1948. *FNRS II* was designed for a working depth of 4,500 m. The principles of *FNRS II* were designed into the *Trieste* that made a successful dive to 3,333 m in 1954.

After efforts spanning several years, it was decided to design a submersible that would reach the seabed at the deepest part of the ocean (Challenger Deep, in the Marianas Trench of the western Pacific). A pressure sphere of diameter 2007 mm consisting of two forged steel hemispheres that were bolted together along a flange was built (Figure 1.6b, p.10). To withstand the hydrostatic pressure of 110 MPa, the vessel was built with a wall thickness of 127 mm. It was thus overdesigned and could withstand considerably more than the rated pressure. The weight of the sphere

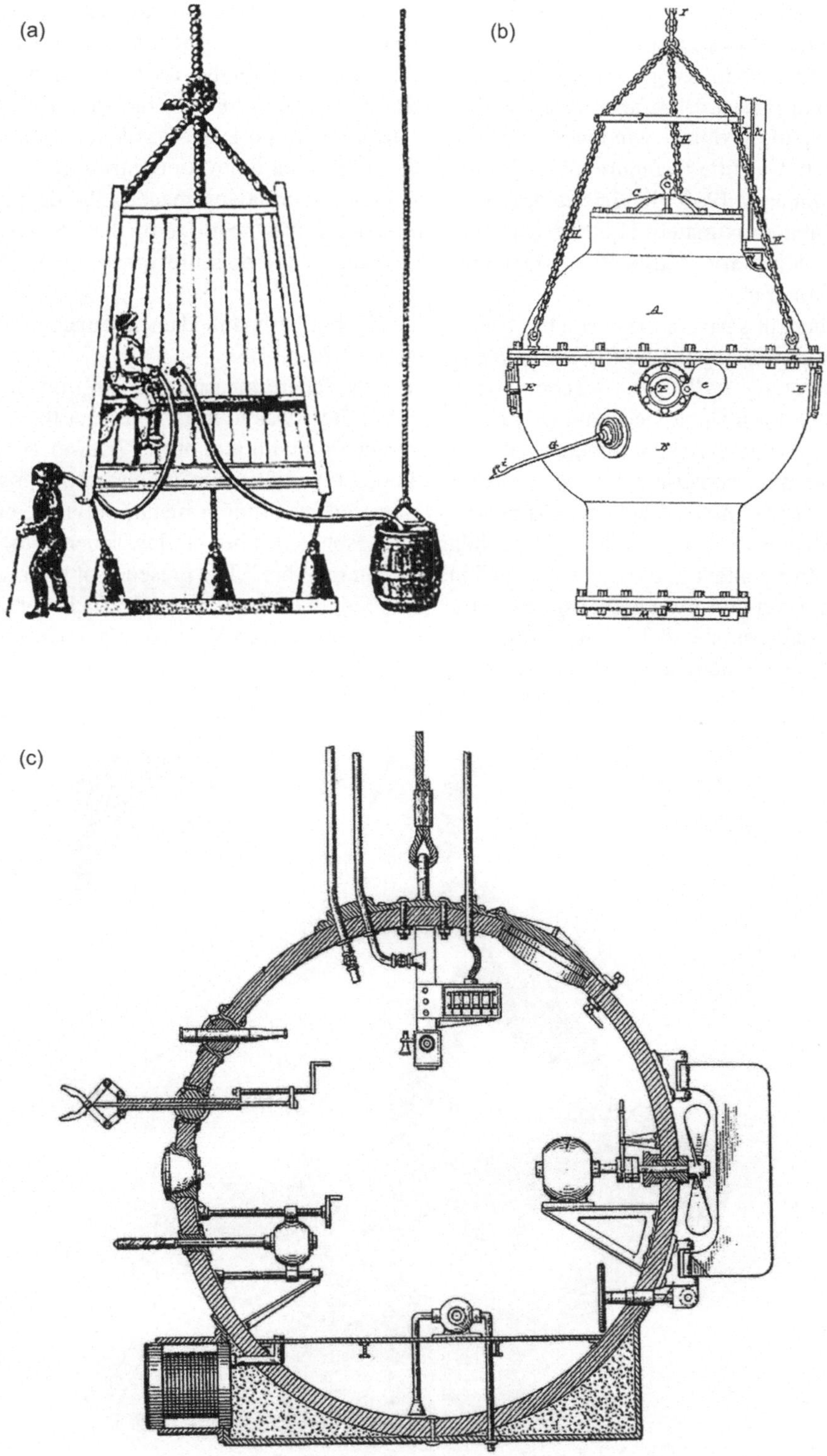

FIGURE 1.4 Diving bells [10]. (a) Halley's diving bell, (b) Richards & Alcott (1849), (c) Lemon's salvage apparatus (1922).

was 14.25 metric tonnes in air and 8 metric tonnes in water (giving it an average specific gravity of 14.25/(14.25–8) = 2.28 times that of seawater). A ballasting arrangement was necessary because it was not possible to design a sphere large enough to hold a person that could withstand the necessary external pressure, yet also have metallic walls thin enough for it to be neutrally buoyant. A conical-shaped plexiglass window, in addition to the viewing port that was located in way of the access hatch, was fitted, (Figure 1.6c); this is another important innovation introduced by Piccard.

On 23 January 1960, the *Trieste* accomplished its mission having reached the deepest part of the ocean at approximately 11,000 m [21]. It had previously been acquired by the US Navy and had undergone necessary modifications, including the fitting of the pressure sphere mentioned in the previous paragraph.

Since then, this part of the ocean has been reached by both unmanned and manned submersibles, one of the latest being the *Deepsea Challenger* (Figure 1.7a, p.11).

The *Deepsea Challenger* was equipped with a pilot sphere measuring 1,100 mm in diameter, large enough for a single occupant (Figure 1.7b, p.11). Steel was used to construct the outer walls of the pilot sphere, having a thickness of 64 mm (compared with that of the *Trieste*). A number of components were constructed using a newly developed type of composite material whose generic name is *syntactic foam*. Syntactic foams are synthesised by filling a metal, polymer, or ceramic matrix with hollow spheres called microballons or cenospheres or non-hollow spheres (for example, perlite). In this context, the term "syntactic" means "put together." The presence of hollow particles results in a lower density, a higher specific strength (strength divided by density), a lower coefficient of thermal expansion, and, in some cases, radar or sonar transparency. A manufacturing method for low-density syntactic foams is based on the principle of buoyancy.

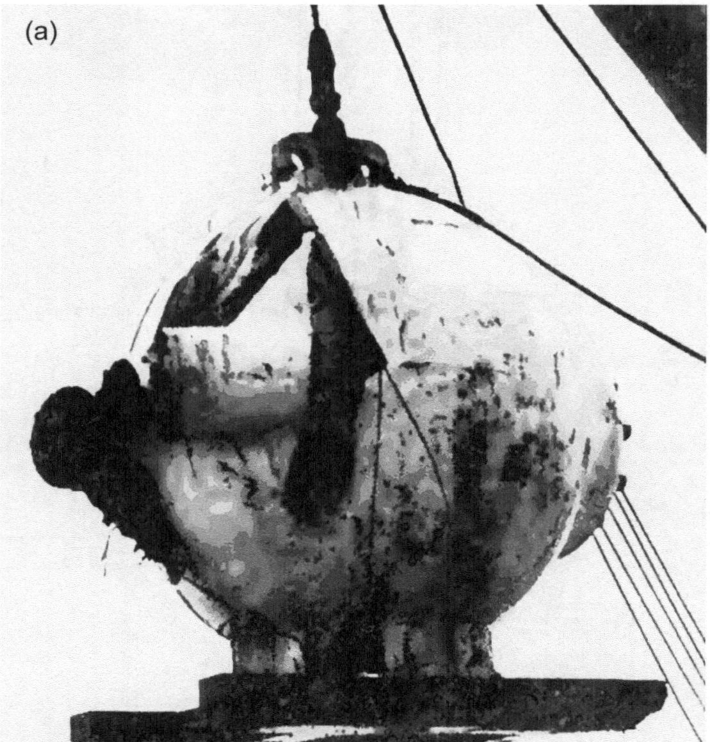

FIGURE 1.5 Beebe's bathysphere [3]. (a) The bathysphere suspended prior to a dive, (b) Schematic view of interior.

(b)

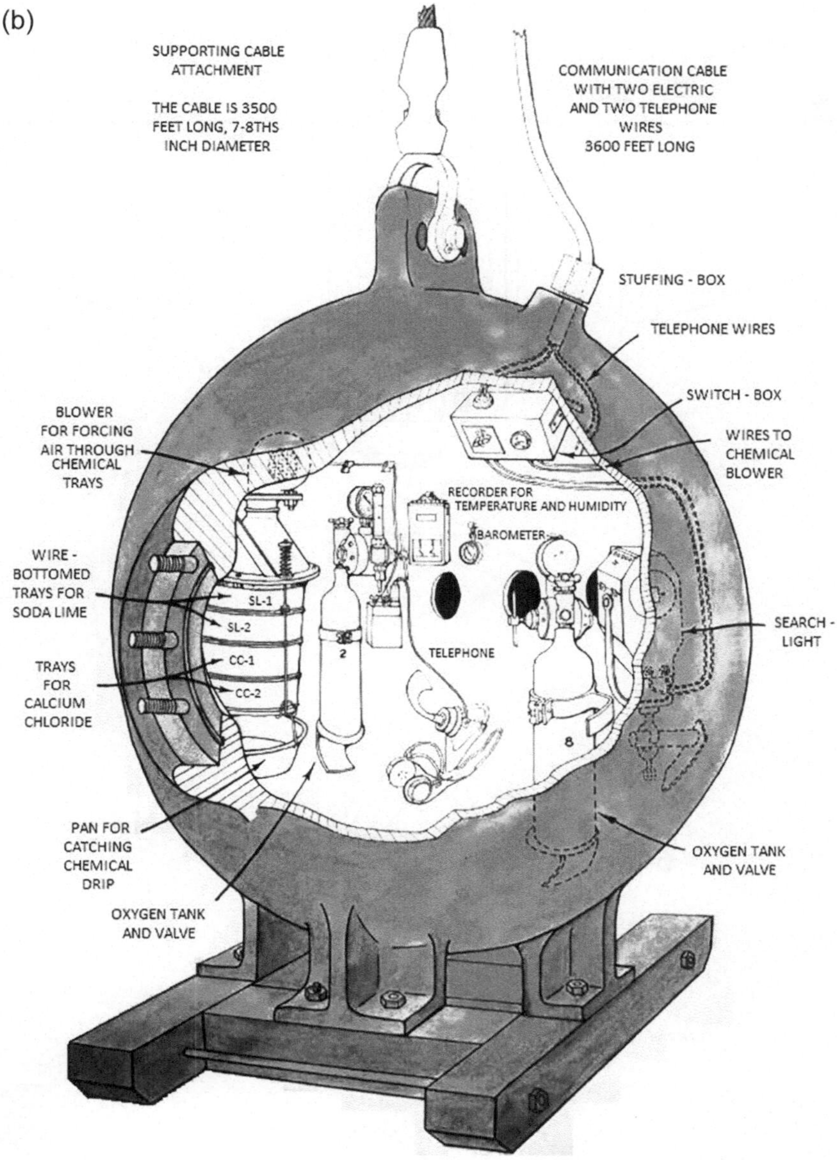

FIGURE 1.5 (*CONTINUED*) Beebe's bathysphere [3]. (a) The bathysphere suspended prior to a dive. (b) Schematic view of interior.

A special-purpose syntactic foam was developed for use in the case of the *Deepsea Challenger* that is on the one hand capable of withstanding the design pressure and secondly is more homogeneous and possesses greater uniform strength than other commercially available syntactic foams. Its specific density of about 0.70 ensures that it will float in water. It is composed of very small hollow glass spheres suspended in an epoxy resin. The foam was used to incorporate the thruster motors as part of the infrastructure, and these were mounted within the foam. It was therefore not necessary to use a steel skeleton for mounting various pieces of equipment.

A third successful attempt at reaching the deepest part of the ocean was achieved in 2020 by the Chinese-designed and built submersible *Fen Dou Zhe* ("Striver"), fitted with a titanium pressure hull (Figure 1.8). This reached a depth of 10,909 m in the Marianas Trench.

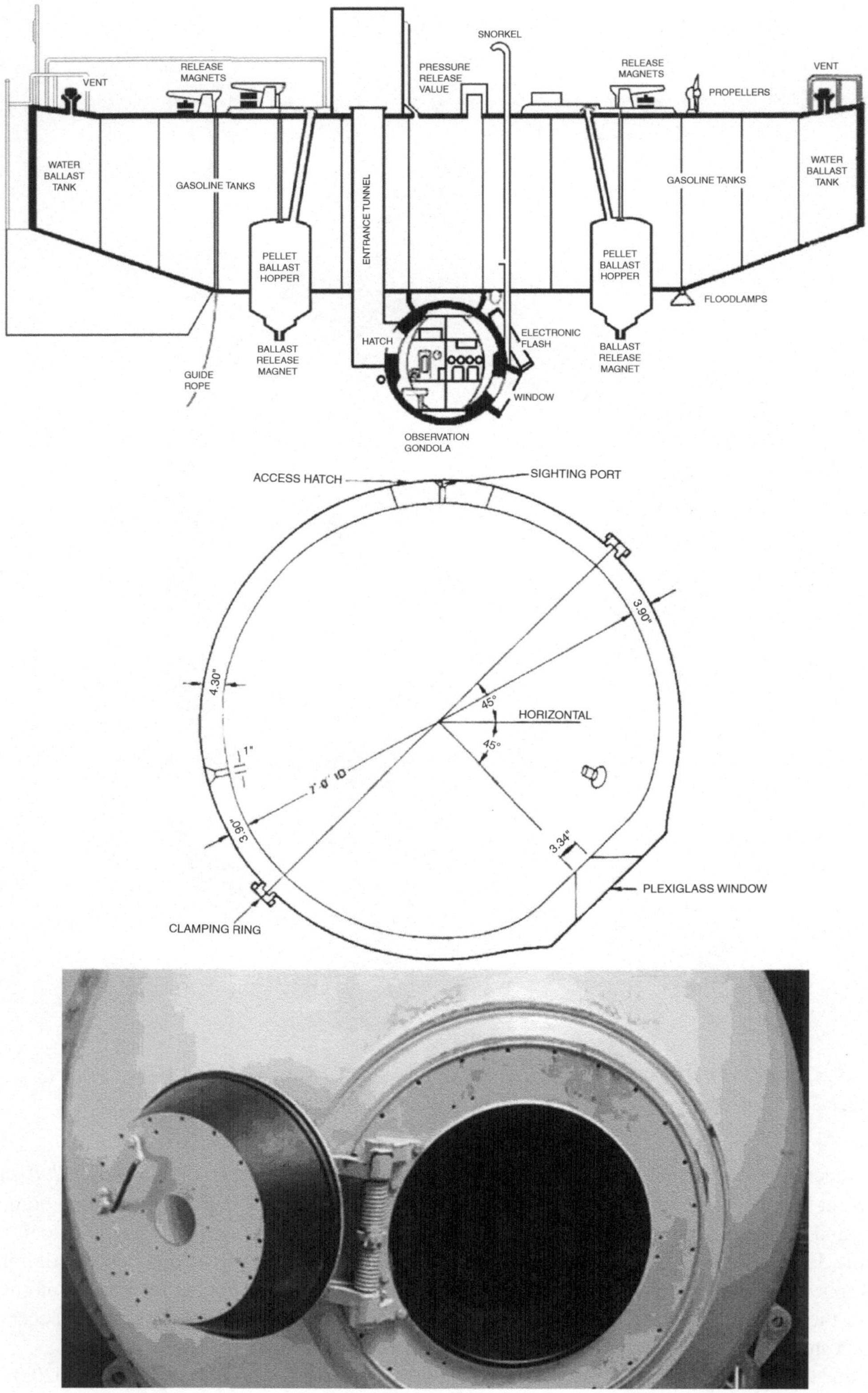

FIGURE 1.6 Bathyscaphe *Trieste* (1959) [21]. (a) General arrangement showing the ballast tanks and the pressure sphere, (b) section, (c) access hatch.

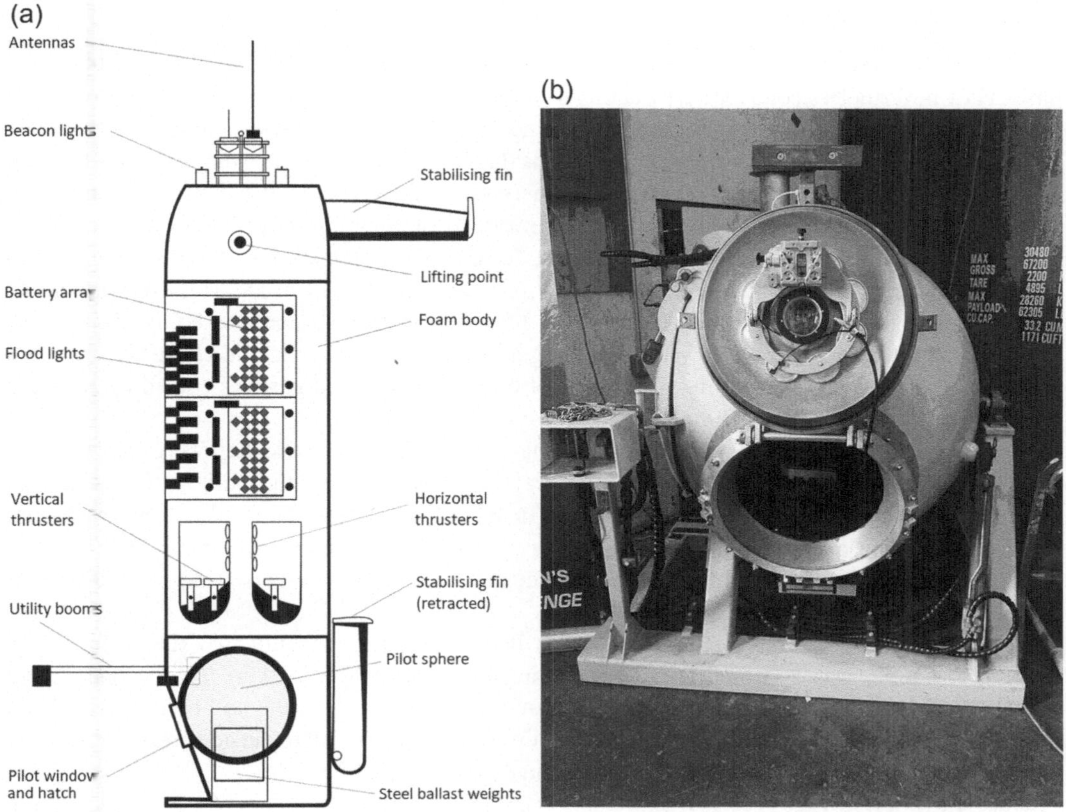

FIGURE 1.7 The *Deepsea Challenger*. (a) General arrangement. (b) Full-scale model of the pilot sphere.

FIGURE 1.8 The *Fen Dou Zhe* deep-diving submersible [26].

1.3.3 The Evolution of Submarines until the End of the 19th Century

The history of submarine development includes a number of designs that were proposed but never built. They nevertheless contributed by acting as points of departure for subsequent stages of development. This is particularly true for the Renaissance with Figure 1.9a showing a model of a sketch by Leonardo da Vinci. Another notable proposal is that by William Bourne, who proposed a design of a completely enclosed boat that could be submerged and rowed below the surface (Figure 1.9b). The proposed vessel consisted of a wooden framework covered with waterproofed leather.

Some years later, Buonaiuto Lorini proposed designs of equipment that would enable men to work underwater (Le Fortificationi, 1609). Giovanni Borelli included in De Motu *Animalium II*, published in 1685, the drawing shown in Figure 1.9c. He is also considered to be the first person to propose a self-contained underwater breathing apparatus. He designed and constructed a brass helmet with a glass window having a diameter of 60cm. The exhaled gas was cooled by seawater after passing through copper tubing.

There exist also designs that led to vessels that did not fulfil the task they were intended for. An example is the *Rotterdam Ship* (Figure 1.10a) that was designed and built by Louis de Son but did not succeed in diving below the surface. In 1620, Cornelis van Drebbel's design was more successful. He had a submersible built in the form of a boat, operated underwater by 12 rowers (Figure 1.10b). The oars were waterproofed, and the craft successfully traversed the river Thames at depths of 4–5 m for several hours [6].

Between 1690 and 1692, the physicist Denis Papin designed and built two submarines. The first design (1690) was a strong and heavy square metal box (Figure 1.11a). The second design (the "kettle"), which he built in 1692, differed from the original in that the hull had an oval shape (Figure 1.11b). It was fitted with a centrifugal air pump and was resistant to outside pressure because, thanks to the pump, air was permitted to enter and exit the hull.

During the period 1720–1721, Nikonov in St. Petersburg built a test model of a submarine in the form of a barrel (Figure 1.12a, p.15), and this was tested successfully. The results of the tests on a subsequent series of full-scale vessels capable of carrying four persons were unsuccessful, however, and the project was abandoned.

In 1775, Bushnell in the United States designed and built a vessel large enough to contain a single occupant. It was made of two hollowed-out pieces of oak, held together using iron hoops, and sealed with tar to ensure watertightness (Figure 1.12b, p.15). The aim of this vessel, named the *Turtle*, was to attack enemy warships undetected and to attempt to sink them by attaching a bomb to their underwater surface. The vessel performed as expected but did not succeed in its ultimate aim, for reasons unrelated to its design.

The *Nautilus* (Figure 1.13, p.15) was built by American Robert Fulton for the French Navy during the Napoleonic wars. It incorporated many features that are encountered in the submarines of today, and it had an ellipsoidal form. The shell plating was made of copper sheets to which internal iron frames were attached to provide sufficient strength and stiffness. This was thus the first time that iron was being used in the construction of a vessel operating below the sea surface.

In 1837, Rodriguez Labandera designed and built the *Hipopótamo*, the first submarine of Latin America in Guayaquil, Ecuador, for the Peruvian government, the aim being to attack and sink enemy ships (Figure 1.14). Its hull was made of wood with caulked planks, and it was waterproofed with bitumen. It was fitted with paddle wheels driven by human-powered pedals

In 1850, Wilhelm Bauer designed a submarine that he called the *Brandtaucher* (*Incendiary Diver*) having a length of 8 m, a beam of 2.2 m, and a depth of 2.6 m (Figure 1.15). The hull was made of riveted sheet iron and was fitted with a propeller. The vessel sank during trials, was later recovered, and is now on display at the German Armed Forces Museum in Dresden.

The French *Le Plongeur* was the first submarine to be machine rather than human-powered (1864) (Figure 1.16, p.17). The need for space to install the machinery and compressed air bottles necessitated a substantial increase in size, compared to previously built vessels.

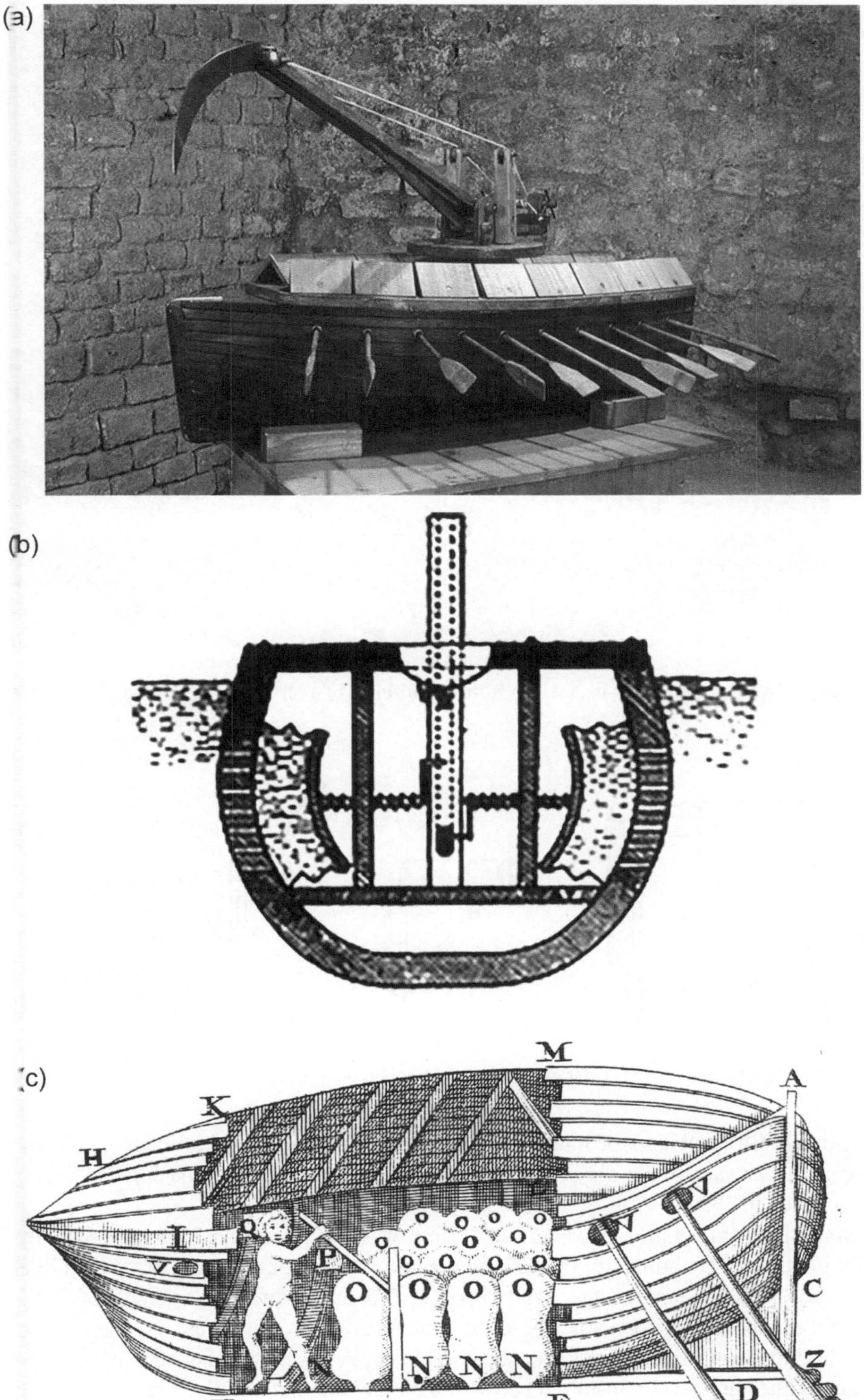

FIGURE 1.9 Proposals for submarines never built. (a) Leonardo da Vinci (1515), (b) William Bourne (1576)[6], (c) Giovanni Borelli.

FIGURE 1.10 Two Dutch inventions. (a) The Rotterdam ship, (b) a replica of the Drebbel submerged boat.

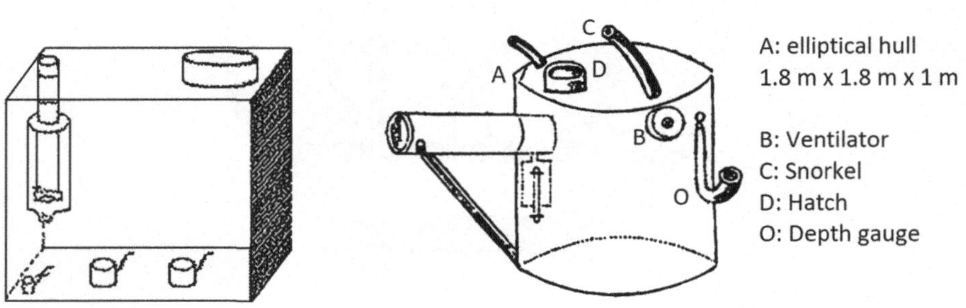

FIGURE 1.11 Papin's designs. (a) First design and (b) second design.

However, this led to problems of longitudinal instability and poor control characteristics, so that *Le Plongeur* never saw active service and was used as a floating storage facility until she was decommissioned in 1935.

The first successful engine-powered submarine was designed and built in Spain by Narcís Monturiol i Estarriol. He built and, in 1867, successfully tested the *Ictineu I* (from the Greek words "ichthys" and "nafs", meaning "fish" and "boat"). The submarine had a length of 7 m and was equipped with appendages for gathering coral and other finds of interest. It was *double-hulled*, with each hull made of olive wood staves sheathed in copper. It was propelled using foot power via two pedals. Ballast tanks were installed between the inner and outer hulls. The vessel descended to 30 m, but the effort of propelling it with foot power led the inventor to consider engine power. He developed a steam engine run by a chemical reaction between potassium chlorate, zinc, and manganese dioxide that produced both heat and oxygen. This was installed in *Ictineu II*, the successor

FIGURE 1.12 Early forms of underwater vehicles (18th century). (a) Model of Nikonov's submarine (1721) and (b) Bushnell's Turtle (1775).

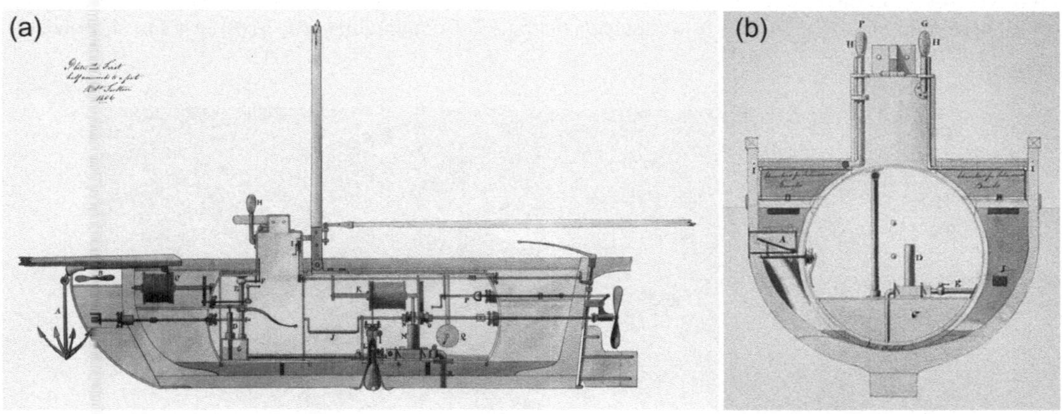

FIGURE 1.13 Fulton's *Nautilus*. (a) The "plunging boat" concept (1807) and (b) section showing crew, ballast, and weapons storage compartments.

to *Ictineo I*, which thus became the first submarine to make successful use of a combustion engine (Figure 1.17).

The *Resurgam*, built in England in 1878, had a length of 5 m and was a hand-powered, one-man-operated vessel. A second, much larger *Resurgam* was built and launched in 1879. This was fabricated using iron plates that were fastened to iron frames, with the central section of the vessel clad with wood that was secured by iron straps. Her length was 14 m, her diameter was 3,000 mm, and she displaced 30 long tons (Figure 1.18).

During the last decade of the 19th century, two vessels were built in the United States by marine engineer Simon Lake (*Argonaut 1* and *Argonaut 2*). *Argonaut 1* had a length of 11 m, was cigar-shaped, and was built of steel (Figure 1.19). Her main attribute, like that of her successor, was a wet diving chamber that allowed a diver to leave and re-enter the submarine. Lake introduced the even keel concept and controlled submergence that was subsequently introduced to all

FIGURE 1.14 The *Hipopótamo*. (1) Rudder blade, (2)Stern compartment, (3) Stern valve, (4) Suction pump, (5) Bilge containing ballast (rocks), (6) Propulsion blades, (7) Commanding officer, (8) Jose Luis Labandera.

FIGURE 1.15 Model of the *Brandtaucher.*

later submarines, in contrast to the use of horizontal rudders used in early designs. The vessel proved a success although not for military use, as the US Navy adopted the Holland designs. It was used instead for undersea exploration and for commercial purposes. A later design by Lake, the *Protector*, was acquired by the Russian Navy.

The French submarine *Morse* was designed to combine the best features of two earlier submarines, the *Gymnôte* and the *Gustave Zédé* (Figure 1.20). The *Gymnôte* was fitted with an electric

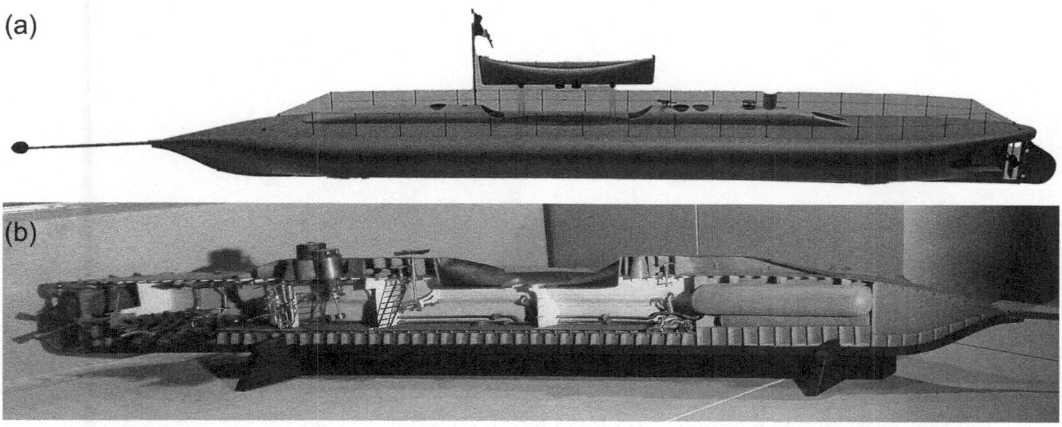

FIGURE 1.16 *Le Plongeur.*

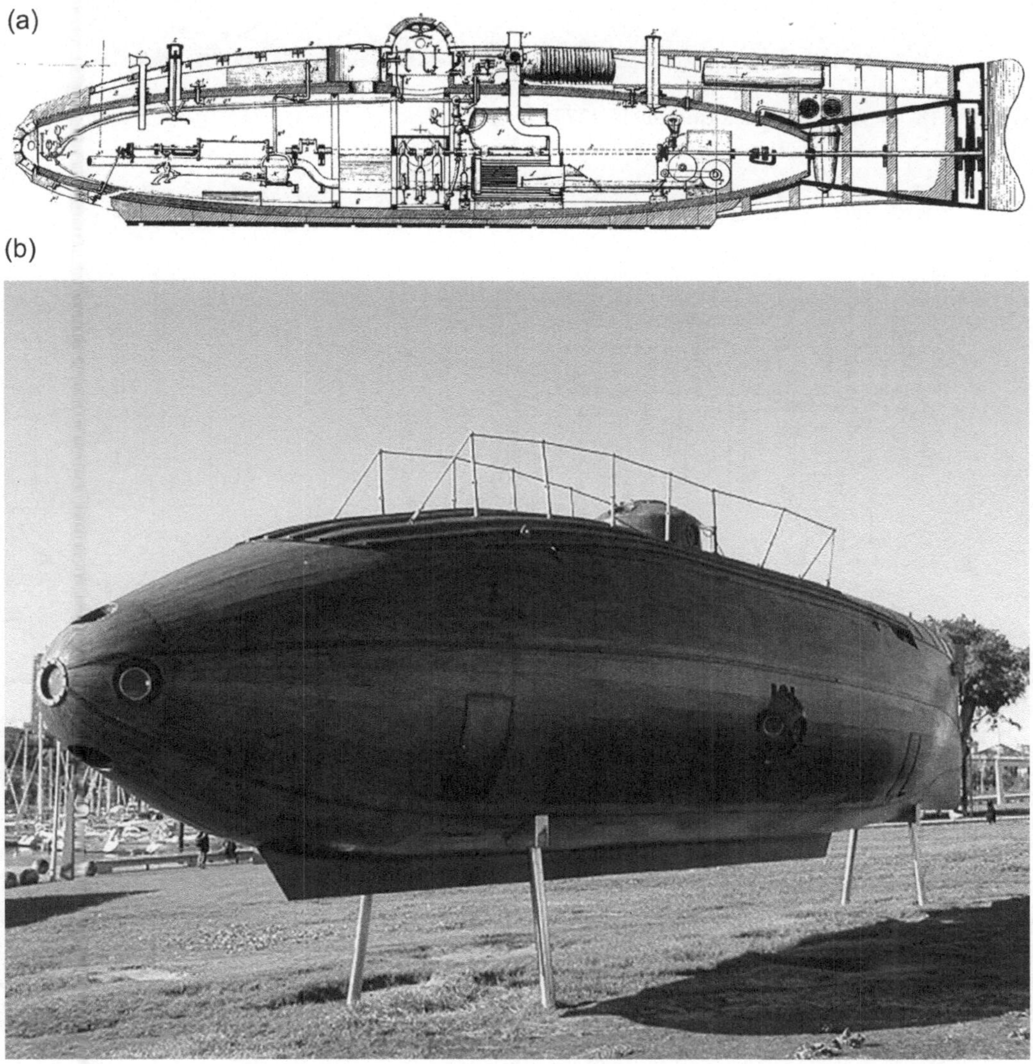

FIGURE 1.17 The Ictineu II. (a) Longitudinal profile and (b) Museum full scale replica.

FIGURE 1.18 A replica of the *Resurgam* II (1879).

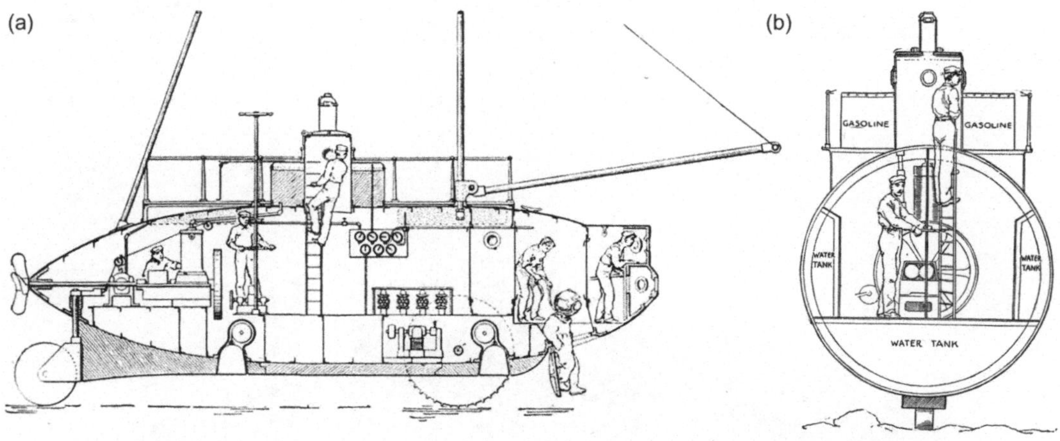

FIGURE 1.19 *Argonaut 1.*

engine, the first periscope and the first naval electric gyrocompass. The *Gustave Zédé* was the first submarine to carry out a successful submerged attack using torpedoes against a surface target. Her hull was made of bronze rather than steel to resist corrosion, which also allowed a magnetic compass to operate inside. She had a different profile fore and aft, unlike other submarines at that time, which had the same hull form at the fore and aft ends. Seventy-six circular ring frames provided sufficient strength to the outer shell plating of the pressure hull. The *Morse* was laid down at Cherbourg and launched in July 1899. She was of a single-hull construction and built of Roma bronze, a copper alloy proposed by her designer M. Romazotti, the aim being for her to be more flexible than an all-steel hull. Furthermore, the copper alloy would interfere less with the vessel's magnetic compass. She was also fitted with a detachable lead keel.

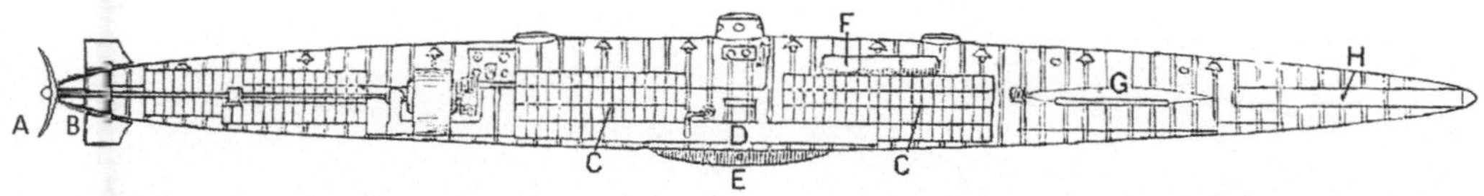

FIGURE 1.20 The French Navy submarine *Morse*.

FIGURE 1.21 *Le Narval.*

The submarines of the time suffered from the fact that the optimum features for operating underwater were disadvantages on the surface. Thus, a circular pressure hull, designed for minimum resistance when submerged, had poor sea-keeping characteristics when the vessel was surfaced; additionally, early submariners were not adequately trained to cope with these conditions. Electric propulsion, although safe and efficient underwater, provided limited range and speed on the surface and thus restricted the submarine to coastal operations. It was these two disadvantages that Laubeuf, the designer of *Le Narval*, sought to overcome (Figure 1.21).

To address the problem of the circular pressure hull's seakeeping characteristics on the surface, a double-hull design was used, enclosing the pressure hull that contained the crew and machinery in a more seaworthy outer hull that was not pressurised. This approach had been followed by Monturiol some years earlier. The space between the inner and outer hulls was used for ballast tanks and fuel tanks, since these can operate at hydrostatic pressure. This led to improved performance on the surface. *Le Narval* proved to be a success, and her design features provided a template for future submarines for many years.

Other advantages claimed for the double-hull type of construction were:

- A greater reserve of buoyancy and, hence, improved seaworthiness.
- Since the external shape was that of a surface vessel, surface stability was similar and much better than that of single-hull vessels.
- The strength of the pressure hull could be increased, since web frames exterior to the pressure hull of appreciable depth could be fitted between the hulls.
- If exterior frames are used, there was more space inside the pressure hull.
- The deck top of the outer hull provided a space for exercise for the crew.

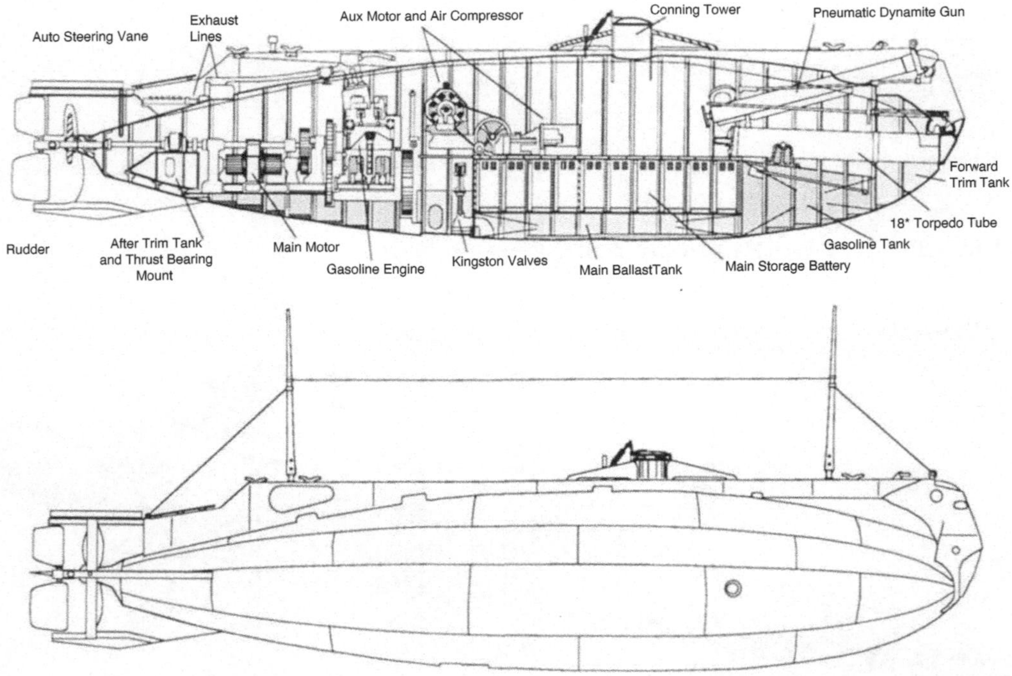

FIGURE 1.22 USS Holland (1898) (a) Longitudinal section (b) Longitudinal profile.

The disadvantage was that the double-hull vessel as then designed took longer to dive.

Towards the end of the 19th century, Holland began to experiment with a series of submarines. The first one to be adopted by the US Navy was the *USS Holland I*, in 1900 (Figures 1.22 and 1.23). In later models, the gasoline engine was replaced by a diesel engine that was used exclusively for another 50 years. Holland's designs were used as the basis for submarine design in Britain some years later.

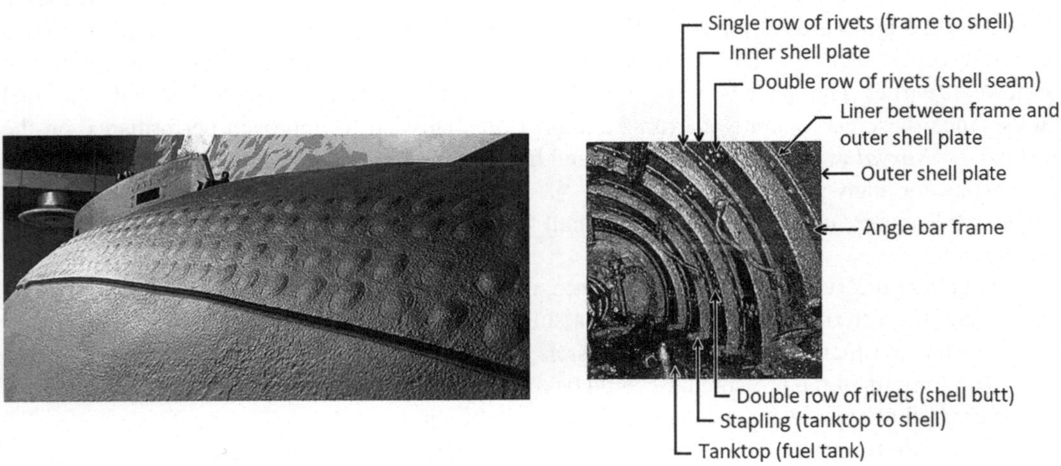

FIGURE 1.23 Structural features of Holland's designs. (a) Reinforcing plate on the "Fenian Ram" and (b) structural connections and arrangements [7] (vessel reconstruction).

1.4 DESCRIPTION AND CLASSIFICATION OF UNDERWATER VEHICLES: SUBMARINE TYPES

In this and the next section, we shall provide a summary of the different types of underwater vehicles currently in use, with an emphasis on pressure hull type and design.

Underwater vehicles can be classified into the following two major groups:

1. Submarines (navy submarines and commercial submarines, including private)
2. Submersibles of all types

Examples of submarines types are shown in Figure 1.24.

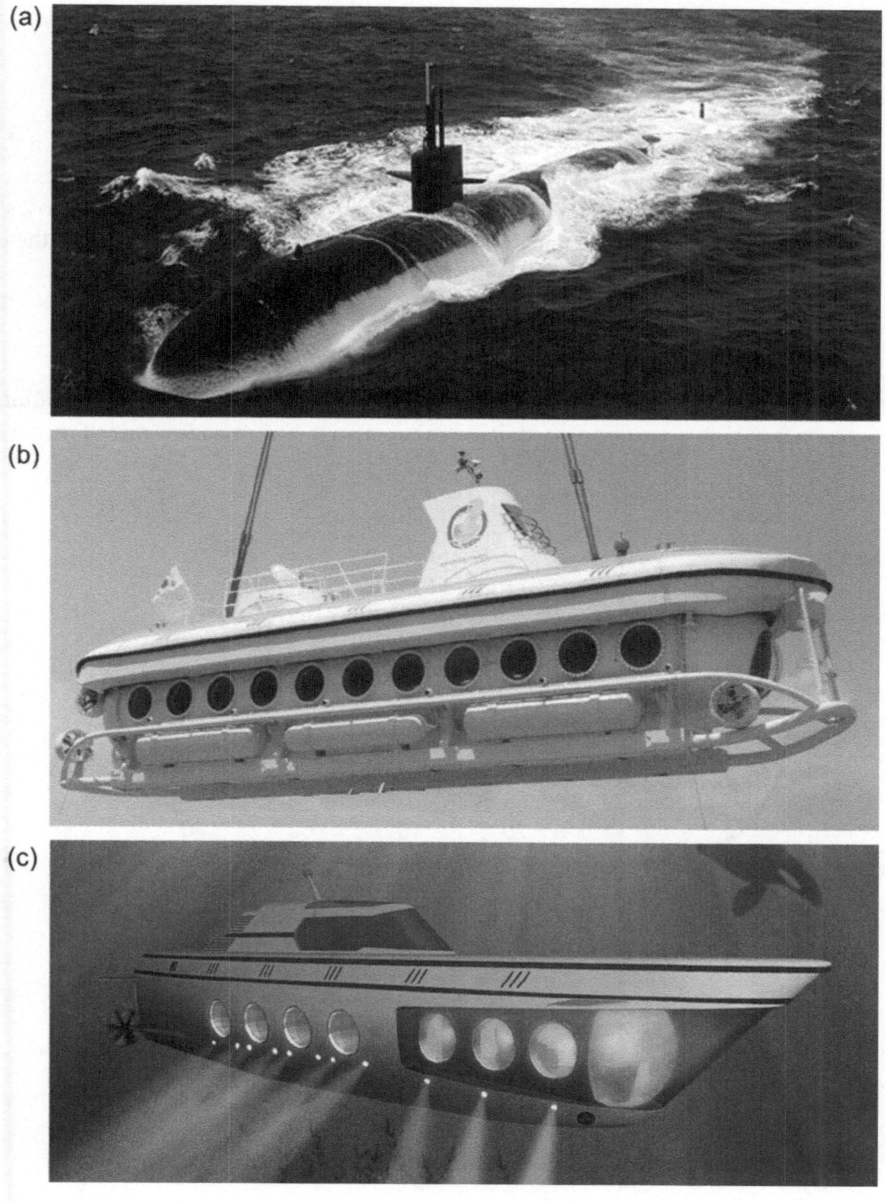

FIGURE 1.24 Types of submarines. (a) Naval (nuclear) submarine, (b) tourist submarine, and (c) private submarine.

1.4.1 Uses of Submersibles

The use of submersibles has expanded significantly as developments in technologies such as materials, automatic control, and communications have facilitated their use in a wide range of activities such as offshore oil and gas, maritime, fisheries, aquaculture, offshore renewable energy, marine science, and marine mining. Figure 1.25 includes a plot of ocean floor area versus depth. It is seen that 99% of the entire ocean floor lies at a depth of less than 6,500 m, and furthermore that, known mineral deposits lie at depths that do not exceed 5,000 m. For this reason, most manned submersibles have been designed to operate at depths not exceeding 6,000 m.

Table 1.1 provides an indicative list of non-naval as well as naval fields in which submersibles have found application. Additionally, submersibles of different types can be used in combination with one another and with surface (mother) ships in order to increase effectiveness, for both naval and commercial activities, as described in [13,27].

1.4.2 Classification of Submersibles

Figure 1.26 includes a breakdown of the different types of underwater vehicles. It is seen that these are broadly classified as *manned* and *unmanned* and that a number of subcategories exist in both groups. A number of these refer to scientific equipment that is used for research purposes that we will not be discussing further in this chapter. We shall instead focus on the strength of the pressure hull forms encountered in the different types of submersibles.

1.4.3 Manned Submersibles

A manned submersible is a manned, non-combatant craft capable of independent operation on and under the sea surface that is equipped with its own propulsive power and means of direct viewing

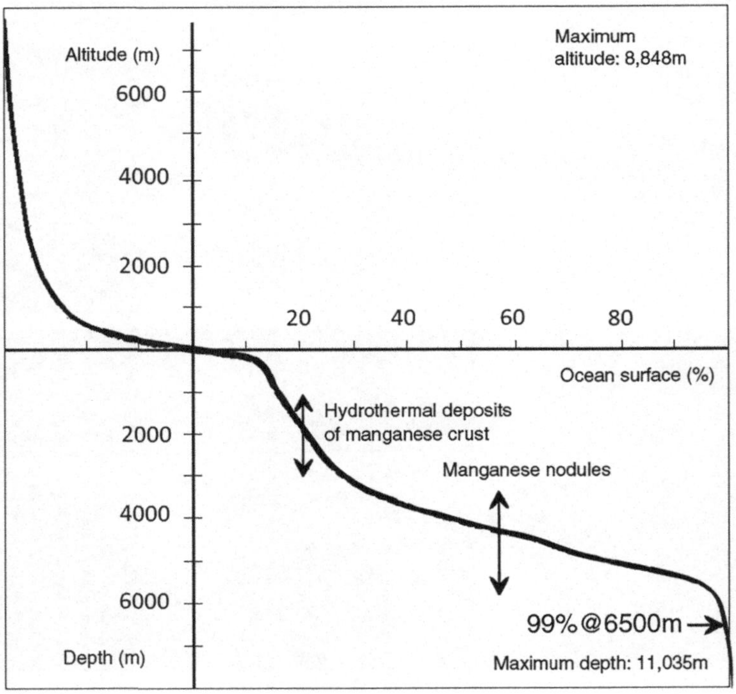

FIGURE 1.25 Ocean floor area versus depth and presence of mineral deposits.

TABLE 1.1
Uses of Submersibles

Field	Specific Activity
Offshore (oil and gas)	Baseline environmental assessment
	Geophysical survey
	Pipeline survey
	Debris/clearance survey
	Emergency response
Environmental monitoring	Water quality
	Ecosystem assessment
	Route survey
	Habitat mapping
Hydrography	Deep sea mining
	Charting
	EEZ survey
	Pre/post dredging survey
Search and recovery	Asset location
	Marine archaeology
Naval operations [16,17]	For support of ship-to-objective manoeuvre
	Countering mines in the sea
	Countering diesel submarines
	Environmental monitoring
	Clandestine operations

for the occupants who are embarked within a dry atmosphere [6].[7] Manned submersibles are classed as (Figure 1.26):

- Human-operated vehicles (HOVs)
- Deep submergence rescue vehicles (DSRVs)
- Atmospheric diving suits (ADSs)
- Rescue bells

In this section, we will provide brief descriptions of these, the most important from the point of view of pressure hull analysis being HOVs and DSRVs.

Alvin was the first manned submersible, designed and built in the United States. It has been used extensively for scientific and military expeditions and has been upgraded several times since it was launched in 1964. Figure 1.27 shows the vehicle in its present condition and the titanium sphere that has been installed during a refit. For the buoyancy material syntactic foam has been used.

The success of *Alvin* led a number of countries to develop and use deep-diving manned submersibles. In 1985, France commissioned the *Nautile*, which is capable of operating at depths of up to 6,000 m (Figure 1.28c). The *Nautile* has been used in seafloor ecology research, searches for hazardous waste as well as in the military field.

In 1987, Russia commissioned *Mir-1* and *Mir-2*, which have been used to obtain environmental data and seafloor topography (Figure 1.28a). Other uses to which they have been put are detection of radiation from a sunken submarine, detection of Second World War sunken warships as well as Arctic surveys. The personnel sphere is fabricated using a maraging steel alloy that has a

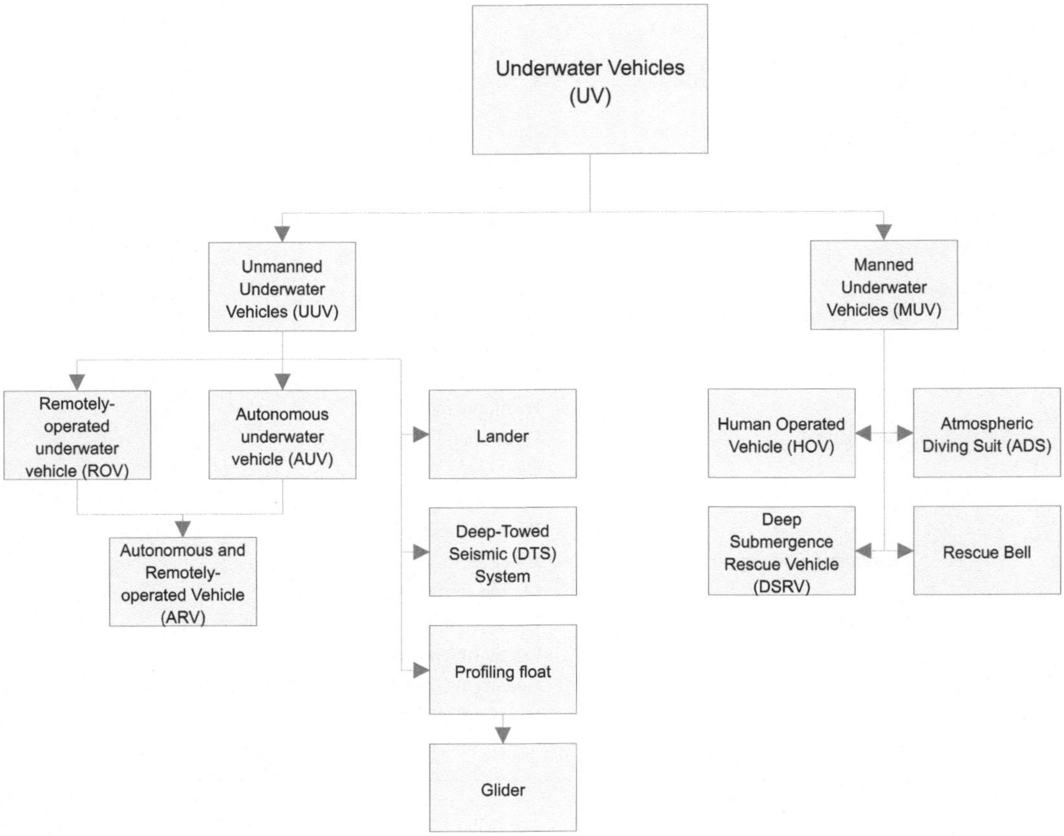

FIGURE 1.26 Classification of submersibles (underwater vehicles).

FIGURE 1.27 The *Alvin*, first manned deep-diving submersible. (a) Vessel launching (b) titanium spherical pressure hull.

10% higher strength/weight ratio than titanium. This alloy contains about 30% cobalt and smaller amounts of nickel and chrome. Two hemispheres were made by casting and machining. They were then bolted together, thus avoiding welded joints. The resulting vessel density is close to the density

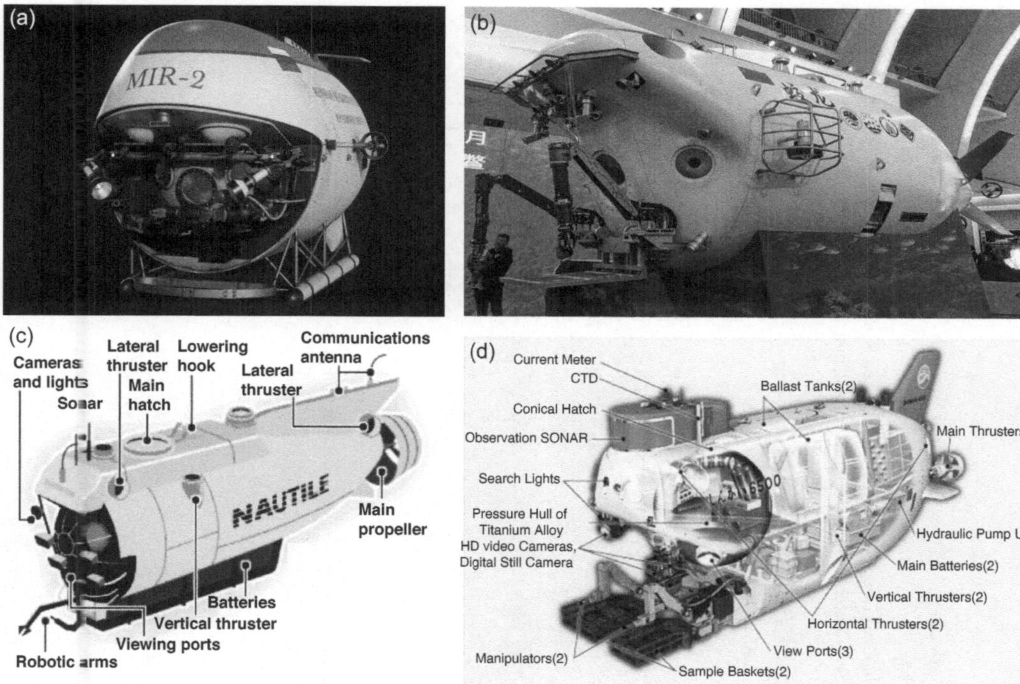

FIGURE 1.28 Deep-diving submersibles. (a) Mir (Russia), (b) Jiaolong (China), (c) Le Nautile (France), (d) Shinkai 6500 (Japan).[8]

of water, making it easier to operate at different depths. Additional buoyancy is provided by use of 8 m^3 of syntactic foam. Both craft are now on display in museums.

Another early manned submersible is the *Shinkai 6500,* a Japanese vessel that entered service in 1990 (Figure 1.28d). The *Shinkai* ("*Deep Diver*") was fitted with a pressure hull made of a lightweight titanium alloy with an internal diameter of 2,000 mm and a thickness of 73.5 mm. The permissible out-of-circularity of the spherical pressure hull was 0.5 mm.

More recently, China deployed the *Jiaolong*, a manned deep-diving submersible that can dive to a depth of over 7,000 m (Figure 1.28b). It has the second-greatest depth range of any crewed research vehicle in the world.

The focus has thus shifted from competition to achieve the greatest depth to scientific research and investigation of resources.

Structural parameters and equipment performance data for manned deep-diving submersibles are included in Table 1.2.

DSRVs were originally designed and commissioned in the United States (Figure 1.29, p.28). The need for such vessels became apparent after the loss of the *USS Thresher* with all lives lost. The primary aim of DSRVs is to reach sunken submarines and to rescue the crew. As such, they are designed for quick deployment in the event of an accident. In addition to search and rescue (SAR) operations, DSRVs have been used in clandestine missions and, in this context, are termed *Deep-Submergence Vessels* (DSVs).

Rescue bells (*chambers*) are also used in submarine SAR operations when conditions permit. The concept was originally introduced in the early 1930s for the US Navy, and the currently used equipment is known as the McCann Submarine Rescue Chamber (Figure 1.30, p.29).

Atmospheric Diving Suits (ADSs) are classed as small one-person submersibles that enable a person to descend to depths up to 700m for many hours by eliminating the physiological dangers to which divers subject to ambient pressure are exposed. The suit is articulated, thereby enabling

TABLE 1.2
Structural Parameters and Equipment Performance of Manned Submersibles [26]

	Alvin	Le Nautile	Shinkai 6500	MIR1/MIR2	Jiaolong	Shenhai Yongshi	Fendouzhe
Country	United States	France	Japan	Russia	China	China	China
Operating organisation	Woods Hole Oceanographic Institution	French Research Institute for the Exploitation of the Sea	Japan Agency for Marine-Earth Science & Technology	Russian Academy of Sciences	National Deep Sea Center	Institute of Deep-Sea Science and Engineering, Chinese Academy of Sciences	Institute of Deep-Sea Science and Engineering, Chinese Academy of Sciences
Status	In operation	In operation	In operation	In operation	In operation	In operation	In operation
Year of constr/modif.	1964/2013	1984	1989	1987	2009/2019	2017	2020
Maximum dive depth (*m*)	6,000	6,000	6,500	6,000	7,000	4,500	11,000
No. of passengers	3	3	3	3	3	3	3
Total length (*m*)	7.0	8.0	9.7	7.8	8.2	9.3	10.2
Total height (*m*)	3.68	3.61	4.1	3.0	3.4	3.9	4.4
Width (*m*)	2.6	2.7	2.8	3.8	3.0	3.0	3.3
Weight (*t*)	20.4	19.5	26.7	18.6	22	20	35
Pressure hull material	Titanium Alloy	Titanium alloy	Titanium alloy	Titanium alloy	Titanium alloy	Titanium alloy	Titanium alloy
Inner diameter of ball cabin (*m*)	1.98	2.1	2.0	2.1	2.1	2.1	1.8
Diameter of observation window (*m*)	127(1) 127 (2) 280 (2)	120(1) 120(2)	120(1) 120(2)	200(1) 120(2)	200(1) 120(2)	200(1) 120(4)	120(3)

(Continued)

TABLE 1.2 (*Continued*)
Structural Parameters and Equipment Performance of Manned Submersibles [26]

	Alvin	Le Nautile	Shinkai 6500	MIR1/MIR2	Jiaolong	Shenhai Yongshi	Fendouzhe
Life support time (h)	72	143	128	82	84	82	87
Maximum speed (km)	3	2.5	2.5	5.0	2.5	2.5	2.5
Payload (kg)	205	200	200	200	220	220	200
Battery/energy (kW-h)	Lithium-Ion/115	Lead-acid/50	Silver-zinc/86.4	Nickel-cadmium/100	Silver-zinc/110	Lithium-Ion	Lithium-Ion
Underwater operat time (hour)	4–5	4–5	4	17–20	12	10	10–13
Year of first operation	1964	1985	1989	1987	2008	2017	2020
Dives per year	110–150	100–115	60	20	26	100	–

FIGURE 1.29 Deep submergence rescue vehicles. (a) USN DSRV-1 Mystic [15] and (b) NATO NSRV-SRV 1 [22].

movement of limbs while maintaining a pressure of one atmosphere, so that decompression is not required during ascent to the surface (Figure 1.30b). The potential structural failure modes of ADS are buckling collapse in compression, leaks, and lockup of joints. Leaks and buckling in compression both cause a reduction in buoyancy. Joint leaks and locking of articulating joints may be reversible when pressure is reduced [20]. ADSs have evolved since the 19th century and have now reached a stage of mature development.

1.4.4 Unmanned Submersibles (Underwater Vehicles)

Unmanned underwater vehicles (UUVs), sometimes known as underwater drones, are submersibles that operate underwater without a human occupant. A modern UUV is defined as a self-propelled submersible whose operation is either fully autonomous (pre-programmed or real-time adaptive

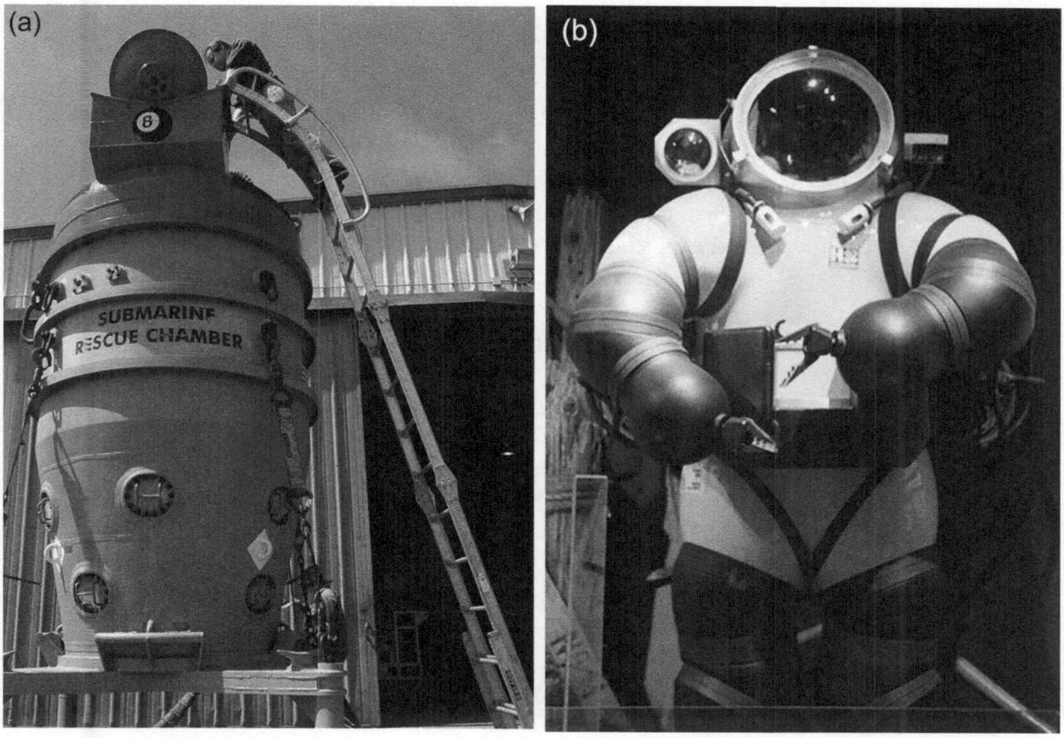

FIGURE 1.30 Rescue chambers and atmospheric diving suits. (a) McCann Submarine Rescue Chamber [24]. (b) Atmospheric diving suit [20].

mission control) or under minimal supervisory control [12]. There are at present a number of ways used to classify them, the most important being:

- Type of vehicle
- Pressure hull form
- Pressure hull material

Table 1.3 includes the UUV classification adopted by the US Navy, for its own purposes. This classification is based on dimensions and performance capabilities. UUVs are classified as portable, lightweight, heavyweight vehicles and large vehicles. The overwhelming majority in existence belong to the first three categories.

In this section, we will provide a brief overview of the types of UUV in operation. Of the UUVs included in the flow chart of Figure 1.26, the two most important ones from the point of view

TABLE 1.3
US Navy UUV Classification [8]

Class	Diameter (inches)	Displacement (pounds)	Endurance (hours)		Payload (cubic feet)
			Low Hotel Load	High Hotel Load	
Man-Portable	3.00–9.00	<100	10–20	<10	<0.25
Lightweight Vehicle (LWV)	12.75	500	20–40	10–20	1.00–3.00
Heavyweight Vehicle (HWV)	21.00	<3,000	40–80	20–50	4.00–6.00
Large	>36.00	20,000	>400	100–300	15.00–30.00

of structural design are remotely operated vehicles (ROVs) and autonomous underwater vehicles (AUVs).

Remotely Operated Vehicles

A *remotely operated underwater vehicle* (ROUV or ROV) is a free-swimming submersible craft used to perform underwater observation, inspection, and physical tasks [23]. It receives instructions from a human controller via a cable that remains attached to it at all times. The uses of ROV include sub-sea infrastructure surveillance, inspection and maintenance, military procedures, search and rescue missions, fish finding and exploring, and obtaining data about regions of the oceans where humans cannot go themselves. A more detailed list follows.

- Surveys (hydrographic surveys, offshore structure inspections and surveys)
- Support of diving operations (offshore industry)
- Military (mine-hunting and mine-breaking)
- Science (ocean and marine life studies)
- Education (construction of a simple ROV by students)
- Media (filming in spaces that are inaccessible or dangerous to divers)
- Hobby (construction of small ROV)

Table 1.4 includes a classification of ROV based on size, mass, and required propulsive power [23].

ROVs generally have two different types of configurations:

- Open or box frame (Figure 1.31)
- Torpedo-shaped (Figure 1.32)

The structure of an open ROV consists of an aluminium or composite frame that supports all mechanical systems and electronic sensors. Torpedo-shaped ROVs, also known as *tow fish*, are suited to being towed by a surface vessel and are equipped with side sonar that is used to detect objects on the sea bed, such as downed aircraft. They can be used for all types of data collection including marine life.

An AUV is designed to accomplish a specific task or tasks and is self-powered so that it functions independently, without continuous input from an operator. It thus requires no communication during its mission. AUVs are used to conduct underwater research and are also used in the military field. Recently, researchers have focused on the development of AUVs for long-term data collection in oceanography and coastal management [4]. The oil and gas industry uses AUVs to make detailed maps of the seafloor before they start building sub-sea infrastructure. Furthermore, the use of AUVs for pipeline inspection and inspection of underwater man-made structures is becoming more common. Scientists use AUVs to study lakes, the ocean, and the ocean floor. In military applications, an

TABLE 1.4
Classification and Use of ROV

Category	Size	Operating Depth and Uses
Micro	<3 kg	In spaces inaccessible to diver (sewer, pipeline, cavity)
Mini	15 kg	As diver alternative (in spaces dangerous for diver)
General	<5 HP	Max. working depth 1,000 m
Inspection class		Equipped with manipulator arms for commercial or industrial use
Light workclass	<50 HP	Max working depth 2,000 m
Heavy workclass	<220 HP	Max. working depth 3,500 m equipped with two manipulator arms
Trenching and Burial	>200 HP	Max working depth 6,000 m. Able to carry out cable laying operations

FIGURE 1.31 Open (box-framed) ROV. (a) *Atalanta* and (b) *Diomède*.

AUV is more often referred to as an unmanned undersea vehicle (UUV). The largest AUVs operate as substitute submarines (Figure 1.33a).

The essential difference between ROVs and AUVs is the type of control technique used. There are advantages and disadvantages to both control techniques. ROVs are real-time controllable, and as a result, manipulation can be executed as necessary, without the need for complex coding. AUVs, on the other hand, have greater operational flexibility since their motion is not limited by the presence of a cable. AUVs and ROVs can be used in tandem, in which case an AUV is used to survey a site and determine an appropriate location for the subsequent deployment of an ROV.

Autonomous and remotely operated vehicles (ARVs) are hybrid underwater vehicles. The ARV can operate as an AUV for self-cruise, and it can also work as an ROV for fine operation via a fibre-optic cable. A typical ARV consists of the underwater vehicle, tether system, and surface support.[9] *Dream Chaser* is a prototype ARV that has been designed in China for research within ocean

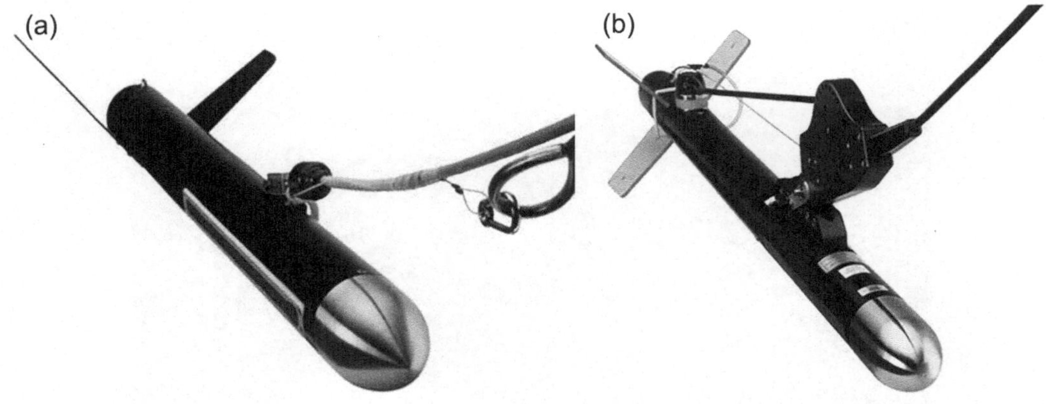

FIGURE 1.32 Torpedo-shaped ROVs (tow-fish).

FIGURE 1.33 Autonomous underwater vehicles. (a) US Navy ORCA XLUUV and (b) REMUS 600 AUV.

trenches (the *hadal*[10] region). The configuration of ARV *Dream Chaser* differs from conventional unmanned submersibles using monohulls (Figure 1.34). Its purpose is to [11]:

1. Perform measurements and data recording of seawater temperature, salinity, depth, etc.
2. Monitor and record environmental data
3. Conduct accurate mapping of underwater terrain

4. Carry out underwater virtual reality imaging
5. Perform sampling of macro-biological trapping, seawater, sediments
6. Carry out large-scale surveying and mapping of the underwater terrain

The entire system is illustrated in Figure 1.34. It is seen that it consists of the following four parts:

1. A submersible (ARV) (Figure 1.35)
2. A simplified tether management system (TMS)
3. A ship-based integrated control system
4. A launch and recovery system (LARS)

As in the case of ROVs, a frame structure is used to support mechanical and sensory equipment, and in this case, aluminium alloy 6061 is used. A FE model of the frame structure was subjected to loads corresponding to operating conditions and distributions of stresses and deformations obtained.

Three pressure hull casings are used to house electronic and other equipment that cannot operate at ambient pressure. Titanium alloy was used, and FE analysis was used to assess strength. Since there is no possibility of exceeding the design depth, the safety factor was reduced from 1.4 to 1.25, thereby leading to a lighter structure. Results of the FE analysis are included in Table 1.5.

A *Deep-Towed Seismic* (DTS) system is an observation system that is towed by a ship and is used for deep-sea geology and geophysics research. Such systems have been developed to obtain high-resolution images of the sea floor structure. By towing the instruments at a altitude of 50–100 m above the seabed rather than from the surface, improved resolution images are obtained. The system involves the use of an array of geological and geophysical instruments that include side-scan sonar, etc. Applications include geotechnical and geological investigations for sediment characterisation, seismic stratigraphy, and geohazard studies.

An overview of the system is shown in Figure 1.36. The mothership, to which a towing cable is attached, operates at a specified towing speed. The towing cable connects the mothership to the towed vehicle and is used to transmit power as well as data. The *sparker*, fitted to the towed vehicle, emits seismic wave signals in the direction of the seabed that are reflected by the strata and are

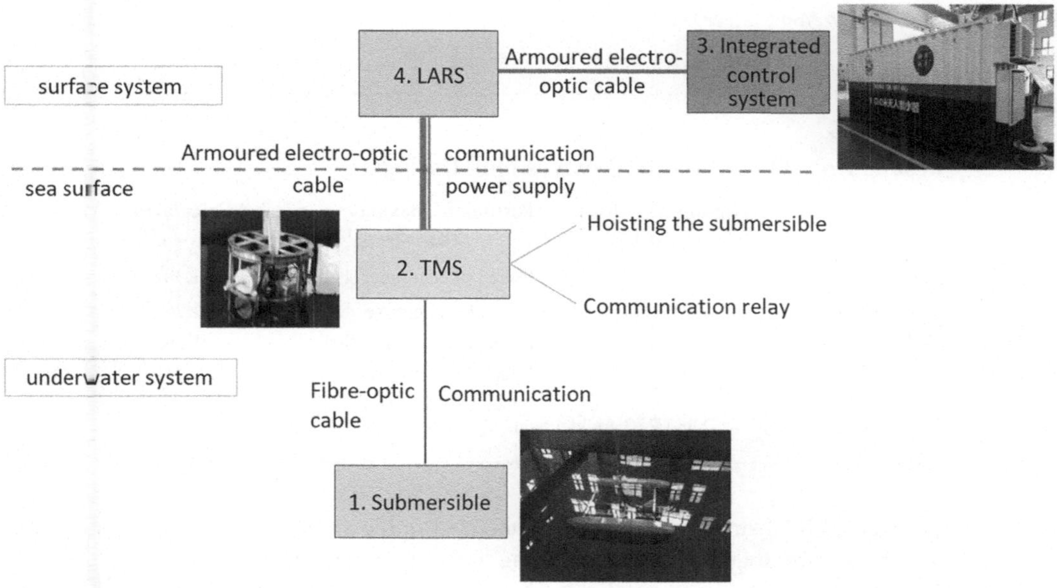

FIGURE 1.34 *ARV Dream Chaser* system operation [11]. (a) Submersible body (schematic) and (b) the submersible during underwater equilibrium tests.

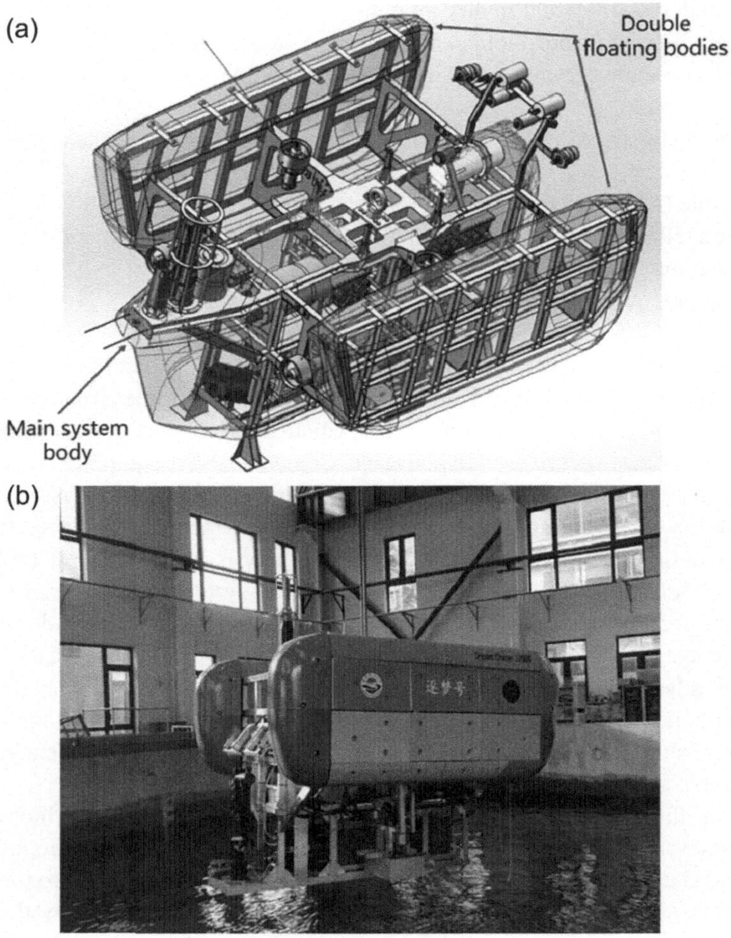

FIGURE 1.35 *ARV Dream Chaser* [11].

TABLE 1.5
Results from FE Analysis of Pressure Hull Casings[a]

Item	Maximum von Mises Stress (MPa)	Position of Maximum von Mises Stress	Calculated Safety Factor
Shell of PC1	615	Round fillet at inner shell	1.30
Head of PC1	634	Edge of seal groove	1.26
Shell of PC2	588	Round fillet at inner shell	1.36
Head of PC2	625	Edge of seal groove	1.28

[a] Pressure hull casings PC2 and PC3 are identical.

subsequently received by hydrophones that are embedded in the seismic array cable. A drogue is used to provide backward drag, thereby maintaining the array cable straight.

Lander is a general term for any autonomous, unmanned oceanographic research vehicle that free falls to the sea floor unattached to any cable and then operates independently[11] (Figure 1.37) [19]. Following completion of its mission, ballast weights are released either by a pre-programmed

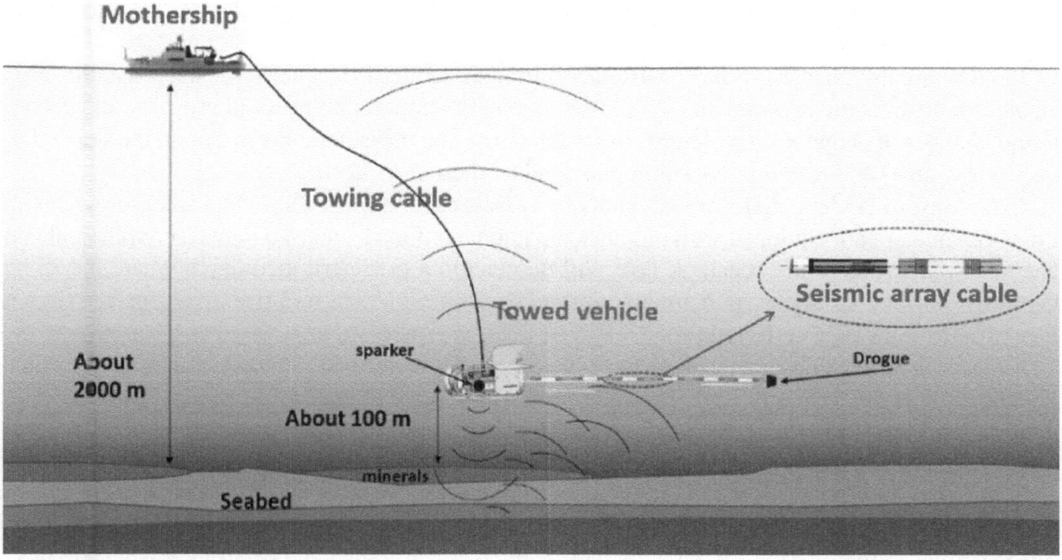

FIGURE 1.36 Deep-towed seismic (DTS) system [28].

FIGURE 1.37 Benthic lander.

timing device or following transmission of a signal from the surface. The lander then floats back up to the surface by virtue of its positive buoyancy. The qualifying term "benthic" is used to denote that the lander operates on the sea bed.

Landers are used because when an artefact is brought to the surface from the deep sea, the large changes in hydrostatic pressure and, to a lesser extent, in temperature result in property and behavioural changes. It is therefore preferable to conduct tests and measurements in situ on the sea bed. It is also less time- and resource-consuming to conduct these tests in situ.

In contrast to landers, *floats* are autonomous vessels used to make subsurface measurements in the ocean and are not equipped with propulsion equipment. Floats are used to measure physical and chemical properties of the ocean. A float will descend to a predetermined depth where it will be neutrally buoyant. Once a certain amount of time has passed, a float will rise to the surface so that it can transmit the data collected.

A profiling float can change its buoyancy in order to move vertically to collect data over a range of depths ("profiles") (Figure 1.38). Profiling floats are capable of making a few hundred profiles to a maximum depth of 2,000 m following which they transmit their data to shore via satellite communication. The deepest-diving models can reach depths of up to 6,000 m.

Underwater gliders carry out the same types of measurements that floats do, but can translate in the horizontal direction because they are fitted with hydroplanes (Figure 1.39). In that sense, their role is similar to that of an AUV, but they have a much greater range and endurance. They are used to obtain sampling over extended periods and can provide data that are much more extensive than a conventional AUV. Additionally, the processing is less costly [18,25].

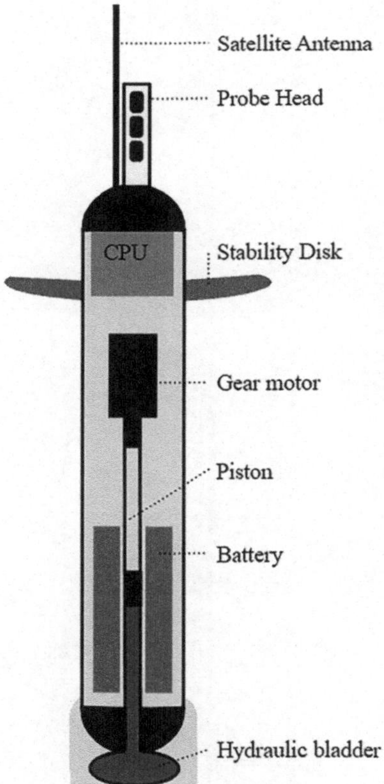

FIGURE 1.38 Profiling float.

FIGURE 1.39 Underwater glider.

NOTES

1 In the 1930s the terms "bathyscaphe" and "bathysphere" were coined by Beebe and Piccard respectively, based on the Greek words vathýs (βαθύς), skáphos (σκάφος), and sphéra (σφαίρα) meaning "deep", "craft" and "sphere" respectively. These terms are no longer in use.
2 These were first fitted to a German U-boat during the Second World War and are now used by navies around the world. The insert in Figure 1.2a shows the layout of voids used on the original (Alberich) tiles.
3 The *Alexandriade* is a 12th century manuscript (MS. Bodl. 264, Bodleian Libr. Oxford Univ.) See also discussion in: https://www.sukellushistoriallinenyhdistys.fi/ancient-diving/.
4 A demonstration of the use of a diving bell was given by two Greeks in Toledo in 1538, as related by Taisnier [5].
5 FNRS: Fonds Nationale de Rechèrche Scientifique (Belgian National Scientific Research Fund).
6 The drawing included in Figure 1.9c is based on a description included in [Bourne].
7 This definition precludes underwater habitats without independent means of propulsion, swimmer delivery vehicles that are not "dry" and diver support or delivery chambers that are tethered to the surface.
8 https://www.jamstec.go.jp/e/about/equipment/ships/shinkai6500.html.
9 The WHOI ARV *Nereus* is described in *Oceanus Magazine*, Vol. 47, No. 3 (www.whoi.edu/oceanus).
10 Hadal (Gk. Αδης = Underworld) region: ocean floor at depths exceeding 6,500 m.
11 The term "lunar lander" was used for moon landings during the period 1969–1972.

REFERENCES

1. Aristotle *Problemata Physika*. In: Ross, W.D. (Ed). (1927). *The Works of Aristotle, Volume VII Problemata, Book XXXII*, Ch. 5 transl. E.S. Forster. Oxford: Oxford University Press, 406 p.
2. ASRANet (1990). *Analysis and Design of Submarine Structures (Online Course)*. London: ASRANet Ltd.
3. Beebe, W. (1934). *Half Mile Down*. New York: Harcourt Brace & Co, 344 p.
4. Blidberg, D.R. (2001). The development of autonomous underwater vehicles (AUV); a brief summary. *Proc. IEEE ICRA, Seoul, Rep. Korea*. Vol. 4, 122–129.
5. Bline, D. (1977). The unique roles and history of manned and unmanned submersibles in oceanography. In: Geyer, R.A. (Ed). *Chap. 3: Submersibles and Their Use in Oceanography and Ocean Engineering, Oceanography Series*. Amsterdam, The Netherlands: Elsevier, Vol. 17, 61–76.
6. Busby, R.F. (1976). *Manned Submersibles*. Alexandria, VA: Office of the Oceanographer of the Navy.
7. Comstock, J.P. (1967). *Principles of Naval Architecture*. Second Revision. Jersey City: SNAME.
8. Department of the Navy (2004). *The Navy Unmanned Undersea Vehicle (UUV) Master Plan*.

9. Edmonds, C., Bennett, M., Lippmann, J., Mitchell, S.J. (2016). *Diving and Subaquatic Medicine*, 5th Ed. Boca Raton, FL: CRC Press.
10. Hoggett R. *Underwater Robotics*. Available at: https://cyberneticzoo.com/underwater-robotics. (Accessed date February 15, 2024).
11. Jiang, Z., Lu, B., Wang, B., Cui, W.-C., Zhang, J.-F., Luo, R., Luo, G., Zhang, S., Mao, Z.-G. (2022). A prototype design and sea trials of an 11,000 m autonomous and remotely-operated vehicle dream chaser. *J. Mar. Sci. Eng.*, Vol. 10, 812. https://doi.org/10.3390/jmse10060812.
12. Laun, A.W. (2013). "Development and validation of a conceptual design program for unmanned underwater vehicles". *M.Sc. Thesis*, Dept. Mech. Eng. Mass. Inst. Tech., Cambridge, MA.
13. Ludvigsen, M., Sørensen, A.J. (2016). Towards integrated autonomous underwater operations for ocean mapping and monitoring. *Ann. Rev. Contr.* Vol. 42, 145–157.
14. Lurton, X. (2002). *An Introduction to Underwater Acoustics: Principles and Applications.* London: Springer Praxis Publ. Ltd, 347 p.
15. Military.com. *Deep Submergence Rescue Vehicle DSRV.* Available at: https://www.military.com/equipment/deep-submergence-rescue-vehicle-dsrv (Accessed date February 15, 2024).
16. National Research Council (2005). *Autonomous Vehicles in Support of Naval Operations.* Washington, DC: The National Academies Press. https://doi.org/10.17226/11379.
17. NATO Combined Joint Operations from the Sea Centre of Excellence. (2012). "Guidance for Developing Maritime Unmanned Systems (MUS) capability." Report 12/029-2, Norfolk, VA.
18. Rudnick, D.L., Davis, R.E., Eriksen, C.C., Frantanoni, D.M., Perry, M.J. (2004). Underwater gliders for ocean research. *Mar. Tech. Soc. J.* Vol. 38, No. 1, 48–59.
19. Tengberg, A. et al. (1995). Benthic chamber and profiling landers in oceanography – A review of design, technical solutions and functioning. *Prog. Oceanog.* Vol. 35, 253–294.
20. Wikipedia. *Atmospheric Diving Suit.* Available at: https://en.wikipedia.org/wiki/Atmospheric_diving_suit (Accessed date February 15, 2024).
21. Wikipedia. *Bathyscaphe Trieste.* Available at: https://en.wikipedia.org/wiki/Trieste_(bathyscaphe) (Accessed date February 15, 2024).
22. Wikipedia. (2024). *NATO Submarine Rescue System.* Available at: https://en.wikipedia.org/wiki/NATO_Submarine_Rescue_Syste (Accessed date February 15, 2024).
23. Wikipedia. *Remotely Operated Underwater Vehicle.* Available at: https://en.wikipedia.org/wiki/Remotely_operated_underwater_vehicle (Accessed date February 15, 2024).
24. Wikipedia. *Submarine Rescue Chamber.* Available at: https://upload.wikimedia.org/wikipedia/commons/b/b3/Submarine_Rescue_Chamber-USN.jpg/ (Accessed date February 15, 2024).
25. Wikipedia. *Underwater Glider.* Available at: https://en.wikipedia.org/wiki/Underwater_glider (Accessed date February 15, 2024).
26. Yang, B., Liu, Y., Liao, J. (2021). Manned submersibles—deep-sea scientific research and exploitation of marine resources. *Bull. Chin. Acad. Sci.* Vol. 5, 622–631.
27. Yang, L., Zhao, S., Wang, X., Shen, P., Zhang, T. (2022). Deep-sea underwater cooperative operation of manned/unmanned submersible and surface vehicles for different application scenarios. *J. Mar. Sci. Eng.* Vol. 10, 909. https://doi.org/10.3390/jmse10070909.
28. Yang, S., Zhu, X.-Q., Ren, H. (2023). Dynamic analysis of a deep-towed seismic system based on a flexible multi-body dynamics frame. *Ocean Eng.* Vol. 279, 114587.

2 Materials for Submarines and Deep-Diving Submersibles

2.1 GENERAL

When considering materials for potential use in underwater vehicles, it has been found that both the environment and the behaviour of the materials give rise to a number of technical problems that are unique to them. In the past, different solutions have been found to some of these problems, and we shall discuss these. Both metallic and non-metallic materials have been used in the construction of submarine and submersible pressure hulls and other structural components. The metals used are steel and titanium alloys, whereas non-metallic materials include composites, glass, acrylic as well as hybrid metal–matrix composites (MMCs) (titanium and composites).

2.2 MATERIALS FOR SUBMARINES AND DEEP-DIVING SUBMERSIBLES

Up to the Second World War, submarines operated at depths that did not exceed 150 m. However, with the development of nuclear propulsion and the use of sonar, the need for them to remain undetected led designers to consider increased operating depths. The ability to remain undetected is an important attribute of operational success. In the case of a metallic pressure hull, increased depth necessitates increased shell plating thickness, improved fabrication techniques as well as improved fatigue strength. This gives rise to problems during the fabrication process that are related to plate forming and welding. It proved difficult to make use of the grades of steel that were available at the time so that advances in submarine propulsion and weaponry took place. The two main protagonists during that period, the United States and the Soviet Union, followed different paths in their efforts to provide solutions to these problems. The United States developed a number of new steel alloys, whereas the USSR chose to use titanium to fabricate the pressure hulls of certain classes of vessels. Each of these paths had its advantages and disadvantages, and it is instructive to discuss these briefly. Improvements in welding techniques were also found necessary.

Figure 2.1 gives the weight-to-displacement ratio for pressure hulls constructed of a variety of materials. The collapse (crushing) depth is plotted along the vertical axis, and the W/Δ ratio is plotted along the horizontal axis. For an efficient submarine platform, the W/Δ ratio ought to lie in the range 0.25–0.35 in order to retain sufficient reserve of buoyancy for non-structural equipmnet and personnel. On this basis, steel can be used for collapse depths of up to 1,000 m. At this depth, the corresponding W/Δ ratio is equal to 0.15 for titanium, aluminium, and glass-reinforced polymer (GRP) sandwich, whereas for carbon fibre, it is 0.08. Of these, titanium is costly, and aluminium is prone to corrosion.

On the basis of naval requirements for submarine operations, the following table was included in a report published several years ago.

The ideal material as specified in Table 2.1 does not exist at this point in time, but since the original publication of this list, a lot of effort has been made in the development of materials and joining techniques. In the case of submarines, the availability, strength, and low cost of steel make it the natural choice for the majority of applications, at least for the foreseeable future.

With regard to deep-diving submersibles, the pressure hull weight is nearly proportional to the material yield strength and inversely proportional to the material specific strength, as seen from (12.5). Therefore, to reach great depths, it is necessary to use materials that have high yield strength

DOI: 10.1201/9781003489153-2

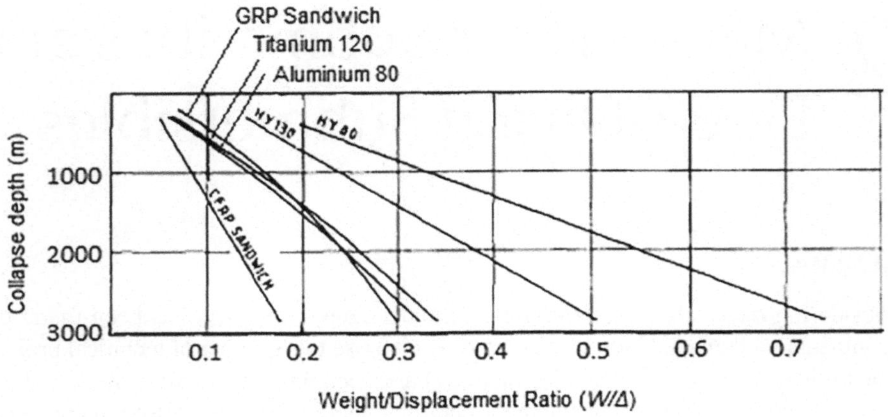

FIGURE 2.1 *W*/Δ ratio for various materials versus collapse depth.

TABLE 2.1
Ideal Characteristics for a Metal Alloy for Submarine Pressure Hulls

Density	As Low as Magnesium (1,740kg/m_s)
Yield strength	2,000 MPa
Moduli	Tension 280 GPa, shear 100 GPa
Toughness	Will not fracture in a brittle manner under severe plastic deformation at –18°C (0°F) or lower
Weldability	95% joint efficiency or greater in as welded condition for yield strength, toughness and fatigue
Formability	Hot-formed or cold-formed to shape without the necessity for subsequent heat treatment
Repairability	Weld repairable under service conditions
Fatigue	Notched fatigue strength of welded structure at least 90% of un-notched fatigue strength of basic metal for low cycle high stress and plastic strain conditions wherein number of peak stress cycles is 20,000 or less.
Corrosion	Not susceptible to stress corrosion. Corrosion fatigue strength equal to air fatigue strength
Stability	Will not creep or change dimensions significantly under operating stresses (75% of yield strength)
Availability	Based on present designs: plates up to 4,000 mm wide, 7,600 mm long, and 100 mm thick. Shapes in the forms of Tees having a mass of up to 220 kg/m
Isotropy	Mechanical properties identical in any plane strength

or else use materials that do not suffer from yielding. A wide variety of materials have been considered for use in submersible design.

The selection of an appropriate pressure hull material is based on the following criteria:

With regard to structural design

- *High Strength-to-Weight Ratio*: the material must be strong and light.
- *High Ductility*: the ability of a material to deform plastically without fracturing.
- *Resistance to Brittle Fracture*: the ability of a material to resist failures with little or no plastic deformation.
- *Fracture Toughness*: the ability of a cracked material to resist catastrophic propagation of cracks.

With regard to fabrication

- *Weldability*: the property of a material which allows it to be welded under normal conditions.
- *Formability*: the ability of a metal to be shaped through plastic deformation.

With regard to performance over time

- *Corrosion Resistance*: the ability of a material to resist its deterioration by chemical or electrochemical action within its environment.
- *Resistance to Stress-Corrosion and Cracking*: the ability of a material to resist failures caused by the combined action of a flaw (e.g. a crack) and tensile stress.
- *Resistance to Low Cycle Fatigue*: the ability of a material to withstand localised fluctuating stress.
- *Creep Resistance*: the ability of a material to withstand permanent deformation over time.
- *Stress Relief Embrittlement*: the process by which localised residual stresses in a metal are reduced by the reduction in the normal ductility when a metal is heated to a suitable temperature and slowly cooled.

Materials of possible interest for submersible pressure hull construction are (Figure 2.2):

- High-strength steels
- Light alloys, including high-strength aluminium and titanium
- Fibre-reinforced polymers (plastics) (FRP)
- MMCs
- Ceramic composites

Unreinforced ceramics, including solid glass, have very high compressive strength, but are too brittle for general application in pressure hulls.[1] MMCs, of which continuous ceramic (alumina) fibre-reinforced aluminium is a leading contender, offer very high specific strength and stiffness, but fabrication methods are not yet sufficiently developed, and costs are too high for these to be

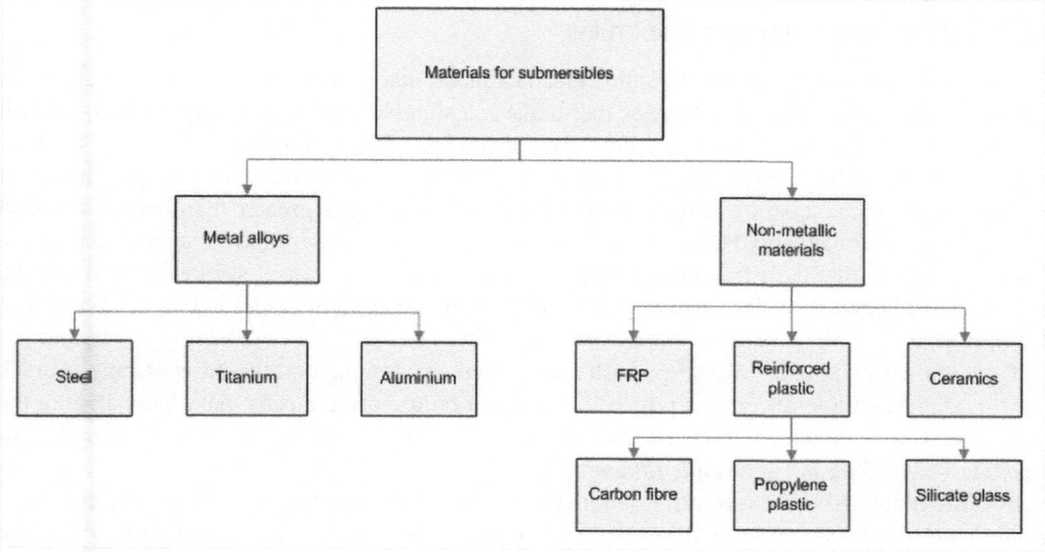

FIGURE 2.2 Materials for submersibles.

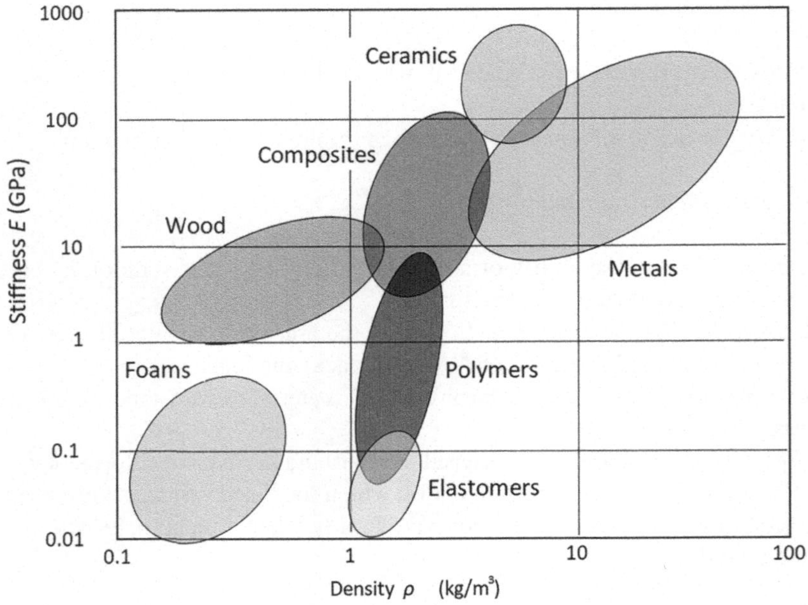

FIGURE 2.3 Stiffness–density diagram for selected groups of materials [12].

viable as a pressure hull option. FRP, therefore, emerges as the strongest alternative to steel or light alloys and offers some major potential advantages over the use of metal alloys [16]. From the point of view of structural performance, FRPs offer not only a high strength-to-weight ratio but are also characterised by high durability, stiffness, damping performance as well as flexural strength. From the point of view of performance over time, they offer resistance to corrosion, wear, and impacts [13].

Figure 2.3 provides a description of material properties (stiffness and density) for a range of materials. The highest stiffness-to-density ratios arise for materials in the upper left-hand region of the diagram.

2.2.1 Titanium in Submarine Structures

We shall consider first the use of titanium. This is a metal that is abundant in the earth's crust and possesses certain mechanical properties that make it a suitable candidate for use in pressure hull construction. Titanium and titanium alloys have a strength that is comparable to many stainless steels at a much lighter weight (higher strength/weight ratio). Additionally, they display excellent corrosion resistance, superior to that of aluminium and sometimes greater than that of stainless steel, and are non-magnetic. However, high-strength grades of titanium have one potential disadvantage, which is that their fracture toughness is much lower than that of steel (less than 60 MPa $m^{0.5}$ and a weld ductility of less than 10%) (Figure 2.4a). Low-strength grades generally do not have this problem.

Titanium has been used in aerospace structures that are designed on the basis of fatigue crack growth rate (defect tolerance). The possibility arises of failure mechanisms that do not arise in the aerospace sector, arising in other fields of application. These include *cold creep*, *sustained load cracking*,[2] *dwell fatigue*,[3] and *ripple fatigue*[4] [7].

Because titanium costs several times more than steel and it is technically difficult and expensive to fabricate marine vessel hulls using it, it has been avoided by naval shipbuilding industries in most countries. A different manufacturing process is required to that of steel, and the bending and shaping of heavy plates made of titanium alloy are much more difficult than in the case of steel.

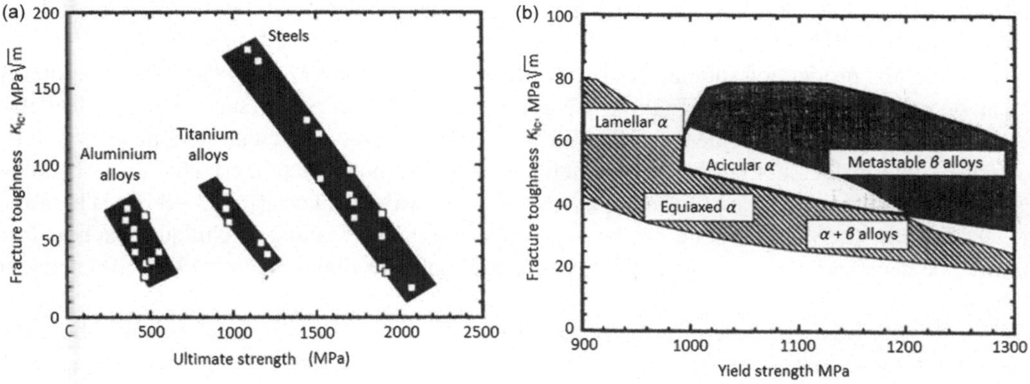

FIGURE 2.4 Fracture toughness as a function of strength. (a) Various structural alloys. (b) Titanium alloys.

Its extreme hardness causes problems when machining and shaping products. Titanium is also sensitive to contaminants, such as chlorine and cadmium. Prior to utilising titanium alloys for marine applications, advances are needed to provide more economical procedures for the fabrication of structural components.

In the 1970s, Soviet nuclear submarines with a titanium pressure hull were deployed. The hull weight needed for extreme operating depths was thus reduced by using this material.

2.2.2 Steels Developed for Submarine Pressure Hulls

Pressure hulls are designed so that they fail in an elasto-plastic mode, and thus, the yield stress has a direct bearing on the collapse load. Therefore, by increasing the yield stress, it is possible to increase the diving capability of a vessel.

A series of steels bearing the acronym HY ("high yield") was developed in the United States for use in the construction of nuclear-powered submarine pressure hulls. These are HY-80, HY-100, and HY-130, with each of the numerals 80, 100, and 130 referring to their yield strength in ksi (80,000 psi, (555 N/mm^2) 100,000 psi (694 N/mm^2), and 130,000 psi (903 N/mm^2)). Their chemical composition is summarised in Table 2.2.

HY-80 and HY-100 are both weldable grades, whereas HY-130 is generally considered unweldable. Modern steel manufacturing methods that can precisely control time/temperature during processing of HY steels have made the cost to manufacture more economical.

In addition to its high strength, HY-80 has good corrosion resistance and good formability. Using HY-80 steel requires careful consideration of the welding processes, filler metal selection, and joint design to account for microstructure changes, distortion, and stress concentrations. In addition to strength, attention to fatigue is required. This requires that the stress does not exceed the material yield stress at any stage of the vessel's life, this being an additional requirement that had to be considered when designing the pressure hull.

TABLE 2.2
Chemical Composition of HY-80, HY-100, and HY-130 Steels

	Percentage Content (%)								
Material	**C**	**Mn**	**P**	**S**	**Si**	**Ni**	**Cr**	**Mo**	**V**
HY-80	0.15	0.31	0.0100	0.018	0.27	2.23	1.30	0.35	–
HY-100	0.15	0.25	0.0075	0.002	0.25	3.12	1.58	0.45	0.015
HY-130	0.12	0.79	0.004	0.003	0.35	4.96	0.57	0.41	0.060

With the use of HY-80 steel, the operating depth of submarines could be increased to 550 m. It was first used in 1953 for the pressure hull of the *USS Albacore*, a small diesel experimental submarine. The first production submarines to use HY-80 steel were the *Permit* class. These reportedly had an operating depth of 400 m. In the UK Navy more recent *Astute* class submarines, a steel alloy designated Q1N that is equivalent to HY-80 has been used. This particular steel has low sulphur content so that it does not suffer from lamellar tearing. Its notch toughness characteristics have also been improved so that its minimum average Charpy value equals 100 J at −84°C. The use of higher-strength steels will depend on developments in joining (welding) techniques such as laser that will be necessary to overcome the fatigue-related problems that arise in the heat-affected zone (HAZ).

HY-100 steel was introduced for the deeper-diving *Seawolf* class of US fast attack submarines where the design dictated that the vessel had to operate at greater depths than previous designs. In addition to yield strength, the alloy had to have a high degree of toughness and impact strength and excellent fabrication properties. The strength increase was obtained by water quenching and ageing heat treatment. This is a low-alloy martensitic steel that contains small additions of chromium and nickel to help with corrosion resistance.

The shell plating of the pressure hull of the *Seawolf* submarines was 50 mm thick, and the submarines were capable of diving to depths of up to 600 m. Collapse depth estimates lie in the range 725–900 m.

Table 2.3 includes the critical mechanical properties of several different types of steel that are used for submarine and submersible pressure hull construction.

With the passage of time, steels of increasing yield strength have been developed and used; however, the use of steels whose yield strength exceeds 800 MPa has been limited because of the following reasons:

1. As yield strength increases, fracture toughness decreases. Therefore, it is necessary to develop higher-quality steels in order to retain the requisite fracture toughness.
2. It is more difficult to fabricate welded structures using these types of steels.

TABLE 2.3
Mechanical Properties of Selected Steels for Use in Submarine Pressure Hulls

Material Class	**Strength Yield Stress** MPa	**Fracture toughness Charpy Energy/Temperature** Joules/degrees C	**Comments**
HY-100, 80 HLES	690	47 J/−84	Common pressure hull steel
HSLA-100	690	81/−84	High-strength low-alloy steel not requiring preheating for $t < 25$ mm
HY-80, Q1N	552	47/80/−84	Common pressure hull steel
HSLA-80	552	81/−84	High-strength low-alloy steel not requiring preheating for $t < 25$ mm
BS EN 10 025 S550 QL1	550	40/−20	High-strength low-alloy steel not requiring preheating for $t < 25$ mm. Economical and suitable when weight saving required
BS EN 10 025 S355 K2G3	355	40/−20	General-purpose suitable for internal and external structures
Lloyds Grade EH36	355	34/−40	General-purpose suitable for internal and external structures

TABLE 2.4
Yield Stress of Steels Used in Submarine Pressure Hull Construction Worldwide

Yield Stress (MPa)	USA	UK	France	Germany	China	Russian Federation	Japan	Australia	Sweden
550	HY80	Q1N		1.394			NS56		
590			60HLES		921A TA5-A	AK-25 48-OT3			
630				1.3974					
690	HY100	Q2N				AB-2	NS70		
700			80HLES					BIS812 EMA	Weldox 700
785					980	AK-29	NS80		
890	HY130		100HLES			AB-3?	NS90		Weldox 900
980						AK-??			
1,078	HY156						NS110		
1,175						AK-??			

Source: http://seesaawiki.jp/w/doramarine.

3. Since high-yield steels permit increases in design stresses, these will be more prone to fatigue cracks. This problem has been encountered in surface ships.
4. At high stresses, the material becomes more prone to stress corrosion cracking.

Data concerning the use of steel in pressure hull construction in various navies are summarised in Table 2.4.

HSLA (*high-strength low-alloy*) steels are made to specific mechanical properties. They have better mechanical properties and corrosion resistance than low-carbon steels and are 20%–30% lighter. Small quantities of carbon are added to improve formability and weldability (0.05%–0.25%). Other attributes such as strength, impact strength, and corrosion resistance are improved by adding a number of metals (manganese, small quantities of copper, and others). For submarine structures, they are suited to non-pressure hull applications because weld cracking results in a reduction in fracture toughness.

Aluminium alloys have been used in submersible construction for structural members, hard containers, hydraulic components, and propellers. Their attractiveness stems from the high strength-to-density ratio, availability, low cost, and ease of fabrication. In submersible construction, series 5,000 alloys, certain 7,000 series alloys, and 6,061 alloy have been used.

2.2.3 Welding of High-Strength Steels

Because the properties of high-strength steels are acquired through heat treatment, the welding process results in a degradation of the properties of the HAZ. For this reason, strict controls of pre-heat, heat input, and temper bead sequences have been developed and imposed in order to minimise the extent, hardness, and brittleness of the HAZ.

In the past, the weld metal could be heat-treated and therefore had to have a higher alloy content to match the strength of the plate. This increased the risks of weld cracking. As a result, the weld metal strength had to be increased even more, to over-match the plate strength, in order to compensate for its lower toughness. This resulted in high-tensile residual stresses that were induced in the tee butt welds from the over-matched weld metals and the pre-heating regime used, which exceeded the through-thickness strength and ductility of the weld metal. Ultrasonic testing revealed that lamellar tearing (cracking) occurred in the pressure hull plating below the frame webs.

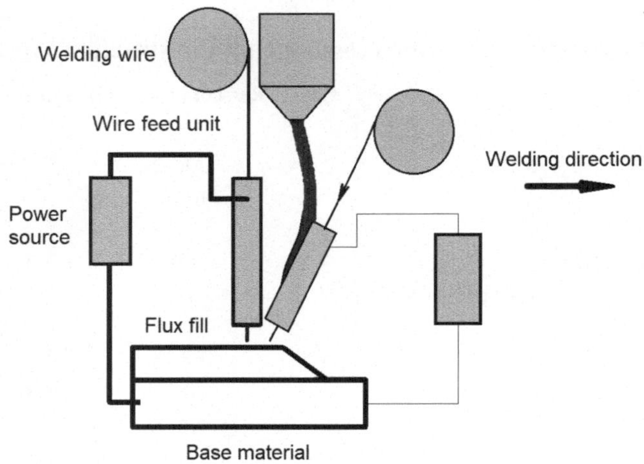

FIGURE 2.5 Tandem welding.

This called for the reduction of the residual stresses in the tee-butt connections. In order to achieve this, lower-strength electrodes, with pro-active control of preheating and deposition techniques, were developed.

Current welding practice for UK submarine manufacture involves twin *tandem* submerged arc welding for rotated sub-unit circumferential butt welds and for frame-to-hull and web-to-table tee butt welds. Pressure hull static circumferential butts and sub-unit vertical seams are welded by a mechanised *flux-core arc welding* (FCAW) process. A semi-automated FCAW process is used for all other welding.

Features of tandem welding:

- In tandem sub-arc welding, each of the two wires is connected to its own power source and fed simultaneously by its own feed unit (Figure 2.5)
- Both AC and DC sources can be used
- The wires are normally large diameter (3–5 mm), and deposition rates are about twice that of single-wire welding

2.3 COMPOSITE MATERIALS

Apart from high-strength steels and light alloys such as high-strength aluminium and titanium, materials of possible interest for pressure hull construction are artificial *composites*. These consist of two or more constituent materials bonded together so that the gross properties of the composite are superior to those of the constituents. Typically, a composite material consists of reinforcing fibres, a matrix that holds the fibres together, coatings added to improve bonding and load transfer at the fibre–matrix interface, and fillers that are added to reduce cost and improve dimensional stability.

Composites are a large group of materials that include fibre-reinforced polymers (FRPs),[5] metal–matrix composites (MMCs), and ceramic composites. They possess a number of important advantages over metal alloys which are:

- Low density (used in upper regions of the hull, thus lowering the CG of the vessel)
- Lower *W*/Δ ratio for pressure hull leading to enhanced endurance
- Ease of fabrication
- Easily repairable and reduced maintenance costs

- Poor conductor of electricity
- Reduced magnetic, acoustic, infrared, and radar signatures
- Good fatigue characteristics

2.3.1 Fibre-Reinforced Polymers: Composition and Properties

FRPs (plastics) form part of the group of *composite materials*. Composites consist of more than one type of material and are generally classified according to their constituents (Figure 2.6). They include a base material known as a *matrix* and filler materials that may be particles, sheets, or fibres, either natural or synthetic. The purpose of including different materials, each of which has different properties to the others, is for them to act in unison so that each overcomes the deficits of the others.

The types of composites that have found application in underwater vehicles are reinforced with fibres that are embedded within the matrix, as shown in Figure 2.7. Unidirectional fibres provide optimum axial strength in a selected direction. Fibres, however, may be arranged to be aligned in several directions.

In fibre-reinforced composites, the matrix is usually a ductile and tough material of low density whose strength is an order of magnitude less than that of the fibre. Its roles are to provide compressive strength, maintain the embedded fibres in the desired orientation, transfer stress to other phases, and protect phases from the environment.[6] Materials used for the matrix are polymers, metals, and ceramics. We shall be focussing on polymers, since they are used in fibre-reinforced composites used in underwater vehicles.

2.3.2 Reinforcing Fibres

Fibres on the other hand are significantly stronger in tension than the matrix and are much stiffer than it. They are also of low density, and their primary role is to bear tensile loads. In general terms, the strength of a composite material depends on fibre strength, length, and shape. It also depends

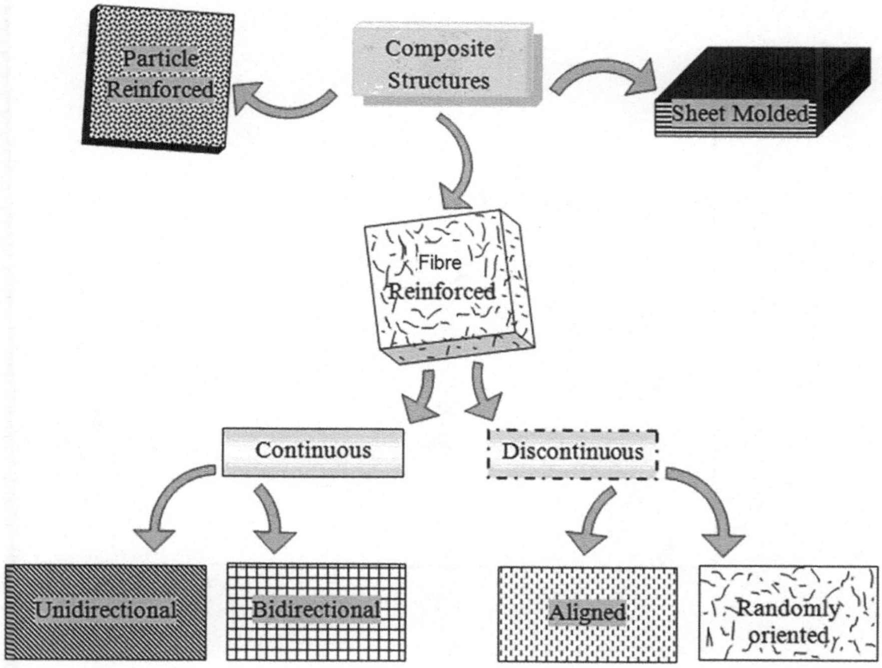

FIGURE 2.6 Classification of composites [1].

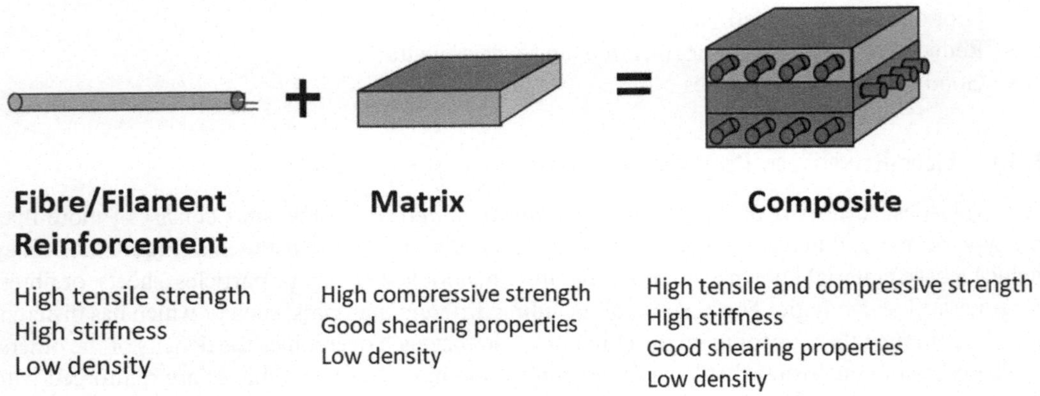

FIGURE 2.7 The basic composition of a fibre-reinforced composite [12].

on the bonding of the fibre to the matrix and the presence of voids, moisture, and flaws within the fibres. Materials used for reinforcing fibres are glass, carbon, organic materials, boron, ceramic, and metals. Figure 2.8 shows the specific strength of a number of different types of reinforcing fibres in use.

The cross section of fibres may be circular, square, or hexagonal, and their diameter lies in the range 0.025–0.15 mm. The length-to-diameter ratio is equal to 100 for chopped fibre and much longer for continuous fibre.

Of the many types of FRPs available, those that are reinforced using glass and carbon fibres can ensure that the weight-to-displacement ratio is low enough to allow the required payload to be carried [6]. Glass fibres are characterised by:

- High tensile strength
- Chemical resistance
- Moisture resistance

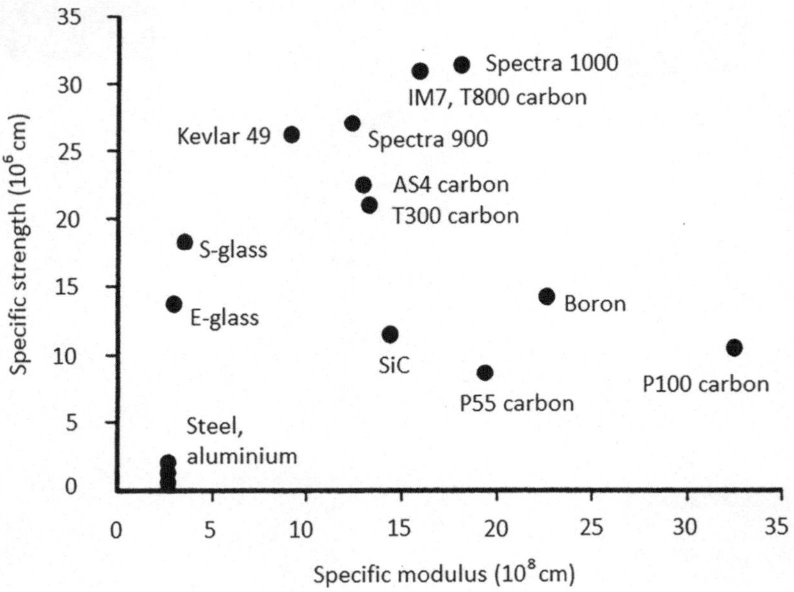

FIGURE 2.8 Strength of different types of fibres used in fibre-reinforced composites [3].

TABLE 2.5
Mechanical Properties of Various Types of Glass Fibre [21]

Fibre Type	Tensile Strength (MPa)	Compressive Strength (MPa)	Modulus of Elasticity (GPa)	Density (gm/cm³)	Thermal Expansion (ìm/m°C)	Softening *T* (°C)	Price ($/kg)
E-glass	3,445	1,080	76.0	2.58	5	846	~2
C-glass	3,300	–	69.0	2.49	7.2	–	–
S-2 glass	4,890	1,600	85.5	2.46	2.9	1,056	~20

- Satisfactory thermal properties
- Satisfactory electrical properties

The most important constituent of glass is silicon dioxide, SiO_2, which is abundant in nature (sand) and is prepared into a white powder. When cooled as fused quartz into a glass, it can be converted into fibres for use in FRP. The process of fibre production involves high temperatures, and in order to reduce the working temperature, specific materials (*fluxing agents*) are added to lower the melting point. As a result, a number of different types of glass fibres are produced, depending on the production process. These include A-glass, C-glass, E-glass, and S-glass[7] (Table 2.5).

The most common type of glass fibre used in FRP is E-glass, which is alumino-borosilicate glass with less than 1% w/w alkali oxides. E-glass is strong, has good heat resistance and high electrical properties. Additionally, a major advantage is its low price. For more critical needs, S-Glass offers higher heat resistance and about one-third higher tensile strength (at a higher cost) than that of E-glass.

Carbon fibre on the other hand is several times stronger than steel (4,000 MPa) yet five times lighter. Compared to 6,061 aluminium, it is seven times stronger, twice as stiff yet 1.5 times lighter. Other attractive properties are thermal stability and non-permeability to X-rays. Its stiffness covers a wide range of values (standard modulus, 207–240 GPa, intermediate modulus 240–340 GPa, and high modulus 340–960 GPa). There is, however, a trade-off between stiffness and strength. Additionally, the cost of carbon fibre greatly exceeds that of glass fibre; therefore, structures made entirely of carbon fibre–reinforced polymers are not easily affordable.

The question of fibre alignment is important and has to be considered for each application. It is possible to orient the fibres in one or more directions, as shown in Figures 2.9 and 2.10. A *lamina* is defined as any arrangement of unidirectional or woven fibres in a matrix, whereas a *laminate* is a stack of laminae arranged with their main reinforcement in at least two directions.

As seen in Figure 2.10, the maximum tensile strength is obtained when the fibres are oriented parallel to the applied load.

2.3.3 The Use of Polymers as a Matrix in a Composite Material

As mentioned previously, the material used for the matrix of composites used in underwater vehicles is some form of *polymer*. A polymer may be either natural or synthetic and is the outcome of the linking of a large number of *monomers*. Monomers are chemical compounds, examples of which that are of relevance to FRP are *epoxides*. These, following *polymerisation*, form *epoxy resin*,[8] which is commonly used as a matrix material [2]. Figure 2.11 shows the structure of the epoxide group that includes two atoms of carbon and one atom of oxygen (known as a *cyclic ether*).

The process of polymerisation involves the joining of a large number of monomers to form a molecule that may extend in three dimensions and is many times larger than each individual monomer that comprises it. It is possible to combine one or more types of monomers in polymerisation. Usually, at least 100 monomer molecules must be joined to form a substance with particular

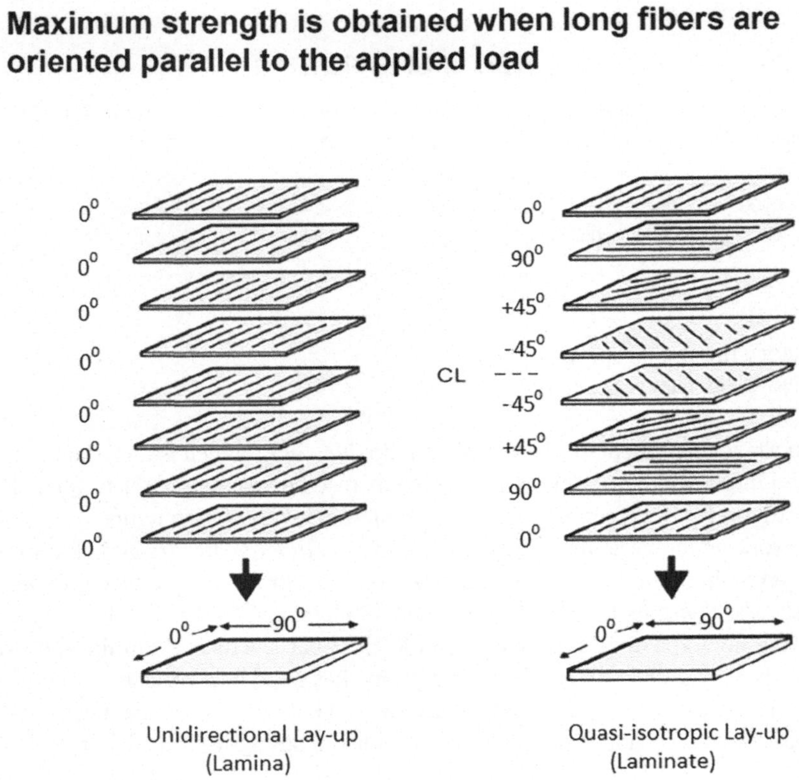

FIGURE 2.9 Fibre orientation [12]. (a) Unidirectional Lay-up (Lamina). (b) Quasi-Isotropic Lay-up (Laminate).

physical properties such as stiffness and strength. What differentiates polymers from other chemical compounds that involve a large number of molecules such as crystals is the formation of stable covalent bonds.[9]

Polymers in which linear polymer chains are cross-linked to form a three-dimensional network structure are called *cross-linked polymers.* The purpose of cross-linking is to improve mechanical properties (such as to increase rigidity) of the material comprising individual polymer chains (Figure 2.12a). The properties of the end product depend on cross-link density so that at the one end of the spectrum at which the cross-link density is non-existent, the resulting material is malleable and, upon heating, can be given any shape, whereas at the other end at which the cross-link density is high, the material is rigid. Polymers of both types are widely used.

The term *curing* is used to denote the process of cross-linking polymer chains. If curing involves the use of heat, a *thermosetting resin* (polymer) results (Figure 2.12b).[10] In addition to heat, curing may involve high pressure or mixing with a catalyst. The heat is not necessarily endothermic as it may be released as a result of the reaction of the polymer with a catalyst. Upon heating, thermosetting resins (*thermosets*) remain solid until the temperature reaches the point where it begins to degrade.

Among the thermosetting resins used for FRP (unsaturated), *polyesters* are by far the most widely used and find extensive use in marine structures (boats, etc). These resins are suitable for practically every moulding process available for thermosets. They offer ease of handling, low cost, dimensional stability, and a balance of good mechanical, chemical, and electrical properties. Polyester can be used in *orthophthalic* and *isophthalic* forms. The orthophthalic form is produced from phthalic acid and has a styrene content ranging from 35% to 45%. It is an inexpensive, general-purpose resin with applications that do not require corrosion and elevated temperature resistance or

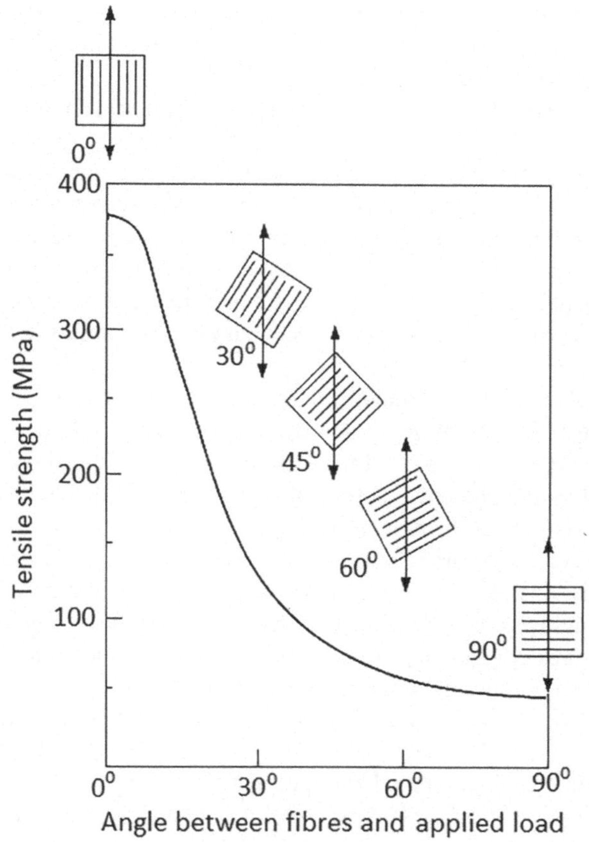

FIGURE 2.10 Effect of fibre angle on tensile strength [12].

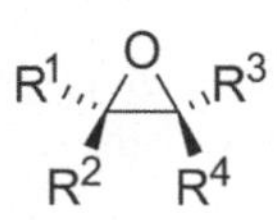

FIGURE 2.11 Structure of the epoxide group [20].

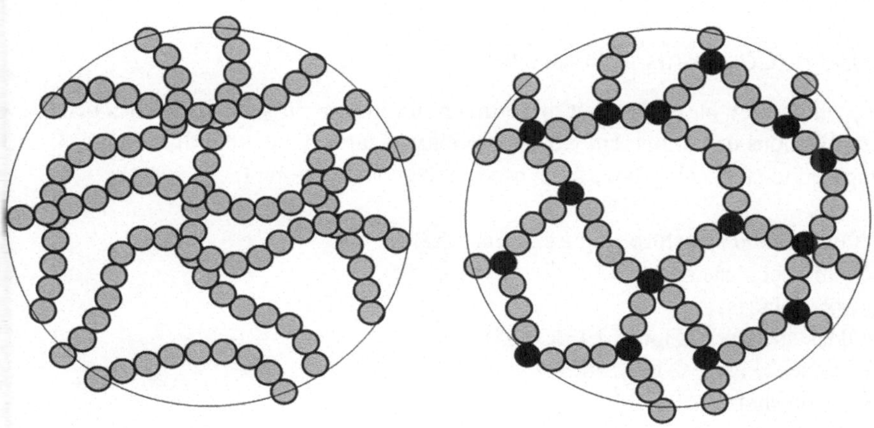

FIGURE 2.12 Polymer chains and cross-linking [22]. (a) Individual linear polymer chains. (b) Polymer chains that have been cross-linked to give a rigid 3D thermoset polymer.

high mechanical properties. The isophthalic form is produced from isophthalic acid and includes 42%–50% styrene. It is corrosion-resistant with better thermal and mechanical properties and is more expensive than orthophthalic polyester.

Other thermosetting resins that can be considered for use in marine applications are *epoxy resins* and *vinyl esters.*

Epoxy resins are low-molecular-weight pre-polymers and are one of the most commercially available classes of thermosetting resins that can react with themselves or with catalysts. Epoxy has two major known types: *bisphenol A*, with lower reported shrinkage and high corrosion resistance, and *bisphenol B*, with a better mechanical property and often has higher market value, the most popular epoxy being bisphenol A. Epoxy resins have better mechanical properties than their polyester and vinyl ester counterparts. They have excellent adhesion properties and are suited for service at higher temperatures.

To combine the favourable thermal and mechanical properties of epoxies as well as the low cost of polyesters, vinyl esters have been developed from styrene and the condensation of methacrylic acid ($CH_2{=}C(CH_3)CO_2H$) with an epoxy. In this process, only the position of the reactive chains differs from that of polyesters, thereby creating an elongated chain that exhibits a higher resistance to impact loads and also to the absorption of water.

Their flammability is a major drawback, and this necessitates the use of protective equipment. Compared to polyester, vinyl esters have a higher resistance to vibration loads, a higher thermal resistance as well as toughness, which results in reduced cracking following impact loads. Although extremely durable, vinyl ester is reported to not bond easily with other materials, and delamination of the repair resin can occur, making repairs more difficult.

Additional factors that influence material choice for underwater vehicles are:

- The effects of immersion (combined with pressure)
- Compressive fatigue
- Creep behaviour

In certain applications, fire resistance, thermal and electrical conductivity, and electromagnetic properties may also be important.

The relative advantages and disadvantages of polyester, vinyl ester, and epoxy are listed in Table 2.6, whereas the mechanical properties of these and other thermosetting resins are listed in Table 2.7.

The mechanical properties of FRP composites are contrasted with those of metal alloys and MMCs in Table 2.8.

2.3.4 Uses of Composites in Submarines

The use to which they have been put up to this point in time in submarines has been restricted to external applications due to fire, smoke, and toxicity hazards. One other drawback is their high cost when compared to steel. They have been used in the construction of:

- Fin fairings and structure (Figure 2.13a, p.55)
- Deck and deck casings
- Control surfaces
- Dry deck shelters[11] (Figure 2.13b, p.55)
- Bow domes (Figure 2.14, p.56)
- Propulsion shaft
- Stern structure
- Engine room (machinery foundations)

These have been installed on board a number of submarines in different navies.

TABLE 2.6
Comparison of Polyester, Vinyl Ester, and Epoxy Resins [11]

Material	Advantages	Disadvantages
Polyester	Easy to handle Cheaper than epoxy and vinyl ester	Moderate mechanical properties High styrene emission High shrinkage after cure Limited time frame after handling
Vinyl ester	High chemical resistance Better mechanical properties than polyester Less sensitive to temperature and humidity than polyester Better bonding and adhesion than polyester	High content of styrene More expensive than polyester High shrinkage after curing Need post-curing for superior mechanical properties
Epoxy	High mechanical and thermal properties Superior resistance to humidity Long handling periods possible Low shrinkage after cure	More expensive than vinyl ester Mixture of components is critical

TABLE 2.7
Mechanical Properties of Thermosetting Resins

Material	Specific Gravity	Modulus of Elasticity (GPa)	Poisson's Ratio	Tensile Strength (MPa)	Tensile Failure Strain (%)	Compressive Strength (MPa)	Heat Distortion Temp. (°C)
Polyester (orthophthalic)	1.23	3.2	0.36	65	2	130	65
Polyester (isophthalic)	1.21	3.6	0.36	60	2.5	130	95
Vinyl ester	1.12	3.4	–	83	5	120	110
Epoxy (DGEBA[a])	1.20	3.0	0.37	85	5	130	110
Phenolic	1.15	3.0	–	50	2	–	120

[a] DGEBA, Bisphenol A (*Di**G**lycidyl **E**ther of **B**isphenol **A***).
Source: Adapted from Table 1 in [16].

2.3.4.1 Sonar Domes

Composite construction is now the material of choice for sonar domes on surface ships as well as submarines. The domes provide a smooth flow around sonar transducers and protect the transducers from impact damage. Domes on submarines are truly massive structures that can measure 10 m across and weigh over 20 tonnes, as they are built to match the outer hull cross section (Figure 2.14b, p.56). The domes are water-filled, so they are not subjected to the hydrostatic pressure that the pressure hull experiences. Even so, demanding performance requirements are imposed on submarine bow dome structures.

The domes must have a high degree of acoustic transparency. To avoid signal distortion, stiffeners are undesirable. Early sonar domes were made with polyester resin and alternate plies of woven roving and mat E-glass. Minimising the void content is also of paramount importance in order to improve both acoustic transparency and overall part strength.

Investigation of material properties indicated that good sonar transmission at various frequencies was achievable with glass-reinforced plastics. When this attribute is coupled with weight-saving,

TABLE 2.8
Properties of Potential Pressure Hull Materials [16]

Material	Specific Gravity	Fibre Volume Fraction	Modulus of Elasticity (GPa)	Compressive Yield Strength (MPa)	Specific Modulus of Elasticity (GPa)	Specific Compressive Yield Strength (MPa)
Steel HY80	7.8	–	207	550	27(1)[a]	71(1)
Steel HY100	7.8	–	207	690	27(1)	88(1.2)
Steel HY130	7.8	–	207	890	27(1)	114(1.6)
Aluminium (L65 alloy)	2.8	–	70	390	25(0.9)	139(2.0)
Titanium (6-4 STOA alloy)	4.5	–	110	830	24(0.9)	184(2.6)
GRP[b] (polyester/E-glass)	1.7	0.34	15	210	8.8(0.33)	124(1.7)
GRP (epoxy/S-glass UD[c]	2.1	0.67	65	1200	31(1.1)	571(8.0)
GRP (epoxy/S-glass filament wound)	2.1	0.67	50	1000	24(0.9)	476(6.7)
CFRP[d] (epoxy/HS carbon UD)	1.7	0.67	210	1200	124(4.6)	706(9.9)
CFRP (epoxy/HS carbon fil. wound)	1.7	0.67	170	1000	100(3.7)	588(8.3)
MMC (6061 alum./ SiC carbon UD)	2.7	0.5	140	3000	52(1.9)	1111(16)
MMC (6061 alum./ HS carbon UD)	2.3	0.5	200	600	87(3.2)	261(3.7)
MMC (6061 alum./ alumina fibre UD)	3.1	0.5	190	3100	61(2.3)	1000(14)

[a] Values in parentheses are normalised with respect to HY80.
[b] Glass fibre–reinforced polymer.
[c] Unidirectional.
[d] Carbon fibre–reinforced polymer.

freedom from maintenance and corrosion, and ease of fabrication of complex shapes, GRP is seen to be preferable to rubber and steel.

2.3.4.2 Periscope Fairings and Antenna Masts (GRP)

Submarine periscope and antenna masts have foil-shaped fairings that have traditionally been made of GRP (Figure 2.13c). The masts are all made of composite materials, but the periscopes have a K-Monel "I-beam" structural core.

The cost per unit weight of composites for submarine use is an order of magnitude greater than that of steel alloys that find their way on to surface ships. Submarine composites are marine composites manufactured to aerospace standards.

In the French *Le Triomphant* class of nuclear submarines, 50% of the exostructure components are made of composite materials (Figure 2.15, p.57). These include the outer deck, sonar dome, propulsion nozzle, rudders, and fairings. The reasons for this are better resistance to corrosion, weight gain, and acoustic transparency (Table 2.9, p.57).

The following table provides a comparison of the properties of candidate materials for pressure hulls, both metallic and non-metallic.

FIGURE 2.13 Submarine components made of composite materials. (a) X-type rudder installed in Columbia class submarine (USN) [5]. (b) Dry deck shelter (HMS Spartan). (c) Periscopes and radar equipment.

(a)

(b)

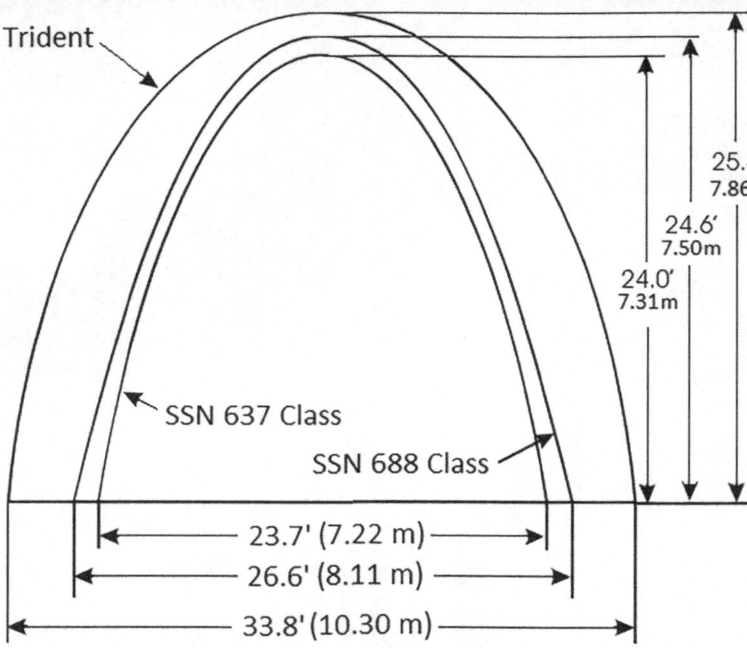

	Surface Area sq-ft	Weight lbs
SSN 637 Class	1388	26,260
SSN 588 Class	1519	28,177
Trident	2282	43,700

FIGURE 2.14 Bow dome of Los Angeles class submarine. (a) Bow dome of Los Angeles class submarine [4]. (b) Bow dome sizes.

FIGURE 2.15 Use of composites (shaded regions) in the hull structure of the *Le Triomphant* class of submarines [9].

TABLE 2.9
Summary of Differences Between Steel and Composites Construction

Property	Steel Construction	Composite Construction
Weight	High	Allows significant reduction in structural weight
Corrosion	Rusts in marine environment resulting in high maintenance cost	Very durable in marine environment, little maintenance
Combustibility	Non-combustible, will not contribute to fire or generate toxic fumes	Combustible, surface must be protected in fire hazard areas
Thermal conductivity	High, must be insulated to prevent fire propagation and to control infrared signature	Low, inherent insulation more than sufficient
Electrical conductivity	High, inherently provides electromagnetic shielding	Low, must embed conductive layer if electromagnetic shielding is needed

The following figure gives an idea of the profile of cylindrical pressure hulls fabricated in both metallic and non-metallic materials (Figure 2.16).

With regard to the end closures, special problems arise in the fabrication of hemispherical or tori-spherical dome closures forming the ends of a composite pressure hulls (Figure 2.17). In a spherical shell, optimum design requires uniform thickness and isotropic (membrane and flexural) elastic rigidities and strength.

2.3.4.3 The Future

The combined use of different materials has also been studied, and it has been found that titanium can be successfully used for the skeletal structure of a composite hull (MMC – Table 2.10).

The potential use of composites in submarines is illustrated in Figure 2.18. This includes the fore and aft structure, propeller shafts, propeller, machinery foundations, HP air tanks, piping, valves, flooring, and partitions, as well as the pressure hull.

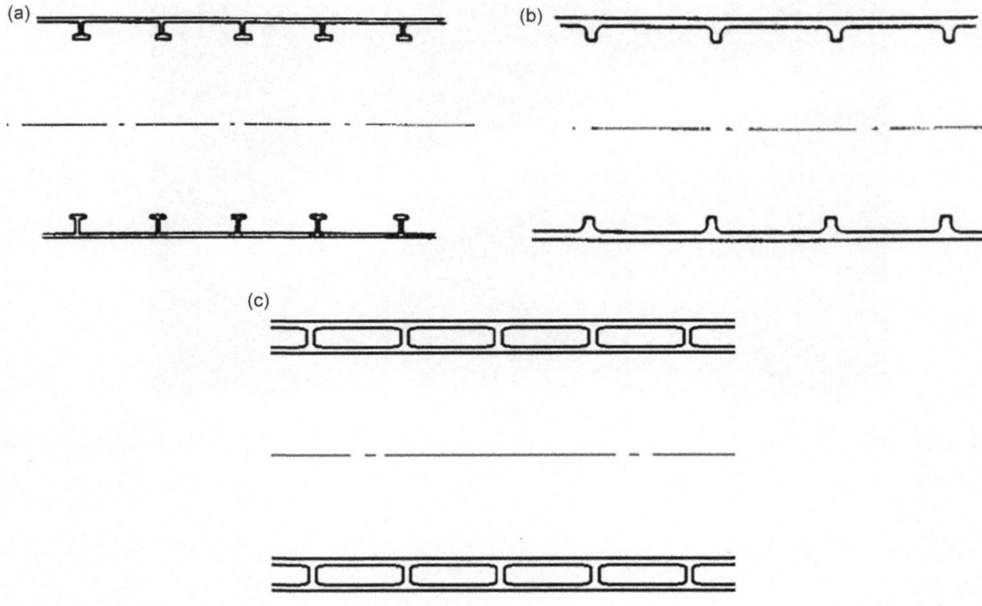

FIGURE 2.16 Cylindrical pressure hulls in various materials [16]. (a) Steel ring-stiffened cylinder. (b) Composite cylinder with integral ring stiffeners. (c) Composite sandwich cylinder.

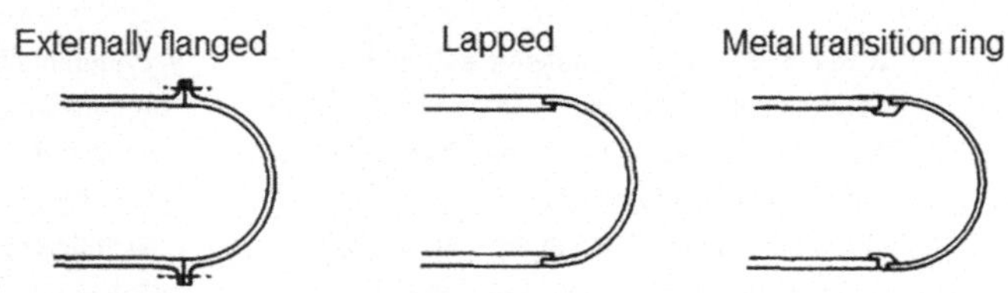

FIGURE 2.17 Dome/cylinder joint configurations.

TABLE 2.10
Properties of Some Candidate Pressure Hull Materials [16]

Material[a]	Specific Gravity	Fibre Volume Fraction	Elasticity Modulus (GPa)	Compressive Yield Strength (MPa)	Specific Modulus of Elasticity (GPa)	Specific Compressive Yield Strength (MPa)
Steel HY80 (Q1N)	7.8	–	207	550	27(1)	71(1)[b]
Steel HY100 (Q2N)	7.8	–	207	690	27(1)	88(1.2)
Steel HY130 (Q3N)	7.8	–	207	890	27(1)	114(1.6)
Aluminium (L65 alloy)	2.8	–	70	390	25(0.9)	139(2.0)
Titanium (6-4 STOA alloy)	4.5	–	110	830	24(0.9)	184(2.6)
GRP (polyester/E-glass woven roving)	1.7	0.34	15	210	8.8(0.33)	124(1.7)
GRP (epoxy/S-glass unidirectional)	2.1	0.67	65	1,200	31(1.1)	571(8.0)
GRP (epoxy/S-glass filament wound)	1.7	0.67	50	1,000	24(0.9)	476(6.7)

(Continued)

TABLE 2.10 (*Continued*)
Properties of Some Candidate Pressure Hull Materials [16]

Material[a]	Specific Gravity	Fibre Volume Fraction	Elasticity Modulus (GPa)	Compressive Yield Strength (MPa)	Specific Modulus of Elasticity (GPa)	Specific Compressive Yield Strength (MPa)
CFRP (epoxy/HS carbon unidirectional)	1.7	0.67	210	1,200	124(4.6)	706(9.9)
CFRP (epoxy/HS carbon filament wound)	1.7	0.67	170	1,000	100(3.7)	588(8.3)
MMC (6061 alum./SiC fibre UD)	2.7	0.5	140	3,000	52(1.9)	1,111(16)
MMC (6061 alum./HS carbon UD)	2.3	0.5	200	600	87(3.2)	261(3.7)
MMC (6051 alum./ alumina fibre UD)	3.1	0.5	190	3,100	61(2.3)	1,000(14)

[a] GRP, glass-reinforced polymer; CFRP, carbon fibre–reinforced polymer; MMC, metal–matrix composite.
[b] Values in parentheses are normalised with respect to HY80.

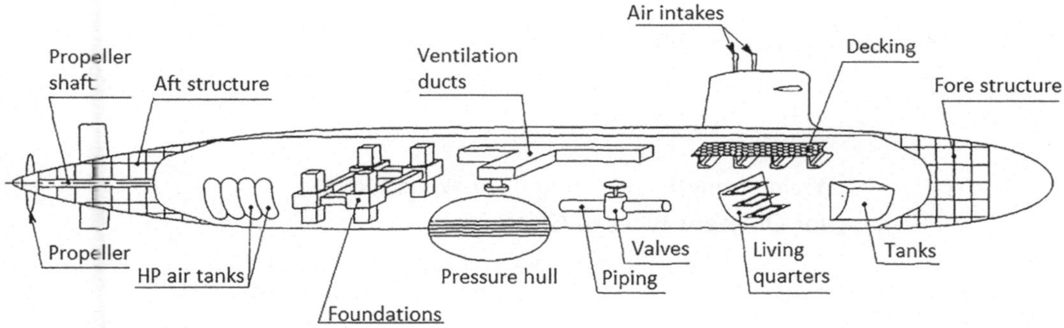

FIGURE 2.18 Potential use of composite in submarines [9].

2.4 USES OF GLASS AND ACRYLIC IN UNDERWATER VEHICLES

2.4.1 Glass

Ever since research began in the second half of the 20th century on the design of deep-diving submersibles, glass has been one of the materials that has been considered for use because of its particular mechanical properties and also because for many uses, its transparency is desirable.

In the 1960s, research was carried out on deep-sea glass domes, but the results were not successful because even though the domes were manufactured to satisfactory standards, incompatibility with mounting systems led to the glass cracking near the interface, leading to dome failure. This was due to the different degrees of contraction of the two materials; this and related problems have since been resolved.

The following table includes data on the yield strength and the strength-to-weight ratio of different types of glass. The density is equal to 2,600 kg/m^3 throughout.

The strength of glass depends on the number of impurity defects (Figure 2.19). Ordinary glass has a nominal strength of 70 MPa, whereas synthetic glass has yield strength of 500 MPa due to fewer impurity defects. Glass fibres obtained by drawing from synthetic glass have yield strength of 4,000 MPa for 10 cm gauge length [8].

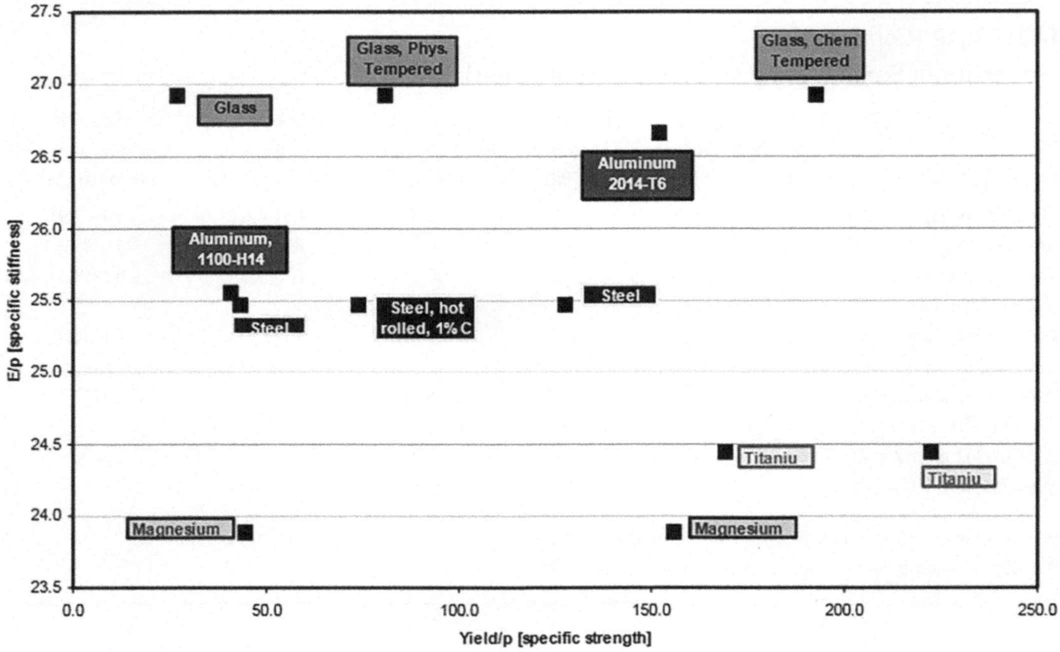

FIGURE 2.19 Specific strength and stiffness of glass and other materials [8].

TABLE 2.11
Yield Strength and Strength-to-Weight Ratio for Different Types of Glass

Type	Yield Strength (MPa)	Strength/ Weight Ratio
Glass	70	0.027
Glass, physically tempered	210	0.081
Glass, chemically tempered	500	0.192
Glass fibres	4,000	1.538

The strength of a particular grade of glass depends on the surface condition so that a protected surface maintained in excellent condition is necessary to maintain the prescribed strength. Flaws act as stress concentrations and reduce strength. Measures to improve the surface condition are fire-polishing and the use of lacquers.

Comparing the data in Table 2.11 with that for other materials in Table 2.10, we note that the strength-to-weight ratio for all types of glass is significantly higher than for all types of metal alloys. This renders glass, in principle at least, a prime candidate for use in underwater vehicles. Other advantages that glass has when compared to metal alloys are

- Transparency, rendering it suitable for visual observations and the use of cameras (Figures 2.20, 2.21)
- Ability to drill and spot face for O-ring seal penetrations
- Invisibility to radio frequencies and magnetic fields
- Electrically non-conductive

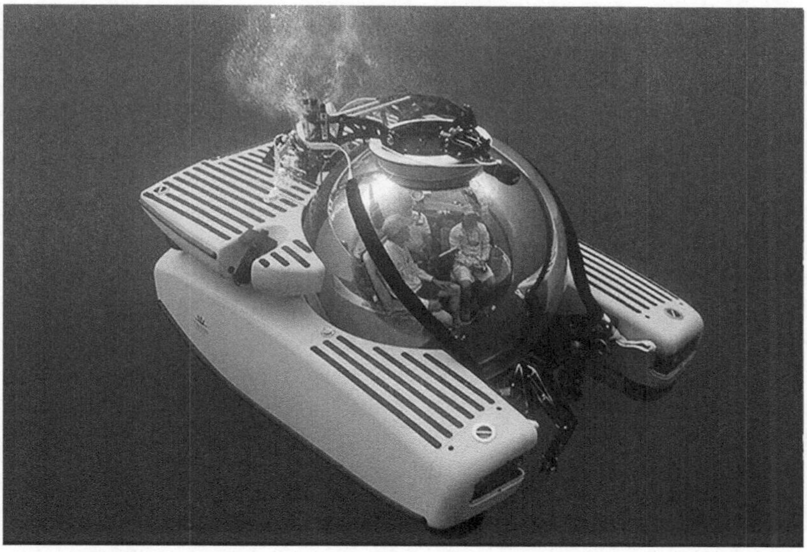

FIGURE 2.20 Spherical pressure hull [14].

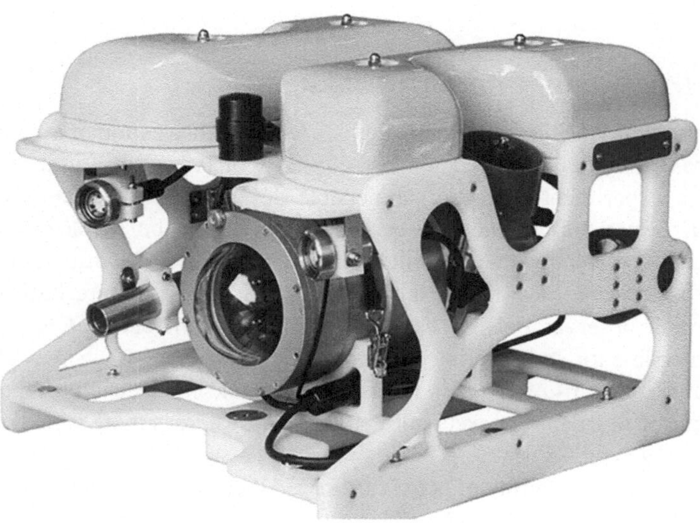

FIGURE 2.21 Camera housing [14].

- Corrosion-resistant
- Non-polluting
- Low cost

The major disadvantage of glass is its *brittleness,* and as a result, it is subject to damage from impact and *spalling*, that is, fragments are broken off as a result of impact or cyclic loading. This is the major reason for which it has not until now found its way into the fabrication of the pressure hulls of large-size submersibles. It is, however, extensively used to house components in all the main classes of undersea vehicles such as AUVs, landers, ROVs, and manned submersibles. Glass housings are made using borosilicate and BK-7[12] and have the dual advantage of acting as both buoyancy chambers and pressure-resistant housings for small and lightweight autonomous instrumentation packages. Glass spheres are fabricated having a diameter in the range 250–450 mm and are used as flotation spheres and instrument housings for depths up to full ocean depth. They are generally fabricated by mating two hemispheres.

Early on, a list of desired properties for glass suitable for use in underwater vehicles was prepared [10]. This list is given below.

- High compressive working strength (>700 MPa) and tensile strength (>175 MPa)
- Elastic constants (modulus of elasticity and Poisson ratio) compatible with those of metals selected for coupling at joints
- Low thermal expansion for ease of residual stress reduction in manufacturing and for low temporary stresses in production and end use. The thermal expansion ought to match that of adjoining metals.
- High transparency and minimal light scattering
- Low density
- Adequate chemical durability for the marine environment

The type of glass currently used in underwater vehicles is *borosilicate* glass that is used for instrument housings that can be deployed at any depth in the ocean. The main constituents of borosilicate glass are silica and boron trioxide, whereas other additives that are used are soda ash and alumina. Borosilicate glass has a very low coefficient of thermal expansion and can withstand thermal shock better than common glass.[13]

2.4.2 Acrylic (Polymethyl Methacrylate)

From early on, acrylic has been used in submersibles for the construction of viewing ports, (bathyscaphe *Trieste*), based on its transparency and the quality of light transmittance. An early report gives details of an experimental acrylic submersible designed and fabricated by the US Navy (*NEMO* – *N*aval *E*xperimental *M*anned *O*bservatory) [15]. This submersible performed numerous dives up to 186 m, and it was found that visibility was excellent with no shape distortion (Figure 2.22).

Subsequent developments have led to the use of acrylic as a pressure hull material for submersibles, underwater observatories, ROVs, and hyperbaric chambers. In Chapter 12, the design of an acrylic submersible for an operating depth of 400 m is described, whereas at present, acrylic submersibles operate to depths of 1,000 m. In principle, acrylic can be used for the pressure hull of submersibles operating at a depth in excess of 2,400 m [18], whereas there is no limit to the operating depth of acrylic viewing ports.

The term acrylic refers to a family of petroleum-based thermoplastic synthetic polymers that are termed polymethyl methacrylate (PMMA).[14] These are often classified as a type of glass in that they are non-crystalline vitreous substances.

The properties that make acrylic suitable for use in underwater vehicles are:

- High optical clarity. It transmits up to 92% of visible light (3 mm thickness)
- Lightweight
- Excellent dimensional stability
- High impact resistance
- Good weatherability and UV resistance
- Low water absorption (0.20% for immersion lasting 24 hours)

Its mechanical properties are given in the following table.

It is seen from Table 2.12 that acrylic has a low modulus of elasticity. This is an advantage when acrylic is used as a structural material for underwater vehicles because it permits localised yielding, leading to a redistribution of stresses. This is important because when used in submersibles,

FIGURE 2.22 *DSV-5 NEMO* on museum display.

TABLE 2.12
Mechanical Properties of Acrylic (PMMA)

Property	Value	Unit
Compressive strength	103	MPa
Poisson's ratio	0.4	–
Modulus of elasticity	3.06	GPa
Material density	1,170–1,200	kg/m^3

it interfaces with other materials. Acrylic plastic, unlike glass, flows plastically and cracks before failing catastrophically, thereby providing an additional measure of safety [17].

A particular distinguishing feature of acrylic, compared to metal alloys, is that, being a thermoplastic polymer, its material properties are time- and temperature-dependent, that is, it is a *viscoelastic* material. Viscoelastic materials exhibit both viscous and elastic characteristics when subjected to external loading. The differences between a linear elastic and a viscoelastic material are illustrated in Figure 2.23.

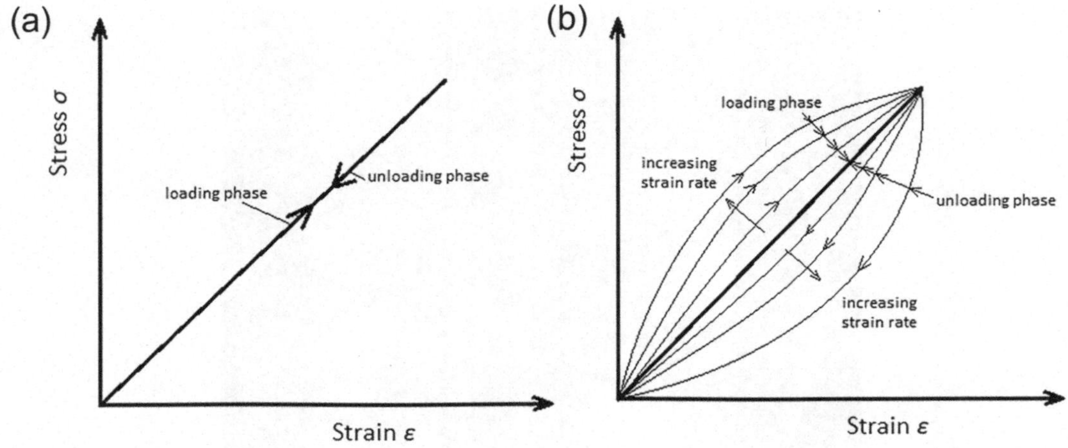

FIGURE 2.23 Comparison of linear elastic and viscoelastic materials. (a) Linear elastic material. (b) Viscoelastic material.

It is seen that in the case of a linear elastic material such as a metal alloy, the stress–stress curve is linear and follows the same path during both the loading (stretching) and unloading (relaxing) phases (Figure 2.23a). The slope is equal to the modulus of elasticity, E.

Viscoelastic materials on the other hand follow different paths in the loading and unloading stages. This phenomenon is called *hysteresis* and is illustrated in Figure 2.23b, with the arrow heads indicating the direction of loading and unloading. At very low strain rates, the behaviour approximates the behaviour of the linear elastic solid, but as the strain rate is increased, hysteresis effects become more accentuated. It is thus seen that different stress–strain curves are obtained for different strain rates. If the stress is maintained constant in a viscoelastic material, the strain increases with time. This phenomenon is termed *creep.*

Thermoelastic polymers (and acrylic) exhibit viscoelastic behaviour at ambient conditions. In particular,

- At low temperatures and high strain rates, elastic behaviour is observed
- At high temperatures and low strain rates, viscous behaviour is oberved
- At intermediate temperatures and strain rates, viscoelastic behaviour is observed

Figure 2.24 presents tensile and compressive yield stress versus time curves for acrylic (PMMA). These were obtained from creep data (strain–time curves) that were then transformed into a family of isochronous stress–strain curves from which yield stress values were determined and plotted as shown in Figure 2.24 [17].

The dependence of acrylic plastic behaviour on temperature is shown in Figure 2.25, and it is observed that all properties shown have a linear inverse relationship with temperature. The variation of collapse depth of model acrylic plastic hulls with temperature is clearly shown as curve B.

Other conditions (limit states) that need to be considered are fatigue and impact loading. With regard to fatigue, acrylic plastic has a well-defined S-N curve with a knee at 10,000 cycles and a stress range of 45 MPa. An endurance limit exists so that for stress ranges below this level fatigue fracture does not occur. Since, however, the number of cycles to which an acrylic pressure hull is expected to be subjected to throughout its lifetime is low, fatigue is not expected to be a problem. The question of impact strength is much more important because acrylic has a much lower strength than metal alloys. This needs to be compensated for by design innovations such as the use of a structural cage [17].

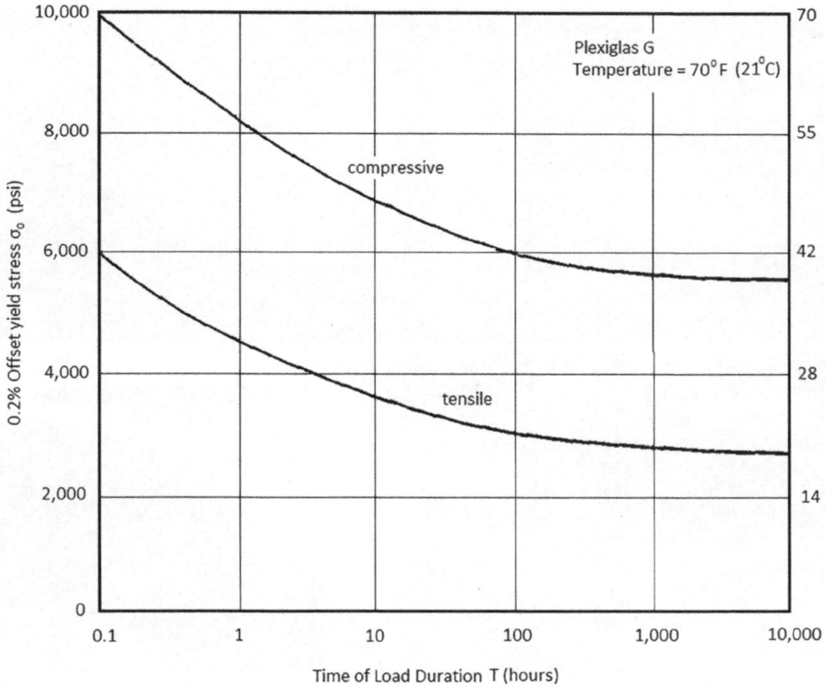

FIGURE 2.24 Yield stress–time curves for acrylic plastic [17].

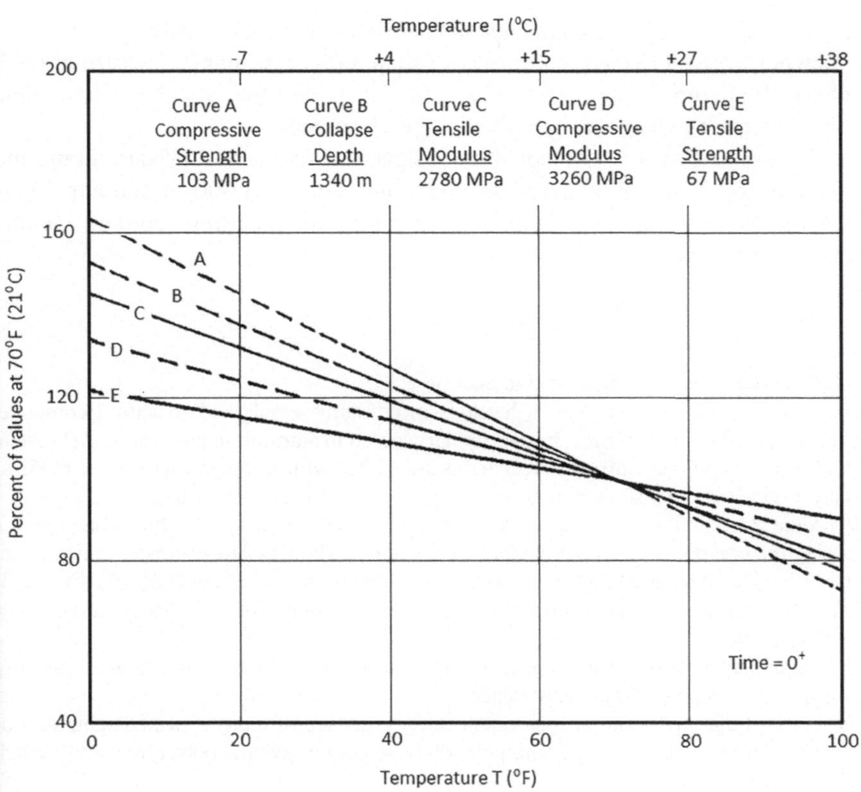

FIGURE 2.25 Properties of acrylic plastic (Plexiglas G) as a function of temperature [17].

FIGURE 2.26 Uses of acrylic in underwater vehicles. (a) Viewing ports, (b) submersible pressure hull, (c) The *Triton* 3300 [19], and (d) The *Triton Titanic Explorer* [19].

Compared to glass, its major disadvantage is that it scratches more easily.

Acrylic viewports are fabricated in a variety of shapes (flat disc, conical frustrum, double bevel, and hemispherical) (Figure 2.26a) and are used in private and tourist submarines, submersibles, decompression and diving chambers as well as research vessels.

Figures 2.26c and d show two submersibles designed and built by a private submarine builder. The first of these, the *Triton* 3300/6 has an operating depth of 1,000 m and can accommodate six persons, whereas the second, the *Titanic Explorer*, has an operating depth of 4,000m and can accommodate two persons.

NOTES

1 Uses of glass and acrylic are discussed in Section 2.4.
2 *Sustained load cracking* occasionally develops in pressure vessels and structural components under stress for sustained periods of time. It is particularly noted in aluminium pressure vessels, such as diving cylinders. It is not a manufacturing defect and is associated with certain alloys such as 6,351 aluminium alloy and service conditions (overstressing due to excessive internal pressure).
3 *Dwell fatigue* refers to the reduction in the fatigue life of an aircraft gas turbine component as a result of exposing the component (turbine blade) to a constant high mean stress during cruising, between the increase of the load during take-off and the reduction on landing. The "cold" of cold dwell fatigue refers to the fact that this phenomenon occurs at temperatures of around 100°C or less, in a relatively cold part of the engine.
4 *Ripple fatigue* may arise over time in pressure transducers fitted in aircraft hydraulic systems in which pressure spikes (*pump ripple*) are experienced.
5 A *polymer* is a large molecule, or macromolecule, composed of many repeated subunits (monomers). Synthetic polymers include epoxy resin, polyethylene, polypropylene, polystyrene, polyvinyl chloride, and others (see Chapter 10).
6 By *phase*, we denote the different load-bearing constituents that comprise a composite material. These are the matrix and fibres in each orientation.

7 The letters are used to distinguish between different primary properties. Thus, E (electrical) denotes low electrical conductivity, S (strength) denotes high strength, and C (chemical) denotes high chemical durability.
8 The term *resin* dates from the early years of the plastic industry and was used to denote naturally occurring amorphous solids.
9 In a covalent bond, two atoms are connected to each other by the sharing of two or more electrons.
10 The other main group of fibre-reinforced polymers are *thermoplastic* (*thermo-softening*) polymers. In these, the cross-links between polymer chains are weak or non-existent, and this leads to the material being more malleable at a structural level than thermosetting polymers. Thermoplastic polymers have low melting points so that if heated soften and are thus easily moulded into various shapes. Examples are *polystyrene* and *polytetrafluoroethylene* (PTFE).
11 A *dry deck shelter* is a removable module that can be fitted to a nuclear submarine to allow special force personnel access and entry.
12 BK7 is a high-quality optical glass that is used whenever the additional benefits of fused silica are not required. It is a relatively hard material with extremely low bubble and inclusion content, while providing excellent transmittance throughout the visible spectrum.
13 Borosilicate glass is commercially available under a number of trade names, the most well-known being *pyrex*.
14 It is also known by its trade names *perspex, plexiglas,* etc.

REFERENCES

1. Altenbach, H., Altenbach, J., Kissing, W. (2018). *Mechanics of Composite Structural Elements*. Springer; Singapore.
2. Encyclopaedia Britannica. (2024). *Polymerization*. Available at: https://www.britannica.com/science/polymerization (Accessed 15 February 2024)
3. Erklig, A. (2005). *Introduction to Composite Materials Lecture Notes*. ME 429 Dept Mech. Eng. Gaziantep Univ., Turkiye.
4. Global Security. (2024). SSN-773-p01.jpg. Available at: https://www.globalsecurity.org/military/systems/ship/images/ssn-773-p01.jpg. (Accessed 15 February 2024)
5. Globe Composite. (2024). *The "X-Factor" in Columbia-Class Submarine Design*. Available at: https://www.globecomposite.com/blog/x-factor-columbia-class-submarine-design. (Accessed 15 February 2024)
6. Graham, D. (1995). Composite pressure hulls for deep ocean submersibles. *Compos. Struct.* Vol. 32, 331–343.
7. Kostrivas, A., Smith, L.S., Gittos, M.F. (2003). "Sustained load cracking in titanium alloys." *10th World Conference on Titanium Hamburg*, Wiley-VCH, Hamburg.
8. Lehman, R. *The Mechanical Properties of Glass. Glass Engineering 150:312*. Dept. Ceram. Mat. Eng. Rutgers Univ, New Brunswick, NJ.
9. Lemière, Y. (1992). L' evolution des materiaux composites dans la construction des sous-marins. *La construction navale en composites, IFREMER, Actes de colloques*. Vol. 15, 43.
10. National Materials Advisory Board (1970). "Massive glass as a naval-structural material." Report NMAB-262, Washington, DC.
11. Nodehi, M. (2022). Epoxy, polyester and vinyl ester based polymer concrete: A review. *Inno. Infrastruct. Solut.* Vol. 7, 40–64.
12. Pourang, H. (2018). *Introduction to Composite Materials*. Iran University of Science and Technology, Iran.
13. Rajak, D.K., Pagar, D.D., Menezes, P.L. Linul E. (2019). Fiber-reinforced polymer composites: manufacturing, properties, and applications. *Polymers* Vol. 11, 1667–1704.
14. Rayotek Scientific Inc. (2024). *Deep Sea, Downhole & Marine Molded Glass Dome & Molded Glass Sphere*. Available at: https://rayotek.com/marine-molded-glass-domes-spheres.htm (Accessed 15 February 2024)
15. Rockwell, P.K., Migliore, H.J. (1972). "Operational evaluation of NEMO, an acrylic-hulled submersible." Technical Report R-778. Naval Civil Engineering Laboratory, Port Hueneme, CA, USA.
16. Smith, C.S. (1991). Design of submersible pressure hulls in composite materials. *Mar. Struct.* Vol. 4, 141–182.

17. Snoey, M.R., Katona, M.G. (1971). "Structural analysis of a full-scale spherical acrylic plastic pressure hull." Technical Report R-716. Naval Civil Engineering Laboratory, Port Hueneme, CA.
18. Stachiw, J.D. (2003). *Handbook of Acrylics for Submersibles, Hyperbaric Chambers, and Aquaria.* Flagstaff, AZ: Best Publishing Company, 1054 p.
19. Triton Submarines. (2024). *Triton 4000/2.* Available at: https://tritonsubs.com/subs/gullwing/?dc=ultradeep (Accessed 15 February 2024)
20. Wikipedia. (2024). *Epoxide.* Available at: https://en.wikipedia.org/wiki/Epoxide (Accessed 15 February 2024).
21. Wikipedia. (2024). *Glass Fiber.* Available at: https://en.wikipedia.org/wiki/Glass_fiber (Accessed 15 February 2024).
22. Wikipedia. (2024). *Thermosetting Polymer.* Available at: https://en.wikipedia.org/wiki/Thermosetting_polymer. (Accessed 15 February 2024).

3 Cylindrical Shells under External Pressure

3.1 INTRODUCTION

In this chapter and the ones that follow, we shall consider one of the most important structural types encountered in underwater vehicles, and in particular in submarines: the ring-stiffened cylindrical shell. This is an important element of many different kinds of underwater vehicles as well as fixed structures: submarines, submersibles as well as offshore platforms. In all of these, a substantial part of the shell plating has a cylindrical form and is stiffened in some way. Similarities are limited to this because loading, materials used, and dimensions differ.

Figure 3.1 provides an overall view of the major components of a submarine, including the cylindrical pressure hull, the fore and aft ends, the sail fin, and a transverse bulkhead.

Before developing the equations of equilibrium of cylindrical shells, we shall provide a general introduction to the structural layout of submarines, following which we will provide an introduction to the theory of shell structures that is relevant to underwater vehicles, namely that of *shells of revolution*. We will then be in a position to consider the strength of cylindrical shells and, following that, of other structural types used in underwater vehicles.

3.1.1 A General Description of Submarine Structures

Figure 3.2 shows a longitudinal section of the structure of a modern submarine. We note that, apart from the external casing that is in direct contact with the ocean environment, there exists an additional casing to which stiffeners are attached. This casing contains all habitable and operational spaces, whereas the external casing provides the required hydrodynamic form to the vessel and encloses a number of ballast tanks and other appendages as well as the pressure hull. Since the hydrostatic pressure inside these tanks is the same as the external pressure, they do not need to be strengthened to withstand the hydrostatic pressure. It ought to be added that not all submarines have this type of layout as in certain cases, external ring stiffeners are used. Also, it is possible that part of the pressure hull is not enclosed by an outer casing.

The form of the stiffened shell that is subjected to hydrostatic pressure is circular, and the vessel has a cylindrical form over a significant proportion of its length. The fore and aft ends consist of domes that have a variety of shapes (hemispherical, torispherical). In the case of torispherical domes, the end consists of a toroidal ring whose form is reminiscent of a truncated cone. This acts as a transition piece between the cylindrical shell and a spherical cap. The use of the toroidal ring facilitates the use of a cap that has a diameter that is much larger than that of the cylindrical shell. This contributes to the effective exploitation of interior spaces and the installation of various weapons systems (torpedo tubes). We note that the cylindrical shell is fitted with stiffeners that have the form of *ring frames*. Ring frames are closely spaced and are used over the full length of the vessel. The combined strength of the stiffened cylindrical shell and transverse bulkheads is sufficient to withstand the hydrostatic pressure that acts at the operating depth of the submarine.

The form and arrangement of the stiffening system are such that the structure can bear the principal external loading that in the case of the submarine is the external hydrostatic pressure. This remains approximately constant over the whole external surface area of the structure.[1] There is no need to provide stiffening in the longitudinal direction since the axial loading is comparatively small and can be borne by the cylindrical shell alone.[2]

DOI: 10.1201/9781003489153-3

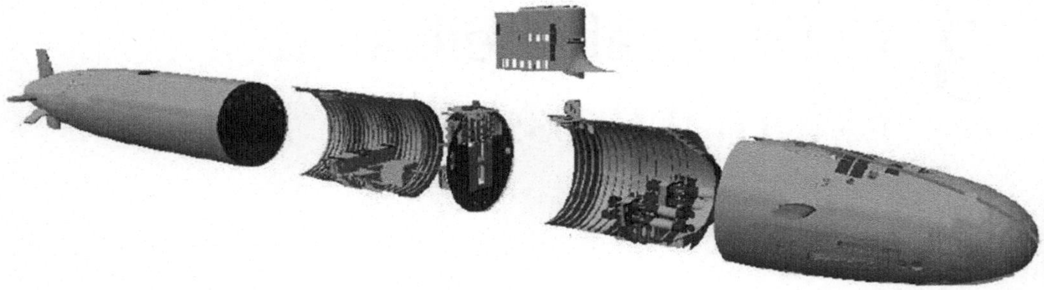

FIGURE 3.1 The *Virginia class* nuclear submarine.

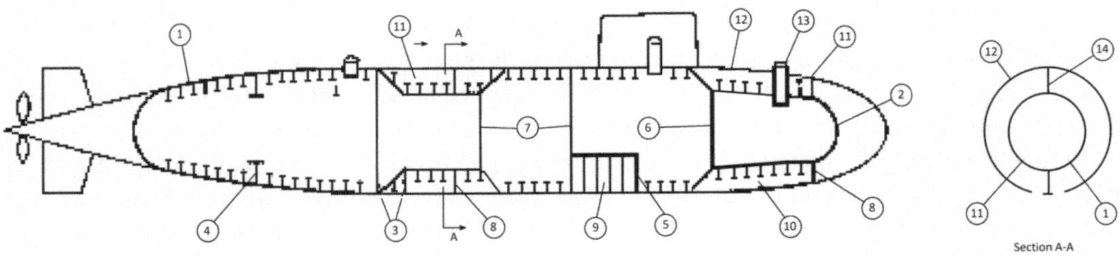

1	• Outer plating and ring frames	8	• Transverse bulkhead
2	• Forward end	9	• Inner ballast tank
3	• Conic section	10	• Outer ballast tank
4	• Ring frame	11	• Main ballast tank
5	• End bulkhead of ext. ballast tank	12	• Outer shell plating
6	• Support bulkhead	13	• Manhole
7	• Subdivision bulkhead	14	• Ballast tank side plating

FIGURE 3.2 Longitudinal section of a submarine pressure hull (*Sturgeon* class) [9].

In comparison to surface ships, the plating of the pressure hull has a larger thickness (>25 mm), and this enables the vessel to dive to the specified operating depth. The ring frames are either cold-formed or built-up, and their spacing lies in the range 0.1–0.2R, where R is the radius of the pressure hull. For optimum (minimum) hydrodynamic resistance, the length-to-diameter ratio ($L/2R$) is approximately equal to 6.

Apart from conventional submarines, smaller vessels are constructed mainly for research purposes (deep-diving submersibles). The pressure hull for these vessels is in many cases spherical and does not require stiffening, as the sphere is the most efficient shape to bear the external hydrostatic pressure; the thickness-to-diameter ratio ($t/2R$) is selected to be sufficiently large. These structures are designed to fail by yielding.

A variety of shapes that are used for the ends of the cylindrical structure may be hemispherical, torispherical, ellipsoidal, or even flat. Figure 3.3d shows a toroidal ring that connects the cylindrical shell to a spherical cap. The ring has a circular cross section in the plane of the vessel and a curvature in a plane perpendicular to the former with a radius smaller than that of the vessel. The advantages of this geometry are that:

1. Smaller stress concentrations arise in way of the connection of the domed end to the cylindrical shell.

2. The loss of volume in comparison to the spherical cap is reduced.
3. A large near-flat surface area is available in way of the vessel axis that is suitable for the installation of penetrations (torpedo tubes).

The disadvantage to the use of such a cross section lies in the difficulties (both theoretical and experimental) in assessing its structural response.

Other major structural items that contribute to the strength of the pressure hull are *transverse bulkheads*. Several types of bulkheads are used, each of which fulfils a specific purpose. *Pressure bulkheads* act as boundaries to tanks that are exposed to external hydrostatic pressure that is equal to that withstood by the stiffened pressure hull and are thus designed to carry the same maximum load. *Holding bulkheads* and *subdivision bulkheads* are subjected to maximum hydrostatic pressure only after flooding of adjacent compartments. For this reason, they can be designed to withstand a lower hydrostatic pressure.

The different levels of a stiffening arrangement of a typical bulkhead are indicated in Figure 3.3a and b. We note that three distinct levels of stiffening have been used. Primary stiffening includes one or two horizontal, parallel beams that extend over the full breadth of the vessel, and often the main deck is used. Secondary stiffening is provided by vertical stiffeners that are connected to the horizontal beams and extend over the full depth of the pressure hull. The horizontal beams that constitute the tertiary level subdivide the bulkhead plating into near-square elements and provide elastic support to it. Such an arrangement ensures that the plating and its stiffening system work in tandem to the maximum possible extent. The load distribution and transmission mechanism to the various members is then highly indeterminate.

In addition to bulkheads, the internal volume is subdivided by *decks*. These may be attached directly to the pressure hull in which case they provide support and thus prevent displacements from occurring at these locations. On the other hand, in such a situation, they are subjected to compressive loads and are liable to buckle. In order to avoid this from occurring, decks can be supported by the shell structure such that they are not subjected to compressive loads. An additional benefit is that possible vibration excitation arising from equipment fitted to the deck is not transmitted to the outer shell structure.

Other sources of concern are *penetrations* of the pressure hull (Figure 3.3e). These are necessary for the vessel's operation and for access by personnel. Circular holes give rise to stress concentrations that in the case of a cylindrical shell are more serious than in the case of flat plates. Circular penetrations are either fitted with cylindrical supports through the pressure hull or else with a coaming around the edge of the shell plating.

3.1.2 Load Transfer in Submarine Pressure Hull Structures

For present purposes, we shall ignore the presence of structures external to the pressure hull (*exo-structures*), and we shall focus on the pressure hull itself and the internal members that are designed to contribute to its structural integrity.

Hydrostatic pressure acts on the pressure hull, which consists of stiffened cylindrical sections, conical sections, the end closures as well as transverse bulkheads. These are attached to internal members such as ring frame stiffeners that are arranged at regular, closely spaced intervals, transverse bulkheads, and decks (Figure 3.4a). The load carried by the outer shell is thus transferred to these internal members (Figure 3.4b). The cylindrical and conical sections are stiffened with ring frames between consecutive transverse bulkheads, and thus, load transfer takes place between the shell plating and the ring frames. Since both the ring frames and the bulkheads are stiffer than the shell plating, the plating bends about the lines of attachment to these. The end closures are generally unstiffened because of their inherent strength and stiffness due to their double curvature.

The cylindrical and conical sections of the pressure hull are supported by transverse bulkheads as well as horizontal decks. The transverse bulkheads thus ensure that the pressure hull retains

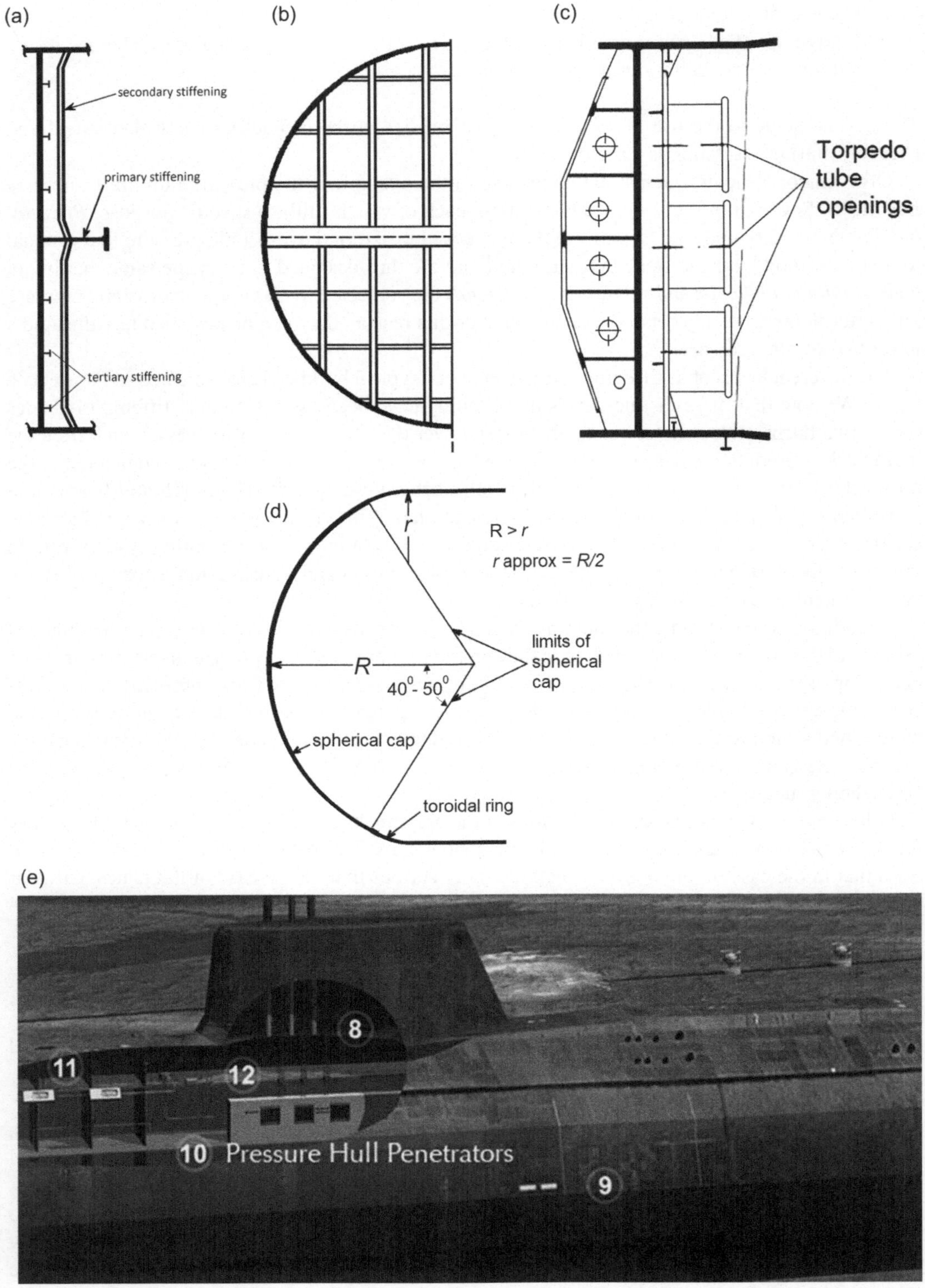

FIGURE 3.3 Major structural components of submarine pressure hull structures. (a) Transverse bulkhead (profile view). (b) Transverse bulkhead (plan view). (c) Flat end closure. (d) Torispherical dome end, (e) Pressure hull penetrations.

its form. If a deck extends across the full breadth of the section and contributes to supporting the pressure hull, it also contributes to form retention, and it is thus subjected to edge compression athwartships.

In the axial direction, as will be shown, the stresses that arise as a result of hydrostatic pressure are not critical. However, they give rise to biaxial stress fields in the cylindrical shell plating and in longitudinally effective decks.

The critical structural components are thus seen to be the stiffened cylindrical shell and the pressure bulkheads, upon which the integrity of the pressure hull depends (Figure 3.4). In order to provide additional support to the pressure bulkheads, various arrangements have been used, an example of which is shown in Figure 3.3c.

3.1.3 Some Observations on the Behaviour of Elements of Curved Structures

In the study of underwater vehicles, we need to extend the analysis of flat and stiffened plates to take account of the curvature that arises in the various component of the structure. In order to do so, we

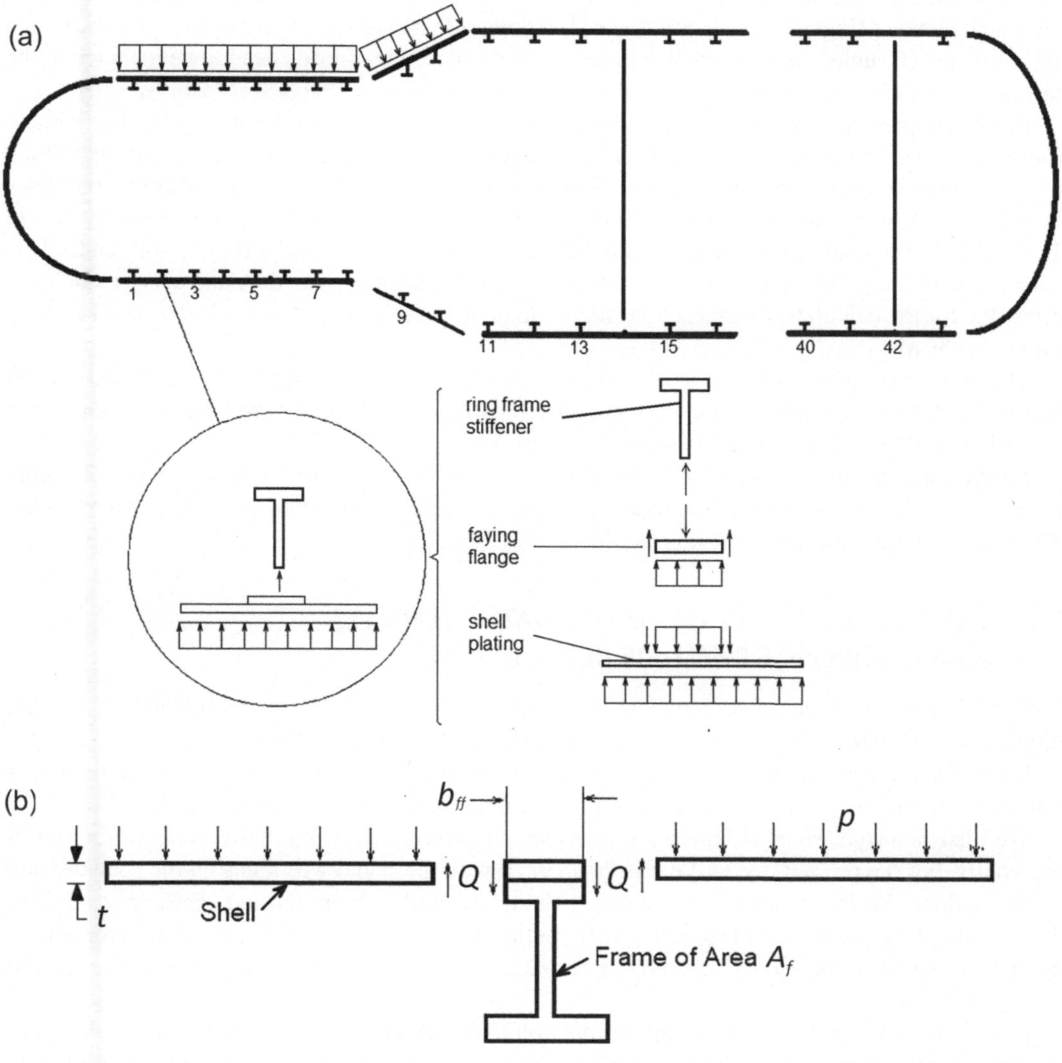

FIGURE 3.4 Load transfer in a submarine pressure hull. (a) Loads acting on pressure hull. (b) Load transfer from plating to ring frame.

have to turn to *shell theory*, which deals with thin structures whose curvature differs in one or more than one direction. A thin flat plate is thus a shell with zero curvature and can thus be treated as a special case of the general theory of shells. Cylindrical and spherical shells have constant curvature in one and two directions, respectively.

One of the earliest thin shell theories was presented by Love and is described in his book [8]. Since then, alternative formulations (Novozhilov,[3] Donnell–Mushtari–Vlasov) have been proposed, and these have been applied to a variety of problems. Applications of shell theory are described in a large number of books that deal with applications in the civil, mechanical, aeronautical, and marine fields, such as [1] and [2].

Cylindrical sections that have a constant diameter constitute a special sub-group of shells since they have a constant curvature in one direction only. If subjected to internal pressure alone, they can be studied using the *membrane theory of shells,* which accounts for the symmetrical deformation of the section (increase in radius and length of the section). Submarine pressure hulls are, however, subjected to external pressure, and for this reason, they can also fail by buckling. In membrane theory, the loads acting on a shell element consist of in-plane forces that are normal to and tangential to the sides of the element, as in the case of flat plates. In bending theory, bending and twisting moments act on a shell element (Figure 3.12c, p.80). In order therefore to derive the equations of equilibrium of a cylindrical shell subjected to external pressure, both of these sets of loads need to be included in the formulation, leading to the *generalised equations of equilibrium of cylindrical sections.*

In this chapter, we shall present the background and assumptions of the theory of elastic buckling of thin-walled unstiffened cylindrical shells. The equation of equilibrium of an unstiffened cylinder subjected to external hydrostatic pressure was originally derived and solved by von Mises in 1914 [11]. A later, extended solution allowing for axial compression as well was given by him in 1929 [16]. These solutions concern cylindrical shells that are simply supported at both ends. This calculation is a first step in the study of stability of real cylindrical structures and acts as a point of reference for the calculation of the actual failure load, analogous to that of the Euler buckling load for elastic beams.

The question that is of interest to the submarine designer is the strength of a cylindrical section that is fitted with ring frames, that is, of a *stiffened cylindrical shell of finite length.* This question was addressed by von Sanden and Günther in 1920–1 [7].

Before we consider the strength of cylindrical shells, we need to consider the properties of a more general category of shells that comprises all types of shells encountered in underwater vehicles. These are termed *shells of revolution.*

3.2 THE PRESSURE HULL OF UNDERWATER VEHICLES AS A COMPOSITE SHELL OF REVOLUTION

Figure 3.5 shows the types of components of a submarine pressure hull that are welded to each other along circumferential boundaries such that a watertight envelope is formed.

Each of these components is a surface that is symmetrical about the axis of the pressure hull. A number of these components are used in arrangements also encountered in submersibles.

We also note that ring stiffeners are provided in certain cases internally and in other cases externally. For the present, we will not be concerned with these and will focus on the pressure hull surfaces alone. Surfaces that are symmetrical about a central axis are termed *shells of revolution.* To describe their geometrical properties, differential geometry is used, whereas to determine their strength properties, the theory of shells of revolution is employed, this being part of the general theory of shells.

In this section, we shall provide an introduction to theory of shells of revolution that will act as the basis for the derivation of the equations of equilibrium of the different types of shells encountered in underwater structures.

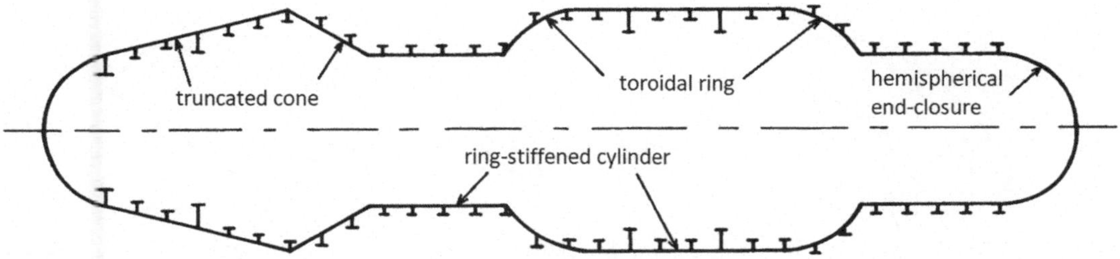

FIGURE 3.5 The components of the pressure hull of an underwater vehicle.

In the structural analysis of a large number of land-based as well as marine structures, the elementary structural types that we deal with are planar (beams, rectangular plates, orthogonally stiffened plates), and these are in general assembled into two and three-dimensional arrangements. In the case of surface ships, we encounter grillages, transverse frames, double bottoms, and superstructures of various forms. At an elementary level, the component members of these structural types may be treated as one-dimensional or as two-dimensional, whose thickness is at least one order of magnitude smaller than that of the other dimensions. This permits the use of thin-walled structures theory and the consequent simplifications that make it possible to avoid having to consider three-dimensional stress fields.

In contrast to these, the distinguishing feature of an underwater vehicle is its curvature, which may be present in one or two directions, so that we speak of a *shell*. Figure 3.5 includes the types of shells that are encountered in underwater structures. These are the cylindrical shell, truncated cone, hemispherical end-closure, and toroidal ring. In the case of underwater structures, the thickness is at least one order of magnitude smaller that the other dimensions, enabling the application of thin-walled structures theory.

3.2.1 Geometrical Properties of Shells

A shell-type structure extends in three-dimensional space and is characterised by the form of its surface and its thickness. An elementary geometrical description is based on a definition of the *middle surface*, which bisects the thickness at all points, and the thickness at every point of the surface (Figure 3.6a). If we select a tangent to the middle surface at a particular position, the normal to this tangent is termed the *shell normal* at this point. A *normal section* at a point is a plane that contains the shell normal at that point. Every normal section bisects the middle surface in a plane curve (Figure 3.6b). The curvature at the point of intersection with the tangent plane is denoted as

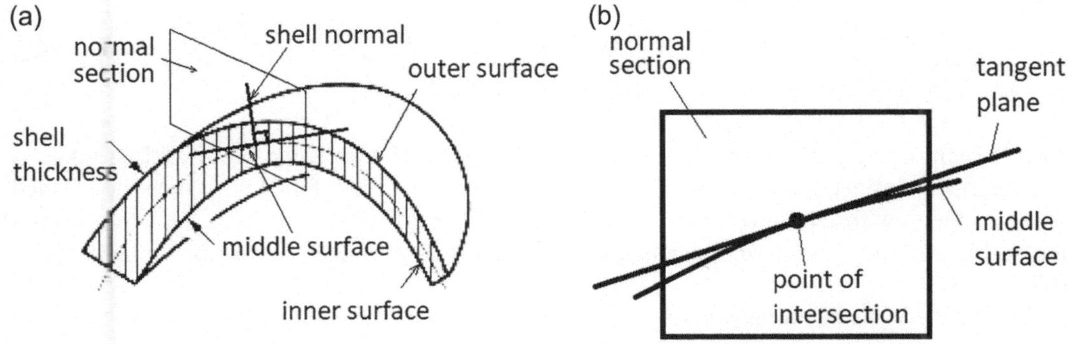

FIGURE 3.6 Geometrical parameters of shells. (a) Middle surface parameters [12]. (b) Definition of normal curvature.

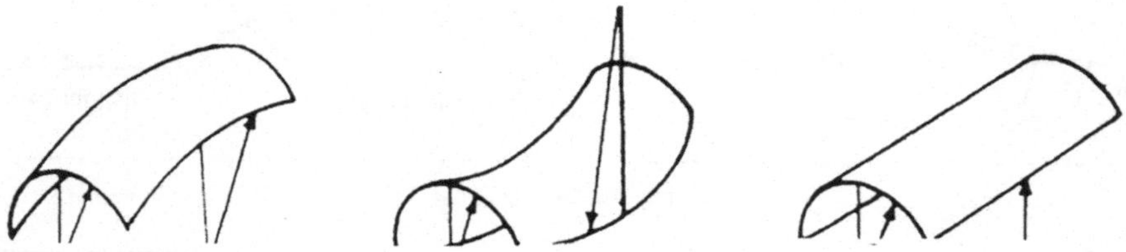

FIGURE 3.7 Classification of surfaces. (a) Synclastic (sphere) – $K>0$, (b) Anticlastic – $K<0$, (c) Cylinder – $K=0$.

the *normal curvature*. It is possible to specify two normal planes that are orthogonal to each other for which the normal curvatures assume maximum and minimum values, these being the *principal curvatures* of the shell surface at that point. We define the *Gaussian curvature K* at a point of a smooth surface as the product of the principal curvatures at that point $K=\kappa_1\kappa_2$. The sign of K depends on the form of the surface, as seen in Figure 3.7.

It is of interest to compare the manner in which a planar structure, such as a rectangular flat plate or a stiffened panel, bears externally applied loads compared to a shell of the type encountered in underwater structures. In the case of lateral pressure loading, a flat plate requires support along its boundaries. The load that is borne by the plate is transmitted to the boundaries in its entirety, and a self-equilibrating force system is set up. The nature of the internal stress field in this case depends on the slenderness of the plate and on the nature of the boundary conditions [14], so that bending stresses as well as possibly membrane stresses are induced. In the case of a shell of revolution that is subjected to axisymmetric loading, in principle, the shell surface does not need supports, that is, it is a *self-supporting* (*self-equilibrating*) structure. The stresses that are induced in this case are membrane stresses alone, and these are the most efficient means of withstanding the applied loading.

The case of edge compression is much more complicated, and in order to gain a preliminary understanding of the effect of curvature, we will compare the buckling strength and the post-buckling response of a flat panel with that of an equivalent curved panel (Figure 3.8).

A perfect flat panel buckles elastically at a stress $\sigma_{cr,p}$ given by[4]

$$\sigma_{cr,p}=\frac{\pi^2 E}{3(1-\nu^2)}\left(\frac{t}{b}\right)^2 \tag{3.1}$$

following which it continues to carry higher loads, albeit with a reduced stiffness. In the presence of imperfections, the stiffness is reduced, although it remains positive so that the plate retains post-buckling strength.

In the case of the curved panel, the elastic buckling stress, $\sigma_{cr,c}$, is given by [14]:

$$\sigma_{cr,c}=\underbrace{\frac{\pi^2 E}{3(1-\nu^2)}\left(\frac{t}{b}\right)^2}_{\text{stable}}+\underbrace{\frac{E}{4\pi^2}\left(\frac{b}{r}\right)^2}_{\text{unstable}} \tag{3.2}$$

This is seen to be much higher than that of the flat plate. However, the post-buckling range is characterised by steep unloading until a minimum value is reached, following which the buckled panel is able to withstand further loading. The effect of initial distortions has a marked effect on the buckling strength of the curved panel, σ_i, significantly more so than in the case of the flat plate (Figure 3.8).

In Figure 3.9, the buckling stress is plotted against b^2/Rt, this being a measure of the curvature of the panel (R=radius of curvature of the panel). This parameter is proportional to the maximum deviation of the panel from its chord divided by its thickness. It is seen that although a curved panel has a higher buckling strength than an equivalent flat panel and that the difference increases with

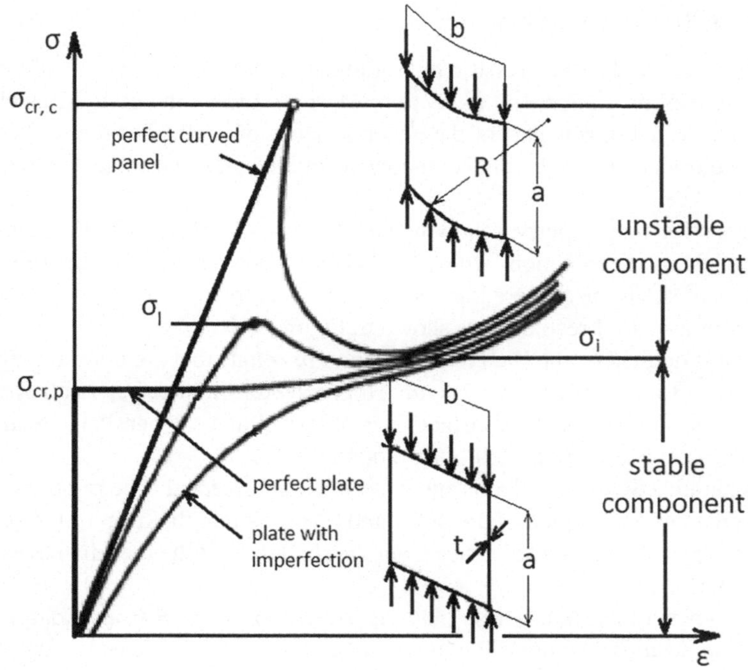

FIGURE 3.8 Comparison of compression strength of flat panel with that of curved panel.

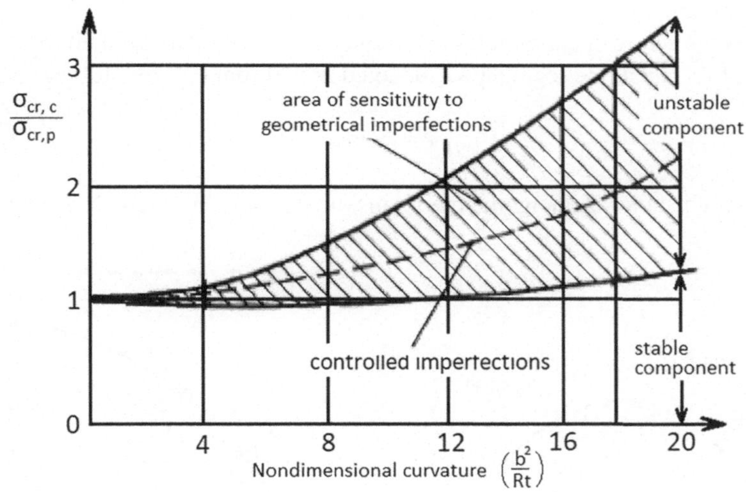

FIGURE 3.9 Effect of initial distortions on buckling strength of curved panel.

non-dimensional curvature b^2/Rt, there is significant variability in the buckling strength as a result of initial distortions. This variability is directly related to the unstable component, that is, to the presence of curvature.

A first conclusion concerning the design of curved panels and more generally shells is that in order to avoid elastic buckling, either the induced stresses have to be significantly lower than the elastic buckling stress, as in the case in aeronautical and space structures, or else inelastic design is required, that is, the material strength has to be exploited, as is the case in underwater structures. Furthermore, it is important to ensure that initial distortions lie within prescribed limits in order to reduce the variability in buckling strength.

3.2.2 Shells of Revolution: Geometry

In Section 3.2.1, we noted that the pressure hull of underwater vehicles consists of a number of different types of shells of revolution. In this section, we shall discuss the basic ideas of the theory of shells of revolution, which forms part of the general theory of shells, in order to be able to derive the equations of equilibrium of individual components of interest when subjected to external hydrostatic pressure.

We start by considering geometry, following which we shall outline both the membrane and the bending theory of shells of revolution. We will then be in a position to apply the results obtained to the different types of shells encountered in underwater structures.

The geometry of a shell of revolution is shown in Figure 3.10.

It is seen that it is defined by the shell *axis*, the *apex*, *parallels* that are normal to the shell axis at increasing distances from the apex and *meridians* that extend from the apex in all directions and lie normal to the parallels along the shell surface. It is observed that the parallels are all circular, this being the characteristic feature of a shell of revolution (SOR).

In order to generate a SOR, a shell axis and a meridian are selected. The rotation of the meridian about the axis generates the surface of the SOR, and it is seen that the form of the surface depends on the shape of the meridian and its distance from the shell axis. If the meridian does not extend up to the apex, the SOR is *truncated.*

Figure 3.10b shows a meridian whose plane is located an angle θ from a datum meridian. In order to specify the position of a point P lying on its middle surface, it is necessary to introduce an additional coordinate, and to this end, we select the angle φ between the normal to the shell at P, AP, and the shell axis. We see that a *spherical coordinate system* (θ, φ) is used, hence the use of the terms meridian and parallel. In Figure 3.10b, the distance from point P to the shell axis is r, and r_φ is the radius of curvature of the meridian (line segment OP=OB).

Point O is thus the centre of curvature. The distance from point P to the shell axis along a normal to the meridian at P is then line segment AP, denoted as r_a. From Figure 3.10b, we have

$$r = r_a \sin\phi \tag{3.3}$$

Point B is located a distance ds along the meridian where $ds = r_\varphi d\varphi$, where $d\varphi$ is angle POB.

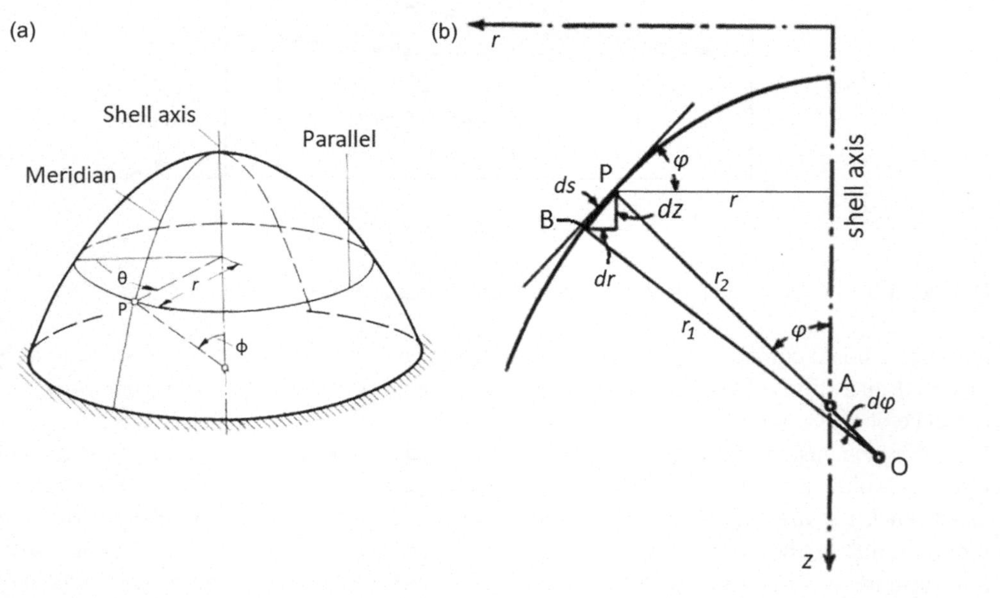

FIGURE 3.10 Geometry of a shell of revolution. (a) Geometry of an SOR. (b) Geometry of a meridian [2].

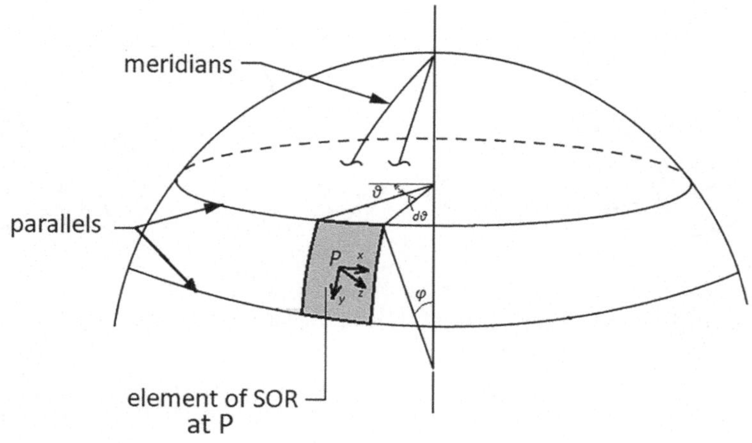

FIGURE 3.11 Element of SOR.

We now consider an infinitesimally small element of the shell surface at P, bounded by pairs of meridians and parallels. We define a local cartesian coordinate system (*x*, *y*, *z*) at P, as shown in Figure 3.11. The *x*-axis is in the direction of the tangent to the parallel, the *y*-axis is in the direction of the tangent to the meridian, and the *z*-axis is in the direction of the normal to the shell middle surface at P (positive inwards).

3.2.3 Stresses in Thin Shells: The Membrane Theory of Thin Shells

It was mentioned previously that the pressure hull of underwater vehicles can be treated as a thin shell. Thin shell theory is based on the following assumptions (Love–Kirchhoff):

1. The shell thickness is negligibly small in comparison with the least radius of curvature of its middle surface ($r \gg t$)
2. The strains and displacements that arise within the shell are small. As a consequence, products of deformations may be neglected, leading to a set of geometrically linear equations.
3. The points that lie on a normal to the middle surface before loading is applied lie on a straight line following application of the loading (Euler–Bernoulli assumption for beams)
4. A normal to the middle surface before loading is applied is also a normal to the middle surface following application of the load. That is, the deformations resulting from normal shear are neglected.
5. The direct stress normal to the middle surface can be neglected. This, however, is not valid in the vicinity of concentrated loads
6. Out-of-plane deformations are small compared to the shell thickness (*very thin shell theory*). Otherwise we need to treat the shell as a three-dimensional continuum (thick shell analysis).

The internal stress resultants that arise along the sides of a shell element from the application of generalised loading of the shell are shown in Figure 3.12. Figure 3.12a shows the stress resultants that are normal to and tangential to the edges of the element along a meridian and a parallel. Figure 3.12b shows the distribution of forces acting normal to the middle surface and consists of transverse shear forces per unit width Q_ϑ and Q_φ. Figure 3.12c shows the bending (M_ϑ, M_φ) and twisting moments per unit width ($M_{\vartheta\varphi}$, $M_{\varphi\vartheta}$) that act along the edges of the element.

Along a meridian, the distribution of normal stress resultants per unit width is termed N_ϑ, so that the total force acting along the length $r_\varphi d\varphi$ of an element is $N_\vartheta r_\varphi d\varphi$. Since this varies over the element

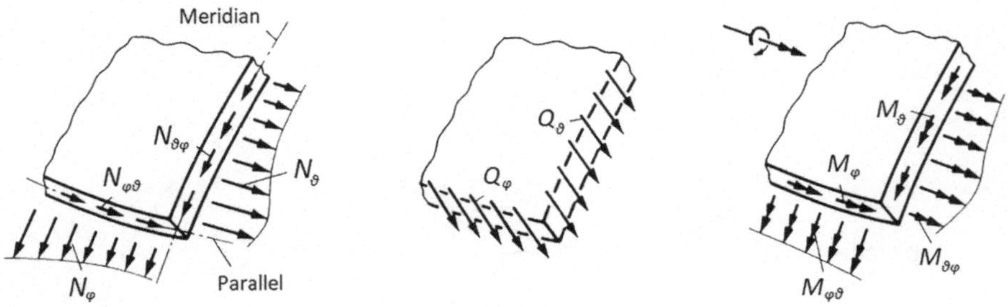

FIGURE 3.12 Internal forces in a shell element [12]. (a) In-plane normal and shear forces. (b) Transverse shear forces. (c) Bending and twisting moments.

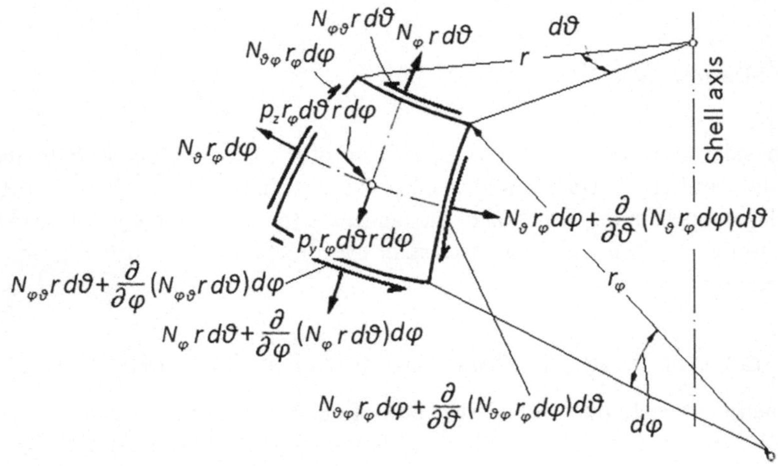

FIGURE 3.13 Internal force system in an SOR (membrane theory) [12].

width, the normal stress resultant acting along the opposite side is equal to $N_\vartheta r_\varphi d\varphi + \partial/\partial\theta(N_\vartheta r_\varphi d\varphi)\, d\theta$, as shown in Figure 3.13. Correspondingly, along the parallels bordering the element, the normal forces are $N_\varphi r d\theta$ and $N_\varphi r d\theta + \partial/\partial\varphi(N_\varphi r d\theta) d\varphi$. The shear stress resultants per unit width acting tangentially to meridians and parallels are termed $N_{\vartheta\varphi}$ and $N_{\varphi\vartheta}$, respectively. These also vary across the width of the element so that along opposite sides, the shear stress resultants are incremented, as shown in Figure 3.13.

The total number of variables is 10, so that in total 10 equations are required to describe the generalised internal force system of a thin shell. Equilibrium conditions about the three orthogonal axes provide us with six equations, and we conclude that the stress distribution is statically indeterminate. To resolve this difficulty, we make use of the basic assumption made in the analysis of pin-jointed struts, which is that in way of their supports, bending moments and shearing forces do not arise. In the case of a thin shell, this means that throughout the surface:

$$M_\vartheta = M_\varphi = M_{\vartheta\varphi} = M_{\varphi\vartheta} = Q_\vartheta = Q_\varphi = 0 \tag{3.4}$$

The internal stress system thus consists of the normal and shear stress resultants along each edge alone (N_ϑ, N_φ, $N_{\vartheta\varphi}$, $N_{\varphi\vartheta}$). This stress system constitutes the *membrane theory of shells* (Figure 3.13). In this chapter, we shall encounter cases in which membrane theory can and cannot be applied.

It will become apparent that it is important to distinguish between the system of stress resultants that arises in the case of a general shell and that of a shell of revolution and also to understand the effect that rotationally symmetric loading has as opposed to non-symmetric loading. These issues arise in the case of SOR encountered in underwater vehicles, and in each case, appropriate equations of equilibrium are used.

Having focused on the internal force system for membrane theory, it is necessary to consider the cases in which the membrane theory of shells does not provide a true representation of the internal stress distribution. Bending moments and in-plane shearing forces, which are not included in membrane theory, arise in way of supported edges, unless the geometry of the shell and the nature of the applied loads ensure that these are zero. Furthermore, they will not be equal to zero in way of concentrated loads and loads distributed over part of the surface, so that in these cases, membrane theory is not applicable. However, the boundary bending moments and shear forces attenuate rapidly so that they can be disregarded over a large part of the shell surface. In the case of underwater structures, since the applied load consists of uniform external pressure, membrane theory can in principle be applied for regions of the shell not adjacent to the boundaries or stiffening systems, provided no other loads arise.

3.2.4 Membrane Theory of SOR Subjected to Rotationally Symmetric Loading

As seen in the previous section, the internal stress field that arises as a result of the membrane theory assumptions, shown in Figure 3.13, consists of in-plane normal stress resultants (N_ϑ, N_φ) and tangential shear stress resultants ($N_{\vartheta\varphi}$, $N_{\varphi\vartheta}$) along the element edges. These vary in the direction of meridians and parallels.

When rotationally symmetric loading is applied, the circumferential force $N_\vartheta r_\varphi d\varphi$ varies along each meridian but retains a constant value along each parallel (Figure 3.14). As a result, the in-plane shear forces along parallels are zero, and by the principle of complementary shear stresses, there will be no in-plane shear loads acting along the meridians ($N_{\vartheta\varphi}=N_{\varphi\vartheta}=0$). Rotationally symmetric loading thus gives rise to N_ϑ and N_φ alone, these being the normal stress resultants per unit length. Additionally, the load component remains constant in the y-direction. The equations of equilibrium for an SOR that is subjected to rotationally symmetric loading are thus derived in terms of these two variables alone. Equilibrium is considered:

a. In the direction of the tangent to the meridian (y-axis)
b. Normal to the middle surface (z-axis)

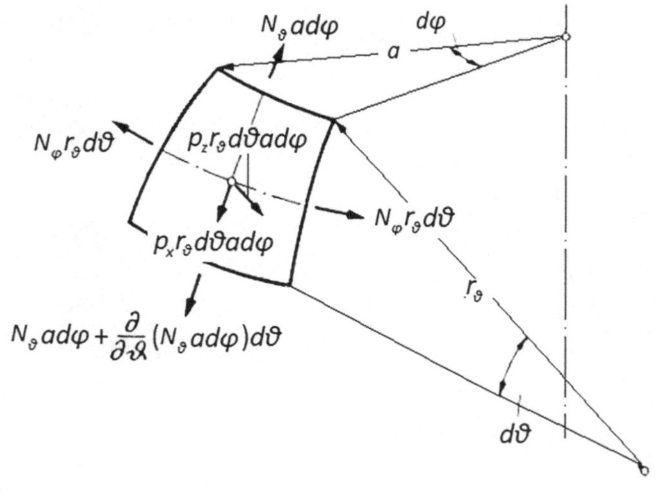

FIGURE 3.14 Internal force system in an SOR (membrane theory) for rotationally symmetric loading [12].

Along the y-axis, we have:

$$\frac{\partial}{\partial\varphi}(N_\varphi r d\vartheta)d\varphi + \frac{\partial}{\partial\vartheta}(N_{\vartheta\varphi} r_\varphi d\varphi)d\vartheta - N_\vartheta r_\varphi d\varphi d\vartheta \cos\varphi + p_y r_\varphi d\varphi r d\vartheta = 0 \tag{3.5}$$

In order to understand the need for the third term in this equation, consider Figure 3.15. This figure shows a section of the shell in the plane of a parallel. It is seen that the circumferential force $N_\vartheta r_\varphi d\varphi$ acting at the ends has a net resultant in the radial direction equal to $N_\vartheta r_\varphi d\varphi d\vartheta$. This force can be resolved into a component in the direction of the tangent to the meridian, given by $N_\vartheta r_\varphi d\varphi d\vartheta \cos\varphi$, and a second one at right angles, as shown in Figure 3.15b.

The component in the direction of the tangent to the meridian is the third term in the above equation. In the direction perpendicular to the middle surface of the element, for equilibrium:

$$\frac{\partial}{\partial\vartheta}(N_\vartheta r_\varphi d\varphi)d\vartheta + \frac{\partial}{\partial\varphi}(N_{\vartheta\varphi} r d\vartheta)d\varphi + N_{\vartheta\varphi} r_\varphi d\varphi \frac{r d\vartheta}{r_\vartheta \tan\varphi} + p_x r_\varphi d\varphi r d\vartheta = 0 \tag{3.6}$$

The presence of the third term in (3.6) is justified on the basis of Figure 3.16. It is seen that the tangential shear stress resultants form an angle with each other, denoted as $d\varepsilon$. As a result, a component $N_{\vartheta\varphi} r_\varphi d\varphi$ acts in the x-direction.

The length element $r d\vartheta$ of the parallel circle must be equal to $r_\varphi \tan\varphi d\varepsilon$, hence

$$d\varepsilon = \frac{r d\vartheta}{r_a \tan\phi} \tag{3.7}$$

This leads to the third term in (3.6).

The last condition to be considered is equilibrium in the z-direction (normal to the middle surface):

$$N_\varphi r d\vartheta d\varphi + N_\vartheta r_\varphi d\varphi d\vartheta \sin\varphi + p_z r_\varphi d\varphi r d\vartheta = 0 \tag{3.8}$$

Equilibrium of the element requires $N_{\vartheta\varphi} = N_{\varphi\vartheta}$. Substituting this relationship and equation (3.4) in equations (3.6), (3.7), and (3.9), and dividing all terms by $d\vartheta d\varphi$, we obtain the equations of equilibrium of a shell of revolution subjected to rotationally symmetric loading:

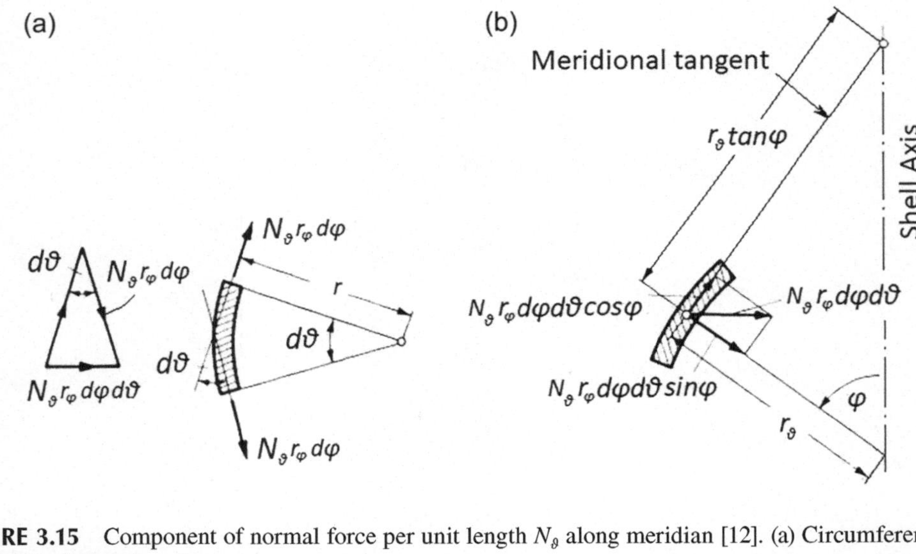

FIGURE 3.15 Component of normal force per unit length N_ϑ along meridian [12]. (a) Circumferential section. (b) Meridional section.

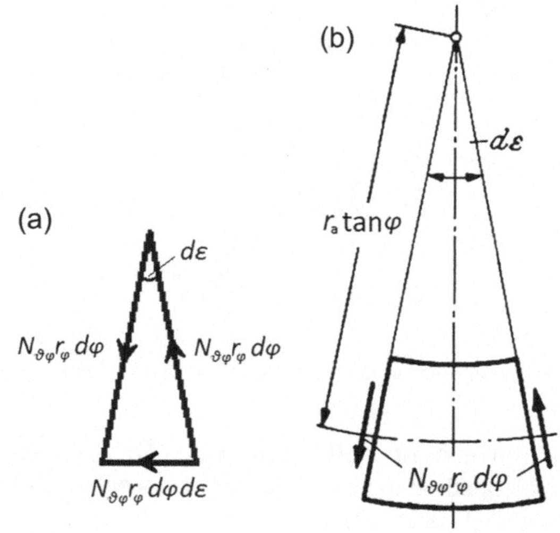

FIGURE 3.16 Projection of shell element including tangential shear forces in the direction of the shell normal [12]. (a) Resultant arising from the action of $N_{\vartheta\varphi}r_\varphi d\varphi$. (b) Definition of angle between tangential shears.

$$\left.\begin{aligned}
&\frac{\partial N_\vartheta}{\partial \vartheta}r_\varphi + \frac{\partial}{\partial \varphi}(N_{\vartheta\varphi}r_a \sin\varphi) + N_{\vartheta\varphi}r_\varphi \cos\varphi + p_x r_a r_\vartheta \sin\varphi = 0 \\
&\frac{\partial}{\partial \varphi}(N_\varphi r_a \sin\varphi) + \frac{\partial N_{\vartheta\varphi}}{\partial \vartheta}r_\varphi - N_\vartheta r_\varphi \cos\varphi + p_y r_\varphi r_a \sin\varphi = 0 \\
&N_\varphi r_a + N_\vartheta r_\varphi + p_z r_\varphi r_a = 0
\end{aligned}\right\} \tag{3.9}$$

Equation (3.9a) ensures equilibrium in the direction of the tangent to the meridian, equation (3.9b) in the direction of a parallel, and equation (3.9c) normal to the middle surface of the shell. This system, consisting of two differential equations and one algebraic equation, is sufficient to determine the three unknowns N_ϑ, N_φ, and $N_{\vartheta\varphi}$, and hence, the system is statically determinate.

3.2.5 Bending Theory of SOR Subjected to Rotationally Symmetric Loading

As mentioned earlier, the membrane theory does not provide an accurate picture of shell response in way of boundaries and also in the vicinity of concentrated loads. In order to succeed in this, it is necessary to introduce the bending moments and transverse shear forces induced in the shell. This leads to the *bending theory of shells* that together with the membrane theory provides a complete description of thin shell response. We shall limit our discussion to the case of SOR subjected to rotationally symmetric loading, which in the case of underwater vehicles is the most important one. In doing so, we shall provide a simplified version of the bending theory of thin shells.

For an SOR subjected to rotationally symmetric loading, the membrane forces that act are shown in Figure 3.14. The bending theory of SOR subjected to rotationally symmetric loading requires inclusion of the following:

1. Bending moments M_ϑ, M_φ along the edges of the element
2. The transverse shear force Q_φ that varies along a meridian

These are shown in Figure 3.17. The transverse shear force Q_ϑ and the twisting moments, $M_{\vartheta\varphi}$, $M_{\varphi\vartheta}$ are equal to zero because of rotational symmetry. For the same reason, M_ϑ retains a constant value in the circumferential direction.

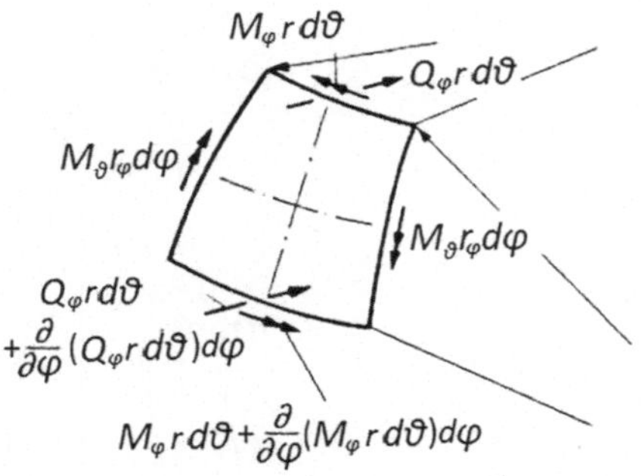

FIGURE 3.17 Bending forces in an SOR subjected to rotationally symmetric loading [12].

We shall now obtain the equations of equilibrium for an SOR subjected to rotationally symmetric loading in accordance with bending theory. To do so, we consider equilibrium in the direction of the coordinate axes and rotational equilibrium about these axes.

In the direction of the tangent to the meridian, equilibrium is satisfied if [12]:

$$\frac{\partial}{\partial\varphi}(N_\varphi r d\vartheta)d\varphi - N_\vartheta r_\varphi d\varphi d\vartheta\cos\varphi - Q_\varphi r d\vartheta d\varphi + p_y r_\varphi d\varphi r d\vartheta = 0 \tag{3.10}$$

The first two terms and the last term originate from membrane theory, whereas the third term arises because the transverse shear forces $Q_\varphi r d\vartheta$ acting along each parallel lie at an angle $d\varphi$ to each other and a resultant acts in the negative y-direction.

In the circumferential direction, equilibrium is satisfied because no resultant force acts in this direction (rotational symmetry). In the z-direction, equilibrium is satisfied provided:

$$N_\varphi r d\vartheta d\varphi + N_\vartheta r_\varphi d\varphi d\vartheta\sin\varphi + \frac{\partial}{\partial\varphi}(Q_\varphi r d\vartheta)d\varphi + p_z r_\varphi d\varphi r d\vartheta = 0 \tag{3.11}$$

Rotational equilibrium about the x- and z-axes is satisfied because of rotational symmetry. For rotational equilibrium about the y-axis, we require:

$$\frac{\partial}{\partial\varphi}(M_\varphi r d\vartheta)d\varphi - Q_\varphi r d\vartheta r_\varphi d\varphi + M_\vartheta r_\varphi d\varphi\frac{r d\vartheta}{r_a\tan\varphi} = 0 \tag{3.12}$$

In this equation, the third term is the resultant of the bending moments $M_\vartheta r_\varphi d\varphi$ acting on meridians at an angle $rd\vartheta/(r_a\tan\varphi)$.

Dividing (3.10–3.12) by $d\vartheta d\varphi$ and setting $r=r_a\sin\vartheta$, we obtain the equations of equilibrium of the bending theory for an SOR subjected to rotationally symmetric loading:

$$\left.\begin{aligned}
&\frac{\partial}{\partial\varphi}(N_\varphi r_\varphi\sin\varphi) - N_\vartheta r_\varphi\cos\varphi - Q_\varphi r_\vartheta\sin\varphi + p_y r_\varphi r_a\sin\varphi = 0\\
&N_\varphi r_\vartheta\sin\varphi + N_\vartheta r_\varphi\sin\varphi + \frac{d}{d\varphi}(Q_\varphi r_a\sin\varphi) + p_z r_\varphi r_a\sin\varphi = 0\\
&\frac{\partial}{\partial\varphi}(M_\varphi r_a\sin\varphi) - Q_\varphi r_\varphi r_a\sin\varphi + M_\vartheta r_\varphi\cos\varphi = 0
\end{aligned}\right\} \tag{3.13}$$

These equations contain five unknowns (N_ϑ, N_φ, M_ϑ, M_φ, Q_φ), and the problem is thus statically indeterminate. The number of unknowns can be reduced to three by expressing N_ϑ, N_φ, M_ϑ, and M_φ in terms of the displacements v and w. The new system of equations is then expressed in terms of

v, w, and Q_φ. One of these equations can be used to eliminate Q_φ, leading to two equations in terms of v, w.

These two equations can be considerably simplified by introducing new variables [10]. The first of these is the angle of rotation of a tangent to a meridian, α_1, defined by:

$$\alpha_1 = \frac{1}{r_\varphi}\left(v + \frac{dw}{d\varphi}\right) \tag{3.14a}$$

The second variable, α_2, is given by:

$$\alpha_2 = r_a Q_\varphi \tag{3.14b}$$

Following manipulation, the two equations of equilibrium, expressed in terms of α_1, α_2, assume the following form [15]:

$$\left.\begin{aligned}
&\frac{r_a}{r_\varphi^2}\frac{d^2\alpha_1}{d\varphi^2} + \frac{1}{r_\varphi}\left[\frac{d}{d\varphi}\left(\frac{r_a}{r_\varphi}\right) + \frac{r_a}{r_\varphi}\cot\varphi - \frac{r_a}{r_\varphi t}\frac{dt}{d\varphi}\right]\frac{d\alpha_1}{d\varphi}\\
&\quad - \frac{1}{r_\varphi}\left[\frac{r_\varphi}{r_a}\cot^2\varphi - v - \frac{v}{t}\frac{dt}{d\varphi}\cot\varphi\right]\alpha_1 = Et\alpha_2\\
&\frac{r_a}{r_\varphi^2}\frac{d^2\alpha_2}{d\varphi^2} + \frac{1}{r_\varphi}\left[\frac{d}{d\varphi}\left(\frac{r_a}{r_\varphi}\right) + \frac{r_a}{r_\varphi}\cot\varphi + 3\frac{r_a}{r_\varphi t}\frac{dt}{d\varphi}\right]\frac{d\alpha_2}{d\varphi}\\
&\quad - \frac{1}{r_1}\left[v - \frac{3v\cot\varphi}{t}\frac{dt}{d\varphi} + \frac{r_\varphi}{r_a}\cot^2\varphi\right]\alpha_2 = -\frac{\alpha_1}{D}
\end{aligned}\right\} \tag{3.15}$$

These two second-order equations when integrated provide a solution to the problem of bending of a SOR subjected to rotationally symmetric loading. They can be used to solve problems involving any type of SOR, such as a sphere and a cone. If the shell thickness t is constant, the terms involving $dt/d\varphi$ vanish so that the terms $d\alpha_2/d\varphi$ and $d\alpha_1/d\varphi$ have the same coefficients.

In order to obtain a solution, they are first converted to a simplified form through the use of a differential operator and from this two independent fourth-order equations are obtained for each of the two variables α_1, α_2. These can then be solved for the geometry of the SOR in question. However, in order to avoid this complex mathematical analysis, numerical solutions have been obtained and are discussed in [3]. The analytical approach followed in the case of conical shells that is typical of this approach is outlined in Section 7.2.3.

In the following sections of this chapter, we shall consider the elastic stability of thin cylindrical shells, which make up a major part of submarine pressure hull structures. We will present the development of the relevant equations of equilibrium by considering a cylindrical shell element, and in doing so, the coordinate axis system and notation will be redefined in a manner that is suited to structural geometry.

3.3 ELASTIC STABILITY OF UNSTIFFENED CYLINDRICAL SHELLS

Before we determine the elastic buckling pressure of an unstiffened cylindrical shell, it is helpful to have a qualitative understanding of the general response of cylindrical shells subjected to external pressure.

3.3.1 Introduction

A perfect cylindrical section (that is, one with zero out-of-circularity and no residual stresses) whose material is linear elastic that is subjected to radial pressure suffers only circumferential (membrane) compression. The load–deflection relationship can be linear or nonlinear. The characteristic response of one particular stiffened cylindrical shell is shown in Figure 3.18a. This figure includes two curves each of which corresponds to out-of-plane deflection at two positions: in way of the attachment of the cylindrical shell to a ring frame and at the midspan position, the latter being the one of interest for present purposes. We note that initially behaviour is linear in both cases. Divergence from linearity increases as the load is increased and approaches the elastic buckling pressure, p_e, calculated using small-deflection theory. The line segment $p = p_e$ is asymptotic to the deflection curve at each point of the shell surface, and this response is termed *asymmetric asymptotic buckling.*

In real structures, asymmetric buckling is of secondary importance because it occurs at pressures that are much higher than the *bifurcation buckling pressure, p_{bcr}.* The term *bifurcation* denotes that a structural member, when subjected to a critical (bifurcation) load, deflects in a direction normal to the axis of loading. By doing so, the load that initially and prior to reaching the bifurcation load is purely compressive, thereby inducing compressive stresses alone, gives rise to both compressive and bending stresses, as the member deflects sideways. The response of the member at and following the bifurcation load is characteristic of the form of the structure and the manner in which it is supported. Figure 3.18b shows the two different types of response (AC, AD) that are possible following bifurcation. In the first case, the member is able to sustain additional loads beyond the bifurcation load, whereas in the second, this not the case. Examples of both types of structures exist. *Stable post-buckling* behaviour is exhibited by compressed plating, whereas *unstable behaviour* is exhibited by curved panels and shell-type structures, as illustrated in Figure 3.18b.

When bifurcation buckling occurs, the load–deflection curve is seen to consist of two distinct parts, before and after the critical pressure, p_{bcr} is reached (Figure 3.18b). Bifurcation occurs when two or more equilibrium paths intersect, these being characteristic of the form and scantlings of the member. For $p < p_{bcr}$, there exists a one-to-one relationship between load and deflection. For $p > p_{bcr}$, curve AB is possible only in theory because any disturbance will result in dynamic response. In the case of curve AC, the shell would be subjected to severe vibration about any point of equilibrium that corresponds to $p > p_{bcr}$. As an example, we need only remind ourselves of the behaviour of a compressed column, which can carry a load higher than the Euler load if it is supported at a nodal

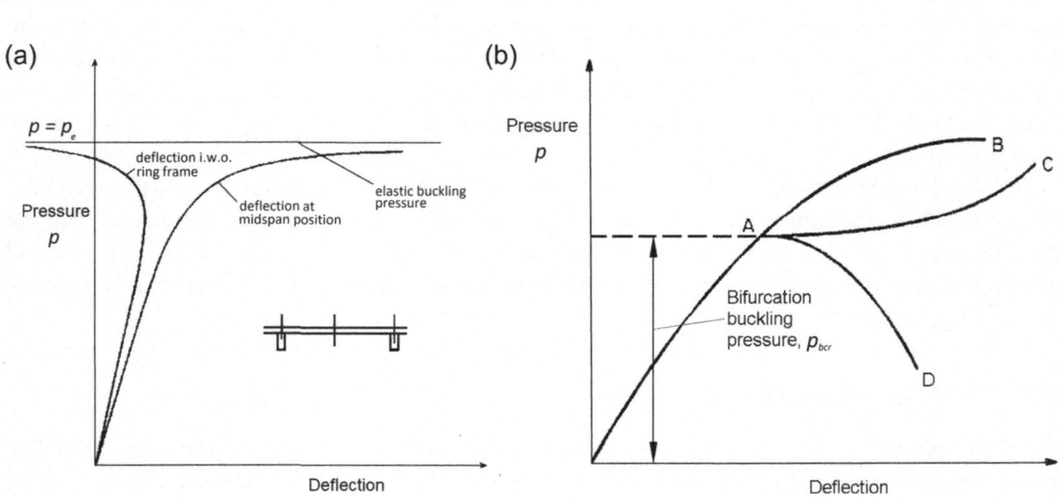

FIGURE 3.18 Buckling of a cylindrical shell under external pressure [6]. (a) Asymmetric asymptotic buckling. (b) Bifurcation buckling.

point, such as the mid-length position. If the member was perfect and the load acted at the centre of gravity, the load carried by the intermediate support would in theory be zero and in practice very small. Upon withdrawal of the support, intense vibration about the equilibrium position would arise in the post-buckling range. If path AD is followed, the behaviour becomes unstable and the form of the shell changes, as in the case of an oblate dome and a prolate dome (Figure 3.19a and b). In this case, the surface becomes concave when buckling takes place (*snap buckling*). Following the adoption of an alternative buckled form, the structure is once again able to carry additional loads.

The response also depends on the boundary conditions. Thus, if a hemi-spherical cap is simply supported on a smooth flat surface, bifurcation buckling will always arise, whereas if it is clamped, asymptotic buckling will occur. Clamping causes deviations from spherical symmetry so that axi-symmetric buckling is asymptotic. Thus, if axial symmetry is disturbed, bifurcation buckling will occur.

From the mathematical point of view, bifurcation buckling is classed as an eigenvalue problem and is formulated as a set of homogeneous linear differential equations with homogeneous boundary conditions.

True bifurcation buckling as described above, however, does not occur in real structures because of the presence of initial geometrical imperfections (distortions) and also because the load centre does not coincide with the centroid of the member cross section. As a result, lateral deflections increase as soon as the load is applied, and the buckling mode develops continuously (Figure 3.20).

In the case of cylindrical shells, the buckling behaviour depends to a considerable extent on cylinder length. Thus, cylinders may be classed as short, medium, and long, depending on the value of the length parameter L/R. Short cylinders ($L/R \leq 1$), which are of interest when considering submarine cylindrical shells, are primarily influenced by the bending stiffness of the shell in the longitudinal direction. Medium-length cylinders are independent of longitudinal bending, the buckling resistance depending on the membrane stiffness of the shell. The buckling strength of long cylinders is independent of both longitudinal and membrane stiffness and approaches that of a ring or a tube [4].

In the sections that follow, we will develop the theory necessary to obtain an expression for the elastic buckling strength of a cylinder of finite length subjected to external hydrostatic pressure. This is the problem of primary interest in the case of submarine pressure hulls since these are strengthened using closely spaced ring frames. The first step consists of developing kinematic relations that enable both membrane and bending action to be accounted for in the cylindrical shell. The case of an infinitely long shell subjected to arbitrary loading will be considered next. The general form of the shell equations obtained will then be modified accordingly in order to consider the case of the finite-length cylindrical shell subjected to external uniform pressure. The solution of the specialised form of the equations will then be given, leading to the required formula for the elastic buckling strength.

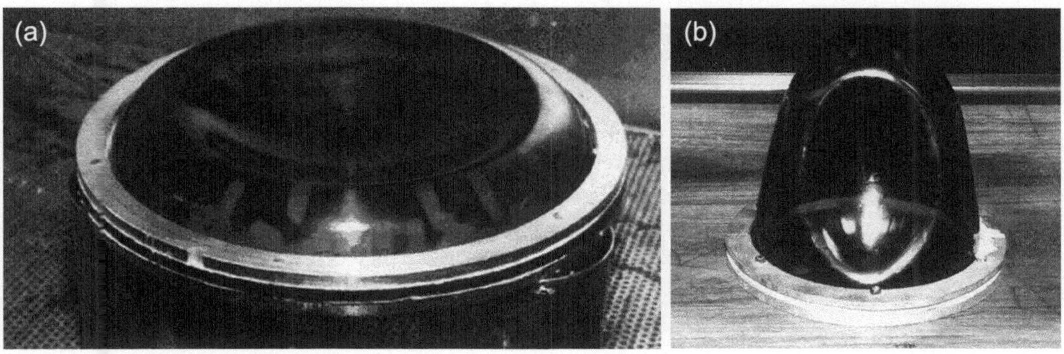

FIGURE 3.19 Dome Buckling.[5] (a) Oblate dome. (b) Prolate dome.

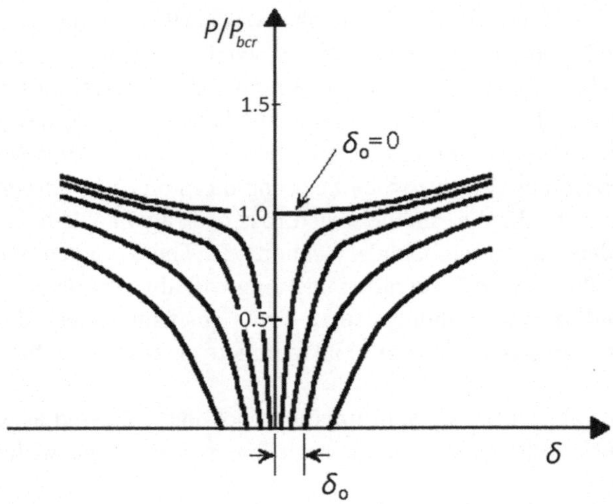

FIGURE 3.20 Load–deflection behaviour of a member with initial distortions (δ_0) exhibiting stable post-buckling behaviour.

3.3.2 Kinematic Relations for Membrane and Bending Behaviour of Cylindrical Shells

We consider an element of a thin cylindrical shell, shown in Figure 3.21. The boundaries of this element are planes that are perpendicular to the mid-surface and include the corresponding principal curvatures. The reference axes are taken so that axes Ox and Oy are tangential to the axes of principal curvature, whereas axis Oz is perpendicular to the mid-surface of the shell element. The radii of the principal curvatures rx and ry lie in planes Oxz and Oyz, respectively. The thickness t is assumed to be constant and small compared to the radii of curvature.

When the element bends, we assume that linear segments such as AD and BC shown in Figure 3.21 that were originally perpendicular to the mid-surface remain linear and perpendicular to it following deformation.

In the case of the curved member shown in Figure 3.22, prior to bending, the radius of curvature measured from O to the neutral surface is equal to r_x and the reference length of an element along the neutral surface is equal to l_0. The length of a fibre located at a distance z from the neutral surface is l.

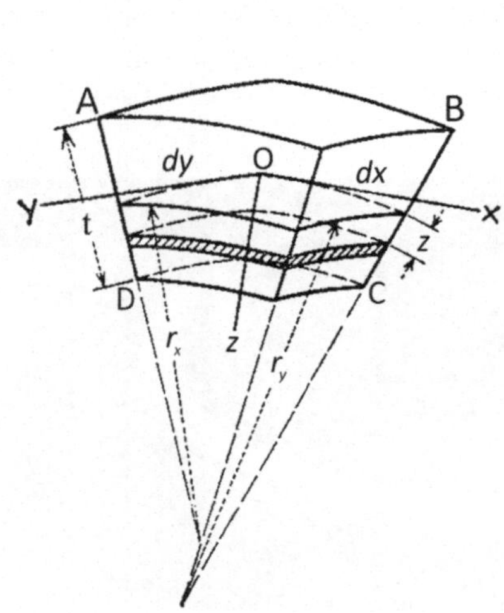

FIGURE 3.21 Shell surface element [14].

Following bending, the radius of curvature changes, as the centre of rotation translates from O to O'. The cross section then distorts through an angle φ_b, as shown by the shaded region, so that the length of the fibre at a distance z from the neutral surface changes from l to $l+\delta$. The angle φ_b represents the change in slope that is caused by the applied bending moment so that:

$$\phi_b = l_0\left(\frac{1}{r_x'} - \frac{1}{r_x}\right) \tag{3.16}$$

where r_x' is the final radius of curvature. The length l_0 is measured along the neutral surface whose position is not known. The axial elongation of the fibre at a distance z from the neutral axis is then

$$\delta = z\phi_b = zl_0\left(\frac{1}{r_x'} - \frac{1}{r_x}\right) \tag{3.17}$$

The axial strain ε_x is then

$$\varepsilon_x = \frac{\delta}{l} = z\left(\frac{l_0}{l}\right)\left(\frac{1}{r_x'} - \frac{1}{r_x}\right) \tag{3.18}$$

From the geometry of Figure 3.22, it is evident that $l_0/r_x = l/r$. Therefore, (3.18) can be written

$$\varepsilon_x = z\left(\frac{r_x}{r}\right)\left(\frac{1}{r_x'} - \frac{1}{r_x}\right) \tag{3.19}$$

Since $r = r_x + z$, (3.19) can be written

$$\varepsilon_x = -\frac{z}{1-\dfrac{z}{r_x}}\left(\frac{1}{r_x'} - \frac{1}{r_x}\right) \tag{3.20a}$$

$$\text{Similarly, } \varepsilon_y = -\frac{z}{1-\dfrac{z}{r_y}}\left(\frac{1}{r_y'} - \frac{1}{r_y}\right) \tag{3.20b}$$

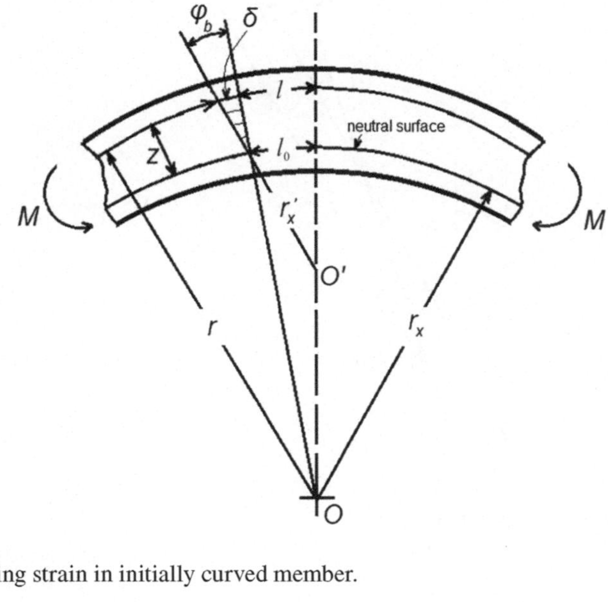

FIGURE 3.22 Bending strain in initially curved member.

Expressions (3.20a and b) are the pure bending strain–displacement relations for a shell that has curvature in the two orthogonal directions.

If, apart from rotating, the sides of the element translate in a direction parallel to their original orientation because of stretching of the neutral surface, and if the corresponding strains in the longitudinal and circumferential directions are ε_{xm} and $\varepsilon_{\theta m}$ respectively, the total strain ε_x is given by (Figure 3.23):

$$\varepsilon_x = \frac{l_2 - l_1}{l_1}$$

where l_1 is the length of the element along the neutral axis prior to stretching, and l_2 is the length of the same fibre following stretching. From Figure 3.23, we have

$$\frac{l_1}{r_x - z} = \frac{ds}{r_x}$$

where ds is the length of a shell element located a distance z from the neutral axis,

$$\text{Hence } l_1 = ds\left(1 - \frac{z}{r_x}\right)$$

When stretching takes place l_1 extends to l_2 so that:

$$l_2 = ds(1 + \varepsilon_{xm})\left(1 - \frac{z}{r_x'}\right)$$

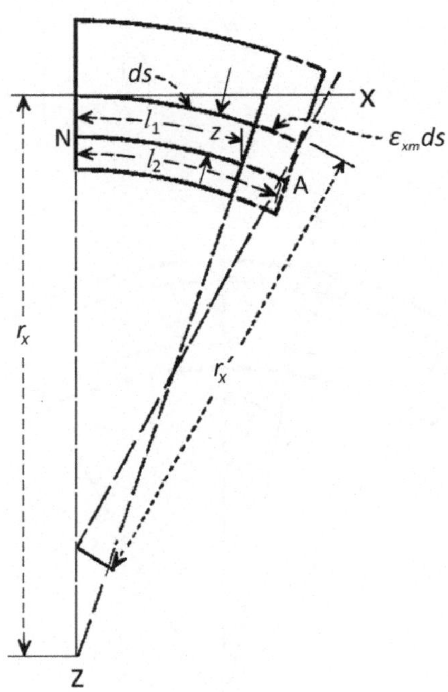

FIGURE 3.23 Shell element in stretching [14].

We obtain for the longitudinal strain ε_x:

$$\varepsilon_x = \frac{\varepsilon_{xm}}{1-\dfrac{z}{r_x}} - \frac{z}{1-\dfrac{z}{r_x}}\left[\frac{1}{(1-\varepsilon_{xm})r_x'} - \frac{1}{r_x}\right] \tag{3.21a}$$

The circumferential strain ε_θ is then:

$$\varepsilon_\theta = \frac{\varepsilon_{\theta m}}{1-\dfrac{z}{r_y}} - \frac{z}{1-\dfrac{z}{r_y}}\left[\frac{1}{(1-\varepsilon_{\theta m})r_y'} - \frac{1}{r_y}\right] \tag{3.21b}$$

For $t \gg r_x$, r_y, we can ignore the quantities z/r_x, z/r_y since z/r_x, $z/r_y \gg 1$. Furthermore, we will ignore the effect of the stretching of the neutral surface on the shell curvature, this being a second-order effect. Expressions (3.21a) and (3.21b) then become:

$$\varepsilon_x = \varepsilon_{xm} - z\left(\frac{1}{r_x'} - \frac{1}{r_x}\right) = \varepsilon_{xm} - \kappa_x z \tag{3.22a}$$

$$\varepsilon_\theta = \varepsilon_{\theta m} - z\left(\frac{1}{r_y'} - \frac{1}{r_y}\right) = \varepsilon_{\theta m} - \kappa_y z \tag{3.22b}$$

where κ_x and κ_θ are changes in the corresponding curvatures. For a plane stress state (σ_z=0), the stresses in the shell element are:

$$\sigma_x = \frac{E}{1-\nu^2}\left[\varepsilon_{xm} + \nu\varepsilon_{\theta m} - z(\kappa_x + \nu\kappa_\theta)\right] \tag{3.23a}$$

$$\sigma_\theta = \frac{E}{1-\nu^2}\left[\varepsilon_{\theta m} + \nu\varepsilon_{xm} - z(\kappa_\theta + \nu\kappa_x)\right] \tag{3.23b}$$

Substituting these relations in the expressions for the loads per unit length that act on the element OABC leads to the following expressions, which are analogous to those in thin plate theory:[6]

$$\left.\begin{aligned}
N_x &= \int_{-t/2}^{+t/2} \sigma_x\, dz = \frac{Et}{1-\nu^2}(\varepsilon_{xm} + \nu\varepsilon_{\theta m}) \\
N_\theta &= \int_{-t/2}^{+t/2} \sigma_\theta\, dz = \frac{Et}{1-\nu^2}(\varepsilon_{\theta m} + \nu\varepsilon_{xm}) \\
M_x &= \int_{-t/2}^{+t/2} z\sigma_x\, dz = -D(\kappa_x - \nu\kappa_\theta) \\
M_\theta &= \int_{-t/2}^{+t/2} z\sigma_\theta\, dz = -D(\kappa_\theta - \nu\kappa_x)
\end{aligned}\right\} \tag{3.23}$$

D is the *flexural rigidity* of the shell, defined as in the case of thin plating, as follows:

$$D = \frac{Et^3}{12(1-\nu^2)} \tag{3.24}$$

A more general case of deformation of the shell element involves, apart from the normal stresses σ_x and σ_θ, shear stresses that act along the sides of the element. If these shear stresses are denoted by τ_{xy} and τ_{xz}, we have:

$$Q_x = \int_{-t/2}^{+t/2} \tau_{xz}\, dz \tag{3.25a}$$

$$N_{x\theta} = \int_{-t/2}^{+t/2} \tau_{x\theta} dz \tag{3.25b}$$

$$M_{x\theta} = -\int_{-t/2}^{+t/2} z\tau_{x\theta} dz \tag{3.25c}$$

For the sides perpendicular to axis Oy, the following relations hold:

$$Q_\theta = \int_{-t/2}^{+t/2} \tau_{\theta z}\, dz \tag{3.26a}$$

$$N_{\theta x} = N_{x\theta} \tag{3.26b}$$

$$M_{\theta x} = -M_{x\theta} \tag{3.26c}$$

The relationship between the shear stress and the twisting moment is determined as in the case of flat plating. Thus,

$$\tau_{x\theta} = -2Gz\kappa_{x\theta} \tag{3.27a}$$

$$M_{x\theta} = D(1-\nu)\kappa_{x\theta} \tag{3.27b}$$

If in addition to rotation, shear deformation takes place in way of the middle surface of the shell,

$$\tau_{x\theta} = (\gamma - 2z\kappa_{x\theta})G \tag{3.28a}$$

$$N_{x\theta} = \int_{-t/2}^{+t/2} \tau_{x\theta} dz = \frac{\gamma Et}{2(1+\nu)} \tag{3.28b}$$

$$M_{x\theta} = \int_{-t/2}^{+t/2} \tau_{x\theta} z dz = D(1-\nu)\kappa_{x\theta} \tag{3.28c}$$

The above expressions can be used to obtain the equations of equilibrium of thin cylindrical shells in which both membrane and bending effects arise.

3.3.3 Infinitely Long Cylindrical Shell Subjected to Arbitrary External Loading

In this section, we shall obtain the equations of equilibrium of an unstiffened cylindrical shell subjected to arbitrary external loading, following the derivation given in [14]. This is the first step towards the derivation of expressions required in submarine pressure hull analysis, in which the loading consists of external hydrostatic pressure that acts on a cylindrical shell having a finite length equal to the frame spacing. This problem will be considered in the next chapter.

3.3.3.1 Kinematic Relations for a Cylindrical Shell under Generalised Loading

In order to derive the equations of equilibrium, we need to consider the distortions arising in an element as a result of the application of external loading, that is, we need to develop the relevant kinematic relations. The presentation in this section follows that in [14] and [15] and has been based on the theory of cylindrical shells developed by Love [8].

Let us consider the cylindrical section of Figure 3.24a, which has a constant diameter $2R$, is of infinite length, and is subjected to arbitrary external pressure loading p per unit area. The applied loading results in deformations so that the directions of components of the neutral surface of element OABC no longer coincide with their initial lines of action. The translations in the direction of the coordinate axes Ox, Oy, and Oz are u, v, and w, respectively. In order to develop the equations of equilibrium of the cylindrical shell under generalised loading, we will consider the deformations of element OABC. The forces and moments that act on the middle surface of the element are shown in Figure 3.24b and c, respectively. For the deformed element, Ox is parallel to the original axis of the cylinder (generatrix), Oz is perpendicular to the plane of the deformed element, and Oy is perpendicular to plane Oxz.

We will first obtain expressions for the rotations of sides BC and AB with respect to sides OA and OC. If the translations u, v, and w are small, we can calculate the angles that arise as a result of these deformations by applying superposition of the angles that result from each of these translations.

We consider rotation of side BC with respect to side OA. This rotation has components in the directions of the three coordinate axes Ox, Oy, and Oz. Rotation of sides OA and BC with respect to axis Ox is a consequence of translations v and w. Since v corresponds to the rotation of sides OA and BC in the direction of the shell's circumference, and since the radius is equal to R, the corresponding rotation of OA with respect to axis Ox is equal to v/R, (Figure 3.25a), whereas that of side BC is equal to

$$\frac{1}{R}\left(v+\frac{\partial v}{\partial x}dx\right) \tag{a}$$

Hence, the angle of rotation of side BC with respect to side OA about axis OA due to translation v is

$$\frac{1}{R}\frac{\partial v}{\partial x}dx \tag{b}$$

Translation w results in the rotation of side OA about axis Ox through an angle $\partial w/R\,\partial\theta$, (Figure 3.25b), whereas side BC rotates through

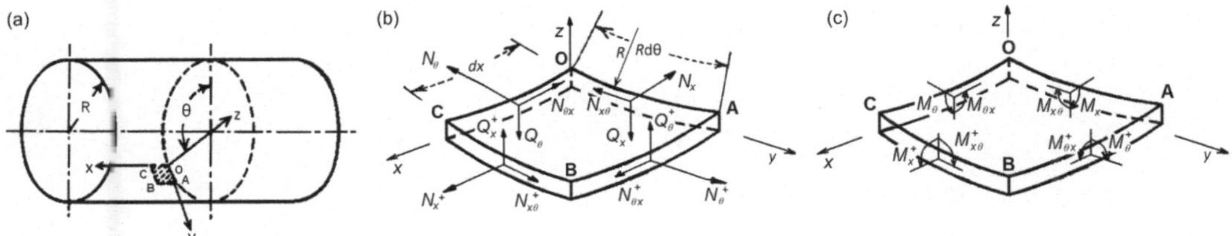

FIGURE 3.24 Forces acting on the middle surface of a cylindrical shell element [14]. (a) Coordinate axis system and shell element OABC. (b) Forces N_x, N_y, N_{xy}. (c) Moments M_x, M_y, M_{xy}.

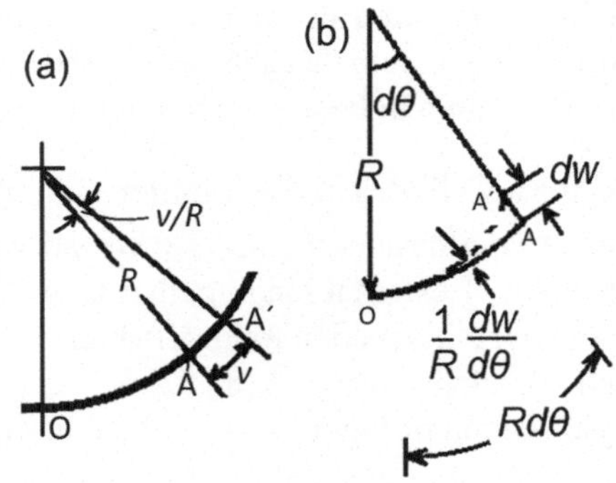

FIGURE 3.25 Rotations of side OA due to translations v and w. (a) Rotation of side OA due to meridional translation v. (b) Distortion of OA due to radial translation w.

$$\frac{\partial w}{R\partial\theta}+\frac{\partial}{\partial x}\frac{\partial w}{R\partial\theta}dx \tag{c}$$

Hence, the angle of rotation of side BC with respect to side OA about axis O_x due to translation w is

$$\frac{\partial}{\partial x}\frac{\partial w}{R\partial\theta}dx \tag{d}$$

Adding the expressions for the rotations of side BC with respect to side OA due to translations v and w, we obtain the total rotation of side BC with respect to side OA about axis Ox:

$$\frac{1}{R}\left(\frac{\partial v}{\partial x}+\frac{\partial^2 w}{\partial x\,\partial\theta}\right)dx \tag{e}$$

The rotation of side BC about the Oy axis with respect to side OA is

$$-\frac{\partial^2 w}{\partial x^2}dx \tag{f}$$

Similarly, the rotation of the Oz axis of side BC with respect to side OA equals:

$$\frac{\partial^2 v}{\partial x^2}dx \tag{g}$$

Expressions (e), (f), and (g) are the three components of the rotation of side BC with respect to side OA.

We shall now obtain the corresponding relations for the rotation of side BC with respect to side OC. The angle subtended by the sides of element OABC due to the initial curvature is equal to $d\theta$. Following deformation that involves translations v and w, this angle changes and the angle of rotation of side OC with respect to axis Ox becomes

$$\frac{v}{R}+\frac{1}{R}\frac{\partial w}{\partial\theta} \tag{h}$$

The corresponding rotation of side AB is

$$\frac{v}{R}+\frac{1}{R}\frac{\partial w}{\partial\theta}+\frac{\partial}{\partial\theta}\left(\frac{v}{R}+\frac{\partial w}{R\partial\theta}\right)d\theta \tag{i}$$

The angle $d\theta$ becomes

$$d\theta + d\theta \frac{1}{R}\left(\frac{\partial v}{\partial \theta} + \frac{\partial^2 w}{\partial \theta^2}\right) \tag{j}$$

During the rotation of the shell surface element, the generatrix rotates about axis Oy through $-\partial w/\partial z$ and about axis Oz through $\partial w/\partial x$. The rotation of part of the generatrix at a distance of $Rd\theta$ from the first rotates about axis Oy through

$$-\frac{\partial w}{\partial x} - \frac{\partial^2 w}{\partial \theta\, \partial x} d\theta \tag{k}$$

The rotation of the same element in the plane tangential with the shell surface is

$$\frac{\partial v}{\partial x} + \frac{\partial(\partial v/\partial x)}{\partial x} d\theta \tag{l}$$

The component of the above angle of twist about Oy is then:

$$-\frac{\partial v}{\partial x} d\theta \tag{m}$$

Adding expressions (k) and (l) gives the total angle of twist:

$$-\left(\frac{\partial^2 w}{\partial \theta\, \partial x} + \frac{\partial v}{\partial x}\right) d\theta \tag{n}$$

Translations v and w result in a rotation of side AB about axis Oz. Translation v corresponds to a rotation of side OC of $\partial v/\partial x$ and of side AB of

$$\frac{\partial v}{\partial x} + \frac{1}{R}\frac{\partial}{\partial \theta}\frac{\partial v}{\partial x} R d\theta \tag{o}$$

so that the relative rotation is

$$\frac{1}{R}\frac{\partial}{\partial \theta}\frac{\partial v}{\partial x} R d\theta \tag{p}$$

Side AB rotates through $\partial w/\partial x$ in the plane of the coordinate axes system. The component of this angle with respect to axis Oz is then equal to

$$-\frac{\partial w}{\partial x} d\theta \tag{q}$$

Adding (p) and (q) gives the total angle of rotation of side AB about axis Oz with respect to side OC:

$$\left(\frac{\partial^2 v}{\partial \theta\, \partial x} - \frac{\partial w}{\partial x}\right) d\theta \tag{r}$$

3.3.3.2 Equilibrium of a Cylindrical Shell Element

We are now in a position to write down the equations of equilibrium for element OABC. As in the case of thin flat plating, this is done by considering equilibrium in the three orthogonal directions.

All forces acting on the shell element are considered, and their projections along the coordinate axes are added.

In the direction of axis Ox, the force components are $(\partial N_x / \partial x)dxRd\theta$ and $(\partial N_{\theta x} / \partial\theta)d\theta dx$. In the direction of axis Oz, the components parallel to N_θ, as a result of angle (q) are

$$-N_\theta\left(\frac{\partial^2 v}{\partial\theta\,\partial x}-\frac{\partial w}{\partial x}\right)d\theta dx \tag{s}$$

The force component parallel to $N_{x\theta}$ is

$$-N_{x\theta}\frac{\partial^2 v}{\partial x^2}dxRd\theta \tag{t}$$

Finally, the components of forces that are parallel to Q_x and Q_θ along Ox as a result of rotations (f) and (k) are:

$$-Q_x\frac{\partial^2 w}{\partial x^2}dxRd\theta-Q_\theta\left(\frac{\partial^2 w}{\partial\theta\,\partial x}+\frac{\partial v}{\partial x}\right)d\theta dx \tag{u}$$

The applied load to the shell element is a lateral pressure of magnitude p, whose component in the direction of the axis Ox is zero. Adding all components, the following equation of equilibrium in the x-direction is obtained:

$$\frac{\partial N_x}{\partial x}dxRd\theta+\frac{\partial N_{\theta x}}{\partial\theta}d\theta dx-N_y\left(\frac{\partial^2 v}{\partial\theta\,\partial x}-\frac{\partial w}{\partial x}\right)d\theta dx$$
$$-N_{x\theta}\frac{\partial^2 v}{\partial x^2}dxRd\theta-Q_x\frac{\partial^2 w}{\partial x^2}dxRd\theta-Q_\theta\left(\frac{\partial^2 w}{\partial\theta\,\partial x}+\frac{\partial v}{\partial x}\right)d\theta dx=0-Q_\theta\left(\frac{\partial^2 w}{\partial\theta\,\partial x}+\frac{\partial v}{\partial x}\right)d\theta dx=0 \tag{3.29}$$

In a similar manner, two additional equations for the shell surface element are obtained by considering equilibrium in the y and z directions. The three equations of equilibrium in directions x, y, and z are then:

$$\left.\begin{aligned}
&R\frac{\partial N_x}{\partial x}+\frac{\partial N_{\theta x}}{\partial\theta}-RQ_x\frac{\partial^2 w}{\partial x^2}-RN_{x\theta}\frac{\partial^2 v}{\partial x^2}-Q_\theta\left(\frac{\partial v}{\partial x}+\frac{\partial^2 w}{\partial x\partial\theta}\right)-N_\theta\left(\frac{\partial^2 v}{\partial x\partial\theta}-\frac{\partial w}{\partial x}\right)=0\\
&\frac{\partial N_\theta}{\partial\theta}+R\frac{\partial N_{x\theta}}{\partial x}+RN_x\frac{\partial^2 v}{\partial x^2}-Q_x\left(\frac{\partial v}{\partial x}+\frac{\partial^2 w}{\partial x\partial\theta}\right)+N_{\theta x}\left(\frac{\partial^2 v}{\partial x\partial\theta}-\frac{\partial w}{\partial x}\right)-Q_\theta\left(1+\frac{1}{R}\frac{\partial v}{\partial\theta}+\frac{1}{R}\frac{\partial^2 w}{\partial\theta^2}\right)=0\\
&R\frac{\partial Q_x}{\partial x}+\frac{\partial Q_\theta}{\partial\theta}+N_{x\theta}\left(\frac{\partial v}{\partial x}+\frac{\partial^2 w}{\partial x\partial\theta}\right)+RN_x\frac{\partial^2 w}{\partial x^2}+N_\theta\left(1+\frac{1}{R}\frac{\partial v}{\partial\theta}+\frac{1}{R}\frac{\partial^2 w}{\partial\theta^2}\right)+N_{\theta x}\left(\frac{\partial v}{\partial x}+\frac{\partial^2 w}{\partial x\partial\theta}\right)+pR=0
\end{aligned}\right\} \tag{3.30}$$

This system of equations describes the general behaviour of a thin cylindrical shell subjected to arbitrary loading. Since deformations of the neutral surface are not allowed, it is analogous to the equations of equilibrium of thin flat plating in the linear elastic small deflection range.

If in these equations we substitute from (3.25) to (3.26), assuming small rotations of sides BC and AB with respect to OA and OC, respectively, the equations are recast as follows:

$$\left.\begin{aligned}
&R\frac{\partial M_{x\theta}}{\partial x}-\frac{\partial M_\theta}{\partial \theta}-RM_x\frac{\partial^2 v}{\partial x^2}-M_{\theta x}\left(\frac{\partial^2 v}{\partial x\partial \theta}-\frac{\partial w}{\partial x}\right)+RQ_\theta=0\\
&\frac{\partial M_{\theta x}}{\partial \theta}+R\frac{\partial M_x}{\partial x}+RM_{x\theta}\frac{\partial^2 v}{\partial x^2}-M_\theta\left(\frac{\partial^2 v}{\partial x\partial \theta}-\frac{\partial w}{\partial x}\right)-RQ_x=0\\
&M_x\left(\frac{\partial v}{\partial x}+\frac{\partial^2 w}{\partial x\partial \theta}\right)+RM_{x\theta}\frac{\partial^2 w}{\partial x^2}+M_{\theta x}\left(1+\frac{1}{R}\frac{\partial v}{\partial \theta}+\frac{1}{R}\frac{\partial^2 w}{\partial \theta^2}\right)-M_\theta\left(\frac{\partial v}{\partial x}+\frac{\partial^2 w}{\partial x\partial \theta}\right)+R(N_{x\theta}-N_{\theta x})=0
\end{aligned}\right\}\tag{3.31}$$

These equations are now expressed in terms of in-plane forces N_x, N_θ, $N_{x\theta}$ and moments M_x, M_θ, $M_{x\theta}$. If the middle surface of the shell deforms as a result of the action of the applied loading, appropriate large deflection kinematic relations have to be substituted in (3.31).

3.3.4 Finite-Length Cylindrical Shell Subjected to Constant Hydrostatic Pressure

In this section, we shall provide the derivation of the equations of equilibrium of a cylindrical shell of finite length subjected to external hydrostatic pressure, this being the basic problem of submarine cylindrical pressure hull stability. The equations of equilibrium for this particular case are derived following [14]. Solving these equations leads to a formula for the critical elastic buckling pressure of a cylindrical shell of finite length. This formula was originally obtained by von Mises [16] and constitutes the cornerstone of submarine cylindrical pressure hull strength analysis.

In the case of a cylindrical shell of finite length, it is necessary to allow for the boundary conditions at the ends, and therefore, recourse has to be made to the equations of equilibrium for generalised loading (3.31). When the external loading is a uniform pressure, all forces apart from N_θ are small. As a result, the terms that include products of forces with angles of line of action are disregarded. The equations of equilibrium (3.31) then assume the following form:

$$\left.\begin{aligned}
&R\frac{\partial N_x}{\partial x}+\frac{\partial N_{\theta x}}{\partial \theta}-N_\theta\left(\frac{\partial^2 v}{\partial x\,\partial \theta}-\frac{\partial w}{\partial x}\right)=0\\
&\frac{\partial N_\theta}{\partial \theta}+R\frac{\partial N_{x\theta}}{\partial x}-Q_\theta=0\\
&R\frac{\partial Q_x}{\partial x}+\frac{\partial Q_\theta}{\partial \theta}+N_\theta\left(1+\frac{1}{R}\frac{\partial v}{\partial \theta}+\frac{1}{R}\frac{\partial^2 w}{\partial \theta^2}\right)+pR=0
\end{aligned}\right\}\tag{3.32}$$

In equations (3.31), it is assumed that the bending and twisting moments are small. As a result, we may ignore the products of these terms with the angles of line of action. From the first two equations of (3.31) we obtain:

$$\left.\begin{aligned}
Q_x&=\frac{1}{R}\frac{\partial M_{\theta x}}{\partial \theta}+\frac{\partial M_x}{\partial x}\\
Q_\theta&=\frac{1}{R}\frac{\partial M_\theta}{\partial \theta}-\frac{\partial M_{x\theta}}{\partial x}
\end{aligned}\right\}\tag{3.33}$$

Substituting these in (3.32), we obtain:

$$\left.\begin{aligned}&R\frac{\partial N_x}{\partial x}+\frac{\partial N_{\theta x}}{\partial \theta}-N_\theta\left(\frac{\partial^2 v}{\partial x\,\partial \theta}-\frac{\partial w}{\partial x}\right)=0\\&\frac{\partial N_\theta}{\partial \theta}+R\frac{\partial N_{x\theta}}{\partial x}-\frac{1}{R}\frac{\partial M_\theta}{\partial \theta}+\frac{\partial M_{x\theta}}{\partial x}=0\\&\frac{\partial^2 M_{\theta x}}{\partial x\,\partial \theta}+R\frac{\partial^2 M_x}{\partial x^2}+\frac{1}{R}\frac{\partial^2 M_\theta}{\partial \theta^2}-\frac{\partial^2 M_{x\theta}}{\partial x\,\partial \theta}+N_\theta\left(1+\frac{1}{R}\frac{\partial v}{\partial \theta}+\frac{1}{R}\frac{\partial^2 w}{\partial \theta^2}\right)+pR=0\end{aligned}\right\}\tag{3.34}$$

To allow for the stretching of the middle surface that arises during buckling, the terms N_θ and p are replaced by $N_\theta(1+\varepsilon_{xm})$ and $p(1+\varepsilon_{xm})(1+\varepsilon_{\theta m})$, respectively, in equations (3.34b) and (3.34c). This is necessary because N_θ and p are large compared to the other stress resultants. The membrane strains are

$$\varepsilon_{xm}=\frac{\partial u}{\partial x}\qquad \varepsilon_{\theta m}=\frac{1}{R}\frac{\partial v}{\partial \theta}-\frac{w}{R}$$

in the longitudinal and circumferential directions, respectively, as shown in Section 4.3.2. Equations (3.34) can then be written as follows[7]:

$$\left.\begin{aligned}&R\frac{\partial N_x}{\partial x}+\frac{\partial N_{\theta x}}{\partial \theta}+pR\left(\frac{\partial^2 v}{\partial x\,\partial \theta}-\frac{\partial w}{\partial x}\right)=0\\&\frac{\partial N'_\theta}{\partial \theta}+R\frac{\partial N_{x\theta}}{\partial x}-\frac{1}{R}\frac{\partial M_\theta}{\partial \theta}+\frac{\partial M_{x\theta}}{\partial x}=0\\&\frac{\partial^2 M_{\theta x}}{\partial x\,\partial \theta}+R\frac{\partial^2 M_x}{\partial x^2}+\frac{1}{R}\frac{\partial^2 M_\theta}{\partial \theta^2}-\frac{\partial^2 M_{x\theta}}{\partial x\,\partial \theta}+N'_\theta-p\left(w+\frac{\partial^2 w}{\partial \theta^2}\right)=0\end{aligned}\right\}\tag{3.35}$$

These equations can now be expressed in terms of translations (u, v, w) by substituting the force and moment terms using the expressions given in Section 3.3.2 and equations (3.31). Furthermore, the following constants are introduced:

$$\phi=\frac{pR}{t}\frac{(1-v^2)}{E}\qquad \alpha=\frac{1}{12}\left[\frac{t}{R}\right]^2\tag{3.36}$$

The equations of equilibrium (3.35) then become:

$$\left.\begin{aligned}&R^2\frac{\partial^2 u}{\partial x^2}+\frac{1+v}{2}R\frac{\partial^2 v}{\partial x\,\partial \theta}-vR\frac{\partial w}{\partial x}+R\phi\left(\frac{\partial^2 v}{\partial x\,\partial \theta}-\frac{\partial w}{\partial x}\right)+\frac{1-v}{2}\frac{\partial^2 u}{\partial \theta^2}=0\\&\frac{1+v}{2}R\frac{\partial^2 u}{\partial x\,\partial \theta}+\frac{1-v}{2}R^2\frac{\partial^2 v}{\partial x^2}+\frac{\partial^2 v}{\partial \theta^2}-\frac{\partial w}{\partial \theta}+\alpha\left[\frac{\partial^2 v}{\partial \theta^2}+\frac{\partial^3 w}{\partial \theta^3}+R^2\frac{\partial^3 w}{\partial x^2\,\partial \theta}+R^2(1-v)\frac{\partial^2 v}{\partial x^2}\right]=0\\&Rv\frac{\partial u}{\partial x}+\frac{\partial v}{\partial \theta}-w-\alpha\left[\frac{\partial^3 v}{\partial \theta^3}+(2-v)R^2\frac{\partial^3 v}{\partial x^2\,\partial \theta}+R^4\frac{\partial^4 w}{\partial x^4}+\frac{\partial^4 w}{\partial \theta^4}+2R^2\frac{\partial^4 w}{\partial x^2\,\partial \theta^2}\right]=\phi\left(w+\frac{\partial^2 w}{\partial \theta^2}\right)\end{aligned}\right\}\tag{3.37}$$

3.3.4.1 Solution of Equations of Equilibrium of a Finite Length Cylindrical Shell under Hydrostatic Pressure

In order to obtain a solution of this system of equations, the boundary conditions and the deflected form of the shell surface have to be specified (translations u, v, w). If the shell surface is simply supported in way of infinitely stiff ring frames, the boundary conditions will be $w = 0$, $\partial^2 w/\partial x^2 = 0$.

To determine the correct deflected form, it is necessary to make the following assumptions:

1. The effect of radial stiffening on the pre-buckling stress distribution in the cylindrical shell is neglected.
2. The rotational and axial restraint to shell buckling due to the presence of radial stiffening is neglected.
3. The ring frames do not deform during buckling of the cylindrical shell.

On the basis of these assumptions, the following expressions were used to obtain an exact solution of the equations of equilibrium (3.37):

In the longitudinal direction,

$$u = A\sin(n\theta)\sin\left(\frac{\pi x}{L}\right) \tag{3.38a}$$

In the circumferential direction,

$$v = B\cos(n\theta)\cos\left(\frac{\pi x}{L}\right) \tag{3.38b}$$

In the radial direction,

$$w = C\sin(n\theta)\cos\left(\frac{\pi x}{L}\right) \tag{3.38c}$$

where L is the unsupported span, given by $L = L_f - t_w$

Expressions (3.38a–c) ensure that the deformed shape consists of a single halfwave (lobe) between the positions of support (ring frames) (Figure 3.26a) and $2n$ lobes along the circumference of the shell (Figure 3.26b).

Substituting (3.38a–c) in (3.37), the following three homogeneous linear algebraic equations in A, B, and C are obtained:

$$\left.\begin{aligned} &A\left(-\lambda^2 - \frac{1-\nu}{2}n^2\right) + B\left(\frac{1+\nu}{2}n\lambda + n\lambda\phi\right) + C(\nu+\phi)\lambda = 0 \\ &A\left(\frac{1+\nu}{2}n\lambda\right) - B\left[\frac{1-\nu}{2}\lambda^2 + n^2 + n^2\alpha + \alpha(1-\nu)\lambda^2\right] - C(n + \alpha n^3 + \alpha n\lambda^2) = 0 \\ &A(\nu\lambda) - B\left[n + \alpha n^3 + (2-\nu)\alpha n\lambda^2\right] - C\left[1 + \alpha\lambda^4 + \alpha n^4 + 2\alpha n^2\lambda^2 + \phi(1-n^2)\right] = 0 \end{aligned}\right\} \tag{3.39}$$

where $\lambda = \pi R/L$. These relations are satisfied by setting $A = B = C = 0$. This is the trivial solution that corresponds to a symmetric deformation in the radial direction. A solution that corresponds to the shell buckling problem is obtained if $A \neq 0$, $B \neq 0$ and $C \neq 0$. For this to be the case, the determinant of the equations has to be equal to zero. Expanding the 3×3 determinant and retaining linear terms leads to the following characteristic equation:

$$\phi(D + E\alpha + F\phi) = G + H\alpha + K\alpha^2 \tag{3.40}$$

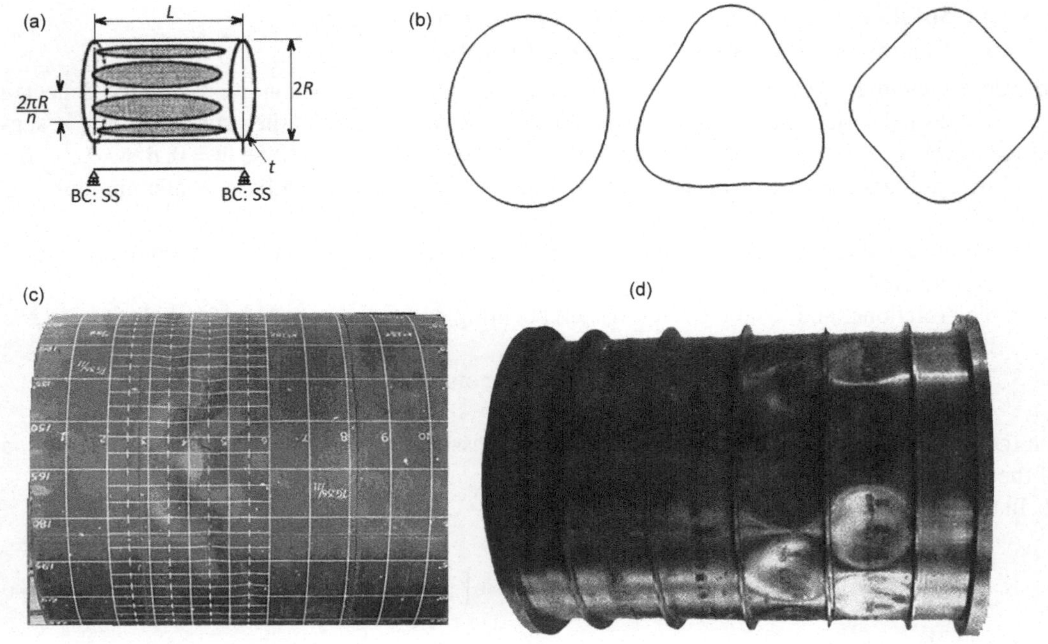

FIGURE 3.26 Interframe elastic buckling of a ring-stiffened cylinder (short). (a) Buckled form of cylindrical shell of medium length simply supported at both ends [4]. (b) Lobes formed during elastic buckling $n = 2$ $n = 3$ $n = 4$. (c) Internal ring stiffeners. (d) External ring stiffeners [13].

where

$$\left.\begin{aligned}
D &= (1-n^2)(n^2+\lambda^2)^2 - \nu\lambda^4 \\
E &= (1-n^2)\left(n^2+\frac{2\lambda^2}{1-\nu}\right)\left[n^2+(1-\nu)\lambda^2\right]+\frac{1+3\nu}{1-\nu}n^4\lambda^2+\frac{2+3\nu-\nu^2}{1-\nu}n^2\lambda^4 \\
&\quad -\frac{2\nu n^2\lambda^2}{1-\nu}-2\nu\lambda^4-\frac{1+\nu}{1-\nu}n^2\lambda^2\left(\lambda^2+n^2\right)^2 \\
F &= -\frac{1+\nu}{1-\nu}\left(1-n^2\right)n^2\lambda^2 \\
G &= -\left(1-\nu^2\right)\lambda^4 \\
H &= -(n^2+\lambda^2)^4+2n^2\left(n^2+\frac{3-\nu}{2}\lambda^2\right)\left[n^2+(2+\nu)\lambda^2\right] \\
&\quad -\left[n^2+(1-\nu)\lambda^2\right]\left[n^2+2(1+\nu)\lambda^2\right] \\
K &= -\lambda^4\left(n^2+\lambda^2\right)\left[n^2(1-\nu)+2\lambda^2\right]
\end{aligned}\right\}$$

If we ignore higher-order terms and substitute in (3.40) for α, φ, and λ, the elastic buckling pressure, $p_m(n)$, of a thin cylindrical shell of finite length is found to be [5,14]:

$$p_m(n) = \left(\frac{Et}{R}\right)\frac{1}{\left[n^2-1+\frac{\lambda^2}{2}\right]}\left\{\frac{1}{\left[\frac{n^2}{\lambda^2}+1\right]^2}+\left(\frac{t}{R}\right)^2\frac{1}{12(1-\nu^2)}\left[n^2-1+\lambda^2\right]^2\right\} \quad (3.41)$$

When buckling takes place, a number of circumferential lobes n are formed, and it is seen from (3.41) that $p_m(n)$ is a function of the number of half-sine waves in the circumferential direction, n. This means that it is necessary to determine the minimum buckling pressure by substituting a range of values for n. To this end, Figure 3.28 was included in von Mises' original paper.

The above formula is a slightly modified version of that derived by von Mises in 1929, in order to give correct answers for low values of n. The term n^2 in the original von Mises equation was replaced by Kendrick by n^2-1 in the first term and in the second expression in the bracket of (3.41) to permit a more accurate determination of the elastic buckling pressure when n obtains low values.

Figure 3.27 shows results for a cylinder whose geometrical and material properties are given in Table 3.1. A variation in frame spacing is seen to give a clear maximum elastic buckling pressure (Figure 3.27a). It is seen in Figure 3.27b that for this particular geometry $15 > n > 12$.

3.3.5 Elastic Buckling Pressure of a Long Cylindrical Shell

In the case of a long shell ($L_f \gg R$), we can ignore second-order terms that include products of λ ($=\pi L_f/R$), since this appears on the denominator. Equation (3.41) can then be written as

$$p_m(n) = \frac{(n^2-1)}{R^3}\frac{Et^3}{12(1-\nu^2)} \quad (3.42)$$

and for $n = 2$,

$$p_m = \frac{3D}{R^3} \quad (3.43)$$

where D is the shell rigidity calculated as for flat plating. Equation (3.43) is known as the *Föppl equation*. Hence, for $n = 2$,

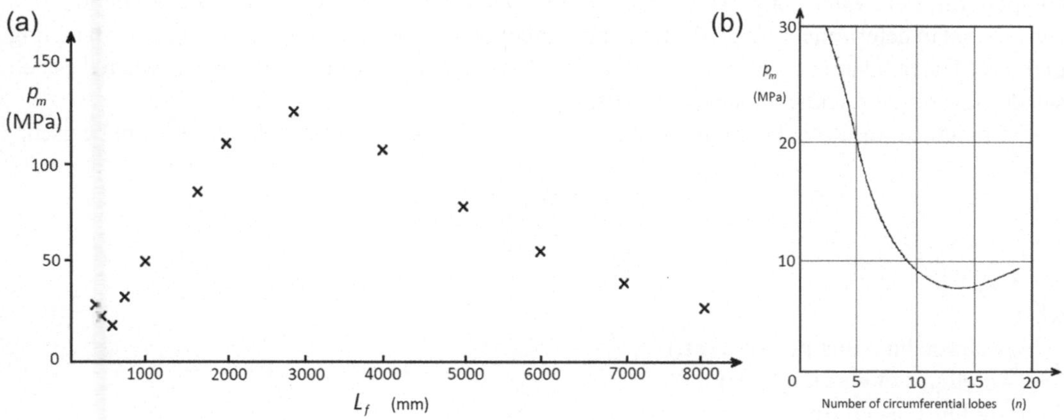

FIGURE 3.27 Elastic buckling pressure p_m and buckling mode n. (a) Effect of frame spacing L_f (for $n = 2$). (b) Number of circumferential lobes, n at buckling.

TABLE 3.1
Ring-Stiffened Cylindrical Section Particulars

Geometry (mm)		Material Properties (MPa)		Critical Pressures (MPa)	
R	2,700	E	207,000	$\sigma_0\ (t/R)$	3.981
L_f	750	v	0.3 (–)	p_{y0}	5.284
t	25	σ_o	430	p_{c3}	3.997
t_f	16	σ_{os}	400	p_{c5}	4.189
t_w	8			p_{c7}	4.457
f	100			p_e	25.51
d	190			p_{c3}[a]	3.895
				p_{c5}[b]	4.116
				p_{c7}[c]	4.252

[a] p_{c3}: Pressure to cause yielding at the outer fibre in the midbay position.
[b] p_{c5}: Pressure to cause yielding at mid-thickness in the midbay position.
[c] p_{c7}: Pressure to cause yielding at mid-thickness (for details on all see Section 5.2).

$$\sigma_{\theta cr} = \frac{E}{1-v^2}\left(\frac{t}{2R}\right)^2 \tag{3.44}$$

where $\sigma_{\theta cr}$ is the critical circumferential compressive stress that results from external hydrostatic pressure loading.

When elastic buckling occurs, a single lobe arises in the longitudinal direction between consecutive ring frames and at least two lobes in the circumferential direction. The elastic buckling pressure depends on:

- The modulus of elasticity of the material (it is directly proportional to it).
- The frame spacing to diameter ratio ($L_f/2R$) (it is inversely proportional to it).
- The thickness to diameter ratio, $t/(2R)$, which is the most important parameter since it is raised to the power 2.5.

In order to determine the lowest elastic buckling pressure from (3.41), it is necessary to calculate and compare different values of pressure that corresponds to different numbers of lobes n. Figure 3.28 can be used to determine directly the number of lobes that correspond to the critical elastic buckling pressure. The number of lobes n is inversely proportional to ($L_f/2R$) and ($t/2R$), that is, when the shell slenderness or the thickness increase n deceases.

For practical (thin) steel cylindrical shells, the following approximate relationship can be used:

$$p_m = \frac{2.6E\left(t/2R\right)^{5/2}}{\left[L/2R\right]-0.45\sqrt{\left(t/2R\right)}} \tag{3.45}$$

where:

p_m = elastic buckling pressure (psi)
E = modulus of elasticity (psi)
R = radius of shell (in)
L = unsupported length of shell plating (in)

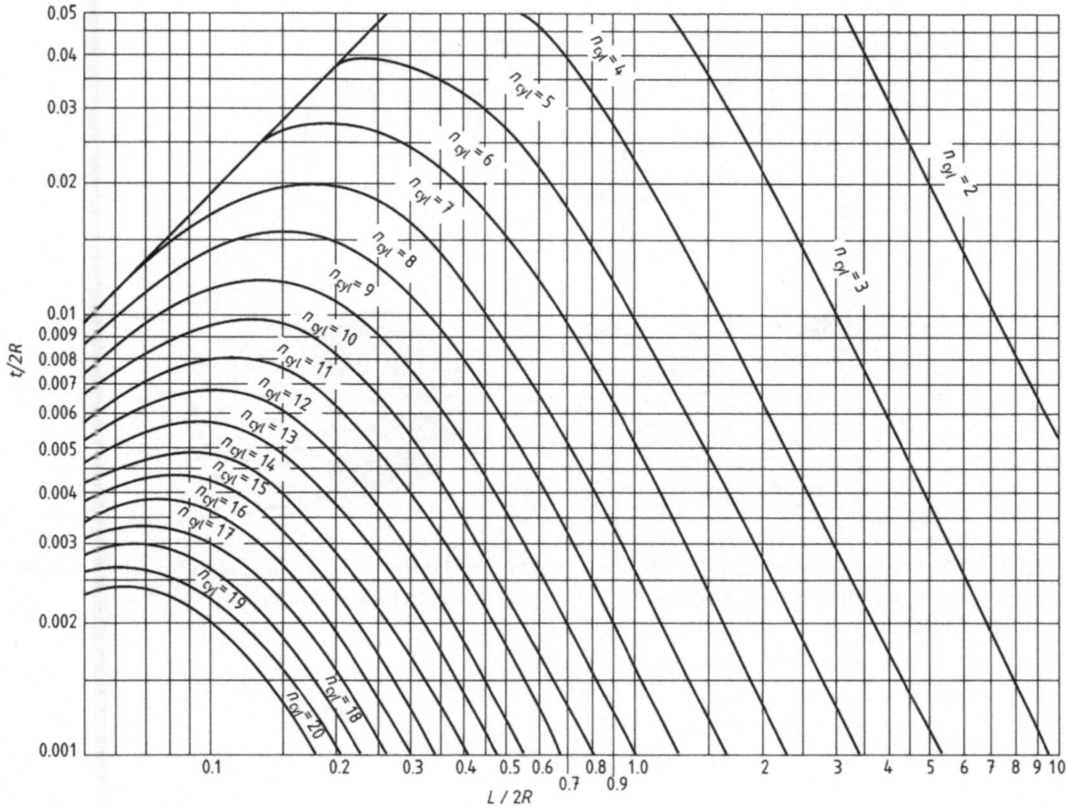

FIGURE 3.28 Number of lobes n present when an unstiffened cylinder loaded in radial and axial compression buckles elastically[8] [16].

This relationship is termed the *Model Basin formula* and was obtained by Windenburg and Trilling [17]. They differentiated von Mises' original 1929 equation with respect to n and set it equal to zero in order to find the minimum buckling pressure. The solution of the resulting equation gives the critical pressure, p_m, which is seen to be independent of n. If the shell is not thin, (3.41) has to be used instead of (3.45).

3.3.6 Pressure Vessels Subjected to Internal Pressure

Let us now consider equilibrium of the two most important generic types of pressure vessels that form the basis of all underwater vehicles – the sphere and the cylindrical shell with end closures. For present purposes, we assume that these vessels are subjected to *internal pressure*, and we shall consider equilibrium conditions in each case.

We consider first a spherical vessel of radius R and wall thickness t. In this case, the internal pressure induces equal biaxial tensile membrane stresses at any point in the shell that can be denoted as σ_θ, σ_φ (Figure 3.29a).

Figure 3.29b shows a free-body diagram consisting of one half of the sphere. The magnitude of stresses σ_φ can be found by considering equilibrium of the half-sphere, assuming that the fluid exerting the pressure has negligible weight. The total force acting on the half-sphere, pA $(= p\pi R^2)$ is balanced by the force acting around the perimeter. This is equal to (stress)(area) = $(\sigma_\varphi)(2\pi Rt)$. Equating,

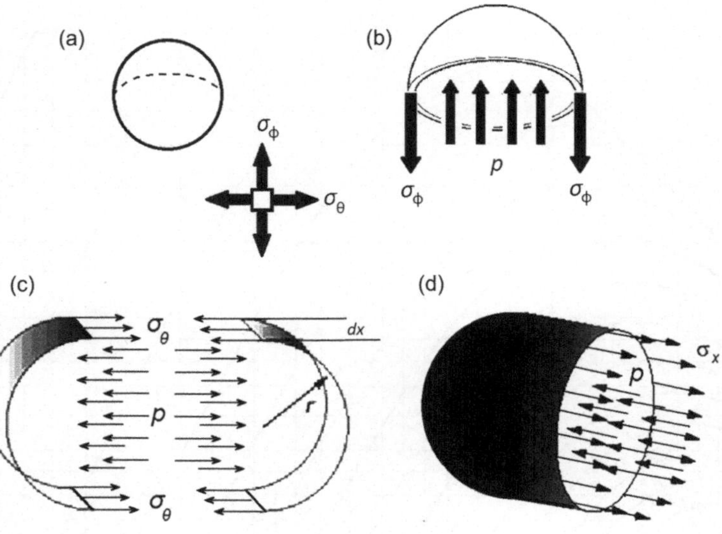

FIGURE 3.29 Vessels subjected to internal pressure. (a) Biaxial stress system in sphere. (b) Free body diagram for sphere. (c) Circumferential stresses in cylinder. (d) Longitudinal stresses in cylinder.

$$p(\pi R^2) = \sigma_\varphi(2\pi Rt)$$
$$\text{Hence,}\quad \sigma_\varphi = \frac{pR}{2t}(\equiv \sigma_\theta) \tag{3.46}$$

Within the surface of the wall, a radial stress, σ_r, is also present in order to balance the internal pressure *p*. When $R/2t$ is large, the radial stresses σ_r can be neglected in comparison to the circumferential membrane stresses (σ_θ, σ_φ).

Let us now consider equilibrium of a cylindrical section of unit length subjected to internal pressure, such that forces act in both the circumferential and longitudinal directions, as shown in Figure 3.29c and d.

In the circumferential direction (Figure 3.29c),

$$p(2R) = 2\sigma_\theta t$$

where σ_θ is the circumferential stress, (identically equal to σ_y, as used in Sections 3.3.2 and 3.3.3.)

$$\text{Hence,}\quad \sigma_\theta = p\left(\frac{R}{t}\right) \tag{3.47}$$

The axial loading of the cylindrical shell is the result of hydrostatic pressure acting on the end closures (Figure 3.29d). Thus, longitudinal tensile stresses σ_x develop around the circumference of the shell of thickness *t*. For equilibrium, the total load is balanced by this stress acting over an area $2\pi Rt$. Since the applied load equals $p\pi R_2$, the resulting longitudinal tensile stress is:

$$\sigma_x = p\left(\frac{R}{2t}\right) \tag{3.48}$$

We note that σ_x is equal to $\sigma_\theta/2$, so that σ_θ is the critical stress. For yielding failure, the critical pressure, p_0, termed the *boiler yield pressure*, is given by:

$$p_0 = \sigma_0 \left(\frac{t}{R} \right) \tag{3.49}$$

The boiler yield pressure is the upper strength limit of the structure and depends on the yield stress of the material and its geometry (R/t). For a biaxial stress field, the von Mises yield criterion can be used:

$$\sigma_x^2 + \sigma_\theta^2 - \sigma_x \sigma_\theta \le \sigma_0^2 \tag{3.50}$$

and since $\sigma_x = \sigma_\theta / 2$,

$$p_{0cr} = 1.155 p_0 \tag{3.51}$$

We now consider the case in which the cylinder is subjected to *external hydrostatic pressure.* Substituting for $p = p_m$ from (3.43), which is Föppl's expression, we obtain the circumferential compressive critical stress, $\sigma_{\theta cr}$, when elastic buckling takes place:

$$\sigma_{\theta cr} = \frac{E}{1 - \nu^2} \left(\frac{t}{2R} \right)^2 \tag{3.52}$$

We note that buckling failure occurs at a very low pressure compared to the boiler yield pressure. We conclude that in the case of the unstiffened cylindrical shell, a significant proportion of the material strength is not utilised. The solution is to reduce the unsupported length by introducing ring frames. This will be considered in the chapters that follow.

NOTES

1 Apart from hydrostatic pressure, naval submarines are required to withstand the loading due to underwater explosions that result in the dynamic response of the structure. This topic will not be dealt with here.
2 We shall see in Section 3.2 that the axial compressive stress is equal to one half the circumferential stress.
3 Complex formulation of equations of equilibrium.
4 This is the Bryan equation for the elastic buckling stress of a simply supported flat plate loaded in edge compression along its short edge. For notation, see Figure 3.8.
5 Ross test results.
6 The sign convention followed is identical to that used in the case of thin flat plates.
7 The term N'_θ represents a minor change in N_θ during buckling, given by $N'_\theta = N_\theta + pR$ [14].
8 The length L corresponds to the frame spacing L_f in the case of a ring-stiffened cylinder.

REFERENCES

1. Calladine, C.R. (1988). *Theory of Shell Structures.* Cambridge: Cambridge University Press, 763 p.
2. Flügge, W. (1960). *Stresses in Shells.* Berlin: Springer Verlag, 552 p.
3. Gill, S.S. (1970). "Basic Principles" Chap. 2. In: Gill, S.S. (Ed). *The Stress Analysis of Pressure Vessels and Pressure Vessel Components.* Oxford: Pergamon Press, 7–71.
4. Greiner R. (2004). "Cylindrical Shells under Uniform External Pressure" Chap. 5. In: Teng, J.G., Rotter, J.M. (Eds). *Buckling of Thin Metal Shells.* London: CRC Press, 154–174.
5. Kendrick, S. (1983). "How Safe are Design Codes?" *Marine Safety Conference University of Glasgow,* Scotland.
6. Kendrick, S.B. (1970). "Externally Pressurised Vessels" Chap. 9. In: Gill, S.S. (Ed). *The Stress Analysis of Pressure Vessels and Pressure Vessel Components.* Oxford: Pergamon Press, 405–511.

7. Labouvie, E.N. (1952). "The Strength of Cylindrical Shells, Stiffened by Frames and Bulkheads, under Uniform External Pressure on all Sides". DTMB Rept. No. T-38. [Transl. of: von Sanden K., Gűnther K. "Űber das Festigkeits problem quersteifter Höhlzylinder unter allseitig gleichmassigen Aussendruck". *Werft und Reederie*, Vol. 1 (1920), Nos. 8, 9, 10, Vol. 2 (1921) No. 17].
8. Love, A.E.H. (1927). *A Treatise on the Mathematical Theory of Elasticity*, 4th Ed. New York: Dover, 672 p.
9. MacNaught, D.F. (1967). "Strength of Ships" Chap. IV. In: Comstock J.P. (Ed). *Principles of Naval Architecture* (revised), New York: SNAME.
10. Meissner, E. (1913). Das Elastizitätsproblem für dünne Schalen von Ringflächen Kugel oder Kegelform (The elasticity problem for thin shells of annular surfaces of spherical or conical shape) *Physikalische Zeits*. Vol. 14, 343–349.
11. von Mises, R. (1914). "Der kritische Aussendruck Zylindrischer Rohre". *Ver Deutsche Ingenieur*, Vol. 58, p 759. (Translated as: "The critical external pressure of cylindrical tubes" USEMB Report 309).
12. Pflüger, A. (1961). *Elementary Statics of Shells*. (transl. Galantay E.) New York: F.W. Dodge, 122 p.
13. Reynolds, T.E. (1962). "Elastic Lobar Buckling of Ring-Supported Cylindrical Shells under Hydrostatic Pressure." DTMB Report 1614, Washington, DC.
14. Timoshenko, S.P., Gere, J.M. (1961). *Theory of Elastic Stability*, 2nd Ed. Tokyo: McGraw Hill Kogakusha, International Student Edition, 564 p.
15. Timoshenko, S.P., Woinowsky-Krieger, S. (1959). *Theory of Plates and Shells*, 2nd Ed. Tokyo: McGraw-Hill Kogakusha, 580 p.
16. Windenburg, D.F. (1933). "The critical external pressure of cylindrical tubes under uniform radial and axial load." Report 366, US Experimental Model Basin, August 1933. (Translation of: von Mises R., Der kritische Aussendruck für allseits belastete Rohre", *Proceedings Stodola Conference*, Zürich, 1929, pp. 418–430.)
17. Windenburg, D.F., Trilling, C. (1934). Collapse by instability of τhin cylindrical shells under external pressure. *Trans ASME*. Vol. 56, 819–825.

4 Interframe Strength of Ring-Stiffened Cylindrical Shells

4.1 IMPERFECTION-FREE CYLINDRICAL SHELL WITH INTERNAL RING FRAMES

In general, the strength of thin-walled structures depends on the nature and magnitude of internal stress fields. If the applied loading gives rise to membrane stresses, failure will occur at a higher load than if only bending stresses arise. This is clear if we examine the distribution of stresses in a circular section. Since submarines have a circular cross section, it is desirable and possible to utilise the membrane stress field that arises when the cylindrical shell is subjected to uniform hydrostatic pressure. To succeed in this, the pressure hull has to be fabricated such that *out-of-circularity* (OOC) is limited to prescribed limits.

Imperfections give rise to local bending which may result in premature failure, either by shell buckling between consecutive ring frames (*interframe* buckling) or by yielding at critical locations of the structure. It has to be noted that OOC arises during the fabrication procedure. Other sources of geometrical imperfections are (a) the presence and (b) the fabrication of ring frames, decks, and bulkheads, which introduce structural asymmetry as well as geometrical imperfections. The construction yard therefore has to adhere strictly to acceptable tolerance levels for geometrical imperfections.[1] We shall return to the question of imperfections in Section 4.2. Before that, we need to consider failure of an imperfection-free, ring-stiffened cylindrical shell.

4.1.1 The Effect of Ring Frames on the Axial Stresses in the Cylindrical Shell

As seen in Section 3.3.6, the circumferential stresses induced in the cylindrical shell are twice as large as the axial stresses that are due to the load carried by the end closures. For this reason, the cylindrical shell is stiffened in the circumferential direction, and equally spaced ring frames are used to this end. Ring frames increase the stiffness of the shell in way of the line of attachment. As a result, the circumferential strains and stresses induced in the shell in way of the ring frames are lower in comparison to the midbay positions. Since the ring frames have a finite rigidity, the induced compressive strains will lead to a reduction in the radius of the section. This, however, will be significantly less than that experienced at the midbay positions.

The fact that the radial stiffness of the shell varies in the axial direction means that bending will take place between successive ring frames. This means that in addition to the membrane stresses that arise in the circumferential and axial directions, bending stresses that vary in the axial direction will be present in the shell. It is therefore necessary to consider the nature of the induced stresses and strains at representative positions along the length of the cylindrical shell.

Consider sections A-A in way of the junction with the ring frame and B-B at the midbay position (Figure 4.1).

In the circumferential direction, in way of section B-B, the compressive circumferential membrane stress is due solely to the action of hydrostatic pressure. In way of section A-A, the circumferential stress in the shell is reduced due to the support provided by the ring frame.

DOI: 10.1201/9781003489153-4

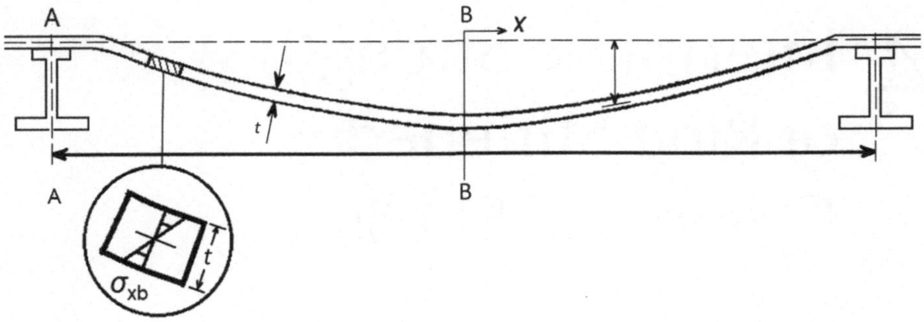

FIGURE 4.1 Cylindrical shell bending between successive ring frames (Insert shows the longitudinal bending stress distribution in the shell plating).

In the axial direction, the action of the hydrostatic pressure in conjunction with the presence of the ring frames leads to bending of the shell, as shown in Figure 4.1. As a result, in way of Section A-A, the total axial stress consists of:

a. The (membrane) compressive stress induced by compression of the end closures, σ_{xm}^{f}
b. The bending stress component, σ_{xb}^{f}, which varies along the length of the shell, between ring frames

The hydrostatic pressure induces in addition a circumferential membrane compressive stress in the shell. Since the axial and circumferential stresses are at right angles to each other, a biaxial stress field results.

We consider section B-B at midbay. In the longitudinal direction, the induced stresses are the result of:

a. Compression of the end closures, giving rise to membrane stresses, σ_{xm}^{m}
b. Bending between the ring frames, σ_{xb}^{m}

In way of the outer surface, the bending strains are compressive, whereas in way of the inner surface, they are tensile. The end result is indicated in Figure 4.2b. It is seen that the maximum axial stresses arise at the outer surface at the midbay position and at the inner surface in way of the ring frame.

4.1.2 Infinitely Long Ring-Stiffened Cylindrical Shell

We shall first consider a cylindrical shell with regularly spaced circular ring frames (spacing = L_f). If in the Föppl equation we replace the flexural rigidity D with EI_{sf}/L_f, where I_{sf} is the moment of

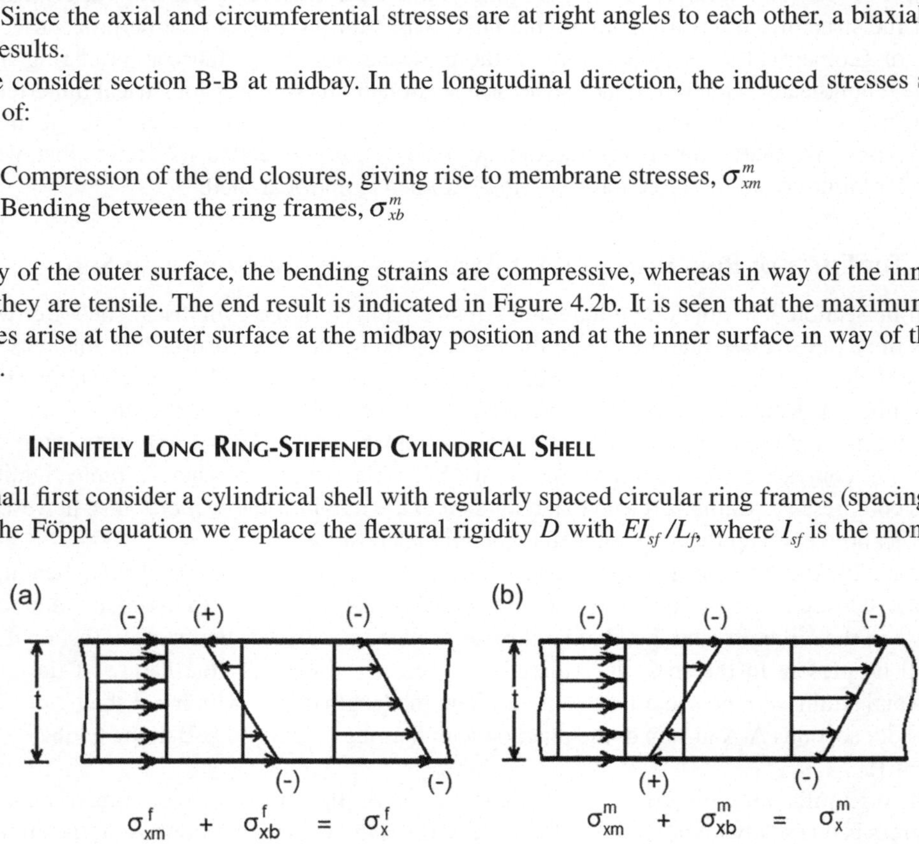

FIGURE 4.2 Longitudinal stresses in the cylindrical shell. (a) In way of the ring frame. (b) At the midbay position.

inertia of the combined shell plating and ring frame, we obtain the *Bresse* formula[2] that is valid for an infinitely long cylinder:

$$p_{fn} = \frac{(n^2 - 1)EI_{sf}}{R^3 L_f} \tag{4.1}$$

Shell bending between frames is not allowed for in this expression. Failure pressure will thus be greater than the actual one, especially when the number of lobes around the circumference increases, although as a rule n has a low value. In using (4.1), it is necessary to recognise that only a part of the shell surface is effective because of the nonlinear variation of the circumferential stress between frames. An acceptable approximation is $0.75L_f$ in lieu of L_f.

To calculate the strength of a long cylindrical shell, we first need to calculate the elastic buckling pressure. If we assume $R = 4{,}000$ mm and $t = 32$ mm, by making use of (3.46), we find $p_{cr} = 0.029$ N/mm². This is a very low pressure, and it corresponds to a water depth of approximately 3 m.

The boiler yield pressure is a measure of the failure strength of the material and is the upper strength limit of a structure. In this particular case, this pressure is 540(32/4,000) = 4.32 N/mm², which corresponds to a depth of water of 430 m.

Clearly then, the structure has to be stiffened in order for it to reach its potential in terms of material strength.

Example 4.1

Calculate the elastic buckling pressure of a ring-stiffened cylindrical shell with radius $R = 4$ m, thickness $t = 32$ mm. The frame spacing is 768 mm, and the combined moment of inertia of the frame and shell is $I_{sf} = 180.10^6$ mm⁴. The web depth equals 250 mm.

Solution

If it is assumed that $n = 2$, the Bresse expression (4.1) gives $p_{f2} = 2.27$ N/mm². This pressure is approximately half the boiler yield pressure, p_0, but it is significantly higher than the elastic buckling pressure of the unstiffened shell.

Example 4.2

Examine the effect of the presence of transverse bulkheads at the ends of the cylindrical shell described in Example 4.1, whose spacing is $L_b = 12$ m.

Solution

a. Cylindrical shell without ring frames
   The elastic buckling pressure of the unstiffened cylindrical shell is found using the Model Basin formula, (3.45), which gives for $L_b = 12$ m $p_m = 0.37$ N/mm². This pressure corresponds to a depth of water of 36 m, which is clearly insufficient.
   The critical buckling mode is determined using Figure 3.28 and for $L_b/2R = 1.5$ and $t/2R = 0.004$, $n = 5, 6$.
b. With ring frames
   In this case, $L = L_f - t_w = 768 - 8 = 760$ mm. The elastic buckling pressure is then $p_m = 0.8185$ N/mm², and from Figure 3.28, we find that $n = 15 - 16$. We note that $p_m = 1.64p_0$, which shows a marked improvement.

The general conclusion from these two examples is that in order to provide sufficient strength to the shell plating and to make rational use of materials, it is necessary to make combined use of ring frames and transverse bulkheads.

4.1.3 Infinitely Long Ring-Stiffened Cylindrical Shell Including Shell Bending Effects

The analysis of an infinitely long,[3] perfect[4] cylindrical shell that is strengthened using regularly spaced ring frames was carried out by von Sanden and Gűnther [3]. They derived and solved the equation of equilibrium of a stiffened cylindrical shell subjected to hydrostatic pressure. In doing so, they analysed an element of the cylindrical shell and recognised that it is subjected to:

a. Radial compression, which gives rise to circumferential compressive membrane stresses, as shown in Figure 4.3a and
b. Bending between consecutive ring frames, because of the higher radial stiffness of the ring frames. This gives rise to longitudinal bending stresses, as shown in Figure 4.2.

A biaxial stress field is thus set up in the cylindrical shell plating. On this basis, the equation of equilibrium was determined to be

$$\frac{d^2M}{dx^2} - \frac{\sigma_\theta t}{R} = p \tag{4.2}$$

The first term on the left-hand side of this equation represents the contribution of longitudinal bending stresses, σ_{xb} (Figure 4.1), whereas the second term represents that of the circumferential compressive membrane stresses σ_θ (Figure 4.3b). Disregarding the first term reduces (3.34) to that of the unstiffened cylindrical shell (3.46). The von Sanden and Günther equation of equilibrium is:

$$\frac{d^4w}{dx^4} + \frac{wt}{R^2 I_{sf}} = \frac{p}{EI_{sf}} \tag{4.3}$$

In 1930, Viterbo [10] introduced a correction to this equation to allow for the correct load that is borne by the ring frame. The load acting on each ring frame per unit length in the circumferential direction comprises the pressure load transmitted to it directly and the pressure load carried by the cylindrical shell and transmitted to it in the form of a transverse shearing force Q_0 on either side.

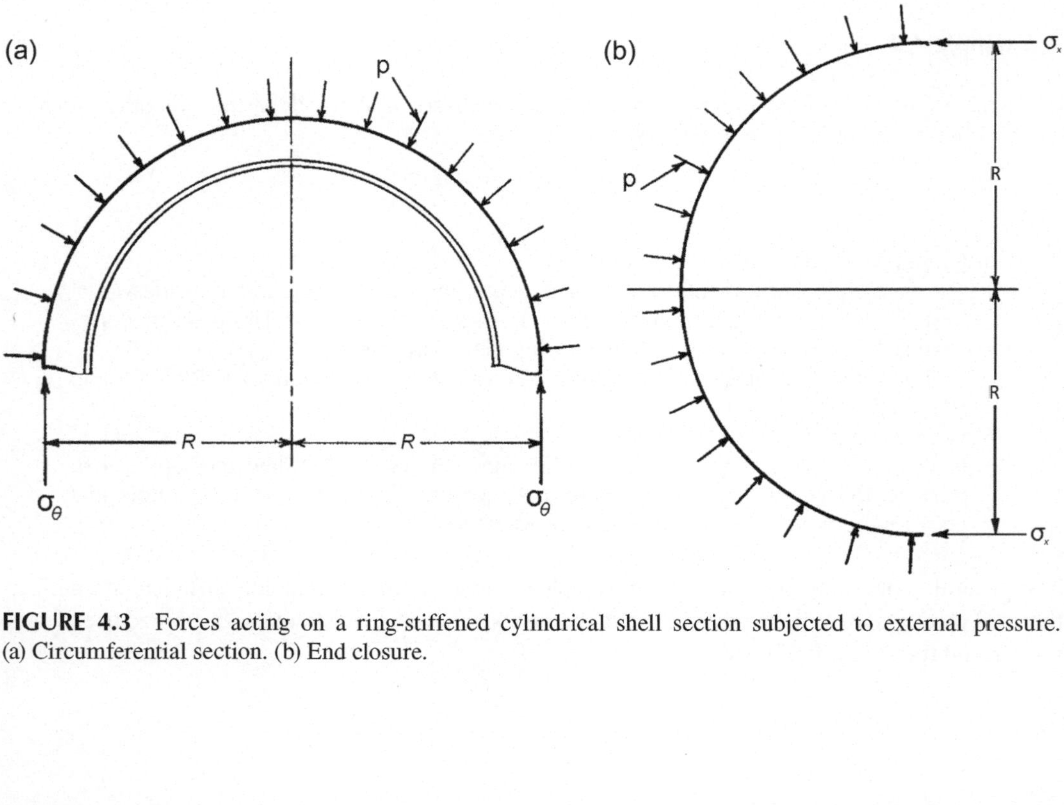

FIGURE 4.3 Forces acting on a ring-stiffened cylindrical shell section subjected to external pressure. (a) Circumferential section. (b) End closure.

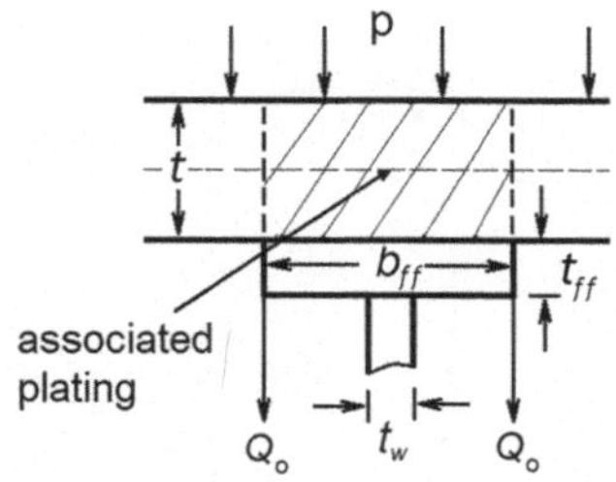

FIGURE 4.4 Shear equilibrium at the cylindrical shell–ring frame intersection [12].

The transverse pressure is resolved into two components, $p\nu/2$ and $p - p\nu/2$. The first component, together with the longitudinal pressure, produces no radial deformation of the frames. Therefore, only the effect of the transverse load $p - p\nu/2$ needs to be considered. Adding, the total load is $pb(1 - \nu/2) + 2Q_0$ where b is the breadth of the associated plating that is taken to be equal to the breadth of the faying flange that is welded to the shell plating, b_{ff} (Figure 4.4).

For equilibrium,

$$2Q_0 = pb\left(1 - \frac{\nu}{2}\right)$$

The equation of equilibrium (4.3) then becomes:

$$\frac{d^4w}{dx^4} + \frac{wt}{R^2 I_{sf}} = \frac{p}{EI_{sf}}\left[1 - \frac{\nu}{2}\right] \tag{4.4}$$

4.1.4 Infinitely Long Stiffened Cylindrical Shell Including Axial Compression

Pulos and Salerno [8] extended this analysis to the study of a stiffened cylindrical shell that is subjected to lateral pressure as well as axial compression resulting from the hydrostatic pressure acting on the end closures (Figures 4.5–4.7).

Figure 4.6 shows an element of a cylindrical shell of radius R and thickness t in isometric view. The shell is subjected to hydrostatic pressure p, which is considered positive when directed toward the generatrix and when producing axial compression. The internal stress field consists of longitudinal unit forces N_x (N/m) and circumferential unit forces N_θ (N/m), and the deflection field consists of radial compression and longitudinal bending deformations.

Figure 4.7 shows in profile view a section of shell plating deformed due to external hydrostatic pressure and axial compression.

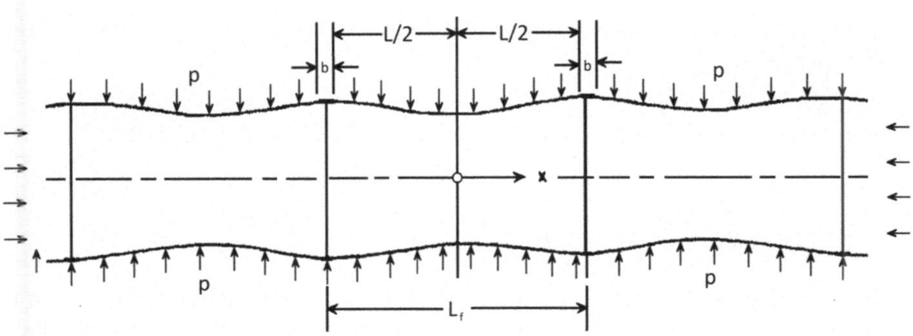

FIGURE 4.5 Coordinate axis system of symmetrically loaded cylindrical shell [8].

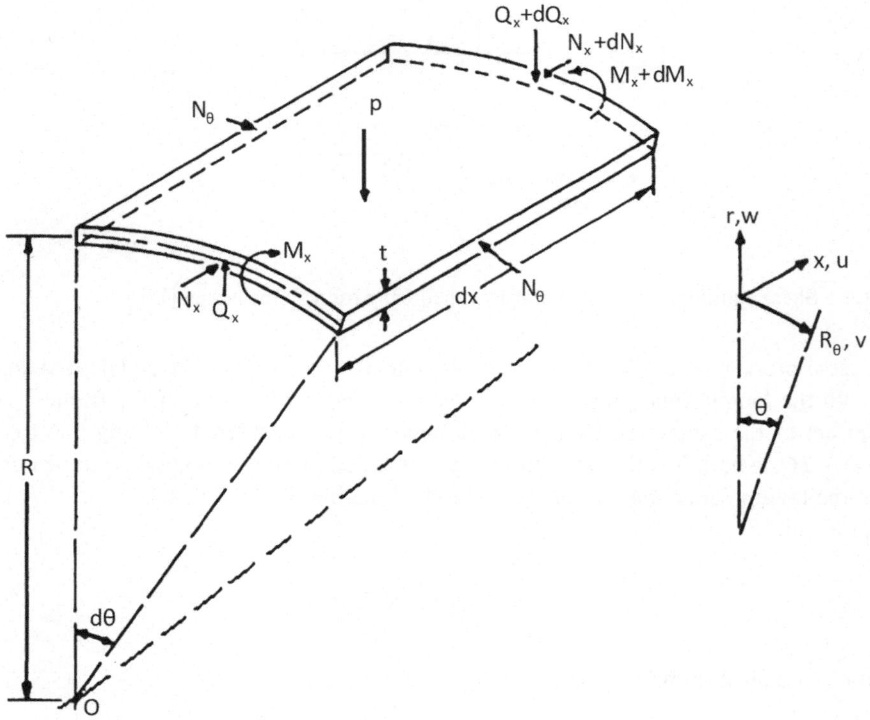

FIGURE 4.6 Element of a cylindrical shell (isometric view) [8].

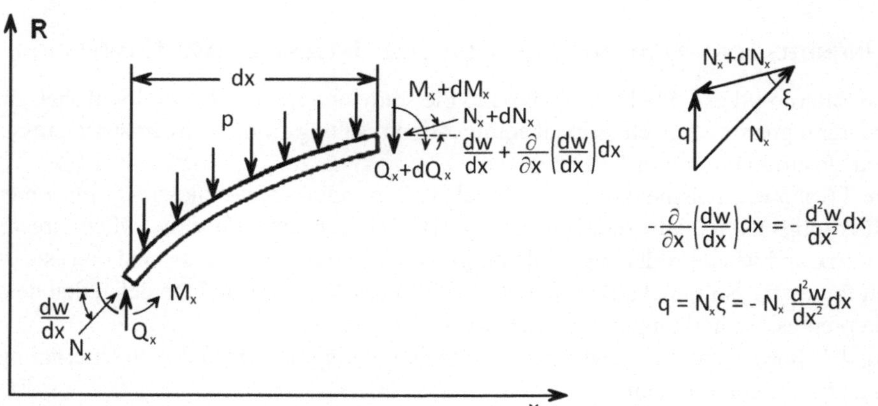

FIGURE 4.7 Forces acting on element of initially bent cylindrical shell (profile view) [8]. (a) Profile view. (b) Axial component of N_x.

From Figure 4.6, for equilibrium of the element of area $Rd\theta dx$ in the radial direction, we have:

$$\frac{dQ_x}{dx} Rd\theta dx + N_x \frac{d^2w}{dx^2} Rd\theta dx - N_\theta d\theta dx + pRd\theta dx = 0$$

It is seen that a net load $-N_x d^2w/dx^2$ per unit length dx acts in the radial direction due to the axial compressive load acting on an initially bent section. The second term thus represents the effect of the contribution of axial (beam-column) compression. The third term represents the contribution of the circumferential force (Figure 4.8). Dividing throughout by $Rd\theta dx$ leads to the following differential equation:

$$\frac{dQ_x}{dx} + N_x \frac{d^2w}{dx^2} - \frac{N_\theta}{R} + p = 0 \tag{4.5}$$

For rotational equilibrium, from Figure 4.8

$$Q_x = \frac{dM_x}{dx} - N_x \frac{dw}{dx} \tag{4.6}$$

Substituting in (4.6) for M_x from

$$M_x = -D\frac{d^2w}{dx^2}$$

and then for dQ_x/dx in (4.5) gives

$$D\frac{d^4w}{dx^4} - N_x \frac{d^2w}{dx^2} + \frac{N_\theta}{R^2} = p \tag{4.7}$$

For plane stress conditions,

$$\varepsilon_x = \frac{1}{Et}(N_x - \nu N_\theta) \tag{4.8a}$$

$$\varepsilon_\theta = \frac{1}{Et}(N_\theta - \nu N_x) \tag{4.8b}$$

Solving for N_θ from (4.8b) gives

$$N_\theta = Et\varepsilon_\theta + \nu N_x \tag{4.9}$$

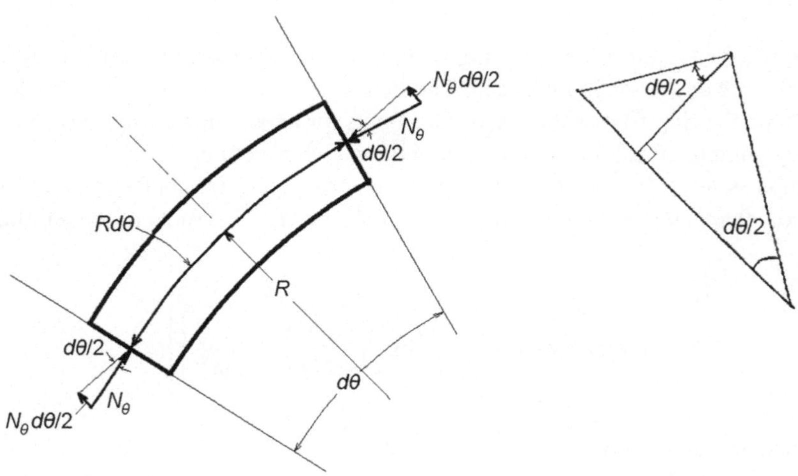

FIGURE 4.8 Circumferential force N_θ.

The circumferential strain of the section, ε_θ, at a distance z from the neutral surface is (Figure 3.23),

$$\varepsilon_\theta = \frac{l_2 - l_1}{l_1} \tag{4.10}$$

where l_1 and l_2 are the initial and final arc lengths at a distance z from the neutral surface following the application of lateral pressure. We have:

$$l_1 = (R - z)\theta \tag{4.11a}$$

$$l_2 = (R - z - w)\theta \tag{4.11b}$$

Substituting, we obtain:

$$\begin{aligned}\varepsilon_\theta &= \frac{(R - z - w)\theta - (R - z)\theta}{(R - z)\theta} \\ &= -\frac{w}{(R - z)} \\ &= -\frac{w}{R}\left[1 + \frac{z}{R} + \frac{z^2}{R^2} + \cdots\right] \\ &\cong -\frac{w}{R}\end{aligned} \tag{4.12}$$

Substituting this in (4.9) and then into (4.7) gives

$$D\frac{d^4w}{dx^4} - N_x\frac{d^2w}{dx^2} + \frac{Et}{R^2}w = -p - \frac{v}{R}N_x \tag{4.13}$$

For external hydrostatic pressure $N_x = pR/2$ so that (4.13) becomes

$$D\frac{d^4w}{dx^4} + \frac{pR}{2}\frac{d^2w}{dx^2} + \frac{Et}{R^2}w = -p(1 - v/2) \tag{4.14}$$

This is the equation of equilibrium of a ring-stiffened cylindrical shell of finite length subjected to hydrostatic pressure and axial compression. The first term in this equation accounts for bending of the shell between the ring frames, the second accounts for axial compression (beam-column effect in shell plating), and the third for deformation in the radial direction.

The equation is an ordinary differential equation with constant coefficients and is solved by assuming a complementary solution of the type $w = f(e^{\lambda x})$. The four roots of the auxiliary equation are:

$$\lambda_1; \lambda_2; \lambda_3; \lambda_4 = \pm\left\{-\frac{pR}{4D} \pm \sqrt{\left[\left(\frac{pR}{4D}\right)^2 - \frac{Et}{DR^2}\right]}\right\}^{1/2} \tag{4.15}$$

The complementary solution is thus

$$w = c_1e^{\lambda_1 x} + c_2e^{\lambda_2 x} + c_3e^{\lambda_3 x} + c_4e^{\lambda_4 x} \tag{4.16}$$

where $c_1 - c_4$ are constants of integration to be determined from the boundary conditions. A particular integral is

$$w = -\frac{pR^2}{Et}(1 - \nu/2) \tag{4.17}$$

This expression represents the radial displacement of a long unstiffened cylinder under hydrostatic pressure. Since from (4.17) $\lambda_1 = -\lambda_2$ and $\lambda_3 = -\lambda_4$ and because

$$e^{\lambda x} = \cosh \lambda x + \sinh \lambda x$$

$$e^{-\lambda x} = \cosh \lambda x - \sinh \lambda x$$

the complete solution of the equation of bending and stretching can be written as

$$w = C_1 \sinh \lambda_1 x + C_2 \cosh \lambda_1 x + C_3 \sinh \lambda_3 x + C_4 \cosh \lambda_3 x - \frac{pR^2}{Et}(1 - \nu/2) \tag{4.18}$$

where $C_1 - C_4$ are new constants of integration.

If the coordinate axis system is chosen so as to take advantage of symmetry, as shown in Figure 4.5, the solution for w has to be an even function of x, and therefore, $C_1 = C_3 = 0$. The deflection w is then given by

$$w = C_2 \cosh \lambda_1 x + C_4 \cosh \lambda_3 x - \frac{pR^2}{Et}(1 - \nu/2) \tag{4.19}$$

The characteristic roots λ_1, λ_3 can be rewritten in the form

$$\lambda_1; \lambda_3 = \sqrt{2}\frac{\theta}{L}\left\{-\frac{p}{p_*} \pm \sqrt{\left[\left(\frac{p}{p_*}\right)^2 - 1\right]}\right\}^{1/2} \tag{4.20}$$

where

$$\theta = \sqrt[4]{3(1-\nu^2)}\frac{L}{\sqrt{Rt}}$$

$$= \frac{1.285L}{\sqrt{Rt}} \quad (\nu = 0.3) \tag{4.21}$$

$$p^* = \frac{2E}{\sqrt{3(1-\nu^2)}}\left[\frac{t}{R}\right]^2 \tag{4.22}$$

θ is termed the *shell slenderness*, whereas p^* is the critical buckling pressure of an unstiffened cylindrical shell subjected to edge compression. The derivation of formula (4.22), based on energy principles, is given in [8]. The solution of equation (4.14) expressed in terms of λ_1, λ_3 in equation (4.20) is seen to depend on the value of p/p^*. For practical purposes, the two limiting cases are:

a. $p/p^* = 0$, which corresponds to zero axial compression (von Sanden and Günther solution) and
b. $p/p^* = 1$, which corresponds to the case in which the applied pressure equals the critical buckling pressure under axial compression alone

We introduce the notation

$$\gamma = \frac{p}{p^*} = \frac{p\sqrt{3(1-v^2)}}{2E}\left(\frac{R}{t}\right)^2 \tag{4.23}$$

so that γ is a measure of the beam-column effect ($\gamma = 0$ implies no such effect).

In the case of a submarine pressure hull, loading consists of both radial and axial compression so that at failure $p/p^* < 1$. In this case, the value of

$$\sqrt{\left[\left(\frac{p}{p^*}\right)^2 - 1\right]}$$

is an imaginary number. The roots λ_1, λ_3 are unequal, complex and are expressed as

$$\lambda_1; \lambda_3 = c \pm ig \tag{4.24}$$

where $i = \sqrt{-1}$. From (4.20) and (4.24),

$$c + ig = \sqrt{2}\frac{\theta}{L}\left\{-\frac{p}{p^*} \pm \sqrt{\left[\left(\frac{p}{p^*}\right)^2 - 1\right]}\right\}^{1/2} \tag{4.25}$$

Squaring both sides of (4.25) and equating real and imaginary parts give

$$c^2 - g^2 = -2\left(\frac{\theta}{L}\right)^2\left(\frac{p}{p^*}\right) \tag{4.26a}$$

$$cg = \left(\frac{\theta}{L}\right)^2\sqrt{\left\{1 - \left(\frac{p}{p^*}\right)^2\right\}} \tag{4.26b}$$

Solution of the system of equations (4.26a and b) gives

$$c; g = \left(\frac{\theta}{L}\right)\sqrt{\left\{1 \mp \left(\frac{p}{p^*}\right)^2\right\}} \tag{4.27}$$

Thus, λ_1, λ_3 become

$$\lambda_1; \lambda_3 = \frac{2\theta}{L}\left(\frac{1}{2}\sqrt{1 + \frac{p}{p^*}} \pm i\frac{1}{2}\sqrt{1 - \frac{p}{p^*}}\right) \tag{4.28}$$

4.1.4.1 Boundary Conditions

Rotation and displacement continuity holds at an intersection of the shell plating with a ring frame and is used in order to determine the two unknown constants C_2, C_4. Since neither the pressure applied to the cylindrical shell nor its geometry varies in the lengthwise direction, the shell–frame intersections will not rotate (Figure 4.9). This condition is expressed as

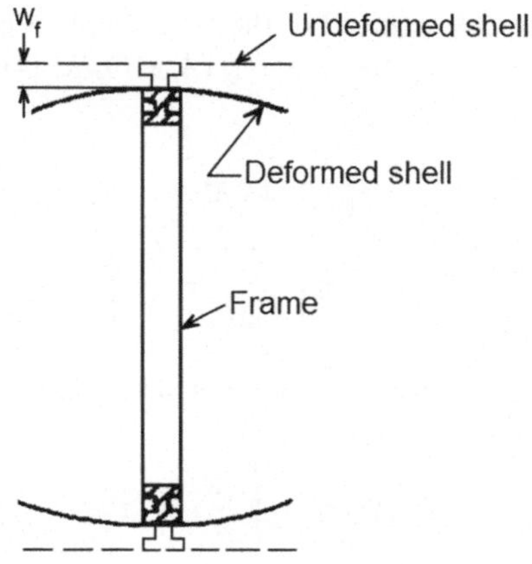

FIGURE 4.9 Deformations in way of cylindrical shell and ring frame intersection [8].

$$\frac{dw}{dx} = 0 \quad at\ x = \pm L/2 \tag{4.29}$$

The second boundary condition concerns the displacement w_f of the cylindrical shell and ring frame in the radial direction. This will depend on the radial stiffness of the intersection which includes the shell plating and the ring frame.

We consider next the shear boundary condition. Allowing for Viterbo's correction, the total load acting on the ring frame is $2Q_0 + pb(1 - \nu/2)$, where Q_0 is the transverse shear transmitted by the shell to the frame, and b is the breadth of the associated plating. The resulting circumferential stress σ_θ in the structure comprising shell plating and the ring frame that is loaded as shown in Figure 4.10 is then

$$\sigma_\theta = -\frac{\left(2Q_0 + pb(1-\nu)\right)}{A_f + bt} R \tag{4.30}$$

The corresponding axisymmetric strain ε_θ is given by

$$\varepsilon_\theta = \frac{w_f}{R} = \frac{\sigma_\theta}{E} \tag{4.31}$$

If we assume that the whole area of the ring frame is concentrated at the radius of the middle surface of the shell plating, the radial displacement w_f is given by

$$w_f = -\frac{\left(2Q_0 + pb(1-\nu)\right) R^2}{E(A_f + bt)} \tag{4.32}$$

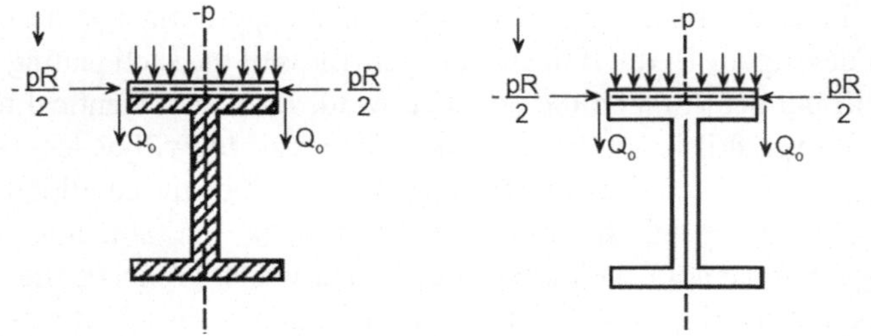

FIGURE 4.10 Loading of cylindrical shell–ring frame intersection [8].

In order to allow for the fact that the centroid of the ring frame does not coincide with that of the shell plating but is located at a radius R_{cg}, the quantity $(A_f + bt)/R^2$ is replaced by $(A_f/R_{cg}R) + (bt/R^2)$. The radial deflection then becomes

$$w_f = -\frac{\left(2Q_0 + \left(pb(1-\nu)\right)\right)}{E\left(\dfrac{A_f}{R_{cg}R} + \dfrac{bt}{R^2}\right)} \tag{4.33}$$

If the frame area A_f is replaced by an effective frame area A_{eff}, defined by

$$A_{\text{eff}} = A_f\left(\frac{R}{R_{cg}}\right)^2 \tag{4.34}$$

the displacement boundary condition (4.32) can be written as

$$w_f = -\frac{\left(2Q_0 + \left(pb(1-\nu)\right)\right)R^2}{E\left(A_{\text{eff}} + bt\right)} \tag{4.35}$$

Since the transverse shear Q_x is related to the radial displacement w by (4.6), for continuity at the intersection of the shell plating with the ring frame, given (4.35), we require at $x = L/2$:

$$\frac{E}{R^2}\left(A_{\text{eff}} + bt\right)w_f = 2D\frac{d^3w}{dx^3} - pR\frac{dw}{dx} = pb(1-\nu/2) \tag{4.36}$$

The constants C_2 and C_4 are determined by substituting (4.19) into boundary conditions (4.29) and (4.36). They are given by (4.53) for the case $p/p^* < 1.0$.

From (4.9) and (4.12), the circumferential membrane stress $\sigma_{\theta m}$ is given by:

$$\sigma_{\theta m} = \frac{N_\theta}{t} = E\frac{w}{z} - \nu\frac{pR}{2t} \tag{4.37}$$

Substituting (4.19) in (4.37) gives

$$\sigma_{\theta m} = -\frac{pR}{t} + \frac{E}{R}\left[C_2\cosh\lambda_1 x + C_4\cosh\lambda_3 x\right] \tag{4.38}$$

The first term in this expression is the membrane stress due to pure radial compression, whereas the second term is the radial component of the interframe shell bending stress.

4.1.4.2 Effective Width of Plating

A value for the effective width of the shell plating is necessary in order to estimate the local and overall buckling strength of the stiffened shell. The definition of effective width is based on an equivalent cross section consisting of the ring frame and an appropriate width of plating that does not fail in a local mode (Figure 4.11b). If the bending stiffness of the shell plating becomes infinitely large, the plating will not deform, and the effective width L_e is then identical to the actual frame spacing, L_f. Since, however, this is not the case, the effective width will be less than L_f.

The determination of the effective width of plating L_e is based on the condition that the circumferential force in the real structure and the equivalent structure has the same magnitude (Figure 4.11a and b). The bending component of the radial force is that which produces the nonlinearity in the circumferential stress distribution, and therefore, the definition of the effective width of plating has to be based solely on this component.

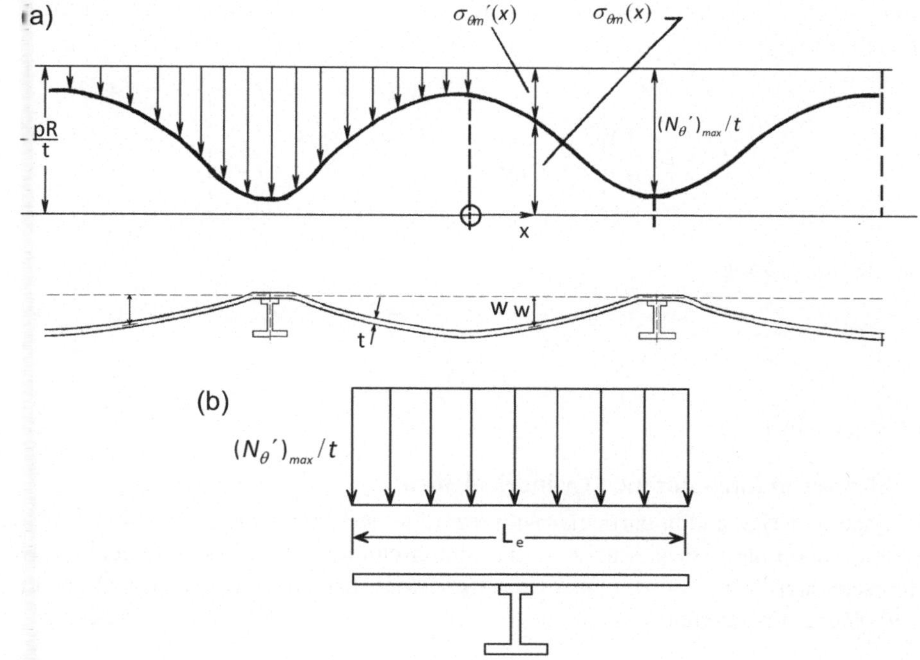

FIGURE 4.11 Circumferential stress distribution and equivalent cross section [8]. (a) Nonlinear distribution. (b) Equivalent linear distribution.

The nonlinear distribution $N'_\theta(x)$ varies in the longitudinal direction, with the peak value $(N'_\theta)_{\max}$ arising in way of the shell–frame intersections (Figure 4.11a). The corresponding circumferential force F_θ is then

$$F_\theta = \int_{-L/2}^{+L/2} \frac{E}{R}\left[C_2 \cosh \lambda_1 x + C_4 \cosh \lambda_3 x\right] dx$$
$$= 2\int_0^{L/2} \frac{E}{R}\left[C_2 \cosh \lambda_1 x + C_4 \cosh \lambda_3 x\right] dx \tag{4.39}$$

For equivalence between the nonlinear and the uniform load distributions

$$F_\theta = L_e (N'_\theta)_{\max} \tag{4.40}$$

where

$$(N'_\theta)_{\max} = N'_\theta(L/2) \tag{4.41}$$

Substituting (4.38) and (4.39) in (4.40) leads to the following expression for the effective width of plating, L_e:

$$L_e = 2\frac{\dfrac{C_2}{\lambda_1}\sinh \lambda_1 \dfrac{L}{2} + \dfrac{C_4}{\lambda_3}\sinh \lambda_3 \dfrac{L}{2}}{C_2 \cosh \lambda_1 \dfrac{L}{2} + C_4 \cosh \lambda_3 \dfrac{L}{2}} \tag{4.42}$$

Substituting for C_2, C_4, λ_1, λ_3 in (4.42) and simplifying yield the following formula for the effective width of shell plating:

$$L_e = L\left(\frac{4}{\theta}\right)\left[\frac{\cosh^2 \eta_1\theta - \cos^2 \eta_2\theta}{\frac{\cosh \eta_1\theta \sinh \eta_2\theta}{\eta_1} + \frac{\cos \eta_1\theta \sin \eta_2\theta}{\eta_2}}\right] \tag{4.43}$$

The quantities η_1, η_2 are given by

$$\eta_1;\eta_2 = \frac{1}{2}\sqrt{1 \mp \frac{p}{p^*}} \tag{4.44}$$

where p^* is given by (4.22).

4.1.4.3 Stresses in Ring-Stiffened Cylindrical Shell

Of importance in design and in particular with regard to buckling is the determination of the positions at which maximum stresses arise. Axial and circumferential stresses therefore have to be considered separately. Since bending gives rise to stresses that vary linearly through the thickness of the shell plating, the maximum stress due to interframe bending in the axial direction is given by

$$\sigma_{xb\max} = \frac{M_x}{I_{sf}}\left(\pm\frac{t}{2}\right) = \left(\pm\frac{t}{2}\right)\frac{D}{I_{sf}}\frac{d^2w}{dx^2} \tag{4.45}$$

where the plus sign is for the outer fibre, and the minus sign is for the inner fibre of the shell plating. Therefore,

$$\sigma_{xb\max} = \pm\frac{E}{1-\nu^2}\frac{t}{2}\frac{d^2w}{dx^2} \tag{4.46}$$

The curvature in the longitudinal direction is obtained from (4.19) and is

$$\frac{d^2w}{dx^2} = C_2\lambda_1^2 \cosh \lambda_1 x + C_4\lambda_3^2 \cosh \lambda_3 x \tag{4.47}$$

Substituting in (4.46), the following expression is obtained for the maximum longitudinal bending stress:

$$(\sigma_{xb})_{\max} = \pm\frac{E}{1-\nu^2}\frac{t}{2}\left[C_2\lambda_1^2 \cosh \lambda_1 x + C_4\lambda_3^2 \cosh \lambda_3 x\right] \tag{4.48}$$

These arise in way of the outer and inner surface fibres of the shell ($\pm t/2$) and vary along the distance between consecutive frames. The maximum total axial stress in the shell plating is then the sum of the longitudinal compressive stress and the maximum compressive bending stress

$$(\sigma_x)_{\max} = -\frac{pR}{2t} \pm \frac{E}{1-\nu^2}\frac{t}{2}\left[C_2\lambda_1^2 \cosh \lambda_1 x + C_4\lambda_3^2 \cosh \lambda_3 x\right] \tag{4.49}$$

The external hydrostatic pressure induces bending towards the generatrix. Therefore, in way of the shell–frame intersection ($x = \pm L/2$), the compressive stresses due to bending will arise in way of the inner fibres. This then is the location of the maximum compressive stress in the stiffened cylindrical shell, which is henceforth denoted as σ_7.

The circumferential membrane stress is given by (4.38). It is seen from (4.38) that it varies along the distance between frames. In addition to this, the Poisson effect gives rise to circumferential bending stresses $\sigma_{\theta b} = \nu\sigma_{xb}$. Since the total circumferential stress σ_θ at any point is the sum of the circumferential membrane and the circumferential bending component,

$$\sigma_\theta = \sigma_{\theta m} + \nu\sigma_{\theta b}$$

$$= -\frac{pR}{t} + \left[\frac{E}{R} \pm \frac{\nu E}{1-\nu^2}\frac{t}{2}\lambda_1^2\right] C_2 \cosh \lambda_1 x + \left[\frac{E}{R} \pm \frac{\nu E}{1-\nu^2}\frac{t}{2}\lambda_3^2\right] C_4 \cosh \lambda_3 x \tag{4.50}$$

In a similar manner, the surface circumferential stresses vary along the length of the cylindrical shell. Maxima arise at midbay ($x = 0$) and at the intersections with frames ($x = \pm L/2$). At these locations, the maximum longitudinal and circumferential stresses are, respectively:

At midbay:

$$(\sigma_x)_{\max} = \underbrace{-\frac{pR}{2t}}_{\text{axial}} \pm \underbrace{\frac{E}{1-\nu^2}\frac{t}{2}\left[C_2\lambda_1^2 + C_4\lambda_3^2\right]}_{\text{bending}} \tag{4.51a}$$

$$(\sigma_\theta)_{\max} = \underbrace{-\frac{pR}{t} + \frac{E}{R}[C_2 + C_4]}_{\text{axial}} \pm \underbrace{\frac{\nu E}{1-\nu^2}\frac{t}{2}\left[C_2\lambda_1^2 + C_4\lambda_3^2\right]}_{\text{bending}} \tag{4.51b}$$

At the intersection with a frame:

$$(\sigma_x)_{max} = \underbrace{-\frac{pR}{2t}}_{\text{axial}} \pm \underbrace{\frac{E}{1-\nu^2}\frac{t}{2}\left[C_2\lambda_1^2 \cosh \lambda_1 \frac{L}{2} + C_4\lambda_3^2 \cosh \lambda_3 \frac{L}{2}\right]}_{\text{bending}} \tag{4.52a}$$

$$(\sigma_\theta)_{max} = \underbrace{-\frac{pR}{t} + \frac{E}{R}\left[C_2 \cosh \lambda_1 \frac{L}{2} + C_4 \cosh \lambda_3 \frac{L}{2}\right]}_{\text{axial}}$$

$$\underbrace{\pm \nu \frac{E}{1-\nu^2}\frac{t}{2}\left[C_2\lambda_1^2 \cosh \lambda_1 \frac{L}{2} + C_4\lambda_3^2 \cosh \lambda_3 \frac{L}{2}\right]}_{\text{bending}} \tag{4.52b}$$

The integration constants C_2 and C_4 are given by:

$$C_{2,4} = \frac{C_{num}}{C_{den}} \tag{4.53}$$

where

$$C_{num} = \mp p(1-\nu/2)\frac{A_{\text{eff}}}{t}(\eta_1 \mp i\eta_2)\left[\sinh \eta_1\theta \cos \eta_2\theta \mp i \cosh \eta_1\theta \sin \eta_2\theta\right]$$

$$C_{den} = 2i\eta_1\eta_2 K'\left[\frac{\sinh \eta_1\theta \cosh \eta_1\theta}{\eta_1} + \frac{\sin \eta_2\theta \cos \eta_2\theta}{\eta_2}\right]$$

$$+ 8i\eta_1\eta_2\left(\eta_1^2 + \eta_2^2\right)\left(\frac{2\theta}{L}\right)^3 D\left[\cosh^2 \eta_1\theta - \cos^2 \eta_2\theta\right]$$

$$K' = \frac{E}{R^2}\left(A_{eff} + bt\right)$$

The minus sign in C_{num} corresponds to the numerator of C_2, and the plus sign corresponds to the numerator of C_4. The quantities η_1, η_2 are given by (4.44).

Substituting in (4.51a and b) and (4.52a and b) for C_2, C_4 we obtain the following expressions for the critical longitudinal and circumferential stresses, respectively [7]:

a. At midbay ($x = 0$)

$$\frac{\sigma_x}{\sigma_{xu}} = \frac{1}{2} \pm \sqrt{\frac{0.91}{1-\nu^2}}\left[\frac{(1-\nu/2)\alpha}{\alpha+\beta+(1-\beta)F_1}\right]F_4 \tag{4.54a}$$

$$\frac{\sigma_\theta}{\sigma_{xu}} = 1 - \left(\frac{(1-\nu/2)\alpha}{\alpha+\beta+(1-\beta)F_1}\right)F_2 \pm \nu\sqrt{\frac{0.91}{1-\nu^2}}\left[\frac{(1-\nu/2)\alpha}{\alpha+\beta+(1-\beta)F_1}\right]F_4 \tag{4.54b}$$

b. At a shell–frame intersection

$$\frac{\sigma_x}{\sigma_{xu}} = \frac{1}{2} \pm \left[\frac{(1-\nu/2)\alpha}{\alpha+\beta+(1-\beta)F_1}\right]\sqrt{\frac{0.91}{1-\nu^2}}F_3F_4 \tag{4.55a}$$

$$\frac{\sigma_\theta}{\sigma_{xu}} = 1 - \left[\frac{(1-\nu/2)\alpha}{\alpha+\beta+(1-\beta)F_1}\right] \pm \nu\left[\frac{(1-\nu/2)\alpha}{\alpha+\beta+(1-\beta)F_1}\right]\sqrt{\frac{0.91}{1-\nu^2}}F_3 \tag{4.55b}$$

In the above expressions:

i. The plus sign corresponds to the stress arising in way of the outer fibres of the shell plating, whereas the minus sign corresponds to the stress arising at the inner fibres.
ii. σ_{xu} is the longitudinal stress in an unstiffened cylinder of infinite length, given by:

$$\sigma_{xu} = -\frac{pR}{t} \tag{4.56}$$

Also:

$$\alpha = \frac{A_{eff}}{L_f t} \tag{4.57a}$$

$$\beta = \frac{b}{L_f} \tag{4.57b}$$

$$F_1 = \left(\frac{4}{\theta}\right)\frac{\left(\cosh^2\eta_1\theta - \cosh^2\eta_2\theta\right)}{\left(\frac{\cosh\eta_1\theta\sinh\eta_1\theta}{\eta_1} + \frac{\cos\eta_2\theta\sin\eta_2\theta}{\eta_2}\right)} \tag{4.58a}$$

$$F_2 = \frac{\left(\dfrac{\cosh\eta_1\theta\sin\eta_2\theta}{\eta_2} + \dfrac{\sinh\eta_1\theta\cos\eta_2\theta}{\eta_1}\right)}{\left(\dfrac{\cosh\eta_1\theta\sinh\eta_1\theta}{\eta_1} + \dfrac{\cos\eta_2\theta\sin\eta_2\theta}{\eta_2}\right)} \tag{4.58b}$$

$$F_3 = \sqrt{\frac{3}{0.91}}\,\frac{\left(-\dfrac{\cosh\eta_1\theta\sinh\eta_1\theta}{\eta_1} + \dfrac{\cos\eta_2\theta\sin\eta_2\theta}{\eta_2}\right)}{\left(\dfrac{\cosh\eta_1\theta\sin\eta_2\theta}{\eta_2} - \dfrac{\sinh\eta_1\theta\cos\eta_2\theta}{\eta_1}\right)} \tag{4.58c}$$

$$F_4 = \sqrt{\frac{3}{0.91}}\,\frac{\left(\dfrac{\cosh\eta_1\theta\sin\eta_2\theta}{\eta_2} - \dfrac{\sinh\eta_1\theta\cos\eta_2\theta}{\eta_1}\right)}{\left(\dfrac{\cosh\eta_1\theta\sin\eta_2\theta}{\eta_2} - \dfrac{\sinh\eta_1\theta\cos\eta_2\theta}{\eta_1}\right)} \tag{4.58d}$$

In the above expressions, θ is given by (4.21) and η_1, η_2 are given by (4.44).

It is seen from (4.54a and b) and (4.55a and b) that a biaxial stress field develops in the cylindrical shell. In order to determine the true collapse mode, it is necessary to calculate the von Mises stress that varies over the interframe spacing in accordance with the magnitudes of its longitudinal and axial components. It is thereby found that failure is governed by both buckling and yielding. The collapse of ring-stiffened cylindrical shells is discussed in Chapter 6.

4.1.5 Infinitely Long Stiffened Cylindrical Shell Including Axial Compression (Approximate Solution)

Failure of pressure hull structures is initiated by yielding rather than elastic buckling, and the collapse pressure is then much lower than the axisymmetric elastic buckling pressure, which includes the beam-column effect. The response is then adequately represented by the linear portion of the pressure–deflection curve, shown in Figure 3.18a.

In an effort to simplify the exact solution of Salerno and Pulos presented in the previous section, Wilson re-derived the basic differential equation and solved it using two methods [12]. In the first, the deflection w was expressed as a Fourier series, and in the second, it was assumed that it takes the form of a power series in pressure, p. It was found that these two solutions are good approximations to the exact solution so that they have been generally used since. Additionally, more accurate expressions were used for the circumferential strain, ε_θ, and the frame constant, K. Wilson's derivation of the equation of equilibrium is outlined below.

A cylindrical shell of length dx and unit circumferential width is considered, as shown in Figure 4.12. The element is at equilibrium under the action of external pressure and the internal longitudinal and circumferential membrane stresses induced (σ_x, σ_θ). For the solution to be valid in the linear elastic small deflection range, firstly, the deflection cannot exceed the shell thickness ($w/t \le 1$), and secondly, the thickness/diameter ratio should be less than 0.05. The axial stress σ_x is given by (3.48), and thus, the corresponding force per unit width N_x acting on the element is

$$N_x = \sigma_x t = pR/2 \tag{4.59}$$

When external pressure acts, the stress is compressive so that a minus sign is required. For equilibrium in the vertical direction,

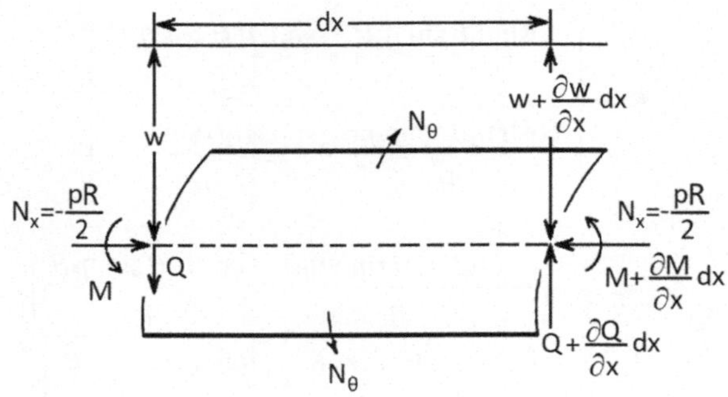

FIGURE 4.12 Forces acting on cylindrical shell element under lateral pressure and axial compression [12].

$$pdx + Q - (Q + dQ) + 2N_\theta(\theta/2)dx = 0$$

or (4.60)

$$pdx - dQ + (N_\theta/R)dx = 0$$

For rotational equilibrium,

$$M + dM = M + Qdx - N_x dw \tag{4.61}$$

Following simplification, equations (4.60) and (4.61) become

$$\frac{dQ}{dx} = p + \frac{N_\theta}{R} \tag{4.62}$$

$$\frac{dM}{dx} = Q - N_x \frac{d^2w}{dx^2} \tag{4.63}$$

Eliminating the shear force Q from (4.62) and (4.63), the equation of equilibrium of the stiffened cylindrical shell is obtained:

$$\frac{d^2M}{dx^2} = p + \frac{N_\theta}{R} - N_x \frac{dw}{dx} \tag{4.64}$$

Where

$$M = \int_{-t/2}^{+t/2} \sigma_x z\, dz \tag{4.65a}$$

$$N_\theta = \int_{-t/2}^{+t/2} \sigma_\theta\, dz \tag{4.65b}$$

$$N_x = pR/2 \tag{4.65c}$$

In these relationships σ_x, σ_θ are the stress components at any position on the stiffened cylindrical shell section. A biaxial stress field thus results, and we note that the response is expressed in terms of internal forces and moments (N_x, N_θ, M). In order to obtain the force–deflection relationship, force–deflection relations have to be substituted in (4.64).

For plane stress state conditions,

$$\sigma_x = \frac{E}{(1-\nu^2)}[\varepsilon_x + \nu\varepsilon_\theta] \tag{4.66a}$$

$$\sigma_\theta = \frac{E}{(1-\nu^2)}[\varepsilon_\theta + \nu\varepsilon_x] \tag{4.66b}$$

The circumferential strain of the section, ε_θ, at a distance z from the neutral surface is

$$\varepsilon_\theta = \frac{l_2 - l_1}{l_1}$$

where l_1 and l_2 are the initial and final arc lengths at a distance z from the neutral surface following the application of lateral pressure. We have

$$l_1 = (R - z)\theta \tag{4.67a}$$

$$l_2 = (R - z - w)\theta \tag{4.67b}$$

Substituting, we obtain:

$$\begin{aligned} \varepsilon_\theta &= \frac{(R-z-w)\theta - (R-z)\theta}{(R-z)\theta} \\ &= -\frac{w}{(R-z)} \\ &= -\frac{w}{R}\left[1 + \frac{z}{R} + \frac{z^2}{R^2} + \cdots\right] \\ &\cong -\frac{w}{R}\left[1 + \frac{z}{R}\right] \end{aligned} \tag{4.68}$$

The longitudinal strain, $\varepsilon_x(z)$, in a fibre located a distance z from the neutral surface is given by the following relationship (Love–Kirchhoff hypothesis):[5]

$$\varepsilon_x(z) = \varepsilon_{xm} - z\frac{d^2w}{dx^2} \tag{4.69}$$

where ε_{xm} is the longitudinal membrane strain of the cross section. From (4.66a) we obtain

$$\sigma_{xm} = \frac{E}{(1-\nu^2)}[\varepsilon_{xm} + \nu\varepsilon_{\theta m}] = -\frac{N_x}{t} \tag{4.70}$$

Hence,

$$\varepsilon_{xm} = \nu\frac{w}{R} - \frac{pR}{2t}\frac{(1-\nu^2)}{E} \tag{4.71}$$

Substituting (4.71) in (4.69), we obtain:

$$\varepsilon_x = \nu \frac{w}{R} - \frac{pR}{2t} \frac{(1-\nu^2)}{E} - z\frac{d^2w}{dx^2} \tag{4.72}$$

Expressions (4.70) and (4.69) enable us to calculate stresses σ_x, σ_θ using (4.66a) and (4.66b):

$$\sigma_x = -\frac{pR}{2t} - \frac{Ez}{(1-\nu^2)}\left[\frac{\nu w}{R^2} + \frac{d^2w}{dx^2}\right] \tag{4.73a}$$

$$\sigma_\theta = -\frac{Ew}{R} - \frac{\nu pR}{2t} - \frac{Ez}{(1-\nu^2)}\left[\frac{w}{R^2} + \nu\frac{d^2w}{dx^2}\right] \tag{4.73b}$$

Substituting (4.73a) and (4.73b) in (4.65a) and (4.65b), the bending moment M and unit force N_θ are found:

$$M = \frac{Et^3}{12(1-\nu^2)}\left[\frac{\nu w}{R} + \frac{d^2w}{dx^2}\right] \tag{4.74}$$

$$N_\theta = -\frac{Etw}{R} - \nu\frac{pR}{2} \tag{4.75}$$

We can now substitute in the equation of equilibrium (4.64) and obtain an explicit load–deflection relationship:

$$\frac{d^4w}{dx^4} + \frac{d^2w}{dx^2}\left[\frac{pR}{2EI} + \frac{\nu}{R^2}\right] + \frac{wt}{R^2I} = \frac{p}{EI}\left[1-\frac{\nu}{2}\right] \tag{4.76}$$

where

$$I = \frac{t^3}{12(1-\nu^2)}$$

If we assume that the circumferential strain, ε_θ, is given by the first term in (4.68), that is, $\varepsilon_\theta = -(w/R)$, the second term in the brackets on the left-hand side of equation (4.76), $(d^2w/dx^2)(\nu/R^2)$, disappears. This term corresponds to the beam-column type loading that is applied to the section as a result of the hydrostatic pressure that acts at the ends of the structure in the axial direction. The equation of equilibrium can then be written as follows:

$$\frac{d^4w}{dx^4} + \frac{d^2w}{dx^2}\left[\frac{pR}{2EI}\right] + \frac{wt}{R^2I} = \frac{p}{EI}\left[1-\frac{\nu}{2}\right] \tag{4.77}$$

At low pressure levels, the load–deflection relationship is practically linear. At higher pressures, the deflection increases nonlinearly, to the extent that as the elastic buckling load is approached, it approaches infinity in an asymptotic manner. However, because this pressure is significantly higher than the collapse pressure, this nonlinearity can be ignored in practical problems, and the equation of equilibrium then reverts to the von Sanden and Günther form, allowing for the Viterbo correction (3.56).

4.1.5.1 Boundary Conditions

To obtain a solution of (4.77), four boundary conditions are required. Two of these are required to ensure symmetry of the deflected form at midspan, a third condition is required to ensure zero rotation at the connection with the end supports, and the fourth is required in order to ensure equilibrium in shear along the ring frame–shell intersection.

At midspan,

$$\frac{dw}{dx} = 0 \tag{4.78a}$$

$$\frac{dw}{d\theta} = 0 \tag{4.78b}$$

At the intersection with the ring frames,

$$\frac{dw}{dx} = 0 \tag{4.78c}$$

The fourth boundary condition is applied to ensure equilibrium in shear. The hydrostatic pressure p gives rise to compression of the cylindrical shell and ring frame arrangement and also longitudinal contraction of the cylindrical shell, due to the Poisson effect. Therefore, the ring frame, to which only radial loading is transmitted, is subjected to a pressure equal to $p(1 - \nu/2)$ and not p, as discussed in Section 4.1.3. The shear equilibrium boundary condition is then

$$2Q_0 + pb\left(1 - \frac{\nu}{2}\right) = Kw \tag{4.79}$$

where K is the stiffness of the ring frame and associated plating, and w is its translation in the radial direction. Since $Q_0 = -EI(dM/dx)$,

$$2EI\frac{d^3w}{dx^3} = Kw - pb\left(1 - \frac{\nu}{2}\right) \tag{4.80}$$

In order to make use of (4.80), it is necessary to determine K. Since the total load supported by the frame per unit circumferential length is $pb(1 - \nu/2) + 2Q_0$, the circumferential stress in the frame is given by

$$\frac{\left[pb(1-\nu/2) + 2Q_0\right]r_c}{A_f + bt}$$

where r_c is the radius of the centroid of the frame and the associated plating. The circumferential strain in the frame is then

$$\frac{\left[pb(1-\nu/2) + 2Q_0\right]r_c}{E(A_f + bt)} = \frac{w}{r_c} \tag{4.81}$$

Thus,

$$2Q_0 = \frac{E(A_f + bt)}{r_c^2}w - pb\left(1 - \frac{\nu}{2}\right)$$

Hence,

$$2EI\frac{d^3w}{dx^3} = \frac{E(A_f + bt)}{r_c^2}w - pb\left(1 - \frac{\nu}{2}\right)$$

Therefore, in (4.80) [11],

$$\begin{aligned} K &= E\left(\frac{A_f + bt}{r_c^2}\right) \\ &\cong E\left(\frac{A_f}{R_g^2} + \frac{bt}{R^2}\right) \end{aligned} \tag{4.82}$$

where R_g is the radius of the centroid of the frame. This expression is valid for ring frames whose depth does not exceed normal proportions.

4.1.5.2 Stress Distributions

Using equation (4.76), Wilson derived the following expressions for the more important stresses whose locations are shown in Figure 4.13. In the axial direction, the stresses in the cylindrical shell vary between maxima in way of the intersection with the ring frames and minima at midbay as a

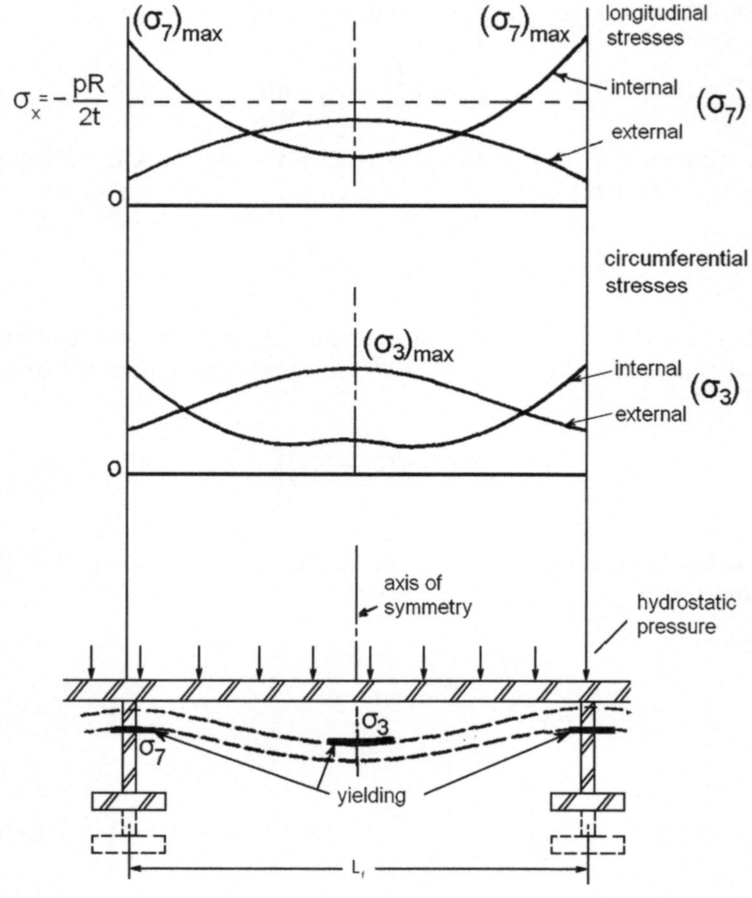

FIGURE 4.13 Longitudinal and circumferential elastic stress fields in a ring-stiffened cylindrical shell [5].

result of bending about the frame supports. The average axial through-thickness stress equals that given by (3.48). In the circumferential direction, the stresses, being predominantly of a membrane nature, are similar on both exterior and interior surfaces. Their variation follows that of the axial stresses because of the Poisson effect.

The stresses shown in Figure 4.13 are given by the following expressions:

i. In the circumferential direction

a At midbay on the external surface of the shell plating:

$$\sigma_3 = \left(\frac{pR}{t}\right)(1+\gamma H) \tag{4.83}$$

b. At midbay on the middle surface of the shell plating (membrane stress):

$$\sigma_5 = \left(\frac{pR}{t}\right)(1+\gamma G) \tag{4.84}$$

c. On the ring frame flange:

$$\sigma_f = \left(\frac{pR}{t}\right)\left[\frac{R}{R_f}\right]\left[\frac{1-(\nu/2)}{1+\dfrac{A}{bt+(2Nt/\xi)}}\right] \tag{4.85}$$

ii. In the longitudinal direction, the maximum stress arises in way of the shell–ring frame intersection, on the inner surface of the shell plating:

$$\sigma_7 = \left(\frac{pR}{2t}\right)\left[1+\sqrt{\frac{12}{1-\nu^2}}\right]\gamma M \tag{4.86}$$

We note that stresses σ_3 and σ_5 are expressed in terms of the circumferential stress σ_θ (3.47), whereas stress σ_7 is expressed in terms of the longitudinal stress, σ_x, (3.48). In calculating σ_θ, σ_x, the end supports are considered fully fixed. The factors γH and γG included in the brackets account for the finite rigidity of the ring frames. The terms that are included in (4.83–4.86) are defined as follows:

θ = shell slenderness (4.21)

β = ring frame stiffness (when $A \to \infty$, $\beta \to 0$) $= \dfrac{2tN}{\alpha(A+bt)}$ where:

N = ring frame bending parameter $= \dfrac{\cosh \alpha L - \cos \alpha L}{\sinh \alpha L + \sin \alpha L}$

$\gamma = \dfrac{A(1-\nu/2)}{(A+bt)(1+\beta)}$ where:

A = shell cross-sectional area $= A_f\left(\dfrac{R}{R_g}\right)^2$

A_f = frame cross sectional area

G = bending parameter at midspan $= \dfrac{-2\left[\sinh\left(\frac{\alpha L}{2}\right)\cos\left(\frac{\alpha L}{2}\right)+\cosh\left(\frac{\alpha L}{2}\right)\sin\left(\frac{\alpha L}{2}\right)\right]}{\sinh \alpha L + \sin \alpha L}$

H = bending parameter at midspan

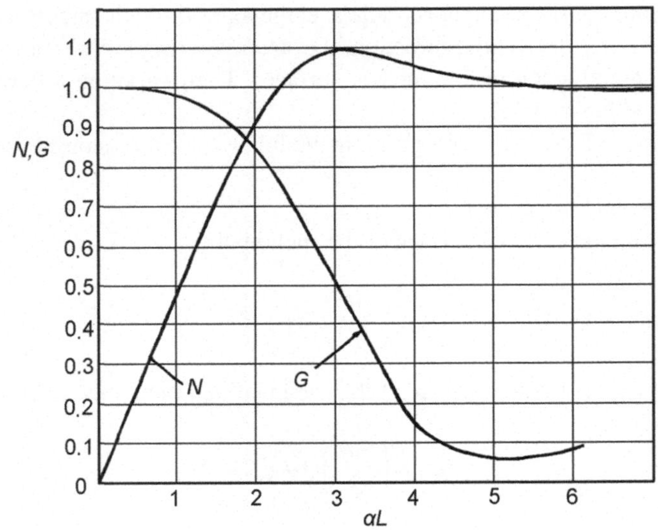

FIGURE 4.14 Ring frame and shell bending parameters *N*, *G* [5].

$$= \frac{-2\left\{\left[1+\sqrt{\left(\frac{3v^2}{1-v^2}\right)}\right]\sinh\frac{\alpha L}{2}\cos\frac{\alpha L}{2}+\left[1-\sqrt{\left(\frac{3v^2}{1-v^2}\right)}\right]\cosh\frac{\alpha L}{2}\sin\frac{\alpha L}{2}\right\}}{\sinh\alpha L+\sin\alpha L}$$

$$M = \text{shell bending parameter in way of ring frames} = \frac{\sinh\alpha L-\sin\alpha L}{\sinh\alpha L+\sin\alpha L}$$

$$\xi^4 = \frac{3(1-v^2)}{(Rt)^2}$$

Figure 4.14 can be used to calculate the bending parameters *N* and *G*. The stress distributions for σ_3 and σ_7 are given by (4.83) and (4.86) and are plotted in Figure 4.13.

Expressions (4.83–4.86) are used to calculate the critical stresses in a ring-stiffened cylindrical shell subjected to external radial pressure and axial compression.

4.2 RING-STIFFENED CYLINDRICAL SHELLS WITH INITIAL DISTORTIONS

In the previous sections of this chapter, we considered imperfection-free structures using a differential approach that led to expressions for deflections and stresses. In real structures, however, a variety of imperfections are present; from the point of view of structural strength, the most important of these are initial distortions and locked-in stresses, both of which are introduced during the forming process and the fabrication process.

4.2.1 Sources of Initial Distortions in the Cylindrical Shell

In the case of the cylindrical shell, the forming process involves bending of flat plating such that it adopts a cylindrical form (Figure 4.15). However, it not possible in practice for the resulting bent plating to be perfectly circular, and the resulting distortions are thus termed OOC of the cylindrical shell. The forming process also induces *cold-bending* stresses in the otherwise stress-free plating, and these have a detrimental effect on collapse strength. This question is discussed in Section 6.7.

FIGURE 4.15 The cold-forming process leading to initial distortions in the form of out-of-circularity of the cylindrical shell.

Other sources of initial distortions are the welding procedures implemented during fabrication. Figure 4.16a and b show two such types of procedures used. In the first of these, vertical flux core arc welding (FCAW) is shown, as used for the circumferential butt joints of the UK *Astute* class submarines, whereas in the second case, submerged arc welding (SAW) is used to attach internal ring frames to cylindrical shell plating (Saab Kockums Shipyard). Fabrication processes involving welding induce initial distortions in the shell plating at an interframe level, and these assume the shapes shown in Figure 4.16c or d. They arise because of the transverse shrinkage of the fillet welds that are used to attach the frames to the shell plating. If the shell is internally stiffened, these distortions are concave inwards, whereas if the shell is externally stiffened, they are concave outwards. It is therefore apparent that when external hydrostatic pressure is applied, strength and other parameters will differ in each case. The various issues that need to be considered are discussed in Section 6.3.

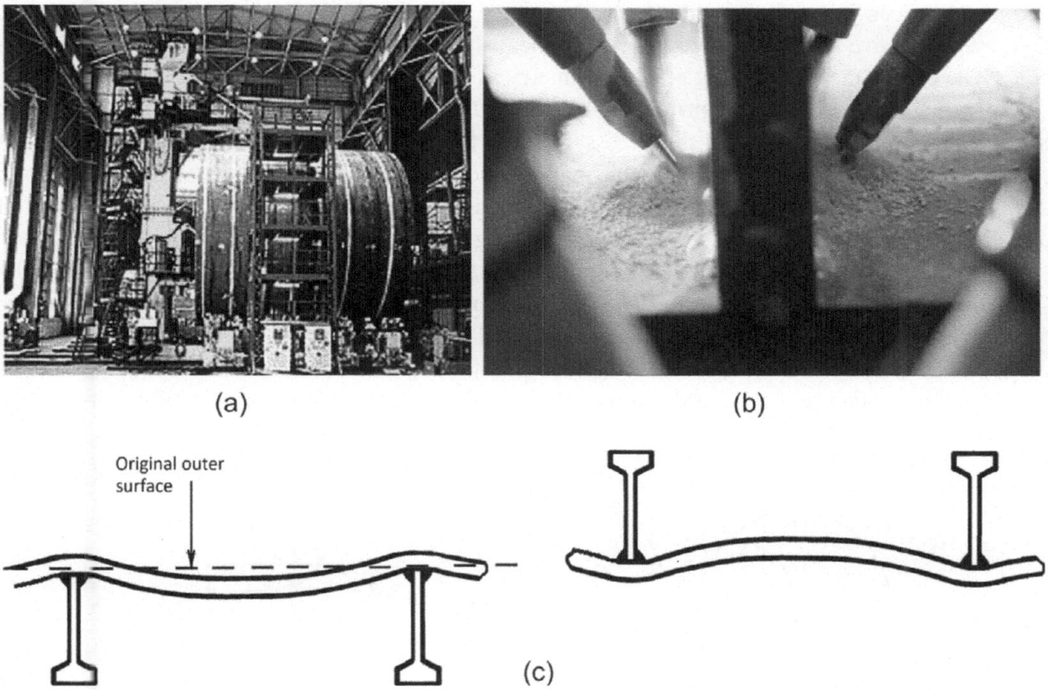

FIGURE 4.16 Welding procedures in submarine fabrication and resulting interframe initial distortions. (a) Rotated FCAW (butt welds). (b) SAW of ring frame to shell. (c) Internally stiffened and externally stiffened shell plating.

Of these, the most important type of initial distortion in ring-stiffened cylindrical shells is OOC, because it has a direct effect on the collapse strength and is ever-present since it is a result of fabrication. OOC of a particular section is defined as the difference between the maximum and minimum values of the diameter. Design codes generally specify that it should not exceed 0.5%*R*. Tests on ring-stiffened cylinders have shown that an OOC in the form of the critical overall elastic buckling mode having an amplitude of 0.05%R results in a 15% reduction in the elasto-plastic overall collapse pressure [1]. Numerical studies have shown similar decreases in overall collapse pressure due to OOC [9].

Other types of geometrical imperfections may also be present, such as *dents* and in the case of spheres and spherical caps, *local flattening*.

4.2.2 Measurement of Initial Distortions

A variety of techniques are used to measure initial distortions. The BS5500 code requires that in the case of OOC, a true record of the shape is obtained following which modal analysis is performed. The method proceeds as follows:

1. Radii are measured at 24 equally spaced intervals (15°). This is done either by swinging an arm internally (Figure 4.17a) or by external measurements with the cylindrical shell rotating on a lathe.
2. Since the point about which the swinging arm rotates does not coincide with the axis of the cylindrical shell, the radial measurements need to be corrected, and this is done by performing Fourier analysis. Details are given in [2].

An alternative to the swinging arm method that is used when it is inconvenient to take radial measurements either internally or externally is to use a chord gauge. This is used to measure curvature at equally spaced intervals around the circumference (Figure 4.17b). The measurements taken are then used to derive the true shape of the cylindrical shell [2]. In addition to these traditional methods, OOC can be measured using photogrammetry as well as laser beams.

Following the recording of primary data, computer-based techniques can be used to refine the raw data in order to provide the necessary input for nonlinear finite element analysis. This is required because an as accurate as possible mathematical model of the initial distortions is required. A *refined model* essentially includes additional values using interpolation in the circumferential direction as well as in the longitudinal direction along each generatrix (Figure 4.18) [6].

Once the shell form is adequately recorded, methods can be implemented to determine the strength of the imperfect structure. Two different approaches to the assessment of the elastic buckling strength of a ring-stiffened cylindrical shell in the presence of initial distortions will be described. The first of these deals with the effect of initial distortions at an interframe level, whereas the second is applied to the general stability problem. In both these methods, the effect of welding

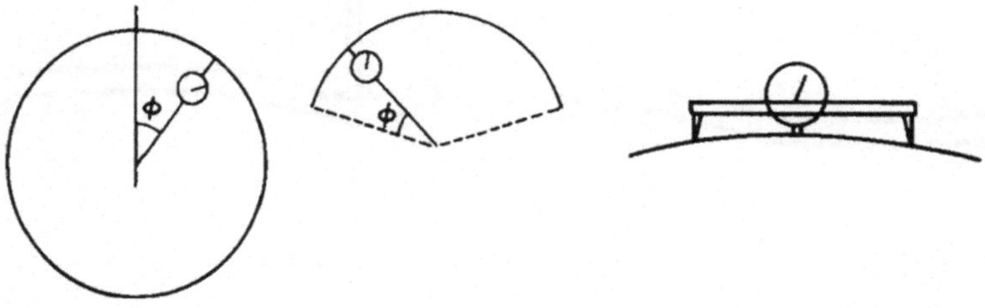

FIGURE 4.17 Traditional methods of measurement of OOC in cylindrical shells [2]. (a) The swinging arm method. (b) The chord gauge method.

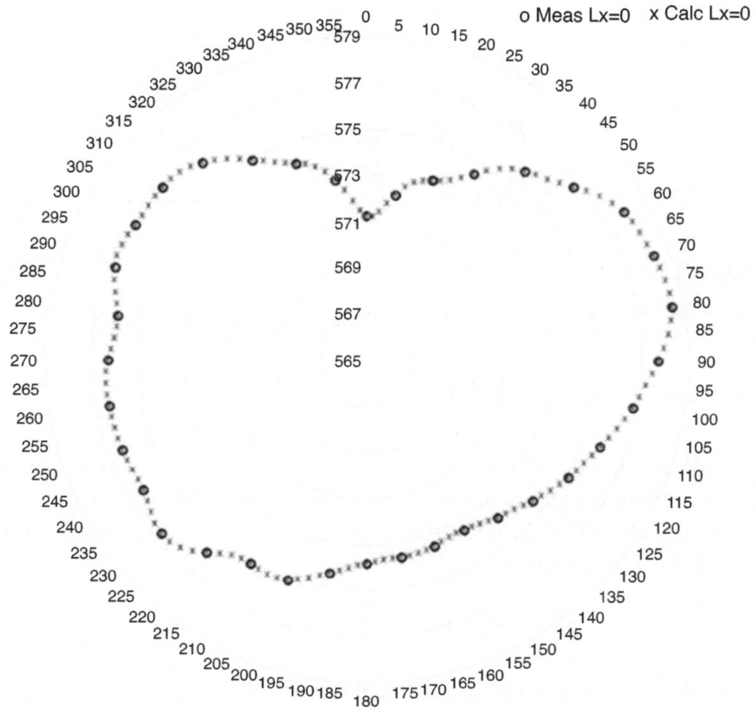

FIGURE 4.18 Refined data for OOC for use in FEA (single section) [6].

residual stresses is ignored. These have an effect on the stress distributions and on the collapse strength of the stiffened shell, and this question will be considered in Section 6.7.

Each of these two approaches provides insight to the structural response as well as useful results. In the first of these, the differential equation of a cylindrical shell with initial distortions of the shell plating between frames was derived and then solved using classical mathematical analysis. In the second approach, the Ritz method was used to determine the elastic overall buckling pressure of a ring-stiffened cylindrical shell. Expressions for the maximum bending stresses and distortions in the shell and ring frame were obtained. In both cases, it was assumed that the cylindrical shell is internally stiffened.

4.3 ANALYSIS OF RING-STIFFENED CYLINDRICAL SHELLS WITH INITIAL DISTORTIONS

In this section, we shall consider the effect that the weld-induced initial distortions have on cylindrical shell response based on the analysis described in [4].

4.3.1 Derivation of the Differential Equation

Measurements have shown that the weld-induced initial distortions in the shell plating can be approximated in a satisfactory manner using simple geometric expressions. To this end, the following expression was used in [4] to represent the interframe initial distortion, $w_0(x)$, of the shell plating:

$$w_0(x) = \frac{4}{L}\left(x - \frac{x^2}{L}\right)(w_0)_{max} \tag{4.87}$$

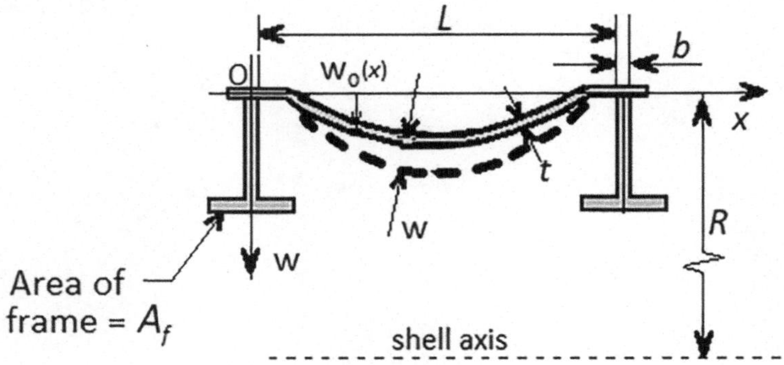

FIGURE 4.19 Shell plating with initial distortions [4].

where x is the co-ordinate measured along a generator from an origin located in way of a ring frame, and $(w_0)_{max}$ is the maximum initial distortion of the shell plating (Figure 4.19).

The moment $M_E(x)$ induced in the shell plating that results from the pressure on the ends of a closed cylinder of radius R is given by:

$$M_E(x) = -N_x(w_i + w(x)) \tag{4.88}$$

where $N_x = -pR/2$, and $w(x)$ is the additional deflection in the radial direction resulting from the applied hydrostatic pressure. Combining (4.87) and (4.88), we obtain:

$$M_E(x) = -N_x\left[\frac{4}{L}\left(x - \frac{x^2}{L}\right)(w_0)_{\max} + w(x)\right] \tag{4.89}$$

where w_i (=$w(x = 0, L)$) is the deflection occurring in way of the shell–frame intersections. The equation of equilibrium of an imperfection-free, ring-stiffened cylindrical shell subjected to radial pressure as well as end compression is (4.14). Rearranging, this assumes the following form:

$$D\frac{d^4w}{dx^4} = p + \frac{1}{R}\nu N_x - \frac{w}{R^2}Et + N_x\frac{d^2w}{dx^2} \tag{4.90}$$

where D is the shell flexural rigidity, defined as in flat plate theory. If we differentiate (4.89) twice and substitute for the last term of (4.90), the following equation is obtained:

$$D\frac{d^4w}{dx^4} = p + \frac{1}{R}\nu N_x - \frac{w}{R^2}Et + N_x\left(\frac{d^2w}{dx^2} - \frac{8(w_0)_{max}}{L^2}\right) \tag{4.91}$$

If we set

$$\alpha^4 = \frac{3(1-\nu^2)}{R^2t^2} \qquad \beta^2 = \frac{pR^3}{2Et}$$

Equation (4.91) becomes

$$\frac{d^4w}{dx^4} + 4\alpha^4\beta^2\frac{d^2w}{dx^2} + 4\alpha^4\left[w - \frac{pR^2}{Et}\left(1 - \frac{\nu}{2}\right) - \frac{8\beta^2}{L^2}(w_0)_{max}\right] = 0 \tag{4.92}$$

We now introduce a change of variable as follows:

$$Z = \frac{pR^3}{Et}\left(1-\frac{\nu}{2}\right)+\frac{8\beta^2}{L^2}(w_0)_{max} - w \tag{4.93}$$

The following simplified form of the equation of equilibrium is thus obtained:

$$\frac{d^4Z}{dx^4}+4\alpha^4\beta^2\frac{d^2Z}{dx^2}+4\alpha^4 Z = 0 \tag{4.94}$$

4.3.2 Solution of the Differential Equation

Equation (4.94) is a linear differential equation whose solution has the form:

$$\begin{aligned} Z = {} & C_1 \sinh K_1x \sin K_2x + C_2 \sinh K_1x \cos K_2x + \\ & + C_3 \cosh K_1x \sin K_2x + C_4 \cosh K_1x \cos K_2x \end{aligned} \tag{4.95}$$

where

$$K_1 = \alpha\sqrt{1-\alpha^2\beta^2} \quad K_2 = \alpha\sqrt{1+\alpha^2\beta^2}$$

The following boundary conditions are assumed:

$$\text{At } x = 0, \quad w = w_i \; \frac{dw}{dx} = 0 \tag{4.96a}$$

$$\text{At } x = L, \quad w = w_i \; \frac{dw}{dx} = 0 \tag{4.96b}$$

Introducing these boundary conditions to (4.5) and its derivative, the following expressions for the constants C_1–C_4 are obtained:

$$C_1 = \left[\frac{K_2 \sin K_2L - K_1 \sinh K_1L}{K_2 \sinh K_1L + K_1 \sin K_2L}\right] Z_0 \tag{4.97a}$$

$$C_2 = \left[\frac{K_2(\cos K_2L - \cosh K_1L)}{K_2 \sinh K_1L + K_1 \sin K_2L}\right] Z_0 \tag{4.97b}$$

$$C_3 = \left[\frac{K_1(\cosh K_1L - \cos K_2L)}{K_2 \sinh K_1L + K_1 \sin K_2L}\right] Z_0 \tag{4.97c}$$

$$C_4 = Z_0 \tag{4.97d}$$

Z_0 is the value of Z at the shell–frame intersection. This is given by:

$$Z_0 = \frac{pR^2}{Et}\left(1-\frac{\nu}{2}\right)+\frac{8\beta^2}{L^2} w_{0\max} - w_i \tag{4.98}$$

When (4.97a–d) are substituted in (4.95), the variable Z is expressed in terms of its value Z_0 at a shell–frame intersection as:

$$Z = \frac{f(x)}{G} Z_0 \tag{4.99}$$

where

$$\begin{aligned} f(x) &= K_2 \sinh K_1 x \cos K_2(L-x) + K_1 \cosh K_1 x \sin K_2(L-x) + \\ &+ K_1 \sin K_2 x \cosh K_1(L-x) + K_2 \cos K_2 x \sinh K_1(L-x) \end{aligned} \tag{4.100a}$$

$$G = K_2 \sinh K_1 L + K_1 \sin K_2 L \tag{4.100b}$$

The parameter Z_0 is obtained by considering equilibrium of radial forces at the shell–frame intersection. From Figure 4.20, we have

$$2D \left.\frac{d^3 w}{dx^3}\right|_{x=0} + \frac{E(A_f + bt)}{R^2} w_i - pb + \nu \frac{pR}{2}\frac{b}{2} + 2\frac{pR}{2}\frac{dw_0(0)}{dx} = 0 \tag{4.101}$$

The value of Z_0 is thereby obtained leading to the following expression for the total deflection of a ring-stiffened cylinder with initial distortions of the form (4.87):

$$\begin{aligned} w &= \frac{\beta^2}{R}(2-\nu) - \frac{2\alpha^2\beta^2}{H}\left[\frac{A_f(1-(\nu/2))}{Rt}\right] f(x) \\ &+ \frac{8\beta^2 (w_0)_{max}}{L^2}\left[1 - \frac{\alpha^2}{H}\frac{(A_f + bt + tL)}{t} f(x)\right] \end{aligned} \tag{4.102}$$

The first two of the three terms in (4.102) represent the deflection of the shell of a cylinder that has no initial distortions (Pulos–Salerno solution), whereas the third represents the contribution to the deflection due to the presence of the initial distortions. We note from the third term that the

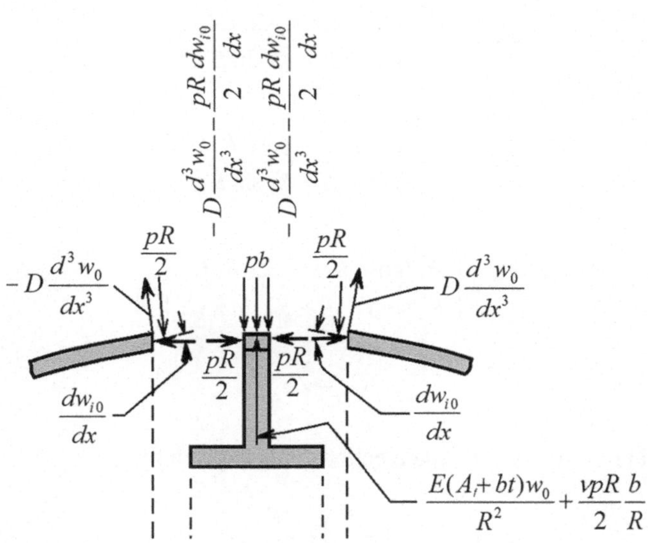

FIGURE 4.20 Forces acting at shell–frame intersection [4].

contribution is directly proportional to the initial distortion. The total deflection is thus the sum of the two quantities so that[6]

$$w = w_{PS} + w_{LS} \tag{4.103}$$

We shall now obtain expressions for the stresses in the circumferential and the longitudinal directions allowing for the presence of initial distortions. In the circumferential direction, the membrane stress $\sigma_{\theta m}$ is given by the first two terms of (4.73b):

$$\sigma_{\theta m} = \frac{Ew}{R} + \frac{\nu pR}{2t} \tag{4.104}$$

Introducing (4.103) into (4.104), the total circumferential membrane stress is:

$$\sigma_{\theta} = \frac{Ew}{R} w_{PS} + \frac{\nu pR}{2t} + \frac{E}{R} w_{LS} \tag{4.105}$$

The last term of (4.105) is due to the presence of initial distortions. Thus, the additional circumferential membrane stress due to initial distortions is given by:

$$\sigma_{\theta m} = \frac{E}{R} w_{LS} \tag{4.106}$$

The bending stress in the longitudinal direction that is due to the initial distortions is given by

$$\sigma_{bxLS} = \pm \frac{Et}{2(1-\nu^2)} \frac{d^2 w_{LS}}{dx^2} \tag{4.107}$$

where

$$\begin{aligned}\frac{d^2 w_{LS}}{dx^2} = -\frac{16\alpha^4\beta^2}{HL^2} w_{LS}\left(\frac{A+bt+tL}{t}\right)&\left[K_1 \cosh K_1 x \sin K_2(L-x)\right.\\ &-K_2 \sinh K_1 x \cos K_2(L-x) - K_2 \cos K_2 x \sinh K_1(L-x) +\\ &\left.+K_1 \sin K_2 x \cosh K_1(L-x)\right]\end{aligned} \tag{4.108}$$

The bending stress in the circumferential direction is then

$$\sigma_{b\theta LS} = \nu \sigma_{bxLS} \tag{4.109}$$

or,

$$\sigma_{b\theta LS} = \pm \frac{\nu}{2} \frac{Et}{(1-\nu^2)} \frac{d^2 w_{LS}}{dx^2} \tag{4.110}$$

The total circumferential stress in the presence of initial distortions is obtained by adding (4.106)–(4.110):

$$\sigma_{\theta} = \frac{E}{R} w_{LS} \pm \frac{\nu}{2} \frac{Et}{(1-\nu^2)} \frac{d^2 w_{LS}}{dx^2} \tag{4.111}$$

The stress σ_{xLS}, in the outer fibres in the longitudinal direction, in the presence of initial distortions, is given by (4.107).

In order to obtain a quantitative assessment of the effect of initial distortions on structural behaviour, an internally stiffened cylindrical shell with the following specifications is considered:

$$L/R = 0.1 \quad A/Lt = 0.5$$

$$t/2R = 0.005 \quad b/L = 0.05$$

The above quantities are defined in Figure 4.19. For this geometry, pressure–strain curves were obtained and are plotted in Figures 4.21–4.23 for a range of values of initial distortions. The range

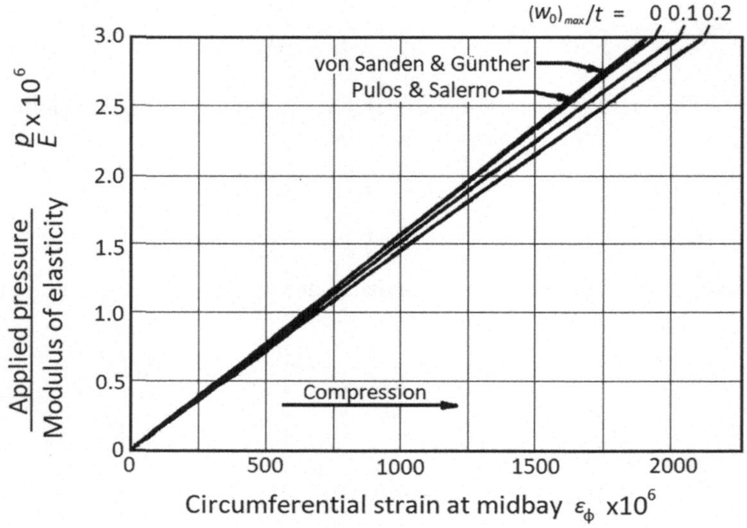

FIGURE 4.21 Pressure versus circumferential strain at midbay [4].

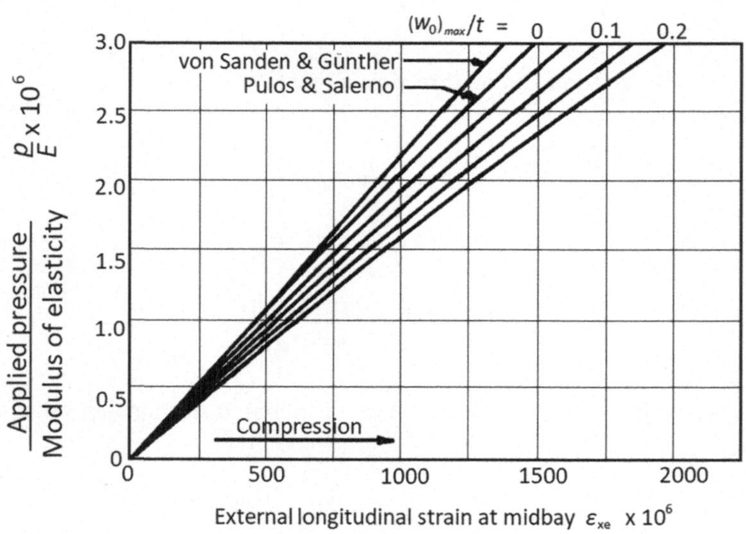

FIGURE 4.22 Pressure versus external longitudinal strain at midbay [4].

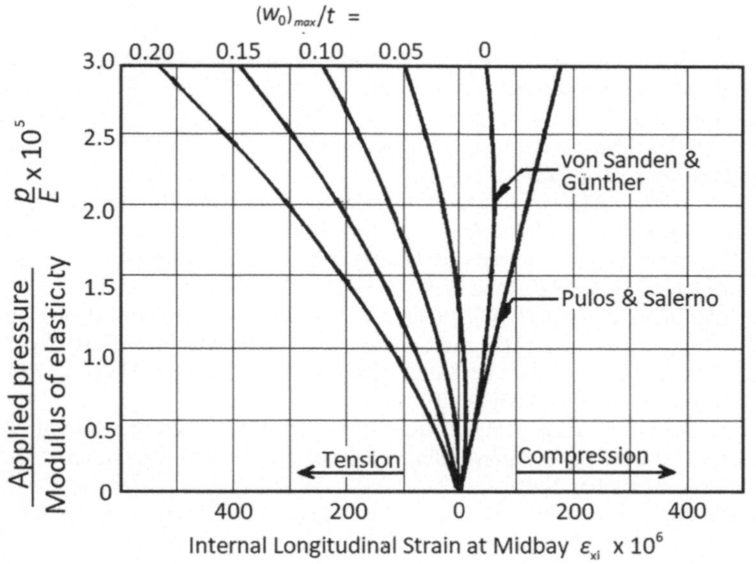

FIGURE 4.23 Pressure versus internal longitudinal strain at midbay [4].

considered corresponds to data originating from test specimens at DTMB and is therefore treated as being representative of real structures.[7] The strains have been calculated at midbay using (a) the von Sanden and Günther theory (ignoring the effect of longitudinal compression); (b) the Pulos and Salerno theory (allowing for the effect of longitudinal compression); and (c) the Lunchick and Short theory (allowing for longitudinal compression and for the presence of initial distortions).

It is seen from Figure 4.21 that in the case of circumferential strains, the beam-column effect allowed for by the Pulos and Salerno solution results in minor differences from the von Sanden and Günther theory. When initial distortions are present, the strains increase as the magnitude of the initial distortion is increased. The discrepancies increase with pressure increases. In an analogous manner, the stresses increase, precipitating earlier yielding.

The effect of initial distortions is, as expected, more pronounced in the case of longitudinal strains. Figures 4.22 and 4.23 illustrate the effect that initial distortions have on external and internal shell strains at midbay. As expected, the effect is more serious in the case of internal strains, and a sign reversal is observed (compression to tension) in Figure 4.23.

NOTES

1 The generally accepted value for maximum permitted out-of-circularity is $5R/1{,}000$ (e.g. BS EN 13445 code).
2 The Bresse formula is based on the Euler column buckling formula.
3 In an infinitely long cylindrical shell, end effects (axial compression) are ignored.
4 That is, no geometric or material imperfections.
5 The Love–Kirchhoff hypothesis corresponds to the Euler-Bernoulli assumption upon which is based the engineering theory of bending of beams (linear increase of strain and stress from the neutral axis in pure bending).
6 The subscripts *PS* and *LS* are used to designate the Pulos–Salerno and Lunchick–Short solutions, respectively.
7 The distortions measured on test specimens are larger than those in full-scale structures because of the scale effect of welding.

REFERENCES

1. Bosman, T.N., Pegg, N.G., Keuning, P.J. (1993). "Experimental and numerical determination of the non-linear overall collapse of imperfect pressure hull compartments". *Proc Warship '93 RINA Internship Symposium Naval Submarines 4*, London.
2. Kendrick, S. (1977). Shape imperfections in cylinders and spheres: Their importance in design and methods of measurement. *J. Strain Analysis*, Vol. 12, No. 2, 117–122.
3. Labouvie, E.N. (1952). "The strength of cylindrical shells, stiffened by frames and bulkheads, under uniform external pressure on all sides". DTMB Rept. No. T-38. [Transl. of: von Sanden, K., Gűnther, K. "Űber das Festigkeits problem quersteifter Höhlzylinder unter allseitig gleichmassigen Aussendruck". *Werft und Reederie*, Vol. 1 (1920), Nos. 8, 9, 10, Vol. 2 (1921) No. 17].
4. Lunchick, M.E., Short, Jr., R.D. (1957). "Behavior of cylinders with initial shell deflection". *Proceeding of the ASME Applied Mechanics Division Conferences*, Amsterdam (also DTMB Report 1150).
5. MacNaught, D.F. (1967). "Strength of Ships" Chap. IV. In: Comstock J.P. (Ed). *Principles of Naval Architecture* (revised), New York: SNAME.
6. Phelps, B.P. (2016). "Refinement of out of circularity and thickness measurements of a cylinder for finite element analysis". Maritime Division Defence Science and Technology Group DST-Group-TN-1521, Australian: Royal Australian Navy.
7. Pulos, J.G. (1963). "Structural analysis and design considerations for cylindrical pressure hulls". DTMB Report 1636, Washington DC.
8. Pulos, J.G., Salerno, V.L. (1961). "Axisymmetric elastic deformations and stresses in ring-stiffened, perfectly circular cylindrical shell under external hydrostatic pressure". DTMB Report 1497, Washington DC.
9. Smith, M.J., MacKay, J.R. (2004). "Overall elasto-plastic collapse of submarine pressure hull compartments" (DRDC Atlantic TM 2004-243). Defence R&D Canada – Atlantic.
10. Viterbo, F. (1930). "Sul problema della robustezza di cilindri cavi rinforzati transversalmente sottopsti da ogni parte a pressione esterna" (On the problem of the strength of transversely reinforced hollow cylinders subjected on all sides to external pressure). *L' Ingegnere* Vol. IV, July 446–456, Aug. 531–540.
11. Wilson, L.B. (1956). "The deformation under uniform pressure of a circular cylindrical shell supported by equally spaced circular ring frames. Part 1 – Asymptotic method". NCRE Report R 337A, Naval Constr. Res. Est., Dunfermline, Scotland.
12. Wilson, L.B. (1966). The elastic deformation of a circular cylindrical shell supported by equally spaced circular ring frames under uniform external pressure. *Trans. RINA*, Vol. 108, 63–72.

5 General Instability of Ring-Stiffened Cylindrical Shells

5.1 INTRODUCTION

The third collapse mode that we shall be dealing with is termed *general instability* (*overall collapse*) (Figure 5.1). This may arise in the case of compartments whose length is large in comparison to the diameter or when the ring frames are weak. Other factors that influence the overall collapse pressure are ring frame spacing and the modulus of elasticity of the material.

From Example 4.1, we conclude that the ring frames contribute to shell strength in a significant way. Additionally, the ring frames need to have sufficient strength in order to avoid overall elastic buckling between successive bulkheads. The effect of frame eccentricity on the overall buckling strength is considered in Section 6.3.2. It is also possible to add further ring frames that would reduce radial deflections when subjected to external pressure.

The first attempt at calculating the elastic overall buckling pressure of a ring-stiffened cylindrical shell was carried out by Tokugawa, who obtained an expression consisting of two terms, the first of which corresponds to failure of the ring frame and attached plating and the second to failure of the shell plating. As a result, this approach came to be known as the "method of split rigidities".

5.2 APPLICATION OF THE STRAIN ENERGY METHOD

In 1954 Bryant obtained an expression for the hydrostatic pressure required for elastic general instability of a cylinder stiffened by uniformly spaced ring frames [4]. This expression has to be treated as being approximate because:

a. Interframe distortions of the shell are ignored.
b. A simplified version of the von Mises formula for elastic buckling of unstiffened cylinders (3.41) is used.
c. It does not account for the interaction of the cylindrical shell with the ring frames.

(a) (b)

FIGURE 5.1 Overall collapse of a stiffened cylindrical shell (*ARE*[1] scale model test results).

DOI: 10.1201/9781003489153-5

In this section, we shall outline the approach followed, which was based on the strain energy method. A comparison of results obtained using this method and the more rigorous method based on the Ritz method, which is discussed in the next section, is included in Section 5.4. Bryant's method is based on the following two assumptions:

1. The strain energy of the ring-stiffened cylinder is the sum of the strain energy of distortion of the cylindrical shell and that of bending of the ring frames.
2. The tangential displacements, v, of the cylinder are identical to those of the ring frame when considered in isolation, for similar radial displacements.

The second assumption is justified given that both the cylindrical shell and the ring frames suffer inextensional bending given by:

$$w = \frac{\partial v}{\partial \theta} \tag{5.1}$$

5.2.1 Cylindrical Shell and Ring Frame Treated Separately

To obtain expressions for the buckling pressure of the ring-stiffened cylindrical shell, we first need to consider the unstiffened cylinder and the ring frame in isolation.

Following buckling, the radial displacements of the cylindrical shell are given by the following expression:

$$w = w_n \sin(n\theta)\sin\left(\frac{\pi x}{L_b}\right) \tag{5.2}$$

For a buckling deformation w_n, the strain energy of the shell increases by the following amount:

$$U_s = \underbrace{U_{s1} w_n^2}_{\text{postbuckling}} + \underbrace{p_{sn} U_{s2} w_n}_{\text{prebuckling}} \tag{5.3}$$

where p_{sn} is the buckling pressure of the shell with n circumferential lobes and one half-sine wave in the axial direction ($m = 1$). The work done during buckling of the shell, W_s, is equal to the product of the external pressure and the decrease in volume of the shell, Vol_s. Thus,

$$W_s = p_{sn}\left(Vol_{s1} w_n^2 + Vol_{s2} w_n\right) \tag{5.4}$$

Rigorous treatments using the strain energy method show that the first-order terms in w_n in (5.3) and (5.4) must be equal to each other. Thus, when conservation of energy $U = W$ is applied, these terms cancel out. We then have:

$$p_{sn} = \frac{U_{s1}}{Vol_{s1}} \tag{5.5}$$

We consider now a ring frame. When subjected to a uniform pressure, the buckled form in the radial direction is described by:

$$w = w_n \sin\theta \tag{5.6}$$

Additionally, a tangential displacement, v, that is proportional to w_n, also arises. The internal strain energy of the ring frame, U_f, is given by:

$$U_f = \underbrace{U_{f1} w_n^2}_{\text{post buckling}} + \underbrace{TU_{f2} w_n}_{\text{pre buckling}} \tag{5.7}$$

The linear term in w_n corresponds to the deformation just prior to buckling. This is proportional to the external radial thrust T multiplied by the mean circumferential compression during buckling.

The work done W_f by the external radial thrust T is then T times the reduction in area of the ring following buckling. It is given by

$$W_f = T\left(A_{f1} w_n^2 + A_{f2} w_n\right) \tag{5.8}$$

As in the case of the unstiffened cylinder, it may be proven that the linear terms in w_n in (5.7) and (5.8) must be equal. Equating U_f and W_f, we obtain:

$$T = \frac{U_{f1}}{A_{f1}} \tag{5.9}$$

The critical buckling pressure of the ring frame attached to shell plating of breadth L_f is then:

$$p_{fn} = \frac{U_{f1}}{A_{f1} L_f} \tag{5.10}$$

and

$$\frac{U_{f1}}{A_{f1}} = \frac{3EZ}{R_f^3} \tag{5.11}$$

where Z is the section modulus, and R_f is the distance from the shell axis to the centre of gravity of the ring frame.

5.2.2 Ring-Stiffened Cylindrical Shell

We consider now a ring-stiffened cylinder fitted with at least 10 uniformly spaced rings. If the overall length between bulkheads is L_b and the frame spacing is L_f, the number of rings is $(L_b/L_f)-1$. During buckling, the cylinder and the ring frames distort, and it is assumed that the overall shape is again given by (5.2). That is, interframe shell distortions are ignored.

We now consider the strain energy of the ring-stiffened cylinder. This has two components. The first corresponds to the strain energy of the cylindrical shell. This is (5.3) modified by substituting the buckling pressure of the ring-stiffened cylinder, p_n, for that of the shell, p_{sn}. We thus have:

$$U_s = \underbrace{U_{s1} w_n^2}_{\text{post buckling}} + \underbrace{p_n U_{s2} w_n}_{\text{pre buckling}} \tag{5.12}$$

The second part of the strain energy corresponds to the presence of circumferential stresses in the ring and shell plating in addition to those stresses already calculated that arise in connection with frame bending.

A ring that is located a distance x_0 from a fixed end deforms into n half-sine waves having an amplitude $w_n \sin(\pi x_0/L_b)$. This deformation concerns the ring itself and the attached plating of width L_f. The strain energy U is then due to the bending of the ring frame as well as that of the attached plating. For the ring frame, the strain energy is given by (5.7) in which the amplitude is w_n and the external pressure p_n is multiplied by L_f for the radial thrust:

$$U_f = U_{f_1} w_n^2 \sin^2\left(\frac{\pi x_0}{L_b}\right) + p_n L_f U_{f2} w_n \sin\left(\frac{\pi x_0}{L_b}\right) \tag{5.13}$$

The total internal strain energy U of the ring-stiffened cylindrical shell is then found by summing over all ring frames and adding that of the cylindrical shell, U_s, given by (5.12):

$$U = U_s + \sum_{i=1}^{N} U_f = \underbrace{w_n^2 \left(U_{s1} + U_{f1} \sum_{i=1}^{N} \sin^2\left(\frac{\pi x_0}{L_b}\right)\right)}_{\text{buckling}} + \underbrace{p_n w_n \left(U_{s2} + U_{f2} L_f \sum_{n=1}^{N} \sin\left(\frac{\pi x_0}{L_b}\right)\right)}_{\text{prebuckling}} \tag{5.14}$$

The work W done by the external pressure p_n is equal to the product of p_n and the reduction in internal volume of the shell. This is given by (5.4), where $p=p_n$.

$$W = p_n \left(Vol_{s1} w_n^2 + Vol_{s2} w_n \right) \tag{5.15}$$

At the critical buckling pressure, p_n, $U = W$. Substituting from (5.14) and (5.15) and observing as previously that the first-order terms in w_n cancel out, we obtain:

$$p_n = \frac{U_{s1}}{Vol_{s1}} + \frac{U_{f1}}{Vol_{f1}} \sum_{n=1}^{N} \sin^2\left(\frac{\pi x_0}{L_b}\right) \tag{5.16}$$

We may assume that the strain energy and volume of the shell and frame included in (5.16) are equal to the corresponding quantities included in (5.5) and (5.10), provided the expression for the moment of inertia of the ring frame includes the attached shell plating of width L_f.

The reduction in area of the ring frame, previously included in (5.8), now becomes:

$$A_{f1} w_n^2 \sin^2\left(\frac{\pi x_0}{L_b}\right) + A_{f2} w_n \sin\left(\frac{\pi x_0}{L_b}\right)$$

For the whole cylindrical shell, a summation of the first term in the above expression is performed for all ring frames. Equating this to the decrease in volume of the entire cylinder, we obtain

$$Vol_{s1} w_n^2 = A_{f1} w_n^2 \sum_{n=1}^{N} \sin^2\left(\frac{\pi x_0}{L_b}\right)$$

Substituting in (5.16) leads to:

$$p_n = \frac{U_{s1}}{Vol_{s1}} + \frac{U_{f1}}{A_{f1}}$$

Or

$$p_n = p_{sn} + p_{fn} \tag{5.17}$$

We see that the buckling pressure of the ring-stiffened cylinder is the sum of p_{sn}, which is the elastic buckling pressure of the unstiffened cylinder that is supported solely by the end bulkheads and p_{fn}, which is the pressure at which a ring frame consisting of the ring and the attached cylindrical shell buckles into n circumferential waves.

The N frames that are fitted between the end bulkheads subdivide this length into $(N+1)$ bays. As a result, a small discrepancy in the estimate of the volume change arises, because the two half-bays adjacent to the bulkheads are neglected. The deflections there are very small compared to the mean value so that the error should be $\ll 1/N$.

The unstiffened cylinder buckles in one half-sine wave along its length ($m = 1$) and in n circumferential waves. We can use the von Mises formula (3.41) to calculate p_{sn}, for which $m = 1$. For practical cases, $t/R \ll 1$ so that the second term in the large bracket of (3.41) may be ignored. Hence,

$$p_{sn} = \frac{Et}{R} \frac{\lambda^4}{\left(n^2 - 1 + \frac{\lambda^2}{2}\right)\left(n^2 + \lambda^2\right)^2} \tag{5.18}$$

A suitable formula for p_{fn} is that obtained by Bresse (4.1). Substituting (5.18) and (4.1) in (5.17), we obtain:

$$p_n = p_{fn} + p_{sn} = \frac{(n^2 - 1)EI_{sf}}{R^3 L_f} + \frac{Et}{R} \frac{\lambda^4}{\left(n^2 - 1 + \frac{\lambda^2}{2}\right)\left(n^2 + \lambda^2\right)^2} \tag{5.19}$$

where

p_{fn} = elastic buckling pressure of ring frame including the attached plating

p_{sn} = approximate elastic buckling pressure of unstiffened shell

$\lambda = \frac{\pi R}{L_b}$

The length L_b is taken between sections that ensure near-perfect circularity (transverse bulkheads or heavy ring frames). Since overall failure can result from insufficient ring frame strength, this can be avoided by a small increase in structural weight.

The von Mises term in (5.19) represents the shear strength of the cylindrical shell in the absence of ring frames. This is negligible in comparison to the Bresse term.

Expression (5.19) has the merit, not only of computational simplicity but also of showing the contribution of the cylindrical shell and the ring frames individually. The effects of changes in bulkhead spacing and ring frame scantlings are thus more clearly apparent. On the other hand, this expression does not cater explicitly for the two cases of internal and external ring frames, since the second term is based on the assumption that the ring frame is merged with the shell plating. The effect that ring frame eccentricity has on the elastic overall response of ring-stiffened shells is discussed in Section 6.3.3.

Example 5.1

Calculate the overall elastic buckling pressure of the cylindrical shell of Example 4.1 for the possible failure modes of the section if the bulkhead spacing is 12 m.

Solution

$$\lambda = \pi(4)/(12) = 1.047$$

$$p_{f2} = 2.27\ \text{N/mm}^2$$

We calculate the two terms of (5.19) (p_{fn}, p_{sn}) and p_n (N/mm²):

n	p_{fn}	p_{sn}	p_n
2	2.27	21.69	23.97
3	6.06	2.40	8.47
4	11.37	0.31	11.68
5	18.19	0.12	18.31

The critical failure mode occurs for $n=3$ (and $p_3=1.7p_0$)

5.3 APPLICATION OF THE RITZ METHOD

A more rigorous approach that is considered to be the best available analytical solution to the question of overall elastic instability of ring-stiffened shells is that by Kendrick [5]. He used the Ritz method to determine the equilibrium state of the structure consisting of the cylindrical shell plating and ring frames. Before describing the implementation of the method, we shall outline its basic features.

The Ritz method is an approximate variational method that makes use of the principle of minimum potential energy to determine the equilibrium state of a structure. The method is approximate because it makes use of assumed displacement fields that do not necessarily coincide with the displacements of the exact solution of the problem in question. Its success lies in its ability to replace differential equations by systems of linear equations that can be solved much more easily.

The first step of the method involves defining an approximate displacement field $\tilde{w}$ that consists of linear combinations of coefficients and shape functions, as follows:

$$\tilde{w}(x) = \sum_{i=1}^{n} a_i \phi_i \tag{5.20}$$

where a_i are *Ritz coefficients*, and ϕ_i are *shape functions*. The Ritz coefficients are initially unknown, and their determination leads to the solution of the problem, whereas the shape functions are approximations to the distorted form of the structure in question. As such, they have to satisfy the geometrical boundary conditions. In principle, if an increasing number of shape functions are used, the solution becomes more accurate and approaches the exact solution.

The second step involves determining the total potential of the structure, Π. This is the sum of the potential of the internal strain energy, U, and that of the external work, W:

$$\Pi = U + W \tag{5.21}$$

The approximate displacement field $\tilde{w}$ is then substituted in the total potential so that this becomes an approximate potential that is expressed in terms of the unknown Ritz coefficients a_i and the assumed shape functions.

The third step involves applying the principle of minimum potential energy, which is a variational principle, according to which at equilibrium the first variation of Π is equal to zero:

$$\delta\Pi = 0 \tag{5.22}$$

From the calculus of variations,

$$\frac{\partial \Pi}{\partial a_1}\delta a_1 + \frac{\partial \Pi}{\partial a_2}\delta a_2 + \frac{\partial \Pi}{\partial a_3}\delta a_3 + \cdots = 0 \tag{5.23}$$

Since each of these terms is individually equal to zero, we have

$$\frac{\partial \Pi}{\partial a_1}\delta a_1 = \frac{\partial \Pi}{\partial a_2}\delta a_2 = \frac{\partial \Pi}{\partial a_3}\delta a_3 \ldots = 0 \tag{5.24}$$

The quantities δa_1, δa_2, etc., are arbitrary so that

$$\frac{\partial \Pi}{\partial a_i} = 0 \quad i = 1, 2, \ldots n \tag{5.25}$$

This is termed the *Ritz condition*. By applying (5.25), a system of equations whose number is equal to the number of Ritz coefficients is obtained, and this enables us to determine the values of the coefficients. This is in essence the solution of the problem. The system of equations is solved by setting the determinant equal to zero. This ensures that a non-trivial solution is obtained.

We now return to the problem is question. We shall be considering the general instability of a ring-stiffened cylindrical shell of finite length, shown in Figure 5.2.

Prior to buckling, the structure is in a state of stable equilibrium, and the induced displacements (u_1, v_1, w_1) are the result of the applied load. When buckling takes place, additional displacements (u_2, v_2, w_2) arise beyond those present prior to buckling. To these additional displacements correspond additional internal strain energy stored in the shell and additional work done by the applied external pressure during buckling. The total potential Π is expressed in terms of the total displacements ($u_1 + u_2$, $v_1 + v_2$, $w_1 + w_2$), which include the pre-buckling displacements and the buckling-related displacements and their derivatives.

The buckling-related displacements are expressed in terms of (5.20) so that they assume approximate values. It is necessary to select appropriate shape functions for the general instability of the ring-stiffened cylindrical shell. To this end, the following expressions are chosen:

$$u_2(x,\theta) = A\cos(n\theta)\cos\left(\frac{m\pi x}{L_b}\right) \tag{5.26a}$$

$$v_2(x,\theta) = B\sin(n\theta)\sin\left(\frac{\pi x}{L_b}\right) \tag{5.26b}$$

$$w_2(x,\theta) = C\cos(n\theta)\sin\left(\frac{m\pi x}{L_b}\right) \tag{5.26c}$$

where *A*, *B*, and *C* are the Ritz coefficients that are in fact the amplitudes of the displacements in each of the three orthogonal directions. Expressions (5.26) correspond to the buckled form shown in Figure 5.2, that is, they represent a smooth half-sine wave along the entire length of the stiffened cylinder. This implies that interframe bending does not arise, that is, the structure is stiffer than it

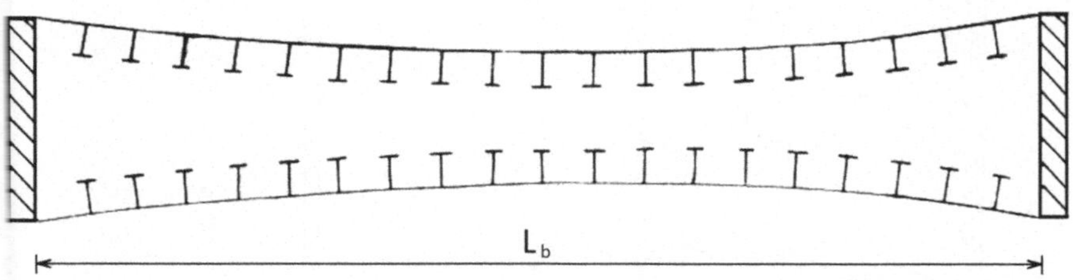

FIGURE 5.2 General instability of a ring-stiffened cylindrical shell (ignoring interframe distortions of shell plating).

actually is. The predicted buckling pressure will in this case be higher than the actual one. In order to allow for the finite bending stiffness of the shell plating between frames, it is necessary to introduce additional terms in (5.26). The buckled form is then as shown in Figure 5.3, and expressions (5.26) become:

$$u_2(x,\theta) = A_1 \cos(n\theta)\cos\left(\frac{\pi x}{L_b}\right) \tag{5.27a}$$

$$v_2(x,\theta) = B_1 \sin(n\theta)\sin\left(\frac{\pi x}{L_b}\right) + B_2 \sin(n\theta)\left(1-\cos\left(\frac{2\pi x}{L_b}\right)\right) \tag{5.27b}$$

$$w_2(x,\theta) = C_1 \cos(n\theta)\sin\left(\frac{\pi x}{L_b}\right) + C_2 \cos(n\theta)\left(1-\cos\left(\frac{2\pi x}{L_b}\right)\right) \tag{5.27c}$$

According to these deflection functions, the entire ring-stiffened cylinder buckles into one half-sine wave in the axial direction between transverse bulkheads so that simple supports are assumed at the ends. The terms involving B_2 and C_2 are introduced to allow for the bending of the shell plating between frames. A zero-rotation condition is assumed in way of the line of attachment of each of the frames to the shell, as seen in Figure 5.3. In this chapter, we shall present the solution of the former problem, illustrated in Figure 5.2; the solution to the latter problem is included in [7].

The deflection functions included in (5.26) have been validated by tests at the David Taylor Model Basin DTMB (Figure 5.4).

We shall now proceed to step 2, that is, the derivation of formulas for the internal strain energy U and of the external work done W.

5.3.1 Strain Energy of a Linear Elastic Body

The strain energy of an isotropic body that extends in three-dimensional space is given by:

$$U = \frac{1}{2}\iiint\left[\overline{\sigma_x}\,\overline{\varepsilon_x} + \overline{\sigma_\theta}\,\overline{\varepsilon_\theta} + \overline{\sigma_z}\,\overline{\varepsilon_z} + \overline{\tau_{x\theta}}\,\overline{\gamma_{x\theta}} + \overline{\tau_{\theta z}}\,\overline{\gamma_{\theta z}} + \overline{\tau_{zx}}\,\overline{\gamma_{zx}}\right]dxdydz \tag{5.28}$$

The bars in the above expression indicate that both extensional and bending components are included. In the case of the types of structures considered in this chapter, and in particular, ring-stiffened cylindrical shells, the following assumptions hold [3]:

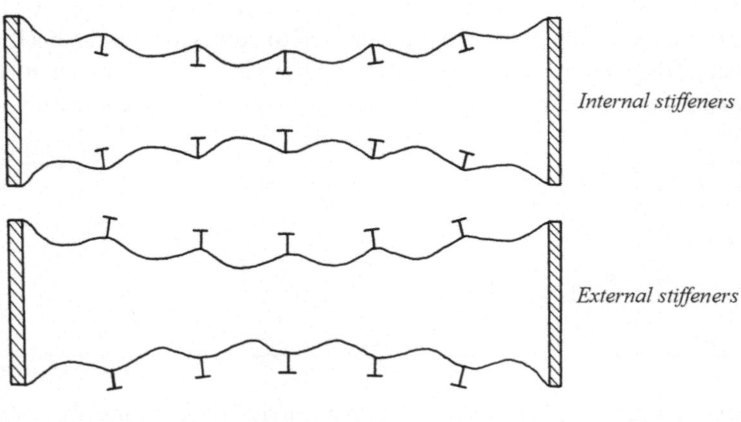

FIGURE 5.3 General instability allowing for interframe bending.

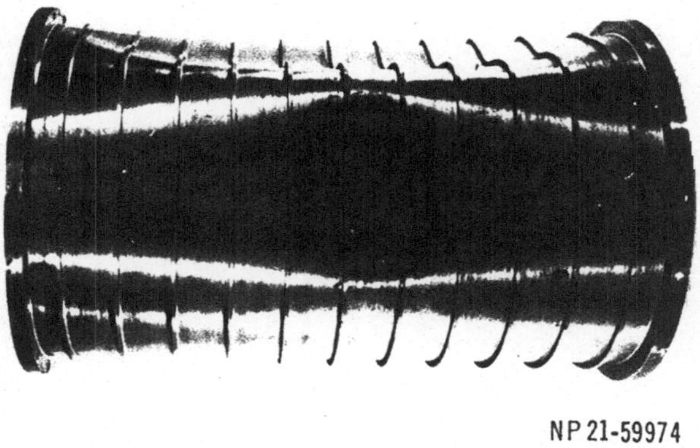

FIGURE 5.4 Externally stiffened test specimen failing by general instability [8].

1. Normals to the undeformed middle surface remain straight, normal, and inextensional during deformation. As a result, transverse normal and shearing strains $\left(\overline{\gamma_{\theta z}}, \overline{\gamma_{zx}}\right)$ may be neglected.
2. Transverse normal stresses $(\overline{\sigma_z})$ are assumed to be negligibly small compared with other normal stress components and may therefore be neglected.

These assumptions were originally introduced by Kirchhoff and form the basis of thin-walled structures theory.

For a ring-stiffened cylindrical shell subjected to external pressure loading, the total potential of the system comprising the structure and its immediate environment is given by:

$$\Pi = U_{sm} + U_{sb} + \sum_{r=1}^{N} (U_{fm})_r + \sum_{r=1}^{N} (U_{fb})_r + W \tag{5.29}$$

where U_{sm}, U_{sb} are the potentials of the membrane and bending strain energy of the shell, U_{fm}, U_{fb} are the potentials of the membrane and bending strain energy of each ring frame, N is the total number of ring frames, and W is the potential of the work done by the pressure load. Each of the terms included in (5.29) has to be determined individually.

5.3.2 Membrane Strain Energy of Cylindrical Shell to First-Order Accuracy

For an unstiffened cylindrical shell, the internal strain energy consists of the corresponding membrane (U_{sm}) and bending strain energy (U_{sb}) terms:

$$U_s = U_{sm} + U_{sb} \tag{5.30}$$

Using cylindrical coordinates, the membrane strain energy, U_{sm}, is given by the following expression, on the basis of (5.28):

$$U_{sm} = \frac{1}{2}\int_0^{2\pi}\int_0^{L_b} \sigma_x\varepsilon_x + \sigma_\theta\varepsilon_\theta + \tau_{x\theta}\gamma_{x\theta} d\theta dx \tag{5.31}$$

where the quantities σ_x, ε_x, etc., are the membrane stresses and strains, etc. For linear elastic analysis,

$$\sigma_x = \frac{E}{1-\nu^2}\left[\varepsilon_x + \nu\varepsilon_\theta\right] \tag{5.32a}$$

$$\sigma_\theta = \frac{E}{1-\nu^2}[\varepsilon_\theta + \nu\varepsilon_x] \tag{5.32b}$$

$$\tau_{x\theta} = \frac{E}{2(1+\nu)}\gamma_{x\theta} \tag{5.32c}$$

Hence,

$$U_{sm} = \frac{Rt}{2}\frac{E}{(1-\nu^2)}\int_0^{2\pi}\int_0^{L_b}\left[\varepsilon_x{}^2 + \varepsilon_\theta{}^2 + 2\nu\varepsilon_x\varepsilon_\theta + \frac{1-\nu}{2}\gamma_{x\theta}{}^2\right]dx\,d\theta \tag{5.33}$$

where L_b is the shell length between fixed end supports (bulkheads).

This is the membrane strain energy of a perfect cylindrical shell expressed in terms of longitudinal, circumferential, and shear components of the membrane strain. We shall now obtain expressions for each of these in terms of displacements (*kinematic relations*).

In the case of a shell that is subjected to external hydrostatic pressure, the primary consequence is a reduction in radius and, hence, also in the shell perimeter. If R_i and R_f are the initial and final radii respectively, the change in length of the perimeter $=2\pi(R_i-R_f)=-2\pi w$ where w is the change in radius. The circumferential membrane strain ε_θ is then given by:

$$\varepsilon_\theta = \frac{(-2\pi w)}{2\pi R_i} = -\frac{w}{R} \tag{5.34}$$

If the application of the load results in both a reduction in radius and a translation, the resulting displacement can be resolved into a component v in the direction of the tangent to the meridian and a component w in the radial direction (Figure 5.5).

Consider the element AB of the circumference shown in Figure 5.5. Following application of the load, points A and B translate to A′ and B′ , respectively. The change in length due to tangential displacements v and $v+(\partial v/\partial\theta)d\theta$ is $(\partial v/\partial\theta)d\theta$. The change in length due to the translation in the radial direction $=Rd\theta-(R-w)d\theta=wd\theta$. We choose to neglect the effect of the difference in radial displacements of A and B as being of higher order. Thus, the total change in the length of AB is:

$$\frac{\partial v}{\partial\theta}d\theta - wd\theta$$

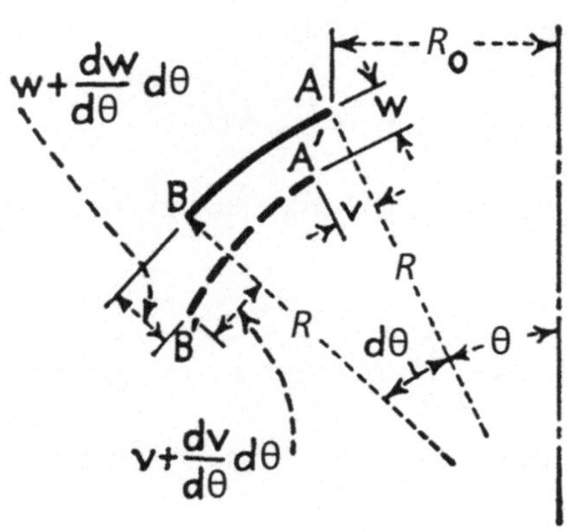

FIGURE 5.5 Shell distortion due to membrane forces [12].

Dividing by the initial arc length, $Rd\theta$, we obtain the circumferential membrane strain ε_θ:

$$\varepsilon_\theta = \frac{1}{R}\left[\frac{\partial v}{\partial \theta} - w\right] \tag{5.35}$$

In the longitudinal direction, in the absence of external longitudinal loads, the strain that is induced results from the Poisson effect. Hence,

$$\varepsilon_x = \frac{\partial u}{\partial x} = \frac{\nu}{R}\left[\frac{\partial v}{\partial \theta} - w\right] \tag{5.36}$$

The expression for the shear strain is identical to that used in plate theory:

$$\gamma_{x\theta} = \frac{\partial u}{\partial s} + \frac{\partial v}{\partial x} = \frac{1}{R}\frac{\partial u}{\partial \theta} + \frac{\partial v}{\partial x} \tag{5.37}$$

where $s = R\theta$.

Substituting (5.35)–(5.37) in (5.33), we obtain the following expression for the membrane strain energy of a shell to first-order accuracy, U_{sm} [5]:

$$U_{sm} = \frac{ERt}{2(1-\nu^2)}\int_0^{2\pi}\int_0^{L_b}\left[\frac{\partial^2 u}{\partial x^2} + \left(\frac{\partial v}{\partial \theta} - w\right)^2 / R^2 + 2\nu\frac{\partial u}{\partial x}\left(\frac{\partial v}{\partial \theta} - w\right) / R + \frac{(1-\nu)}{2}\left(\frac{1}{R}\frac{\partial u}{\partial \theta} + \frac{\partial v}{\partial x}\right)^2\right] dx\, d\theta \tag{5.38}$$

5.3.3 Bending Strain Energy of Cylindrical Shell to First-Order Accuracy

In order to obtain an analogous expression for the bending strain energy of a cylindrical shell, U_{sb}, we need to derive expressions for the curvatures, since the bending strain energy is a function of the changes in the curvatures of a shell element. We therefore need to obtain expressions for the bending and twisting curvature components [12].

In order to estimate the change in curvature due to bending, we consider a shell element mn as shown in Figure 5.6. Before application of a load, the curvature is given by

$$\frac{\partial \theta}{\partial s} = \frac{1}{R}\frac{\partial \theta}{\partial \theta} = \frac{1}{R}$$

Following application of a load, the element translates and rotates to $m_1 n_1$ and the curvature becomes

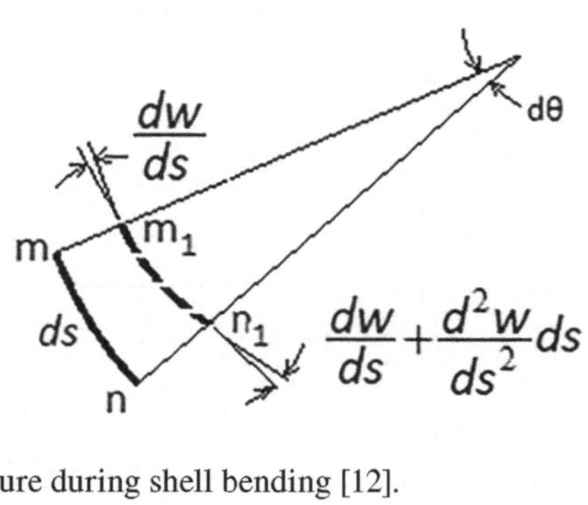

FIGURE 5.6 Change in curvature during shell bending [12].

$$\frac{\partial \theta_1}{\partial s_1} \approx \frac{d\theta + \dfrac{\partial^2 w}{\partial s^2} ds}{(R-w)d\theta}$$

Hence, the change in curvature, κ_θ, is given by:

$$\kappa_\theta = \frac{d\theta + \dfrac{\partial^2 w}{\partial s^2} ds}{(R-w)d\theta} - \frac{1}{R}\frac{d\theta}{\partial\theta} \approx \frac{1}{R^2}\left[w + \frac{\partial^2 w}{\partial \theta^2}\right] \tag{5.39}$$

For zero circumferential membrane strains, we have, from (5.35):

$$\frac{\partial v}{\partial \theta} - w = 0$$

Hence,

$$\kappa_\theta = \frac{1}{R^2}\left(\frac{\partial v}{\partial \theta} + \frac{\partial^2 w}{\partial \theta^2}\right) \tag{5.40}$$

The bending moment producing this change in curvature is

$$M_\theta = -D\kappa_\theta = -\frac{D}{R^2}\left(\frac{\partial v}{\partial \theta} + \frac{\partial^2 w}{\partial \theta^2}\right) \tag{5.41}$$

The bending strain energy per unit area related to pure bending, U_b, is then

$$U_b = \frac{1}{2} M_\theta \kappa_\theta = \frac{D}{2R^4}\left(\frac{\partial v}{\partial \theta} + \frac{\partial^2 w}{\partial \theta^2}\right)^2 \tag{5.42}$$

In addition to (pure) bending strain energy, we need to determine the strain energy due to twisting. Consider the cylindrical shell element OABC shown in Figure 3.24a. During deformation, an element of a generatrix (say OC) rotates through an angle $-\partial w/\partial x$ about the y-axis, as shown in Figure 5.7a. The generatrix CB rotates through

$$-\frac{\partial w}{\partial x} - \frac{\partial^2 w}{\partial x^2} d\theta$$

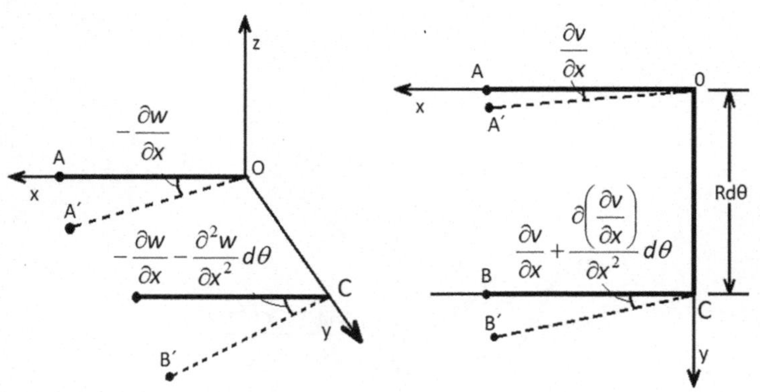

FIGURE 5.7 Rotations of shell element resulting in twisting. (a) Rotation about Oy. (b) Rotation about Oz.

In the plane tangential to the shell surface, the rotation of the element is given by (Figure 5.7b):

$$\frac{\partial v}{\partial x}+\frac{\partial\left(\frac{\partial v}{\partial x}\right)}{\partial \theta}d\theta$$

Because of the angle $d\theta$ between the two generatrices, the latter rotation has a component with respect to the y-axis equal to

$$-\frac{\partial v}{\partial x}d\theta$$

The total angle of twist is:

$$-\kappa_{x\theta}Rd\theta=-\left(\frac{\partial^2 w}{\partial\theta\,\partial x}+\frac{\partial v}{\partial x}\right)d\theta \tag{5.43}$$

The twisting strain energy per unit area, U_t, is:

$$U_t=\frac{1}{2}M_{x\theta}\kappa_{x\theta} \tag{5.44}$$

where

$$M_{x\theta}=-\int_{-t/2}^{+t/2}\tau_{x\theta}z\,dz=\frac{D}{R}(1-\nu)\kappa_{x\theta} \tag{5.45}$$

Substituting (5.45) in (5.44), we obtain

$$U_t=\frac{D}{R^2}(1-\nu)\left(\frac{\partial^2 w}{\partial\theta\,\partial x}+\frac{\partial v}{\partial x}\right)^2 \tag{5.46}$$

Adding expressions (5.42) and (5.46) and integrating over the surface area of the shell, we obtain the following expression for the bending strain energy of a geometrically perfect cylindrical shell [11]:

$$U_{sb}=\frac{D}{2R^4}\iint\left[\left(\frac{\partial v}{\partial\theta}+\frac{\partial^2 w}{\partial\theta^2}\right)^2+2(1-\nu)R^2\left(\frac{\partial^2 w}{\partial\theta\,\partial x}+\frac{\partial v}{\partial x}\right)^2\right]Rd\theta dx \tag{5.47}$$

5.3.4 Membrane and Bending Strain Energy of Shell Plating to Second-Order Accuracy

In the previous section, we obtained expressions for the strain energy of the cylindrical shell to first-order accuracy. However, for reasons that will become clear later, it is necessary to obtain expressions that include second-order terms as well, for each of the strains ε_x, ε_θ [5].

In order to obtain the required expressions, we consider a pair of points (P_1, P_2) lying on the middle surface of the shell along the same meridian a distance δx apart. Using cylindrical coordinates (r, x, θ), the coordinates of these point before and after load application are included in Table 5.1.

The distance between two points (r_1, x_1, θ_1), (r_2, x_2, θ_2) is:

$$\sqrt{(r_1\cos\theta_1-r_2\cos\theta_2)^2+(r_1\sin\theta_1-r_2\sin\theta_2)^2+(x_1-x_2)^2} \tag{5.48}$$

TABLE 5.1
Co-Ordinates of Points (P_1, P_2) Lying on the Shell Middle Surface

Points	Cylindrical Coordinates R	x	θ
P_1	R	x	θ
P_2	R	$x+\delta x$	θ
P_1^1	$R-w$	$x+u$	$\theta+v/R$
P_2^1	$R-w-\left(\frac{\partial w}{\partial x}\right)\delta x$	$x+\delta x+u+\left(\frac{\partial u}{\partial x}\right)\delta x$	$\theta+\frac{1}{R}\left(v+\frac{\partial v}{\partial x}\delta x\right)$

Therefore, the longitudinal strain ε_x is given by:

$$\delta x(1+\varepsilon_x)=\left[\left\{(R-w)\cos(\theta+\frac{v}{R})-(R-w-\frac{\partial w}{\partial x}\delta x)\cos(\theta+\frac{v}{R}+\frac{1}{R}\frac{\partial v}{\partial x}\delta x)\right\}^2\right.$$

$$\left.+\left\{(R-w)\sin\left(\theta+\frac{v}{R}\right)-(R-w-\frac{\partial w}{\partial x}\delta x)\sin\left(\theta+\frac{v}{R}+\frac{\partial v}{\partial x}\delta x\right)\right\}^2+\left(\delta x+\frac{\partial u}{\partial x}\delta x\right)^2\right]^{\frac{1}{2}}$$

$$=\left[(R-w)^2+\left(R-w-\frac{\partial w}{\partial x}\delta x\right)^2-2(R-w)\left(R-w-\frac{\partial w}{\partial x}\delta x\right)\cos\left(\frac{1}{R}\frac{\partial v}{\partial x}\delta x\right)\right.$$

$$\left.+\left(1+2\frac{\partial u}{\partial x}+\frac{\partial u}{\partial x}^2\right)\delta x^2\right]^{\frac{1}{2}}$$

When second-order terms are retained, this becomes

$$1+\varepsilon_x=\sqrt{1+2\frac{\partial u}{\partial x}+\frac{\partial u}{\partial x}^2+\frac{\partial v}{\partial x}^2+\frac{\partial w}{\partial x}^2}$$

Hence,

$$\varepsilon_x=\frac{\partial u}{\partial x}+\frac{1}{2}\left(\frac{\partial v}{\partial x}^2+\frac{\partial w}{\partial x}^2\right) \tag{5.49}$$

This is the expression for the longitudinal membrane strain ε_x of a perfect cylindrical shell that includes both first- and second-order terms.

To derive the expression for the circumferential membrane strain ε_θ, we consider a pair of points (P_1, P_3) lying on the same parallel at an angular distance $\delta\theta$ from each other. Following imposition of the load, P_1 translates to P_1^1 and P_3 to P_3^1. The original and new coordinates of these points are included in Table 5.2.

TABLE 5.2
Co-ordinates of Points (P_1, P_3) Lying on the Shell Middle Surface

	Cylindrical Coordinates		
Points	***R***	***x***	***θ***
P_1	R	x	θ
P_3	R	x	$\theta+\delta\theta$
P_1^1	$R-w$	$x+u$	$\theta+v/R$
P_3^1	$R-w-\left(\frac{\partial w}{\partial\theta}\right)\delta\theta$	$x+u+\left(\frac{\partial u}{\partial\theta}\right)\delta\theta$	$\theta+\delta\theta+\frac{1}{R}\left(v+\frac{\partial v}{\partial\theta}\delta\theta\right)$

Applying (5.48) in the circumferential direction, ε_θ is found from:

$$R\delta\theta(1+\varepsilon_\theta)=\left[\left\{(R-w)\cos(\theta+\frac{v}{R})-(R-w-\frac{\partial w}{\partial\theta}\delta\theta)\cos(\theta+\delta\theta+\frac{v}{R}+\frac{1}{R}\frac{\partial v}{\partial\theta}\delta\theta)\right\}^2\right.$$

$$+\left\{(R-w)\sin\left(\theta+\frac{v}{R}\right)-(R-w-\frac{\partial w}{\partial\theta}\delta\theta)\sin\left(\theta+\delta\theta+\frac{v}{R}+\frac{\partial v}{\partial\theta}\delta\theta\right)\right\}^2$$

$$\left.+\left(\frac{\partial u}{\partial\theta}\delta\theta\right)^2\right]^{\frac{1}{2}}$$

$$=\left[(R-w)^2+\left(R-w-\frac{\partial w}{\partial\theta}\delta\theta\right)^2-2(R-w)\left(R-w-\frac{\partial w}{\partial\theta}\delta\theta\right)\cos\left(\delta\theta+\frac{1}{R}\frac{\partial v}{\partial\theta}\delta\theta\right)+\left(\frac{\partial u}{\partial\theta}\delta\theta\right)^2\right]$$

When first- and second-order terms are retained, this reduces to

$$1+\varepsilon_\theta=\sqrt{1+\frac{2}{R}\left(\frac{\partial v}{\partial\theta}-w\right)+\frac{1}{R^2}\left(w^2+\frac{\partial u}{\partial\theta}^2+\frac{\partial v}{\partial\theta}^2+\frac{\partial w}{\partial\theta}^2-\frac{4}{w}\frac{\partial v}{\partial\theta}\right)}$$

Hence

$$\varepsilon_\theta=\frac{1}{R}\left(\frac{\partial v}{\partial\theta}-w\right)+\frac{1}{2R^2}\left(\frac{\partial u}{\partial\theta}^2+\frac{\partial w}{\partial\theta}^2-2w\frac{\partial v}{\partial\theta}\right)\tag{5.50}$$

The membrane strains ε_x, ε_θ to second-degree accuracy are given by expressions (5.49) and (5.50). In addition to these, we require an expression for the shear strain $\gamma_{x\theta}$. Since the shear strains are zero prior to buckling, the first-order expression is sufficient:

$$\gamma_{x\theta}=\frac{1}{R}\frac{\partial u}{\partial\theta}+\frac{\partial v}{\partial x}\tag{5.51}$$

We shall now consider shell displacements more closely. As mentioned previously, the displacements of the perfect shell (u_p, v_p, w_p) that result from the application of the external load can be expressed as the sum of the displacements that arise prior to buckling (u_1, v_1, w_1) and the buckling-related displacements (u_2, v_2, w_2) so that:

$$u_p=u_1+u_2\tag{5.52a}$$

$$v_p = v_1 + v_2 \tag{5.52b}$$

$$w_p = w_1 + w_2 \tag{5.52c}$$

The pre-buckling displacements (u_1, v_1, w_1) consist of a uniform longitudinal and a uniform radial contraction, that is,

$$u_{1x} = \text{const} \quad u_{1\theta} = 0 \tag{5.53a}$$

$$v_1 = 0 \tag{5.53b}$$

$$w_1 = \text{const} \tag{5.53c}$$

Substituting (5.53) in (5.52) and subsequently in (5.48)–(5.50), these become:

$$\varepsilon_x = \frac{\partial u_1}{\partial x} + \frac{\partial u_2}{\partial x} + \frac{1}{2}\left(\frac{\partial v_2}{\partial x}^2 + \frac{\partial w_2}{\partial x}^2\right) = \varepsilon_{x1} + \varepsilon_{x2} \tag{5.54a}$$

$$\varepsilon_\theta = -\frac{w_1}{R} + \frac{1}{R}\left(\frac{\partial v_2}{\partial \theta} - w_2 + \frac{1}{2R}\left(\frac{\partial u_2}{\partial \theta}^2 + \frac{\partial w_2}{\partial \theta}^2 - 2w_2\frac{\partial v_2}{\partial \theta}\right)\right) = \varepsilon_{\theta 1} + \varepsilon_{\theta 2} \tag{5.54b}$$

$$\gamma_{x\theta} = \frac{1}{R}\frac{\partial u_2}{\partial \theta} + \frac{\partial v_2}{\partial x} = \gamma_{x\theta 2} \tag{5.54c}$$

The strains are now expressed in terms of pre-buckling (ε_{x1}, $\varepsilon_{\theta 1}$) and buckling-related components (ε_{x2}, $\varepsilon_{\theta 2}$, $\gamma_{x\theta 2}$). Substituting (5.54) in (5.47), we obtain:

$$\begin{aligned}
U_{sm} &= \frac{Rt}{2}\frac{E}{(1-\nu^2)}\int_0^{2\pi}\int_0^{L_b}\left\{\varepsilon_{x1}^2 + \varepsilon_{\theta 1}^2 + 2\nu\varepsilon_{x1}\varepsilon_{\theta 1} + \frac{1-\nu}{2}\gamma_{x\theta 1}^2\right\}d\theta dx \\
&+ \frac{Rt}{2}\frac{E}{(1-\nu^2)}\int_0^{2\pi}\int_0^{L_b}\left\{\varepsilon_{x2}^2 + \varepsilon_{\theta 2}^2 + 2\nu\varepsilon_{x2}\varepsilon_{\theta 2} + \frac{1-\nu}{2}\gamma_{\theta x2}^2\right\}d\theta dx \\
&+ \frac{RtE}{(1-\nu^2)}(\varepsilon_{x1} + \nu\varepsilon_{\theta 1})\int_0^{2\pi}\int_0^{L_b}\varepsilon_{x2}d\theta dx + (\varepsilon_{\theta 1} + \nu\varepsilon_{x1})\int_0^{2\pi}\int_0^{L_b}\varepsilon_{\theta 2}d\theta dx
\end{aligned} \tag{5.55}$$

The total potential Π is minimised with respect to the total displacements. However, it is seen that the first integral in (5.55) that includes the pre-buckling terms is a constant, and it can therefore be omitted. The potential Π is therefore in effect minimised with respect to the buckling-related displacements (u_2, v_2, w_2). The third integral can be written as follows:

$$RN_{x1}\int_0^{2\pi}\int_0^{L_b}\varepsilon_{x2}\, d\theta\, dx + RN_{\theta 1}\int_0^{2\pi}\int_0^{L_b}\varepsilon_{\theta 2}\, d\theta\, dx$$

where N_{x1} and $N_{\theta 1}$ are the longitudinal and circumferential forces per unit length in the shell prior to buckling given by:

$$N_{x1} = -\frac{PR}{2} \tag{5.56a}$$

$$N_{\theta 1} = -pR\left[\frac{tL_f}{A_f + tL_f}\right] \tag{5.56b}$$

The negative sign indicates compression. The membrane strain energy then becomes:

$$\begin{aligned} U_{sm} = {} & \frac{Rt}{2}\frac{E}{(1-\nu^2)}\int_0^{2\pi}\int_0^{L_b}\left\{\left(\frac{\partial u_2}{\partial x}\right)^2 + \frac{1}{R^2}\left(\frac{\partial v_2}{\partial \theta} - w_2\right)^2\right. \\ & \left. + \frac{2\nu}{R}\frac{\partial u_2}{\partial x}\left(\frac{\partial v_2}{\partial \theta} - w_2\right) + \frac{(1-\nu)}{2}\left(\frac{\partial v_2}{\partial x} + \frac{1}{R}\frac{\partial u_2}{\partial \theta}\right)^2\right\} d\theta dx \\ & + \frac{R}{2}N_{x1}\int_0^{2\pi}\int_0^{L_b}\left(\left(\frac{\partial v_2}{\partial x}\right)^2 + \left(\frac{\partial w_2}{\partial x}\right)^2\right) d\theta\, dx \\ & + \frac{N_{\theta 1}}{2R}\int_0^{2\pi}\int_0^{L_b}\left(\left(\frac{\partial u_2}{\partial \theta}\right)^2 + \left(\frac{\partial w_2}{\partial \theta}\right)^2 - 2\frac{\partial v_2}{\partial \theta}w_2\right) d\theta dx \end{aligned} \tag{5.57}$$

In (5.57), the first-order terms in the last two integrals have been omitted because they vanish when θ is integrated from 0 to 2π.

The bending strain energy of the shell, U_{sb}, can be written as follows [10]:

$$\begin{aligned} U_{sb} = {} & \frac{Et^3}{24R(1-\nu^2)}\int_0^{2\pi}\int_0^{L_b}\left\{R^2\left(\frac{\partial^2 w_p}{\partial x^2}\right)^2 + \frac{1}{R^2}\left(\frac{\partial^2 w_p}{\partial \theta^2} + w_p\right)^2\right. \\ & \left. + 2\nu\frac{\partial^2 w_p}{\partial x^2}\left(\frac{\partial^2 w_p}{\partial \theta^2} + w_p\right) + 2(1-\nu)\left(\frac{\partial^2 w_p}{\partial x \partial \theta} + \frac{1}{2}\frac{\partial v_p}{\partial x} - \frac{1}{2R}\frac{\partial u_p}{\partial \theta}\right)^2\right\} d\theta dx \end{aligned} \tag{5.58}$$

By substituting (5.52) and (5.53) in (5.58), it has been shown [5] that the total displacements (u_p, v_p, w_p) can be replaced by the buckling-related displacements (u_2, v_2, w_2):

$$\begin{aligned} U_{sb} = {} & \frac{Et^3}{24R(1-\nu^2)}\int_0^{2\pi}\int_0^{L_b}\left\{R^2\left(\frac{\partial^2 w_2}{\partial x^2}\right)^2 + \frac{1}{R^2}\left(\frac{\partial^2 w_2}{\partial \theta^2} + w_2\right)^2\right. \\ & \left. + 2\nu\frac{\partial^2 w_2}{\partial x^2}\left(\frac{\partial^2 w_2}{\partial \theta^2} + w_2\right) + 2(1-\nu)\left(\frac{\partial^2 w_2}{\partial x \partial \theta} + \frac{1}{2}\frac{\partial v_2}{\partial x} - \frac{1}{2R}\frac{\partial u_2}{\partial \theta}\right)^2\right\} d\theta dx \end{aligned} \tag{5.59}$$

We have thus obtained expressions for the membrane strain energy and the bending strain energy of the shell using second-order expressions for the strains (5.49), (5.50), which have been separated into pre-buckling and buckling-related strains. In the next section, we shall derive corresponding expressions for the ring frames.

5.3.5 Membrane and Bending Strain Energy of a Ring Frame

Each ring frame is welded to the shell plating so that it may be assumed that a clamped condition prevails. The shell plating thus acts as a wide flange of the frame and both bend in unison. The strain distribution can then be assumed to be linear along the stiffener depth, as shown in Figure 5.8. The distance of the stiffener centroid from the middle surface of the shell plating (eccentricity) is denoted as e, and the bending strains at the ring frame centroid and the shell plating middle surface as ε_{fb} and ε_{sb}, respectively. Then we have

$$\varepsilon_{sb} - \varepsilon_{fb} = \frac{e}{\rho} \tag{5.60}$$

where $1/\rho$ is the change in curvature of the section. The membrane and bending strain energies of the ring frame are then:

$$U_{fm} = \frac{EA_f}{2} \varepsilon_{fb}^{\ 2} \tag{5.61a}$$

$$U_{fb} = \frac{EI_f}{2} \left(\frac{1}{\rho} \right)^2 \tag{5.61b}$$

where A_f, I_f are the area and the moment of inertia of the ring frame about its centre of gravity. We can express U_{fm} in terms of the strain of the shell plating middle surface, ε_{sb}, by substituting for ε_{fb} from (5.60):

$$U_{fm} = \frac{EA}{2} \left(\varepsilon_{sb} - \frac{e}{\rho} \right)^2 \tag{5.62}$$

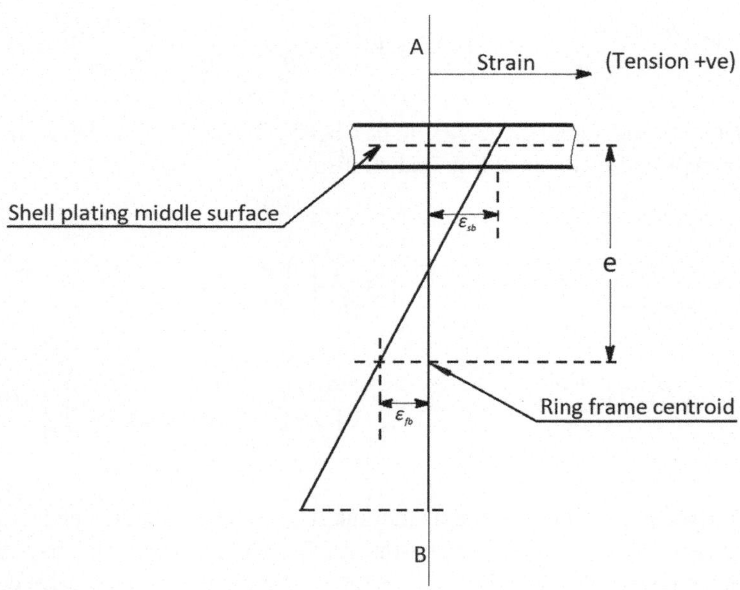

FIGURE 5.8 Strain distribution in ring frame and attached shell plating.

The strain and change in curvature in way of the middle surface of the shell plating in the circumferential direction are:

$$\varepsilon_{sb} = \frac{1}{R}\left(\frac{\partial v}{\partial \theta} - w\right) + \frac{1}{2R^2}\left(\left(\frac{\partial u}{\partial \theta}\right)^2 + \left(\frac{\partial w}{\partial \theta}\right)^2 - 2w\frac{\partial v}{\partial \theta}\right) \quad (5.63a)$$

$$\frac{1}{\rho} = \frac{1}{R^2}\left(\frac{\partial^2 w}{\partial \theta^2} + w\right) \quad (5.63b)$$

Using (5.61b), (5.62), (5.63a), and (5.63b), the membrane and bending strain energies of the ring frame become:

$$U_{fm} = \frac{EA_f}{2R}\int_0^{2\pi}\left\{\left(\frac{\partial^2 w}{\partial \theta^2} + w\right)\left(\frac{e}{R}\right) - \left(\frac{\partial v}{\partial \theta} - w\right) - \frac{1}{2R}\left(\left(\frac{\partial u}{\partial \theta}\right)^2 + \left(\frac{\partial w}{\partial \theta}\right)^2 - 2w\frac{\partial v}{\partial \theta}\right)\right\} d\theta \quad (5.64a)$$

$$U_{fb} = \frac{EI_f}{2R^3}\int_0^{2\pi}\left(\frac{\partial^2 w}{\partial \theta^2} + w\right)^2 d\theta \quad (5.64b)$$

Substituting (5.52), (5.53) for the displacements in (5.64a) and (5.64b) and neglecting constant and zero terms, these become:

$$\begin{aligned} U_{fm} &= \frac{EA_f}{2R}\int_0^{2\pi}\left\{\left(\frac{\partial^2 w_2}{\partial \theta^2} + w_2\right)\left(\frac{e}{R}\right) - \left(\frac{\partial v_2}{\partial \theta} - w_2\right)\right\}^2 d\theta \\ &+ \frac{N_{f\theta 1}}{2R}\int_0^{2\pi}\left(\left(\frac{\partial u_2}{\partial \theta}\right)^2 + \left(\frac{\partial w_2}{\partial \theta}\right)^2 - 2\frac{\partial v_2}{\partial \theta}w_2\right) d\theta \end{aligned} \quad (5.65a)$$

$$U_{fb} = \frac{EI_f}{2R^3}\int_0^{2\pi}\left(\frac{\partial^2 w_2}{\partial \theta^2} + w_2\right)^2 d\theta \quad (5.65b)$$

where $-N_{f\theta 1}$ is the circumferential force in the ring frame prior to buckling. The assumed prebuckling displacement is given by (5.53), and hence, $N_{f\theta 1}$ is given by:

$$N_{f\theta 1} = -pRL_f\left[\frac{A_f}{A_f + tL_f}\right] \quad (5.66)$$

The membrane and bending strain energy of a single ring frame are thus given by (5.65a) and (5.65b), respectively.

5.3.6 Potential of the Work Done

The external work done is equal to the sum of the products of the pressure acting in the radial direction, p, and the resulting deflections and of the pressure P acting in the longitudinal direction and the corresponding longitudinal deflections.

Following deformation, a point on the middle surface of the cylindrical shell having coordinates (r, x, θ) translates to point (r^1, x^1, θ^1). The following relationships hold:

$$r^1 = R - w \tag{5.67a}$$

$$x^1 = x + u \tag{5.67b}$$

$$\theta^1 = \theta + \frac{v}{R} \tag{5.67c}$$

Following deformation, the volume of the material becomes:

$$V^1 = \frac{1}{2}\iint \left(r^1\right)^2 d\theta^1\, dx^1 \tag{5.68}$$

where the integral is taken over the entire deformed surface. Substituting (5.67a–c) in (5.68), the following expression is obtained:

$$V^1 = \frac{1}{2}\int_0^{2\pi}\int_0^{L_b} (R-w)^2\left(1+\frac{\partial u}{\partial x}\right)\left(1+\frac{1}{R}\frac{\partial v}{\partial \theta}\right) d\theta dx \tag{5.69}$$

The change in volume due to displacements u_p, v_p, w_p is given by:

$$V^1 - V = \frac{1}{2}\int_0^{2\pi}\int_0^{L_b}\left\{-2Rw_p + R\frac{\partial v_p}{\partial \theta} + R^2\frac{\partial u_p}{\partial x} - 2w_p\frac{\partial v}{\partial \theta} - 2Rw_p\frac{\partial u_p}{\partial x} + R\frac{\partial v_p}{\partial \theta}\frac{\partial u_p}{\partial x} + w_p{}^2\right\} d\theta dx \tag{5.70}$$

The potential of the pressure in the radial direction (external work done) is the product of the resulting expression for the change (increase) in volume, (5.70), and the radial pressure, p:

$$W_p = -\frac{pR}{2}\int_0^{2\pi}\int_0^{L_b}\left\{2w_p - \frac{\partial v_p}{\partial \theta} - R\frac{\partial u_p}{\partial x} + \frac{2w_p}{R}\frac{\partial v_p}{\partial \theta} + 2w_p\frac{\partial u_p}{\partial x} - \frac{\partial v_p}{\partial \theta}\frac{\partial u_p}{\partial x} - \frac{w_p{}^2}{R}\right\} d\theta dx \tag{5.71}$$

Substituting in (5.71) for u_p, v_p, w_p from (5.52) and (5.53) results in the following expression for the potential of the work done by the radial pressure, after deletion of constant and zero terms:

$$W_p = -\frac{pR}{2}\int_0^{2\pi}\int_0^{L_b}\left\{2w_2\frac{\partial v_2}{\partial x} + \frac{2w_2}{R}\frac{\partial v_2}{\partial \theta} - \frac{\partial v_2}{\partial \theta}\frac{\partial u_2}{\partial x} - \frac{w_2{}^2}{R}\right\} d\theta dx \tag{5.72}$$

The potential of the pressure P acting in the longitudinal direction is equal to the work done during displacement of the boundaries and is given by:

$$W_P = \left(\frac{PR}{2}\right)\int_0^{2\pi}\left\{u_{p(x=0)} - u_{p(x=L_b)}\right\} d\theta \tag{5.73}$$

Substituting for u_p from (5.52a), the above expression becomes:

$$W_P = \left(\frac{PR}{2}\right)\int_0^{2\pi}\left\{u_{1(x=0)} - u_{1(x=L_b)}\right\}d\theta + \left(\frac{PR}{2}\right)\int_0^{2\pi}\left\{u_{2(x=0)} - u_{2(x=L_b)}\right\} d\theta$$

The first integral is a constant, whereas the second is equal to zero. Therefore,

$$W_P = 0 \tag{5.74}$$

The total potential of the work done, W, is thus given by (5.71).

5.3.7 Determination of Ritz Constants

The total potential, Π, is, in accordance with (5.29), the sum of expressions (5.57), (5.59), (5.65a), (5.65b), and (5.72).

The deflection functions (5.26a–c) are substituted into the strain energy expressions. The total potential then becomes a function of shell and frame geometry, the applied pressure, and the deflection parameters. Thus, for a given geometry, the total potential is a function of the deflection parameters alone (A, B and C).

The principle of minimum potential energy is invoked, according to which for equilibrium, the elastic energy must have a stationary value. For this to be the case, any arbitrary variation from this value must be equal to zero. Thus, for the three deflection parameters selected, the following criterion has to be satisfied:

$$\delta\Pi = 0 = \frac{\partial\Pi}{\partial A}\delta A + \frac{\partial\Pi}{\partial B}\delta B + \frac{\partial\Pi}{\partial C}\delta C \tag{5.75}$$

Since δA, δB, etc., are arbitrary and may therefore be non-zero, for equilibrium, the following conditions hold:

$$\frac{\partial\Pi}{\partial A} = \frac{\partial\Pi}{\partial B} = \frac{\partial\Pi}{\partial C} = 0 \tag{5.76}$$

We thus have a system of three linear homogeneous algebraic equations, the solution of which involves calculating the coefficients A, B, and C. For a non-trivial solution, the 3×3 determinant of the coefficients has to be equal to zero, that is:

$$\begin{vmatrix} a_{11} - n^2(A_1 + 2RA_2)\phi_1 & a_{12} - n\lambda\phi_1/2 & a_{13} + \lambda\phi_1 \\ a_{12} - n\lambda\phi_1/2 & a_{22} - \lambda^2\phi_2/2 & a_{23} + n(A_1 + 2RA_2 - 1)\phi_1 \\ a_{13} + \lambda\phi_1 & a_{23} + n(A_1 + 2RA_2 - 1)\phi_1 & a_{33} - \phi_1(n^2A_1 + 2Rn^2A_2 - 1) - \lambda^2\phi_2/2 \end{vmatrix} = 0 \tag{5.77}$$

Where:

$$a_{11} = \lambda^2 + n^2(1+k)(1-\nu)/2$$

$$a_{12} = -n\lambda\{(1+\nu)/2 - k(1-\nu)/2\}$$

$$a_{13} = \lambda\nu - k(1-\nu)n^2\lambda$$

$$a_{22} = n^2 + \lambda^2(1+k)(1-\nu)/2 + R_B$$

$$a_{23} = -n\{1 + \lambda^2 k(1-\nu)\} + R_{BC}$$

$$a_{33} = 1 + k\left\{\lambda^4 + (n^2-1)^2 + 2\lambda^2(n^2-\nu)\right\} + R_C$$

$$A_1 = tL_f / A_f + tL_f$$

$$A_2 = A_f L_f / L_b (A_f + tL_f)$$

$$k = t^2/12R^2$$

N=number of ring frames
m=number of half-waves in longitudinal direction
n=number of waves in circumferential direction

$$R_A = \sum_{r=1}^{N} \sin^2(m\pi r L_f / L_b)$$

$$= (N+1)/2 \quad \text{when} \quad m/(N+1) \neq 1$$

$$= 0 \quad \text{when} \quad m(N+1) = 1$$

$$R_B = 2(1-\nu^2)A_f R_A n^2 / tL_b$$

$$R_C = 2(1-\nu^2)I_f R_A (n^2-1)^2 / tR^2 L_b + \left\{2(1-\nu^2)A_f R_A / tL_b\right\}\left\{1-(e/R)(n^2-1)\right\}^2$$

$$R_{BC} = \left\{2(1-\nu^2)A_f R_A n / tL_b\right\}\left\{1-(e/R)(n^2-1)\right\}$$

$$\lambda = m\pi R / L_b$$

$$\phi_1 = pR\frac{(1-\nu^2)}{Et}$$

$$\phi_2 = PR\frac{(1-\nu^2)}{Et}$$

In the case of uniform pressure, $\phi_1 = \phi_2$. The expansion of the determinant (5.77) gives a cubic equation in ϕ_1. It was found that a good approximation to the value of ϕ_1 is obtained by neglecting the quadratic and cubic terms. The linear expression obtained for ϕ_1 must be minimised with respect to m and n in order to obtain the buckling pressure. The value $m = 1$ represents a single half-sine wave in the longitudinal direction and thus corresponds to the general instability mode shown in Figure 5.2. The value $m = N+1$ corresponds to interframe buckling, shown in Figure 5.9.

The minimum value of the elastic buckling pressure arises for either $m = 1$ or $m = N+1$. The transition from one buckling mode to the other is affected by frame rigidity. In the case of weak frames, the collapse pressure decreases with increases in overall length, L_b, regardless of the number of frames present. When the frames are very strong, collapse is independent of the number of frames and occurs as illustrated in Figure 5.9.

As seen in earlier sections, an important parameter for general instability is L_b/R. Of the two variables, L_b is the one that can be modified for a particular pressure hull. This can be done by

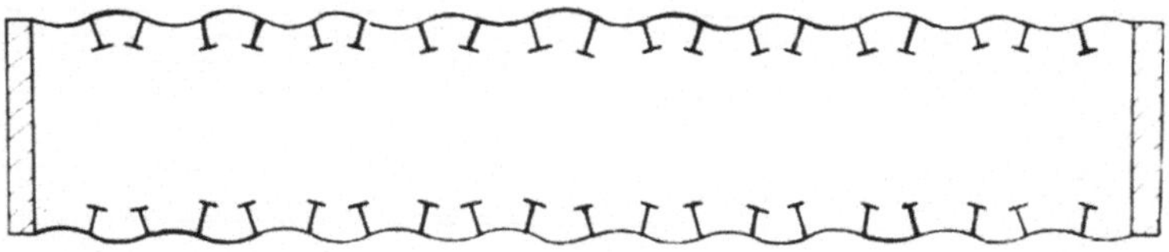

FIGURE 5.9 Local instability (interframe buckling).

introducing additional transverse bulkheads, but this solution gives rise to other problems such as accessibility and system layout. One other way of reducing the effective L_b is by introducing heavy frames of sufficient stiffness and strength at intermediate positions. This enhances overall collapse strength. Indeed, the number of intermediate frames needed, their scantlings as well as those of the ordinary ring frames can be formulated as a problem of classical structural optimisation theory.[2]

Following publication of the method, graphical solutions were developed for ease of use. These were also used to compare results obtained using the Bryant and Kendrick methods. A comparison of results obtained using these two methods is included in the next section.

5.4 GRAPHICAL PRESENTATION OF THE BRYANT AND KENDRICK SOLUTIONS

Tests carried out at DTMB have shown that experimental failure pressures are generally slightly lower than the Bryant solution and slightly higher than the Kendrick solution. In the case of cylinders with light ring frames, the differences are insignificant, but in the case of heavy frames, such as those used in submarine pressure hulls, the differences are larger. It was found that for such cases, the Bryant solution gave failure pressures that are up to 35% higher than the Kendrick solution.

In order therefore to make use of the Kendrick solution and to avoid having to carry out lengthy calculations each time, a set of graphs was developed at DTMB by Reynolds and Ball [9,1]. These graphs relate overall collapse strength to variations in:

- Frame size
- Frame spacing
- Shell radius and thickness
- Compartment length

These are the variables that influence the overall collapse strength of ring-stiffened cylinders. The graphs have been obtained for externally stiffened cylinders with a modulus of elasticity of (30)(10^6) psi (208 GPa). Since the use of internal frames results in slightly higher general instability pressures (Section 6.3.2), these results provide a degree of conservatism.

In order to prepare the graphs, use was made of Bryant's mixed rigidities approach, and it was shown by Reynolds that the Kendrick solution can also be broken down into two terms, as follows:

$$p_{cr} = p_s\left(\frac{L_b}{R},\frac{t}{R},n\right) + p_f\left(\frac{I_e}{L_f R^3},n\right)$$

where

p_{cr} = elastic general instability pressure
p_s = shell pressure factor (to be determined from Figure 5.10)
p_f = frame pressure factor (to be determined from Figure 5.11)
I_{sfe} = moment of inertia of frame and effective length of shell about the centroid of the combined section, given by:

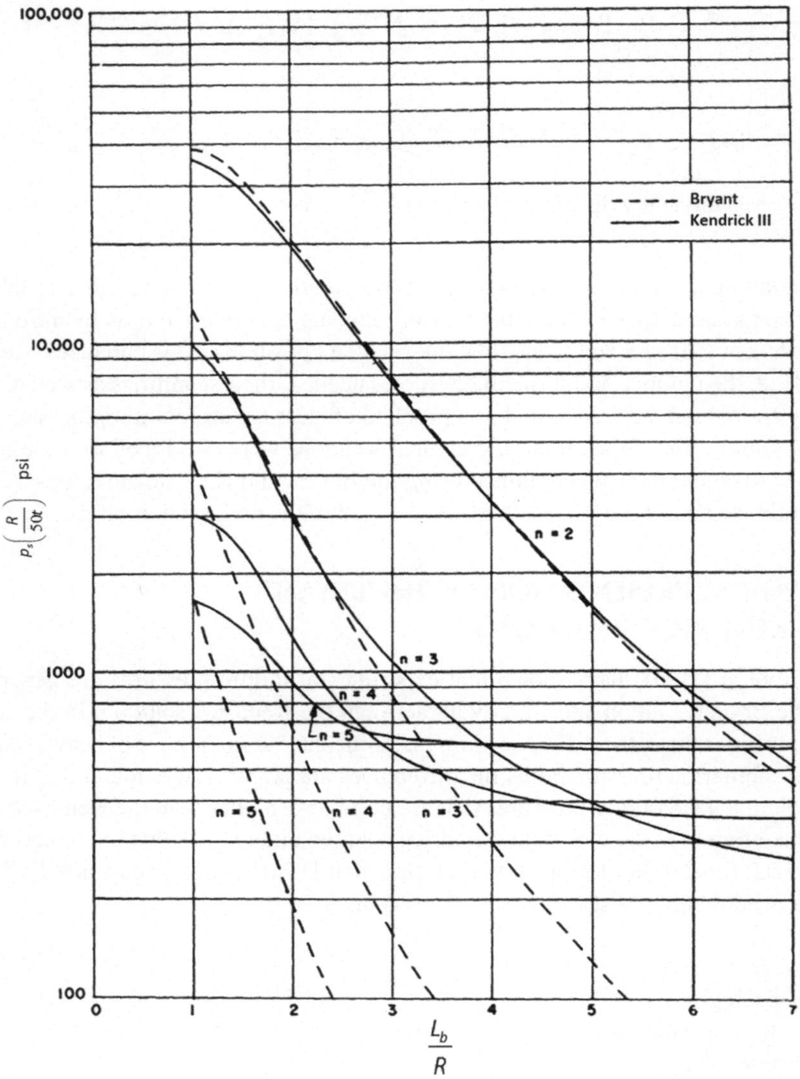

FIGURE 5.10 Shell pressure factor p_s [9].

$$I_{sfe} = \frac{A_f e^2}{1 + \dfrac{A_f}{L_e t}} + I_f + \frac{L_e t^3}{12}$$

I_f = moment of inertia of the frame about its own centroid
e = eccentricity of frame, measured from the midsurface plane of the shell to the frame centroid
L_e = effective length of shell where:

$$L_e = 1.57\sqrt{Rt} \quad \text{provided} \quad L_f > 2\sqrt{Rt}$$

In Figure 5.10, the shell pressure factor p_s corresponding to the modes of collapse n = 2, 3, 4, and 5 is plotted as a function of the relative bulkhead spacing L_b/R. The frame pressure factor p_f is plotted as a function of the relative frame stiffness $I_{sfe}/L_f R^3$. The sum of the two factors is determined for

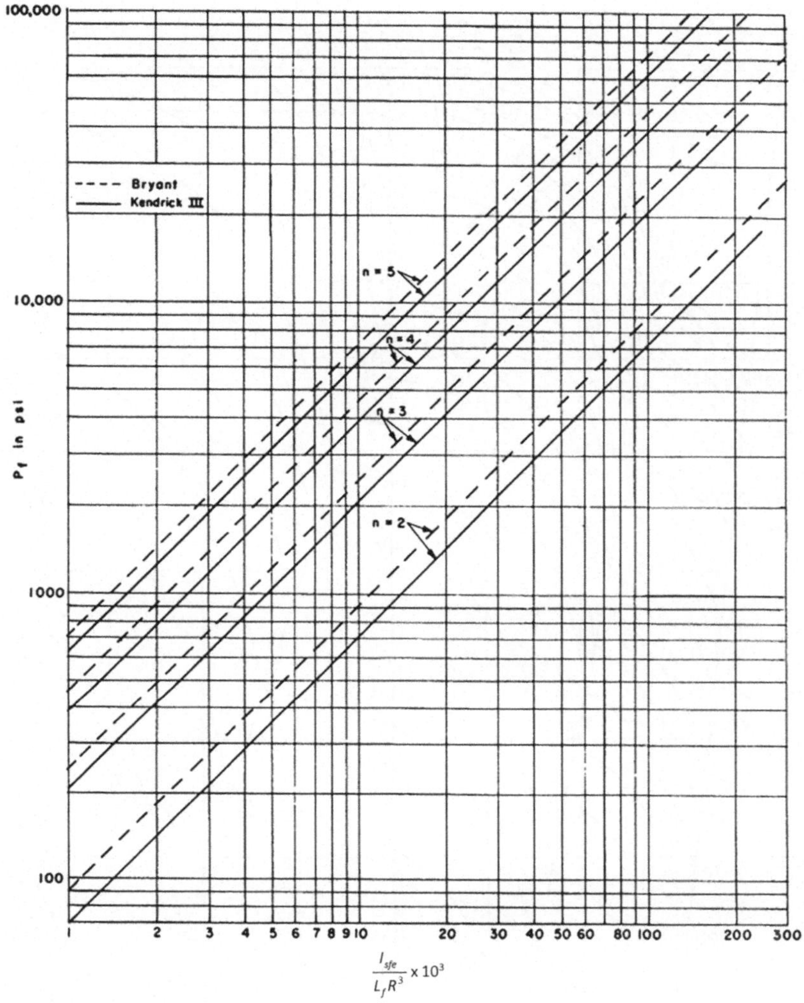

FIGURE 5.11 Frame pressure factor p_f [9].

the modes in question (n=2, 3, 4 and 5), and the lowest value is the critical elastic buckling pressure for general instability.

The data included in Figures 5.10 and 5.11 are combined in Figure 5.12 that has coordinates

$$\left[1-\frac{1}{3}(100\frac{t}{R}-1)\right]\frac{L_b}{R} \quad \text{and} \quad \frac{I_{sfe}}{L_f R^3}$$

In this figure, plots of constant values of p_{cr} appear as families of curves intersecting along straight lines that divide the graph into regions corresponding to n=2, 3, and 4. The buckling pressure is then determined directly once the coordinates are evaluated, this being an approximate value. Following this, Figures 5.10 and 5.11 should be used for a more precise result.

Because of the manner in which the graphs have been constructed, and in particular because of the split into two parts corresponding to shell and frame strength, some differences occur when comparing results with the original solutions of the problem. The average difference was found to be ±2.8% with a maximum difference of ±10%.

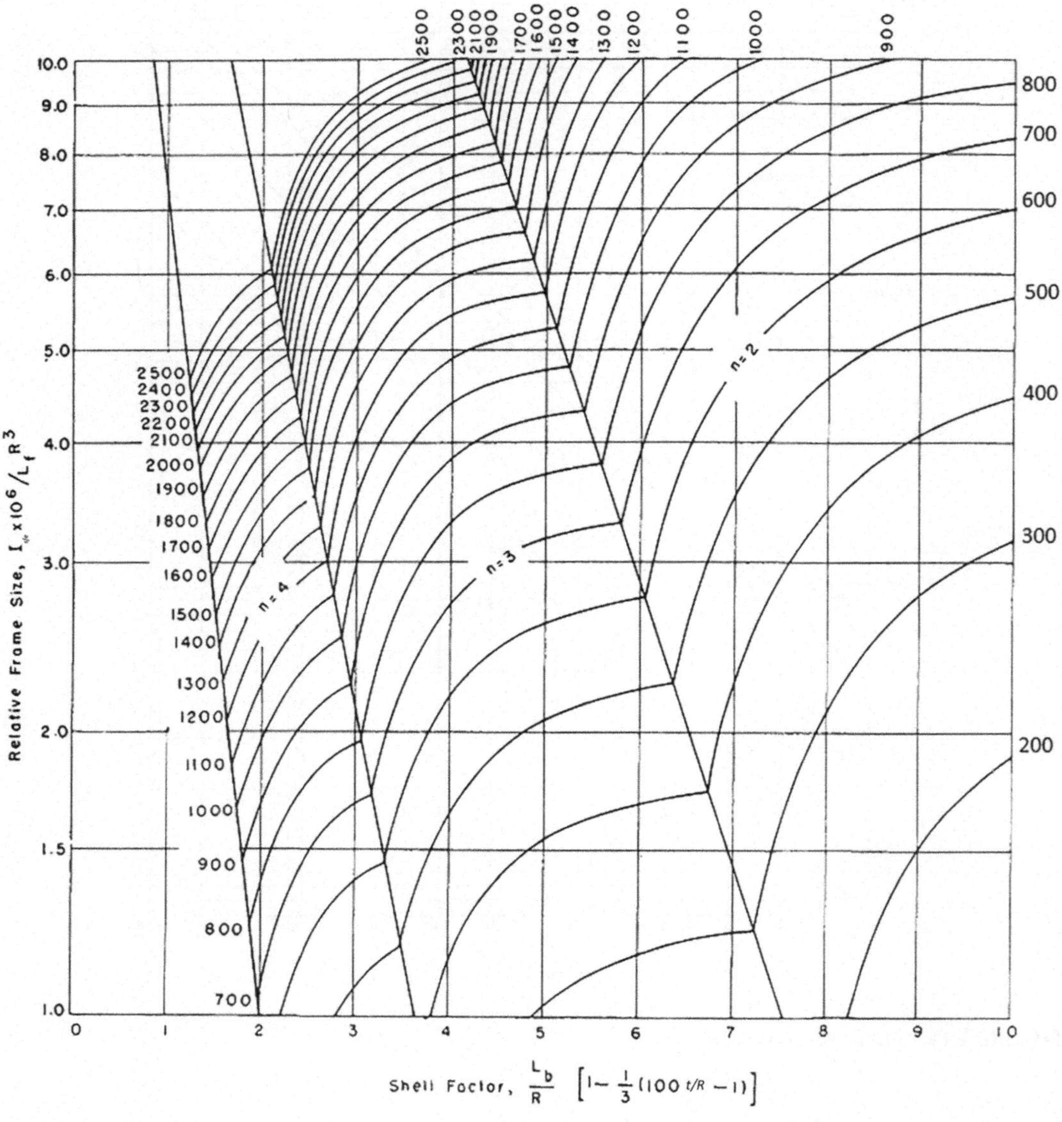

FIGURE 5.12 Curves of constant pressure [9].

Example 5.2

A submarine pressure hull is externally stiffened with 13 ring frames between transverse boundaries. Scantlings are given below:

$R = 96$ in $\quad A_f = 9.63$ in^2
$t = 0.768$ in $\quad I_f = 80.48$ in^4
$L_f = 30$ in $\quad e = 4.49$ in
$L_b = 384$ in

a. Determine the pressure to cause elastic general instability using the Reynolds/Ball graphs
b. Compare the results obtained using the Bryant method, the Kendrick method, and the graphs.

Solution

a. We first determine the effective length: $L_e=13.48$ in

Substituting, $I_{sfe}/L_f R^3=(6.85)(10)^{-6}$

and $\left[1-\frac{1}{3}(100t/R-1)\right]\frac{L_b}{R}=4.26$

From Figure 5.12, the buckling pressure $p_{cr}=1{,}500$ psi (10.42 MPa) for $n=3$.

To use Figure 5.10, we have $L_b/R=4$ giving $p_s(R/1{,}00t)=150$ psi. Hence $p_s=(150)(0.80)=120$ psi. From Figure 5.11, for $n=3$ $p_f=1{,}370$ psi. Hence, $p_{cr}=150+1{,}370=1{,}490$ psi (compared with 1,500 psi found directly from Figure 5.12).

b. The results obtained using the different approaches are included below.

Comparison of results for different methods used to determine the elastic general instability pressure

	Bryant	Kendrick	Graphical Method
p_{cr} (psi)	2,045	1,409	1,490
n	2	3	3

5.5 RING-STIFFENED CYLINDRICAL SHELLS WITH INITIAL DISTORTIONS

The general instability of ring-stiffened cylindrical shells that have initial distortions (out-of-circularity) was studied by Kendrick [6], who extended the analysis carried out for perfect shells, discussed in Section 5.3.

In real structures, initial imperfections, which include initial distortions and locked-in stresses, are always present to some degree, and behaviour is influenced by their presence. It is therefore important to allow for the presence of imperfections in general and for initial distortions in particular. In the case of ring-stiffened cylinders subjected to external radial pressure, the magnitude of the initial out-of-circularities increases as the load is increased. As the buckling pressure is approached, high local bending stresses are induced. These may result in collapse if yielding has already occurred in the shell (that is, the limit of proportionality has been exceeded). It is therefore important to be able to predict the bending stresses that arise in cylindrical shells that have initial out-of-circularity.

The bending stresses have maximum values when the out-of-circularity has the same form as the buckled shape associated with the lowest elastic general instability. For this reason, the out-of-circularity is given by the following expression:

$$w_0(x,\theta)=C_0\cos(n\theta)\sin\left(\frac{m\pi x}{L_b}\right) \tag{5.78}$$

According to the above expression, the initial distortion in the radial direction, w_0, varies in both the circumferential and the longitudinal direction (n sine and m half-sine waves, respectively), C_0 being the amplitude.

The presence of out-of-circularity results in changes to the expressions for the in-plane strains ε_x, ε_θ, and it is therefore necessary to derive these ab initio.

5.5.1 NORMAL STRAINS IN CYLINDRICAL SHELL ALLOWING FOR INITIAL OUT-OF-CIRCULARITY

In order to obtain an expression for the bending strain energy of a cylindrical shell with initial distortions, we need to obtain the corresponding expressions for the normal membrane strains (ε_x, ε_θ). These expressions have to be determined to second-order accuracy. In the longitudinal direction, the initial distortions vary in a sinusoidal manner, as seen from (5.78).

We need to define the displacements u, v, and w_i of the shell middle surface that result from the application of the load. A point lying on the middle surface having cylindrical coordinates (R, x, θ), becomes, following imposition of the load $(R-w_i, x+u, \theta+v/R)$. In the presence of initial distortions w_0 in the radial direction, the initial radius is $R-w_0$, and following imposition of the load this becomes $R-w_0-w_i$. In order to obtain an expression for the longitudinal strain ε_x, we consider the arrangement shown in Figure 5.13.

In Figure 5.13a, we select points P_1 and P_2 that lie on the middle surface of the shell, separated by δx. This figure shows the translations in the radial direction, Figure 5.13b shows the translations in the longitudinal direction, and Figure 5.13c shows the rotation of these points in the circumferential direction. It is seen from Figure 5.13c that the two points are on the same generatrix. Following application of the load, the points translate to P_1' and P_2', respectively, as shown in these sketches.

The distance between two points (r_1, x_1, θ_1), (r_2, x_2, θ_2) is:

$$\sqrt{r_1^2 + r_2^2 - 2r_1r_2\cos(\theta_1-\theta_2) + (x_1-x_2)^2} \tag{5.79}$$

Using this relationship, the following expression is obtained for the strain component ε_x:

$$1+\varepsilon_x = \frac{\sqrt{\left(\frac{\partial w_0}{\partial x}+\frac{\partial w_i}{\partial x}\right)^2 + [1-(w_0+w_i)/R]^2\left(\frac{\partial v}{\partial x}\right)^2 + \left(1+\frac{\partial u}{\partial x}\right)^2}}{\sqrt{1+\frac{\partial w_0}{\partial x}}}$$

When only terms up to second order are retained, this expression becomes:

$$\varepsilon_x = \frac{\partial u}{\partial x} + \frac{1}{2}\left(\frac{\partial w_i}{\partial x}^2 + \frac{\partial v}{\partial x}^2\right) + \frac{\partial w_0}{\partial x}\frac{\partial w_i}{\partial x} \tag{5.80}$$

We shall now determine a corresponding expression for ε_θ. As previously, we select a pair of points P_1, P_3 and specify their positions prior to and following application of the load. Figure 5.14 shows these positions in a transverse section.

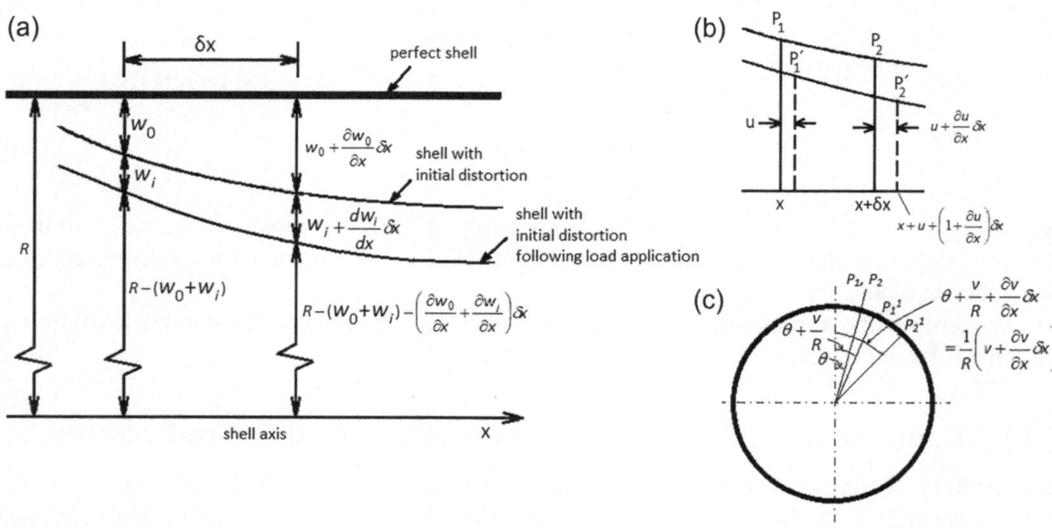

FIGURE 5.13 Intact shell, shell with initial distortions and shell following load action. (a) Radial translations. (b) Longitudinal translations. (c) Rotations in circumferential direction.

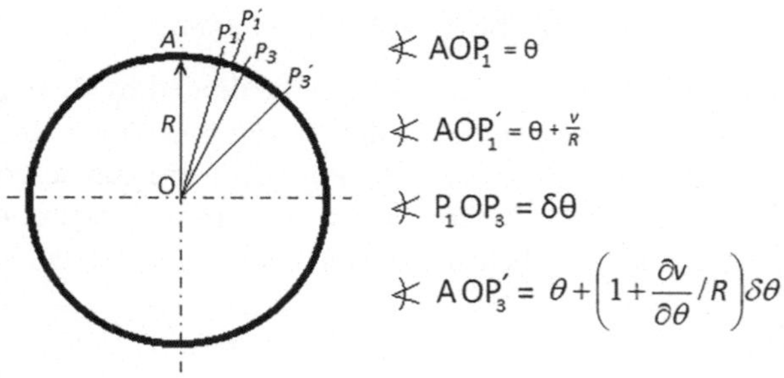

FIGURE 5.14 Rotation of points P_1, P_3 in the circumferential direction.

As previously, P_1 is a point lying in the undeformed middle surface. Point P_3 is also a point in the middle surface at an angular distance $\delta\theta$ from P_1. P_1^1 and P_3^1 are the corresponding points in the deformed surface. The coordinates of these points are included in Table 5.3.

Using (5.79), the exact expression for ε_θ is obtained:

$$1+\varepsilon_\theta = \frac{\sqrt{\left(\frac{\partial w_0}{\partial \theta}+\frac{\partial w_i}{\partial \theta}\right)^2 + [R-(w_0+w_i)]^2\left(1+\frac{1}{R}\frac{\partial v}{\partial x}\right)^2+\left(\frac{\partial u}{\partial \theta}\right)^2}}{\sqrt{\left(\frac{\partial w_0}{\partial \theta}\right)^2+(R-w_0)^2}}$$

Retaining terms up to second order simplifies to:

$$\varepsilon_\theta = \frac{1}{R}\left(\frac{\partial v}{\partial \theta}-w_i\right)+\frac{1}{2R^2}\left(\left(\frac{\partial w_i}{\partial \theta}\right)^2+\left(\frac{\partial u}{\partial \theta}\right)^2-2w_i\frac{\partial v}{\partial \theta}\right)+\frac{1}{R^2}\left(\frac{\partial w_0}{\partial \theta}\frac{\partial w_i}{\partial \theta}-w_0w_i\right) \tag{5.81}$$

Expressions (5.80) and (5.81) are the kinematic relations for an unstiffened cylindrical shell with initial distortions w_0 that is subjected to external radial pressure.

As in the analysis of a perfect ring-stiffened cylindrical shell discussed in Section 5.3, it is necessary to derive expressions for the strain energy of the cylindrical shell and ring frames and the external work done.

TABLE 5.3
Coordinates of Points (P_1, P_3) Resulting from Rotation in Circumferential Direction

	Cylindrical Coordinates		
Points	***R***	***x***	***θ***
P_1	$R-w_0$	x	θ
P_3	$R-w_0-\frac{\partial w_0}{\partial \theta}\delta\theta$	x	$\theta+\delta\theta$
P_1^1	$R-(w_0+w_i)$	$x+u$	$\theta+v/R$
P_3^1	$R-(w_0+w_i)-\left(\frac{\partial w_0}{\partial \theta}+\frac{\partial w_i}{\partial \theta}\right)\delta\theta$	$\theta+\left(1+\frac{1}{R}\frac{\partial v}{\partial \theta}\right)\delta\theta$	$\theta+\left(1+\frac{1}{R}\frac{\partial v}{\partial \theta}\right)\delta\theta$

5.5.2 Strain Energy of Cylindrical Shell and Ring Frames

The analysis of the structure with initial distortions is facilitated by distinguishing between the response of the perfect structure (zero out-of-circularity) with that of the structure with initial distortions. We may thus consider the deflections of the imperfect structure as consisting of two parts: those of the perfect structure and those due to the presence of initial distortions. We can write the deformations u, w_1 of the cylindrical shell that result solely from the application of the load as follows:

$$u_i = u_p + u_3 \tag{5.82a}$$

$$w_i = w_p + w_3 \tag{5.82b}$$

Where

u_i, w_i=deformations due to load action in the presence of initial distortions
u_p, w_p=deformations arising in the absence of initial distortions
u_3, w_3=additional deformations attributed to the presence of initial distortions

The task therefore is to determine each of these deflection components. The buckling-related deformations are assumed to have the following form:

$$u_3 = A\cos(n\theta)\cos\left(\frac{m\pi x}{L_b}\right) \tag{5.83a}$$

$$v_3 = B\sin(n\theta)\sin\left(\frac{m\pi x}{L_b}\right) \tag{5.83b}$$

$$w_3 = C_i\cos(n\theta)\sin\left(\frac{m\pi x}{L_b}\right) \tag{5.83c}$$

According to these expressions, the cylinder buckles into m half-sine waves in the longitudinal direction and n sine waves in the circumferential direction.

A general expression for the membrane strain energy of the cylindrical shell that can be used is (5.33). In the case of the bending strain energy, the expression used in [6] is not (5.59), but the one given below:

$$U_{sb} = \frac{ERtk}{2(1-\nu^2)}\int_0^{2\pi}\int_0^{L_b}\left[R^2\left(\frac{\partial^2 w_i}{\partial x^2}\right)^2 + \left(\frac{\partial^2 w_i}{\partial\theta^2} + w_i\right)^2 / R^2 + \frac{1-\nu}{2}\left(\frac{\partial^2 w_i}{\partial x\,\partial\theta} - \frac{1}{R}\frac{\partial u}{\partial\theta}\right)^2 + \right.$$
$$\left. + \frac{3(1-\nu)}{2}\left(\frac{\partial^2 w_i}{\partial x\,\partial\theta} + \frac{\partial v}{\partial x}\right)^2 + 2\nu\frac{\partial^2 w_i}{\partial x^2}\left(\frac{\partial^2 w_i}{\partial\theta^2} + \frac{\partial v}{\partial\theta}\right) + 2R\frac{\partial u}{\partial x}\frac{\partial^2 w_i}{\partial x^2}\right]dx\,d\theta \tag{5.84}$$

This is based on an expression for the bending strain energy of a cylindrical shell obtained in [2]. The authors of this publication in turn made use of expressions for strains derived by Flügge[3], which they used to obtain an expression for the bending strain energy. The question regarding the correct expression for the bending strain energy of a cylindrical shell had been discussed in several earlier publications because of incompatibilities observed with the corresponding differential equation. It was noted, however, that in general, the resulting errors are of secondary importance because it is the membrane strain energy that dominates behaviour.

With regard to the membrane and bending strain energies of a ring frame, expressions (5.47a) and (5.47b) are valid in this case as well.

5.5.3 Potential of the Work Done

Following deformation, a point on the middle surface of the cylindrical shell having cylindrical coordinates (r, x, θ) translates to point (r^1, x^1, θ^1). The following relationships hold:

$$r^1 = R - (w_0 + w_i) \tag{5.85a}$$

$$x^1 = x + u \tag{5.85b}$$

$$\theta^1 = \theta + \frac{v}{R} \tag{5.85c}$$

Following deformation, the volume of the material is:

$$V^1 = \frac{1}{2} \iint \left(r^1\right)^2 d\theta^1\, dx^1 \tag{5.86}$$

where the integral is taken over the entire deformed surface. Substituting (5.85a–c) in (5.86), the following expression is obtained:

$$V^1 = \frac{1}{2} \int_0^{2\pi} \int_0^{L_b} [R - (w_0 + w_i)]^2 \left(1 + \frac{\partial u}{\partial x}\right)\left(1 + \frac{1}{R}\frac{\partial v}{\partial \theta}\right) dx d\theta \tag{5.87}$$

The change in volume due to the displacements u, v, w_i is given by:

$$V^1 - V = \frac{1}{2} \int_0^{2\pi} \int_0^{L_b} \left\{ -2Rw_i + 2w_0 w_i + w_i^2 + \left(\frac{\partial u}{\partial x} + \frac{1}{R}\frac{\partial v}{\partial \theta}\right)\left(R^2 - 2Rw_0 - 2Rw_i\right) + R\frac{\partial u}{\partial x}\frac{\partial v}{\partial \theta} \right\} dx d\theta \tag{5.88}$$

The term $\partial v/\partial \theta$ can be omitted because material continuity requires that:

$$\int_0^{2\pi} \frac{\partial v}{\partial \theta} d\theta = 0 \tag{5.89}$$

Also, the terms involving $\partial v/\partial \theta$ are not required when determining the potential of the radial pressure. The potential of the pressure in the radial direction (external work done) is then the product of the resulting expression for the change (increase) in volume, (5.88), and the radial pressure, p:

$$W_{p_1} = \frac{p}{2} \int_0^{2\pi} \int_0^{L_b} \left\{ -2Rw_i + 2w_0 w_i + w_i^2 - 2R(w_0 + w_i)\left(\frac{\partial u_i}{\partial x} + \frac{1}{R}\frac{\partial v}{\partial \theta}\right) + R\frac{\partial u_i}{\partial x}\frac{\partial v}{\partial \theta} \right\} dx d\theta \tag{5.90}$$

The potential of the pressure P in the longitudinal direction is equal to the work done during the displacement of the boundaries and is given by:

$$W_{p2} = \left(\frac{PR^2}{2}\right)\int_0^{2\pi}\left\{(u_i)_{x=0} - (u_i)_{x=L_b}\right\}d\theta$$
$$= \left(\frac{PR^2}{2}\right)\int_0^{2\pi}\int_0^{L_r}\frac{\partial u_i}{\partial x}dxd\theta \tag{5.91}$$

The total potential of the work done, W, is the sum of (5.90) and (5.91):

$$W = \frac{p}{2}\int_0^{2\pi}\int_0^{L_b}\left\{-2Rw_i + 2w_0w_i + w_i^2 - 2R(w_0 + w_i)\left(\frac{\partial u_i}{\partial x} + \frac{1}{R}\frac{\partial v}{\partial \theta}\right) + R\frac{\partial u_i}{\partial x}\frac{\partial v}{\partial \theta}\right\}dxd\theta$$
$$+\left(\frac{PR^2}{2}\right)\int_0^{2\pi}\int_0^{L_r}\frac{\partial u_i}{\partial x}dxd\theta \tag{5.92}$$

5.5.4 The Total Potential

The potential of the system comprising of the imperfect ring-stiffened cylindrical shell and the external force field is given by:

$$\Pi = U_{sm}^i + U_{sb}^i + U_{fm}^i + U_{fb}^i + W^i \tag{5.93}$$

The superscript i denotes the presence of initial geometrical imperfections in the cylindrical shell.

If we introduce (5.82a) and (5.82b) to the expressions for the strain energy and the work done, whilst retaining second-order terms in u_3, v, w_3, and w_0 and neglecting second-order terms in p and P, these expressions become [6]:

$$U_{sm}^i = \frac{ERt}{2(1-v^2)}\int_0^{2\pi}\int_0^{L_b}\left[\left(\frac{\partial u_3}{\partial x}\right)^2 + \frac{1}{R^2}\left(\frac{\partial v}{\partial \theta} - w_3\right)^2 + 2v\frac{\partial u_3}{\partial x}\frac{1}{R}\left(\frac{\partial v}{\partial \theta} - w_i\right) + \right.$$
$$\left. + \frac{(1-v)}{2}\left(\frac{1}{R}\frac{\partial u_3}{\partial \theta} + \frac{\partial v}{\partial x}\right)^2\right]dxd\theta \tag{5.94a}$$

$$U_{sb}^i = \frac{ERkt}{2(1-v^2)}\int_0^{2\pi}\int_0^{L_b}\left[R^2\left(\frac{\partial^2 w_3}{\partial x^2}\right)^2 + \frac{1}{R^2}\left(\frac{\partial^2 w_3}{\partial \theta^2} + w_3\right)^2 + \frac{1-v}{2}\left(\frac{\partial^2 w_3}{\partial x\,\partial \theta} - \frac{1}{R}\frac{\partial u_3}{\partial \theta}\right)^2 + \right.$$
$$\left. + \frac{3(1-v)}{2}\left(\frac{\partial^2 w_3}{\partial x\,\partial \theta} + \frac{\partial v}{\partial x}\right)^2 + 2v\frac{\partial^2 w_3}{\partial x^2}\left(\frac{\partial^2 w_3}{\partial \theta^2} + \frac{\partial v}{\partial \theta}\right) + 2R\frac{\partial u_3}{\partial x}\frac{\partial^2 w_3}{\partial x^2}\right]dxd\theta \tag{5.94b}$$

$$U_{fm}^i = \sum_{r=1}^{N}\left(\frac{EA_fR}{2}\right)\int_0^{2\pi}\left\{\left(\frac{\partial^2 w_3}{\partial \theta^2} + w_3\right)\left(\frac{e}{R^2}\right) + \frac{1}{R}\left(w_3 - \frac{\partial v}{\partial \theta}\right)\right\}_{x=rL_f}^2 d\theta \tag{5.94c}$$

$$U_{fb}^i = \sum_{r=1}^{N}\left(\frac{EI_f}{2R^3}\right)\int_0^{2\pi}\left(\frac{\partial^2 w_3}{\partial \theta^2} + w_3\right)_{x=rL_f}^2 d\theta \tag{5.94d}$$

$$
\begin{aligned}
V^i = \frac{p}{2}\int_0^{2\pi}\int_0^{L_b} & \left\{ 2w_0w_3 + w_3^2 - 2R(w_0+w_3)\left(\frac{\partial u_3}{\partial x} + \frac{1}{R}\frac{\partial v}{\partial \theta}\right) + R\frac{\partial u_3}{\partial x}\frac{\partial v}{\partial \theta}\right\} dx d\theta \\
& + \frac{ERt}{2(1-\nu^2)}\int_0^{2\pi}\int_0^{L_b}\left[\left(\frac{\partial u_p}{\partial x} - \frac{\nu w_p}{R}\right)\left\{\frac{\partial^2 w_3}{\partial x^2} + \left(\frac{\partial v}{\partial x}\right)^2 + 2\frac{\partial w_0}{\partial x}\frac{\partial w_3}{\partial x}\right\}\right. \\
& - \left\{\left(\frac{w_p}{R}\right) - \nu\frac{\partial u_p}{\partial x}\right\}\left\{\frac{1}{R^2}\left(\left(\frac{\partial w_3}{\partial \theta}\right)^2 + \left(\frac{\partial u_3}{\partial \theta}\right)^2 - 2w_3\frac{\partial v}{\partial \theta}\right) + \frac{2}{R^2}\left(\frac{\partial w_0}{\partial \theta}\frac{\partial w_3}{\partial \theta} - w_0w_3\right)\right\} \\
& + 2\frac{\partial w_p}{\partial x}\left(\frac{\partial w_3}{\partial x} + \frac{\partial w_0}{\partial x}\right)\left\{\frac{\partial u_3}{\partial x} + \frac{\nu}{R}\left(\frac{\partial v}{\partial \theta} - w_3\right)\right\} \\
& \left. - 2\left(\frac{w_p}{R^2}\right)\left(\frac{\partial v}{\partial \theta} + w_0\right)\left\{\frac{1}{R}\left(\frac{\partial v}{\partial \theta} - w_3\right) + \nu\frac{\partial u_3}{\partial x}\right\}\right] dx d\theta \\
& - \sum_{r=1}^{N}\left(\frac{EA_fR}{2}\right)\int_0^{2\pi}\left[\frac{w_p}{R}\left(1+\frac{e}{R}\right)\left\{\frac{1}{R^2}\left(\left(\frac{\partial w_3}{\partial \theta}\right)^2 + \left(\frac{\partial u_3}{\partial \theta}\right)^2 - 2w_3\frac{\partial v}{\partial \theta}\right)\right.\right. \\
& \left. + \frac{2}{R^2}\left(\frac{\partial w_0}{\partial \theta}\frac{\partial w_3}{\partial \theta} - w_0w_3\right)\right\} + 2\left(\frac{w_p}{R^3}\right)\left(\frac{\partial v}{\partial \theta} + w_0\right) \\
& \left. \times\left\{\frac{\partial v}{\partial \theta} - w_3 - \left(w_3 + \frac{\partial^2 w_3}{\partial \theta^2}\right)\left(\frac{e}{R}\right)\right\}\right]_{x=rL_f} d\theta
\end{aligned}
\tag{5.94e}
$$

5.5.4.1 Deflections of the Perfect Shell

The values of $u_p(x)$, $w_p(x)$ that pertain to the perfect shell can be obtained using the von Sanden and Günther solution, discussed in Section 4.1.3. Kendrick has shown that a sufficiently good approximation can be obtained if these deflections are assumed to consist of uniform displacements in the radial and longitudinal direction in which case interframe bending is ignored. On this basis, the deflections corresponding to uniform radial and longitudinal contractions of an infinite stiffened cylindrical shell may be used.

In order to obtain expressions for these, we assume that they are constant with respect to x and θ in which case the total potential given by (5.93) becomes:

$$
\begin{aligned}
\left(\frac{1}{2\pi}\right)\Pi = & \left\{\frac{ERtL_f}{2(1-\nu^2)}\right\}\left\{\left(\frac{\partial u_p}{\partial x}\right)^2 + \left(\frac{w_p}{R}\right)^2 - 2\nu\frac{\partial u_p}{\partial x}\frac{w_p}{R}\right\} \\
& + \left\{\frac{ERtL_f}{2(1-\nu^2)}\right\}\left(\frac{w_p}{R}\right)^2 + \left(\frac{EA_fR}{2}\right)\left(\frac{w_p}{R}\right)^2\left(1+\frac{e}{R}\right)^2 \\
& + \left(\frac{EI_f}{2R}\right)\left(\frac{w_p}{R}\right)^2 - \left(pR^2L_f\right)\left(\frac{w_p}{R}\right)
\end{aligned}
\tag{5.95}
$$

Minimising (5.95) with respect to w_p/R and $\partial u_p/\partial x$ leads to the following expressions:

$$\left(\frac{w_p}{R}\right) - \nu\frac{\partial u_p}{\partial x} = \varphi_1 - k\left(\frac{w_p}{R}\right) - \left(1-\nu^2\right)\left(\frac{A_f}{tL_f}\right)\left(\frac{w_p}{R}\right)\left(1+\frac{e}{R}\right)^2$$

$$-\left(1-\nu^2\right)\left(\frac{I_f}{R^2tL_f}\right)\left(\frac{w_p}{R}\right) \tag{5.96a}$$

$$\frac{\partial u_p}{\partial x} - \nu\frac{w_p}{R} = -\frac{\varphi_2}{2} \tag{5.96b}$$

where

$$\phi_1 = \frac{pR(1-\nu^2)}{Et} \quad (p = \text{radial pressure})$$

$$\phi_2 = \frac{PR(1-\nu^2)}{Et} \quad (P = \text{longitudinal pressure})$$

Using (5.83a–c) and (5.78), the total potential given by (5.93) is recast as follows:

$$\Pi = \left\{\frac{\pi EtL_b}{4R(1-\nu^2)}\right\}\left(A_{11}A^2 + A_{22}B^2 + A_{33}{C_1}^2 + 2A_{12}AB + 2A_{13}AC_1 + 2A_{14}AC_0\right.$$

$$\left. +2A_{23}BC_1 + 2A_{24}BC_0 + 2A_{34}C_1C_0\right) \tag{5.97}$$

where A_{11}, A_{22}, etc., are coefficients, expressions for which are included in Appendix 2. For the structure to be at equilibrium, the potential V must be a minimum. Therefore, from the calculus of variations,

$$\frac{\partial \Pi}{\partial A} = \frac{\partial \Pi}{\partial B} = \frac{\partial \Pi}{\partial C_1} = 0 \tag{5.98}$$

Substituting (5.97) in (5.98) leads to the following matrix equation, expressed in terms of the deformation parameters A, B, C_1:

$$[\boldsymbol{a}][\boldsymbol{x}] = -C_0[\boldsymbol{c}] \tag{5.99}$$

where

$$[\boldsymbol{a}] = \begin{bmatrix} A_{11} & A_{21} & A_{31} \\ A_{12} & A_{22} & A_{32} \\ A_{13} & A_{23} & A_{33} \end{bmatrix} \quad [\boldsymbol{x}] = \begin{bmatrix} A \\ B \\ C_1 \end{bmatrix} \quad [\boldsymbol{c}] = \begin{bmatrix} A_{14} \\ A_{24} \\ A_{34} \end{bmatrix}$$

Pre-multiplying throughout (5.99) by $[\boldsymbol{a}]^{-1}$ gives the following equation for $[\boldsymbol{x}]$:

$$[\boldsymbol{x}] = -C_0[\boldsymbol{a}]^{-1}[\boldsymbol{c}] \tag{5.100}$$

This matrix equation is solved using the method described in [6] or any other comparable approach to give the minimum buckling pressure.

In the case of uniform pressure, $p = P$ ($\phi_1 = \phi_2$), certain simplifications can be made to the expressions for the coefficients A_{11}, A_{12}, etc., as listed in Appendix 2.

In the case of unstiffened cylinders, it is possible to obtain the exact solution because the assumed deflections satisfy the differential equation of a shell with the assumed initial distortions (5.78) as well as the boundary conditions of simple support at the ends. The general form of the initial distortions of an unstiffened cylinder can then be represented by the following expression:

$$w_0 = \sum_{m=1}^{\infty}\sum_{n=1}^{\infty} {}_{mn}C_0 \cos(n\theta)\sin\left(\frac{m\pi x}{L_b}\right) \tag{5.101}$$

Substituting in (5.82) from (5.83), the total deformations resulting from the applied load are obtained:

$$u_i = u_p(x) + \sum_{m=1}^{\infty}\sum_{n=1}^{\infty} {}_{mn}A\cos(n\theta)\cos\left(\frac{m\pi x}{L_b}\right) \tag{5.102a}$$

$$v = \sum_{m=1}^{\infty}\sum_{n=1}^{\infty} {}_{mn}B\sin(n\theta)\sin\left(\frac{m\pi x}{L_b}\right) \tag{5.102b}$$

$$w_i = w_p(x) + \sum_{m=1}^{\infty}\sum_{n=1}^{\infty} {}_{mn}C_1 \cos(n\theta)\sin\left(\frac{m\pi x}{L_b}\right) \tag{5.102c}$$

Expression (5.102c) can be modified to the following form:

$$w_i = w_p(x) + \sum_{m=1}^{\infty}\sum_{n=1}^{\infty} {}_{mn}C_0\left[\frac{\phi_1}{{}_{mn}\phi_c - \phi_1}\right]\cos(n\theta)\sin\left(\frac{m\pi x}{L_b}\right) \tag{5.103}$$

to allow for the beam-column effect arising at an interframe level. The quantity in brackets is the well-known magnification factor in which ${}_{mn}\phi_c$ is the value of ϕ_1 that makes $|a| = 0$ and thus represents the elastic buckling pressure.

5.5.5 Bending Stresses Due to the Presence of Initial Distortions

For unstiffened cylinders, the maximum circumferential stress due to initial distortions is given by:

$$\sigma_\theta = \sigma_{\theta m} + \sigma_{\theta b} \tag{5.104}$$

where

$$\sigma_{\theta m} = \left[\frac{E}{R(1-\nu^2)}\right]\left[nB - C_1 - \nu\lambda A\right] \tag{5.105}$$

is the membrane stress and

$$\sigma_{\theta b} = \left[\frac{Et}{2R^2(1-\nu^2)}\right]C_1\left[n^2 - 1 + \nu\lambda^2\right] \tag{5.106}$$

is the bending stress.

Comparison of results obtained for the unstiffened and the stiffened cylinder has shown that it is possible in both cases to determine the maximum normal deflection w_3 and hence the maximum bending stress using the following expression:

$$C_1 = C_0 \frac{p}{p_e - p} \tag{5.107}$$

This shows that the deflection increases nonlinearly with load, p. As the load approaches the elastic buckling pressure p_e, the deflection C_1 increases asymptotically towards infinity.

The maximum stresses that are induced in ring-stiffened cylindrical shell arise at midspan between the fixed end supports ($x = L_b/2$), in way of the outer fibres of the frame flange, since these are situated furthest from the centroid of the shell-frame cross section. Kendrick gives the following expression for the maximum flange stress in a cylinder fitted with an odd number of ring frames:

$$\begin{aligned} \sigma_f &= \pm E \left\{ \frac{1}{R} \left(w_3 - \frac{\partial v}{\partial \theta} \right) + \left(\frac{\partial^2 w_3}{\partial \theta^2} \right) \left(\frac{e_f}{R^2} \right) \right\}_{\substack{\theta=0 \\ x=L_b/2}} \\ &= \mp E \left\{ \frac{1}{R}(nB - C_1) + C_1 \left(n^2 - 1\right) \left(\frac{e_f}{R^2} \right) \right\} \end{aligned} \tag{5.108}$$

where e_f is the distance of the frame flange from the middle surface of the shell. For cylinders with an even number of frames, the maximum stress will be less than that given by (5.108), but the difference will be negligible in the case of a large number of frames.

The maximum flange stress is given by

$$\sigma_f = \sigma_{fm} + \sigma_{fb} \tag{5.109}$$

where σ_{fm} is the membrane stress, given by:

$$\sigma_{fm} = \frac{E}{R}(nB - C_1) \tag{5.110a}$$

and σ_{fb} is the bending stress, given by:

$$\sigma_{fb} = E\left(n^2 - 1\right) C_1 \left(\frac{e_f}{R^2} \right) \tag{5.110b}$$

5.6 CONCLUDING REMARKS

Ring-stiffened cylindrical shells used in submarine pressure hulls are generally designed such that the elastic buckling pressure is well above the collapse pressure. Collapse occurs by elasto-plastic yielding, and the most probable location of failure is midway between frames in the circumferential direction, as will become clear in the next chapter. The studies described in this chapter make it possible to quantify the extent to which the overall elastic buckling pressure is reduced as a result of the presence of initial distortions, which, in principle, does not have a direct effect on collapse pressure. This is more directly affected by the presence of welding residual stresses, a topic that is discussed in Chapter 6.

Table 5.4 provides a timeline of the development of the classical theory of unstiffened and ring-stiffened cylindrical shells subjected to external pressure loading.

TABLE 5.4
Development of Elastic Theory for Cylindrical Pressure Hull Strength

Bresse	1859	Infinitely long perfect stiffened cylindrical shell (shell bending ignored)
von Mises	1914	Finite length perfect unstiffened cylindrical shell under hydrostatic pressure
von Sanden & Günther	1920/1	Infinitely long perfect stiffened shell allowing for shell bending between frames
von Mises	1929	Finite length perfect unstiffened cylindrical shell under hydrostatic pressure and axial compression
Viterbo	1930	Correction to von Sanden & Günther solution for ring frame shear load
Pulos & Salerno	1951	Exact solution for interframe strength allowing for axial compression
Kendrick	1953	General instability of ring-stiffened cylindrical shell
Kendrick	1953	General instability of ring-stiffened cylindrical shell with initial distortions
Bryant	1954	Approximate solution for general instability of ring-stiffened shell
Wilson	1956	Approximate solution for interframe strength including end-compression
Lunchick & Short	1957	Effect of initial distortions on interframe strength of ring-stiffened shell
Singer	1965	Effect of ring frame eccentricity on general instability
Kendrick	1970	Ring frame tripping stress

NOTES

1 ARE, Admiralty Research Establishment (UK).
2 Alternatively one can resort to one of the more recent optimisation techniques developed, such as genetic algorithms.
3 The Flügge expressions for the strains are:

$$\varepsilon_x = \frac{\partial u}{\partial x} + (r - R)\frac{\partial^2 w}{\partial x^2}$$

$$\varepsilon_\theta = \frac{1}{R}\frac{\partial v}{\partial \theta} + \frac{r-R}{R}\frac{1}{r}\frac{\partial^2 w}{\partial \theta^2} - \frac{w}{r}$$

$$\gamma_{x\theta} = \frac{1}{R}\frac{\partial u}{\partial \theta} + \frac{r}{R}\frac{\partial v}{\partial x} + \frac{\partial^2 w}{\partial x \partial \theta}\left(\frac{r-R}{R} + \frac{r-R}{r}\right)$$

REFERENCES

1. Ball, W.E. (1962). "Formulas and curves for determining the elastic general-instability pressures of ring-stiffened cylinders." DTMB Report No 1570.
2. Bleich, H.H., DiMaggio, F. (1952). "A strain-energy expression for thin cylindrical shells." Columbia Univ. Dept Civil Eng. Tech Rep. No. 6.
3. Brush, D.O., Almroth, B.O. (1975). *Buckling of Bars, Plates, and Shells*. New York: McGraw-Hill Book Co 379 p.
4. Bryant, A.R. (1954). "Hydrostatic pressure buckling of a ring-stiffened tube." NCRE Rept. 306 Naval Const. Res. Est. Dunfermline.
5. Kendrick, S. (1953). "The buckling under external pressure of circular cylindrical shells with evenly spaced equal strength circular ring frames – Part I". NCRE Report R. 211 Naval Const. Res. Est. Dunfermline, Scotland.
6. Kendrick, S. (1953). "The deformation under external pressure of nearly circular cylindrical shells with evenly spaced equal strength nearly circular ring frames." NCRE Report R. 259 Naval Const. Res. Est. Dunfermline, Scotland.
7. Kendrick, S.B. (1970). Externally pressurised vessels Chap. 9. In: Gill, S.S. (Ed). *The Stress Analysis of Pressure Vessels and Pressure Vessel Components*. Oxford: Pergamon Press, 405–511 pp.
8. Pulos, J.G. (1963). "Structural analysis and design considerations for cylindrical pressure hulls." DTMB Report. No 1639.

9. Reynolds, T.E. (1957). "A graphical method for determining the general-instability strength of stiffened cylindrical shells." DTMB Report No 1106.
10. Salerno, V.L., Levine, B. (1950). "Buckling of circular cylindrical shells with evenly spaced, equal strength circular ring frames: Part II." Polytechnic Institute of Brooklyn PIBAL Report No. 169.
11. Timoshenko, S.P., Gere, J.M. (1961). *Theory of Elastic Stability*, 2nd Ed. Tokyo: McGraw Hill Kogakusha, 564 p.
12. Timoshenko, S.P., Woinowsky-Krieger, S. (1959). *Theory of Plates and Shells*, 2nd Ed. Tokyo: McGraw-Hill Kogakusha, 580 p.

6 Collapse of Ring-Stiffened Cylindrical Shells

6.1 FAILURE MODES OF A SUBMARINE PRESSURE HULL

In Chapters 3–5, we traced the development of the theory of elastic stability of ring-stiffened cylindrical shells subjected to external hydrostatic pressure. In doing so, we provided a theoretical analysis of the response of such structures. However, these results cannot form the sole basis of structural design as it is necessary to consider the effect of material properties, initial imperfections as well as the presence of major internal structural members such as bulkheads and decks, on structural response. Allowing for these enables us to address the question of *collapse strength*, which is an essential step in the design of such structures.

Possible failure modes of a submarine pressure hull are illustrated in Figure 6.1. Two conclusions can be drawn from this figure. Firstly, collapse can be the result of different kinds of external loads, and secondly, it can occur in a variety of modes. In this chapter, we shall focus on the failure modes of ring-stiffened cylindrical shells. These are *interframe shell plating collapse* and *overall collapse* (*general instability*), which takes place between consecutive transverse bulkheads and includes the shell plating and ring frames. There is one other failure mode that is not illustrated in Figure 6.1 and that is *ring frame tripping*, which involves the lateral-torsional distortion of a frame about its line of attachment to the shell plating. This will also be considered in this chapter. The collapse of truncated cones, bulkheads, and decks is considered in Chapter 7 and that of end closures in Chapter 11.

6.2 INTERFRAME COLLAPSE AND GENERAL INSTABILITY COLLAPSE

As in the case of flat stiffened plating, failure in a stiffened cylindrical shell can occur either elastically or elasto-plastically, the actual failure mode depending primarily on the slenderness of the structure and the yield strength of the material. In the case of submarine pressure hulls, interframe collapse is generally elasto-plastic.

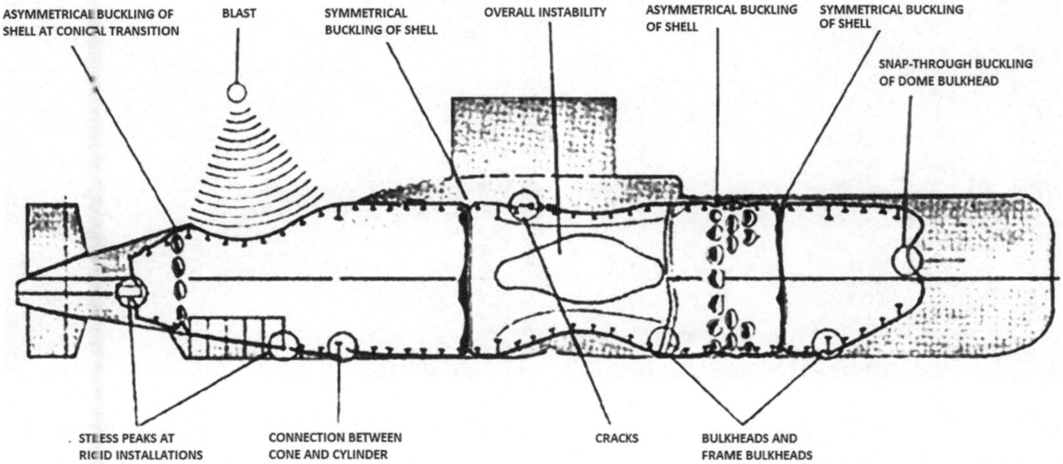

FIGURE 6.1 Failure modes of a submarine pressure hull.

DOI: 10.1201/9781003489153-6

6.2.1 Interframe Collapse

The two possible interframe failure modes are illustrated in Figure 6.2a and b. In the first of these, failure occurs in an *axisymmetric manner* and involves yielding as well as buckling of the shell plating between the frames around possibly the entire circumference and possibly in more than one bay. It is thus a mixed mode, and in this case, the distorted form resembles an accordion pleat. In the second case, the distorted form varies along the circumference, so that it is termed *asymmetric collapse*. Buckling in this case follows yielding, so that it is inelastic. As in the case of axisymmetric failure, it may be observed around the entire circumference and also in more than one bay. It may thus be termed *inelastic asymmetric shell instability between frames.*

We touched upon plastic behaviour of cylindrical shells in Section 3.3.6 in which we considered yielding of an unstiffened cylindrical shell subjected to internal pressure and derived the boiler equation. This gives the pressure required to make the circumferential stress σ_θ equal to the yield stress of the material, ($\sigma_\theta=\sigma_0$), according to the Rankine criterion (uniaxial stress).

The boiler equation, however, does not reflect the strengthening effect of ring frames. In order to allow for their presence in an approximate way, we can replace the actual section by an equivalent orthotropic one whose thickness is such that the area of the frame is included. This is done by replacing t by $t+A_{eff}/L_f$. This is substituted in (3.47) that is solved for the pressure p to give:

$$p_{c1}=\sigma_0\left(\frac{t}{R}\right)\left(1+\frac{A_{eff}}{L_f t}\right) \tag{6.1}$$

where p_{c1} is the pressure to cause mode 1 collapse (yielding in the circumferential direction in a uniaxial stress field). The results obtained in Chapter 4 for a ring-stiffened cylindrical shell subjected to external pressure indicate that in the circumferential direction, the maximum stress occurs on the outer surface of the shell plating at midspan. In the longitudinal direction, the maximum stress occurs at the intersection of the cylindrical shell with the ring frame (Figure 4.13). Which of the two will be larger will depend on the scantlings of the cylindrical shell and ring frames.

Experimental work has shown that for axisymmetric collapse, the circumferential stress is critical. Setting $\sigma_\theta=\sigma_0$ (Rankine yield criterion), the following expression is obtained for the pressure to cause yielding at the outer fibre in the midbay position [17]:

FIGURE 6.2 Interframe failure modes of an externally ring-stiffened cylindrical shell [16]. (a) Axisymmetric collapse. (b) Asymmetric collapse.

$$p_{c3} = \sigma_0\left(\frac{t}{R}\right)\left(\frac{1}{1-a\left[F_2-\nu F_4\sqrt{\frac{0.91}{1-\nu^2}}\right]}\right) \tag{6.2}$$

Where

$$a \equiv \frac{(1-\nu/2)\alpha}{\alpha+\beta+(1-\beta)F_1}$$

α, β are non-dimensional parameters ($\alpha \approx 0.4$, $\beta \leq 0.04$)
F_1, F_2, and F_4 are given by (4.58a), (4.58b), and (4.58d), respectively

In the derivation of (6.2) the beam-column effect has been ignored ($\gamma = 0$).

Alternative expressions are obtained if other failure criteria are used (yielding on the outer surface of the shell plating, yielding through half the shell plating thickness, yielding at mid-thickness). The pressure to cause yielding at mid-thickness in the midbay position is obtained by setting the second term in (4.51a), which is the bending term, equal to zero, leading to:

$$p_{c5} = \underbrace{\sigma_0\left(\frac{t}{R}\right)}_{\text{boilerpressure}}\left(\frac{1}{1-aF_2}\right) \tag{6.3}$$

Expressions (6.2) and (6.3) are based on failure under uniaxial stress conditions, namely in the circumferential direction. Since, however, biaxial conditions exist, the Rankine criterion ($\sigma_\theta = \sigma_0$) ought to be replaced by an appropriate one such as the von Mises criterion. In the case of a thin cylindrical shell, the von Mises yield criterion assumes the following form:

$$\sigma_x{}^2 + \sigma_\theta{}^2 - \sigma_x\sigma_\theta \equiv \sigma_0{}^2 \tag{6.4}$$

If (4.54a) and (4.54b), which give the maximum stresses in the axial and circumferential direction at midbay, are substituted in (6.4), the pressure to cause yielding in the outer fibre, p_{c6}, is obtained:

$$p_{c6} = \sigma_0\left(\frac{t}{R}\right)\left[\frac{3}{4}+a^2\left[F_2^2+F_2F_4(1-2\nu)\sqrt{\frac{0.91}{1-\nu^2}}+F_4{}^2(1-\nu+\nu^2)\left(\frac{0.91}{1-\nu^2}\right)\right]-\frac{3}{2}a\left[F_2-\nu F_4\sqrt{\frac{0.91}{1-\nu^2}}\right]\right]^{-1/2} \tag{6.5}$$

As previously, the pressure to cause yielding at mid-thickness (heart-of-plate), p_{c7}, is obtained by setting the bending stress component equal to zero in (4.54a), (4.54b). Substituting these in (6.4) gives

$$p_{c7} = \frac{\sigma_0\left(\frac{t}{R}\right)}{\sqrt{\left[\frac{3}{4}+a^2F_2^2-\frac{3}{2}aF_2\right]}} \tag{6.6}$$

If the beam-column effect is introduced, the stresses become nonlinear functions of pressure. Equations (6.5) and (6.6) are then transcendental in the pressure since $F_1 - F_6$ are functions of the pressure. It is therefore necessary to resort to an iterative approach to solve these equations [17].

6.2.2 Approximate Solution for Interframe Collapse

In order to avoid having to perform iterative calculations to determine the collapse pressure, Wilson developed an approximate approach that provides accurate enough results for stresses and deflections. This method is described in Section 4.1.5. We shall discuss here how this approach can be used in design.

Geometric imperfections (primarily out-of-circularity) may lead to interframe buckling with a number of lobes forming in the circumferential direction. If the shell plating is relatively thin, elastic buckling will occur, whereas if it is stocky, elastic buckling does not take place, and failure occurs when the compressive membrane stress reaches the yield stress at midspan ($\sigma_5 = \sigma_0$) (Figure 4.13). Elastic shell buckling is similar in nature to column buckling – this can be seen by comparing the load – slenderness equations for these two types of structures (Euler and Bresse equations).

In the case of yielding failure, the following much simpler expression for the collapse pressure, p_{c5}, can be used in pressure hull design. This was obtained by Wilson [22] by solving for the pressure p ($\equiv p_{c5}$) from (4.84) and substituting $\sigma = \sigma_0$:

$$p_{c5} = \left[\frac{\sigma_0 t}{R}\right]\frac{1}{1+\gamma G} \tag{6.7}$$

The parameters γ, G are defined in Section 4.1.5. It is seen from (6.7) that the fraction $1/(1+\gamma G)$ is a measure of the difference with boiler yield strength. In a corresponding manner, for the ring frame, expression (4.85) is used:

$$p_{fy} = p_b\left[\frac{1}{1-\nu/2}\right]\left(\frac{R_f}{R}\right)\left[1+\frac{A}{bt+2Nt/\alpha}\right] \tag{6.8}$$

In the case of sections with sufficiently stiff ring frames, the product γG is slightly negative and thus $p_{c5} \cong p_b$. In any case, a biaxial stress field whose components are a longitudinal stress σ_x and a circumferential stress σ_θ, where $\sigma_\theta \cong 2\sigma_x$, forms prior to collapse. We conclude that structural response depends on the elastic strength as well as on yield strength. Whether elastic or elasto-plastic failure will occur depends on the following [9]:

1. The elastic buckling pressure of the perfect cylindrical shell, p_m
2. The yield pressure of the cylindrical shell, p_{c5}

Example 6.1

A ring-stiffened cylindrical shell has a radius of 4 m and a thickness of 32 mm. The yield stress is 540 N/mm^2 and E=207 kN/mm^2. The ring frames having a cross-sectional area A_f=4,500 mm^2, are spaced 768 mm apart, and have a depth of 250 mm. Their CG lies at a distance of 200 mm from the web toe. Calculate pressures p_{c5} and p_{fy}.

Solution

We require A, the area of the transverse cross section enclosed by the cylindrical shell. This is given by:

$$\frac{A}{A_f} = \left(\frac{R}{R_f}\right)^2$$

Substituting, we obtain

$$A = 4{,}500\left(\frac{4{,}000}{4{,}000-200}\right)^2 = 4{,}986 \text{ mm}^2$$

Shell slenderness:

$$\alpha L = (1.285)(768-8)/\sqrt{(4{,}000)(32)} = 2.73$$

Hence,

$$\alpha = 2.73/760 = 0.0036$$

From Figure 4.14, we obtain:

$$N = 1.072 \quad G = 0.601$$

Hence,

$$\beta = \frac{2tN}{\alpha(A+bt)} = 3.81 \quad \gamma = \frac{A(1-\nu/2)}{A+bt+2Nt/\alpha} = 0.1386$$

where b=0. Substituting in (6.7), (6.8) we obtain:

$$p_{c5} = \frac{(540)(32)}{4{,}000}\frac{1}{(1+(0.1386)(0.601))} = 3.99 \text{ N/mm}^2$$

$$p_{fy} = \frac{4{,}320}{0.85}\left(\frac{3{,}750}{4{,}000}\right)\{1+0.261\} = 4.34 \text{ N/mm}^2$$

We note that the pressure to cause yielding at the midsurface of the plate at midbay, p_{c5}, is much less that the boiler yield pressure p_0 (= 4.32 N/mm^2), whereas the ring frame yield pressure is very close to it. We should not forget the effect of shell bending resulting from out-of-circularity and the consequent reduction in strength.

1. The degree of out-of-circularity
2. The magnitude and distribution of residual stresses arising from cold forming and welding
3. Whether the stress at any point in the structure becomes equal to the yield stress

If $p_m < p_{c5}$, elastic buckling will occur, and in this case, geometrical imperfections are important. If the elastic buckling pressure is slightly higher than p_{c5}, that is, if

$$1.5 \geq \frac{p_m}{p_{c5}} \geq 1.0 \tag{6.9}$$

all of the above factors and especially (4) contribute and elasto-plastic collapse will occur. Finally, for $p_m >> p_{c5}$, pressure p_{c5} determines collapse (6.7). These comments are corroborated by the test data included in Figure 6.3, which includes test data points and a curve approximating the real inter-frame collapse pressure involving elastic buckling and yielding.

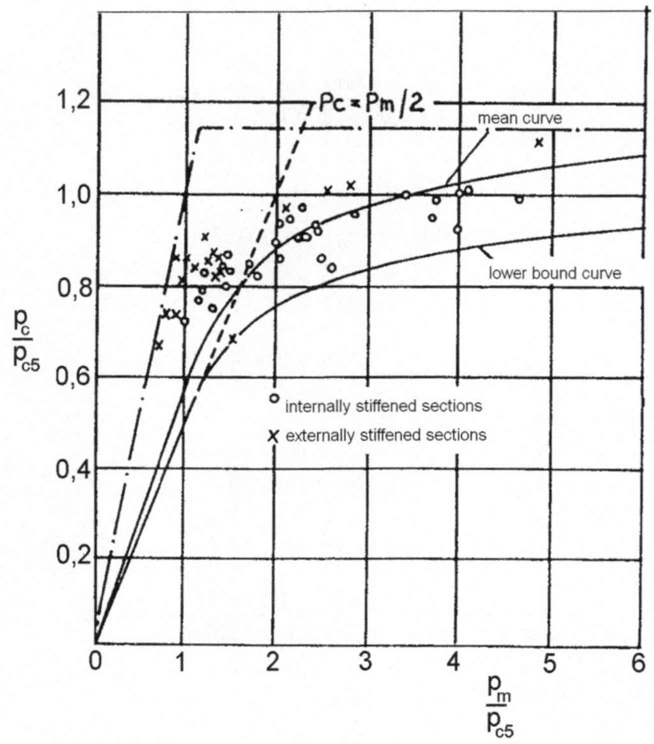

FIGURE 6.3 Interframe collapse pressure (test data).

6.2.3 General Instability Collapse

As far as general instability is concerned, it is necessary to note that for practical geometries, it arises in the inelastic range, that is, yielding precedes collapse. In order to cater for inelastic collapse, early investigators made use of the widely used tangent modulus theory developed by Shanley. Indeed, Lunchick [14,16] obtained the following "modified Bryant formula" that allows for inelastic overall collapse of ring-stiffened cylinders:

$$p_u = \frac{(n^2-1)E_t I_{sfe}}{R^3 L_f} + \frac{E_s t}{R} \frac{\lambda^4}{\left(n^2 - 1 + \dfrac{\lambda^2}{2}\right)\left(n^2+\lambda^2\right)^2 \left[1 - \dfrac{3}{4}\left(1 - \dfrac{E_s}{E_t}\right)\dfrac{\lambda^4}{\left(n^2+\lambda^2\right)^2}\right]} \tag{6.10}$$

In this formula, the elastic modulus E has been replaced by the secant modulus E_s and the tangent modulus E_t, as necessary. An additional term is also present on the denominator of the shell term. Finite element analysis can, however, provide much more accurate results in inelastic stress analysis.

In the sections that follow, we shall consider the collapse strength of cylindrical shells fitted with the stiffening arrangements that are in use. These include:

- Internal ring frames
- External ring frames
- Intermediate heavy frames

6.3 INTERNAL VERSUS EXTERNAL RING FRAMES

In this section, we shall provide a comparison of the relative merits of internal and external ring frames following which we shall consider the failure modes of internal ring frames. We shall consider the question of the position (eccentricity) of ring frames from the points of view of:

- Interframe strength
- Overall elastic buckling strength
- Fabrication issues

6.3.1 INTERFRAME STRENGTH

In the case of externally stiffened cylinders, the additional strains that result from the presence of initial distortions are in the opposite sense to those of internally stiffened cylinders since the initial distortions arise in the outward direction. This indicates that differences in strength arise between internally and externally stiffened cylindrical shells. The issue of ring frame position is considered with reference to axisymmetric yielding at the shell surface at midbay since this is the design failure mode. The von Mises yield criterion was applied since a biaxial stress field is induced, and pressures at which yielding is initiated were calculated. These calculations were performed for a range of values of $(w_0)_{max}/t$ for both internally and externally stiffened cylinders. Results are included in Figure 6.4 so that the effect of yield strength on the pressure to cause yielding can be assessed, for two different ratios of σ_0/E.

It is seen that the effect of weld-induced initial distortions is similar for both levels of yield stress (σ_0/E=0.020, 0.025) and that the pressure to initiate yielding is directly dependent on the magnitude of the yield stress. Ring frame eccentricity has an important effect on yield initiation, as seen in Figure 6.4. In the case of internal framing, increases in the initial distortions result in reductions in the yield pressure. In the case of external framing, the opposite effect is observed, and it is seen that the yield pressures are much higher than for internal framing for the range of initial distortions considered.

We conclude that at an interframe level, for the geometries encountered in submarine pressure hulls, external frames result in increased strength.

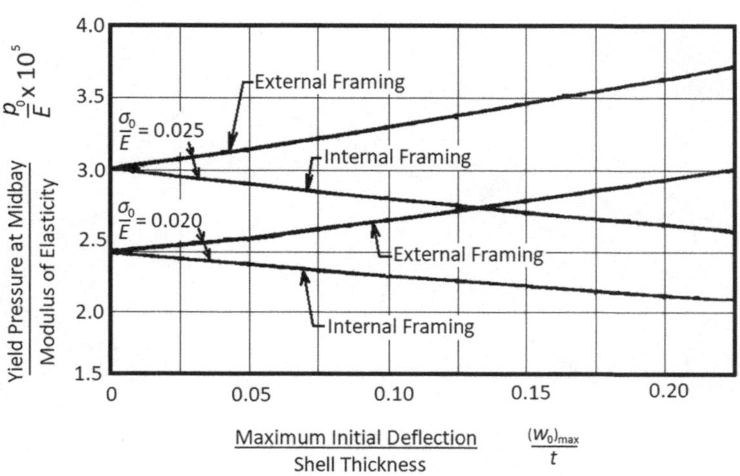

FIGURE 6.4 Effect of magnitude of initial distortion on pressure to initiate yielding (internally and externally stiffened cylinders) [15].

6.3.2 Overall Elastic Buckling Strength

In this section, the effect of the presence of ring frames on the overall elastic buckling strength of a ring-stiffened cylindrical shell will be considered, based on the analysis presented in [2,19]. The question of ring frame eccentricity will be considered in Section 6.3.3. In [2,19], the generalised problem of ring and stringer-stiffened shells is considered, but for present purposes, the approach has been adapted to the ring-stiffened cylinder problem. In order to allow for differences in critical elastic buckling pressure due to the position of the ring frames in an accurate manner, the reference line is chosen as the middle surface of the shell (Figure 6.5). The displacements *u*, *v*, and *w* are expressed in non-dimensional form (see Notation[1]).

The main assumptions made in this treatment are [19]:

1. The ring frames are distributed over the entire surface of the shell.
2. The normal strains $\varepsilon_x(z)$ and $\varepsilon_\theta(z)$ vary linearly in the ring frame as well as in the shell. The normal strains in the ring frame and the shell are equal along their line of contact.
3. The ring frames do not transmit shear. The membrane shear force $N_{x\theta}$ is borne entirely by the shell.
4. The torsional rigidity of the ring frame cross section is added to that of the sheet.

It is seen that a "merged" ring frame approach is followed. This does not permit the determination of ring frame local response (tripping, local buckling, etc). Such failure modes are assumed to arise at pressures higher than the general elastic instability collapse pressure of the cross section.

The following constitutive relations are used:

In the shell,

$$\sigma_x(z) = \frac{E}{1-\nu^2}\left[\varepsilon_x + \nu\varepsilon_\theta - z(\kappa_x + \nu\kappa_\theta)\right] \tag{6.11a}$$

$$\sigma_\theta(z) = \frac{E}{1-\nu^2}\left[\varepsilon_\theta + \nu\varepsilon_x - z(\kappa_\theta + \nu\kappa_x)\right] \tag{6.11b}$$

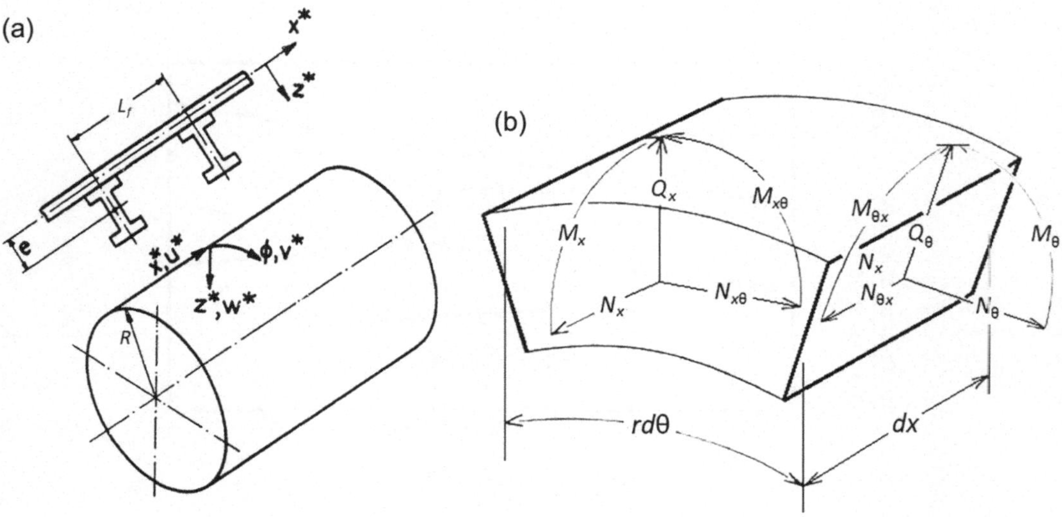

FIGURE 6.5 Notation for ring-stiffened cylindrical shell. (a) Notation. (b) Forces and moments acting on cylindrical shell element.

In the ring frame,

$$\sigma_x(z) = E_f(\varepsilon_x - z\kappa_x) \tag{6.12a}$$

$$\sigma_\theta(z) = E_f(\varepsilon_\theta - z\kappa_\theta) \tag{6.12b}$$

The following linear kinematic (strain–displacement) relations are used, as employed in Donnell's analysis [7].

$$\varepsilon_x = \frac{\partial u}{\partial x} \tag{6.13a}$$

$$\varepsilon_\theta = \frac{\partial v}{\partial \theta} - w \tag{6.13b}$$

$$\gamma_{x\theta} = \frac{\partial u}{\partial \theta} + \frac{\partial v}{\partial x} \tag{6.13c}$$

$$\kappa_x = \frac{\partial^2 w}{\partial x^2} \tag{6.13d}$$

$$\kappa_\theta = \frac{\partial^2 w}{\partial \theta^2} \tag{6.13e}$$

$$\kappa_{x\theta} = \frac{\partial^2 w}{\partial x\,\partial \theta} \tag{6.13f}$$

These relations are identical to those used in flat plate theory, the only difference being the second term in the expression for the circumferential strain, ε_θ. This is due to the change in circumferential dimensions with change in radius.

Substituting from (6.11a and b) in (6.12a and b), we obtain the following relations for the stress resultants:

$$N_x = \frac{Et}{(1-\nu^2)}\left[\varepsilon_x + \nu\varepsilon_\theta\right] \tag{6.14a}$$

$$N_\theta = \frac{Et}{(1-\nu^2)}\left[\varepsilon_\theta(1+\mu) + \nu\varepsilon_x - \chi\kappa_\theta\right] \tag{6.14b}$$

$$N_{x\theta} = N_{\theta x} = \left[\frac{Et}{2(1+\nu)}\right]\gamma_{x\theta} \tag{6.14c}$$

$$M_x = (-D/R)\left[\kappa_x + \nu\kappa_\theta\right] \tag{6.15a}$$

$$M_\theta = (-D/R)\left[\kappa_\theta(1+\eta_0) + \nu\kappa_x - \zeta\varepsilon_\theta\right] \tag{6.15b}$$

$$M_{x\theta} = (+D/R)(1-\nu)\kappa_{x\theta} \tag{6.15c}$$

$$M_{\theta x} = (-D/R)\left[(1-\nu)+\eta_t\right]\kappa_{x\theta} \tag{6.15d}$$

The following terms have been introduced in the above expressions to allow for the presence of ring frames:

ζ allows for the change in bending stiffness due to the eccentricity of the shell ($= E_f A_f eR/L_f D$)
η_0 allows for the increase in bending stiffness of the shell ($= E_f I_f /L_f D$)
η_t allows for the increase in twisting stiffness of the shell ($= G_f I_{ft} /L_f D$)
μ allows for the increase in effective area of the cross section ($= (1-\nu^2)(A_f E_f /L)_f tE$)
χ accounts for the change in extensional stiffness ($= (1-\nu^2)(E_f A_f e)/(EL_f tR)$)

The equations of equilibrium of the stiffened cylindrical shell are [19]:

$$\left.\begin{aligned}
&\frac{\partial N_x}{\partial x}+\frac{\partial N_{x\theta}}{\partial \theta}=0\\
&\frac{\partial N_\theta}{\partial \theta}+\frac{\partial N_{x\theta}}{\partial x}=0\\
&\frac{\partial^2 M_x}{\partial x^2}+\frac{\partial^2 M_\theta}{\partial \theta^2}-\frac{\partial^2 M_{x\theta}}{\partial x\,\partial \theta}+\frac{\partial^2 M_{\theta x}}{\partial x\,\partial \theta}+RN_\theta+N_{x0}R\frac{\partial^2 w}{\partial x^2}+N_{\theta 0}R\frac{\partial^2 w}{\partial \theta^2}=0
\end{aligned}\right\} \tag{6.16}$$

Substituting equations (6.13)–(6.15) in (6.16), the equations of equilibrium are obtained in terms of the displacements, *u*, *v*, and *w*:

$$\left[\frac{Et}{1-\nu^2}\right]\left\{\frac{\partial^2 u}{\partial x^2}+\frac{1-\nu}{2}\frac{\partial^2 u}{\partial \theta^2}+\frac{1+\nu}{2}\frac{\partial^2 v}{\partial x\,\partial \theta}-\nu\frac{\partial w}{\partial x}\right\}=0 \tag{6.17a}$$

$$\left[\frac{Et}{1-\nu^2}\right]\left\{\frac{1+\nu}{2}\frac{\partial^2 u}{\partial x\,\partial \theta}+(1+\mu)\frac{\partial^2 v}{\partial \theta^2}+\frac{1-\nu}{2}\frac{\partial^2 v}{\partial x^2}-\frac{\partial w}{\partial \theta}(1+\mu)\text{-}\chi\frac{\partial^3 w}{\partial \theta^3}\right\}=0 \tag{6.17b}$$

$$\begin{aligned}
&\left(-\frac{D}{R}\right)\left\{\zeta\left(2\frac{\partial^2 w}{\partial \theta^2}-\frac{\partial^3 v}{\partial \theta^3}\right)+\frac{\partial^4 w}{\partial x^4}+(2+\eta_t)\frac{\partial^4 w}{\partial x^2\,\partial \theta^2}+(1+\eta_0)\frac{\partial^4 w}{\partial \theta^4}+\right.\\
&\left.+12\left(\frac{R}{t}\right)^2\left[(1+\mu)\left(w\text{-}\frac{\partial v}{\partial \theta}\right)-\nu\frac{\partial u}{\partial x}\right]+\left(\frac{R^3}{D}\right)p\left[\frac{1}{2}\frac{\partial^2 w}{\partial x^2}+\frac{\partial^2 w}{\partial \theta^2}\right]\right\}=0
\end{aligned} \tag{6.17c}$$

Prior to buckling, the stresses are purely membrane-related, thus:

$$N_{x1} = -pR/2 \tag{6.18a}$$

$$N_{\theta 1} = -pR \tag{6.18b}$$

The boundaries of the cylindrical shell are assumed to be simply supported:

$$\left.\begin{aligned} N_x &= 0 \\ M_x &= 0 \\ v &= 0 \\ w &= 0 \end{aligned}\right\} \text{ at } x = 0,\ L/R \tag{6.19}$$

The following relations are assumed for the deflections:

$$u = A\sin(n\theta)\cos\left(\frac{m\pi R}{L_b}x\right) \tag{6.20a}$$

$$v = B\cos(n\theta)\sin\left(\frac{m\pi R}{L_b}x\right) \tag{6.20b}$$

$$w = C\sin(n\theta)\sin\left(\frac{m\pi R}{L_b}x\right) \tag{6.20c}$$

Substituting expressions (6.20) in the equations of equilibrium (6.17a) and (6.17b) and solving the algebraic equations yield coefficients A and B as functions of C:

$$A = \left(\frac{D_{1n}}{D_{0n}}\right)C = a_n C \tag{6.21a}$$

$$B = \left(\frac{D_{2n}}{D_{0n}}\right)C = b_n C \tag{6.21b}$$

where

$$D_{0n} = \left[\frac{1-\nu}{2}\right](1+\mu)n^4 + \left[1+\mu-\nu\right]m^2n^2\lambda^2 + \left[\frac{1-\nu}{2}\right]m^4\lambda^4$$

$$D_{1n} = -\left[\frac{1+\nu}{2}\right]\chi m\lambda n^4 + (1+\mu)\left(\frac{1-\nu}{2}\right)m\lambda n^2 - \nu\left[\frac{1-\nu}{2}\right]m^3\lambda^3$$

$$D_{2n} = \left[\frac{1-\nu}{2}\right]\chi n^5 + \left\{\chi m^2\lambda^2 - \left(\frac{1-\nu}{2}\right)(1+\mu)\right\}n^3 + \left\{\left[\frac{1+\nu}{2}\right]\nu - (1+\mu)\right\}m^2\lambda^2 n$$

Introducing (6.20) and (6.21) into (6.17c), this becomes:

$$\begin{aligned} &\zeta(-2n^2 - b_n n^3) + m^4\lambda^4 + m^4\lambda^4 + (2+\eta_t)m^2\lambda^2 n^2 + (1+\eta_0)n^4 + \\ &+12\left(\frac{R}{t}\right)^2\left[(1+\mu)(1+b_n n) + \nu m\lambda a_n\right] - \left(\frac{R^3}{D}\right)p\left(\frac{m^2\lambda^2}{2} + n^2\right) = 0 \end{aligned} \tag{6.22}$$

From this equation, the critical elastic buckling pressure for general instability of the ring-stiffened shell, p_{cr}, is obtained. This is the lowest value of p, and this is obtained by selecting the values of n and m that minimise p.

TABLE 6.1
Overall Buckling Pressure for Internally and Externally Ring-Stiffened Cylinders

	Shell Geometry		Overall Buckling Pressure p_{cr} (N/mm²)
R/t=82.17	Unstiffened		0.708
L/R=4.5391	light ring frames	External e/t=−1.653	2.264
	($A_f/L_f t$=0.1471)	Internal e/t=+1.653	2.569
R/t=82.99	Unstiffened		0.687
L/R=4.5384	heavy ring frames	External e/t=−2.817	8.104
	($A_f/L_f t$=0.2948)	Internal e/t=+2.817	9.021

Results from a comparative study are included in Table 6.1. In this table, the overall strength of ring-stiffened cylinders with (a) light ring frames and (b) heavier ring frames is compared with that of the unstiffened cylinder.

For both frame sizes, both internal and external ring frames are considered. In the case of light internal ring frames, the general instability pressure increases by 263% compared to the unstiffened shell, whereas in the case of heavy internal ring frames, the increase is 1,210%. This shows clearly the effect that ring frame size has on the general instability pressure.

When the collapse pressures for internally and externally stiffened shells are compared, it is seen that in the case of light internal ring frames, the collapse pressure is higher than that of external ring frames by 11.5%. In the case of heavier ring frames, the difference increases to 13.5%. A preliminary interpretation of these results is that the higher instability pressure in the case of the internal ring frames is due to their greater stiffness. This is due to their smaller depth-to-radius ratio, when compared to external ring frames. This increased strength is obtained with slightly less material, since for identical cross sections, the external ring frame has an increased length. The question of ring frame eccentricity is considered in more detail in the next section.

6.3.3 The Effect of Ring Frame Eccentricity

The effect of eccentricity is based on the study of the manner in which forces and moments interact in the stiffened shell [19]. To this end, expressions (6.14) and (6.15) are used. Substituting (6.13a–f) in (6.14) and (6.15), these become:

$$N_x = \frac{Et}{(1-\nu^2)}\left[\frac{\partial u}{\partial x} + \nu\left(\frac{\partial v}{\partial \theta} - w\right)\right] \tag{6.23a}$$

$$N_\theta = \frac{Et}{(1-\nu^2)}\left[\left(\frac{\partial v}{\partial \theta} - w\right)(1+\mu) + \nu\frac{\partial u}{\partial x} - \chi\frac{\partial^2 w}{\partial \theta^2}\right] \tag{6.23b}$$

$$N_{x\theta} = N_{\theta x} = \left[\frac{Et}{2(1+\nu)}\right]\left(\frac{\partial u}{\partial \theta} + \frac{\partial v}{\partial x}\right) \tag{6.23c}$$

$$M_x = (-D/R)\left[\frac{\partial^2 w}{\partial x^2} + \nu\frac{\partial^2 w}{\partial \theta^2}\right] \tag{6.24a}$$

$$M_\theta = (-D/R)\left[\frac{\partial^2 w}{\partial \theta^2}(1+\eta_0) + \nu\frac{\partial^2 w}{\partial x^2} - \zeta\left(\frac{\partial v}{\partial \theta} - w\right)\right] \tag{6.24b}$$

$$M_{x\theta} = (+D/R)(1-\nu)\frac{\partial^2 w}{\partial x\,\partial\theta} \tag{6.24c}$$

$$M_{\theta x} = (-D/R)\left[(1-\nu)+\eta_t\right]\frac{\partial^2 w}{\partial x\,\partial\theta} \tag{6.24d}$$

The effect of ring frame eccentricity on the forces and moments in the cross section is seen to be represented by the terms

$$-\chi\frac{\partial^2 w}{\partial\theta^2} \quad \text{and} \quad \eta_0\frac{\partial^2 w}{\partial\theta^2}, \quad -\zeta\left(\frac{\partial v}{\partial\theta}-w\right)$$

in (6.23b) and (6.24b), respectively, for N_θ and M_θ. For long thin shells, M_θ is the most important factor determining the buckling strength. It is seen from (6.24b) that M_θ depends on

$$\frac{\partial^2 w}{\partial\theta^2}\eta_0-\zeta\left(\frac{\partial v}{\partial\theta}-w\right)$$

As the eccentricity e is increased, η_0 also increases, independently of the sign of e, as is clear from the defining expression. The second term in the above expression reduces M_θ whether internal or external ring frames are used.

The eccentricity effect is thus seen to result from the cross-coupling of the bending moments with the membrane forces, the circumferential mid-surface strain, ε_θ, being the major coupling factor. It is made up of two opposing contributions. The first of these, termed the *primary effect*, concerns the manner in which the circumferential membrane stresses in the shell, N_θ, affect the bending stiffness of the shell–frame combination, $\overline{M_\theta}$ (Figure 6.6). The *secondary effect* is the reverse and concerns the effect that the bending strains in the shell–frame combination have on the circumferential membrane stresses in the shell (Figure 6.7).

The total geometric bending stiffness of the shell–ring frame combination is not affected by the position of the ring frame. When buckling occurs, additional membrane forces arise as a result of the initial curvature of the shell. In the case of circumferential membrane forces, N_θ, in the case of outward buckles additional tensile strains and stresses arise, whereas the reverse is the case for inward buckles.

A relationship between the axial and circumferential membrane forces is obtained by differentiating the first two stability equations (6.16a) and (6.16b) with respect to x and θ:

$$\frac{\partial^2 N_x}{\partial x^2} = \frac{\partial^2 N_\theta}{\partial\theta^2} \tag{6.25}$$

Substituting the expressions for the assumed displacements (6.20) into (6.25), the relationship between the membrane forces N_x and N_θ becomes:

$$m^2\lambda^2 N_x = n^2 N_\theta \tag{6.26}$$

As mentioned, N_θ is compressive in a positive (inward) wave and tensile in a negative wave. From (6.26), it is seen that N_x follows N_θ at every point of the shell, and that for long shells, N_x is much larger than N_θ (since for hydrostatic pressure $m=1$ and in long shells, $\beta<1$ while n is always larger than 2).

It should be noted that (6.26) is true only for the classical simple support boundary conditions assumed here (6.19). In Figure 6.6, $\overline{M_\theta}$ represents the geometrical bending stiffness of the cross

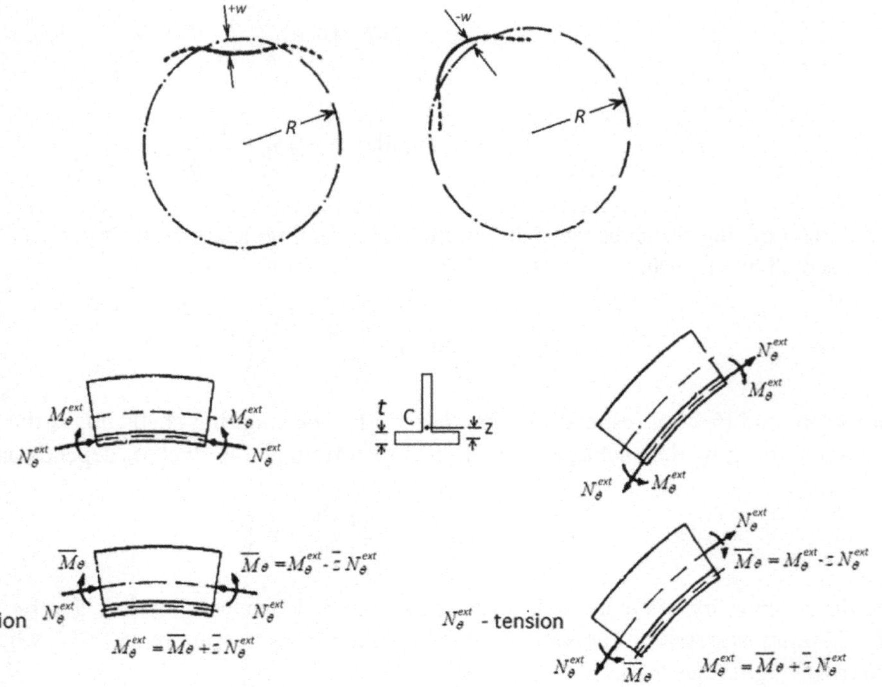

FIGURE 6.6 The primary eccentricity effect (external ring frame) [19].

section of the ring–shell combination and is the bending moment necessary to produce a certain change in curvature.

The circumferential membrane force acting in the shell brings about a change in the bending stiffness of the shell–ring frame cross section. In the case of internal ring frames, the total bending stiffness $\overline{M}_\theta$ is given by:

$$M_\theta^{int} = \overline{M}_\theta - \bar{z}N_\theta^{int} \tag{6.27}$$

where M_θ^{int} is the bending moment necessary to bring about the same change in curvature that $\overline{M}_\theta$ would produce without the membrane force N_θ^{int}. Also,

$$M_\theta^{ext} = \overline{M}_\theta + \bar{z}N_\theta^{ext} \tag{6.28}$$

where M_θ^{ext} is the moment necessary to produce the same change of curvature which would arise without the action of the membrane force N_θ^{ext}. From (6.27) and (6.28), it is seen that the bending stiffness for external stiffening is larger than that for internal stiffening. This is the primary eccentricity effect.

There is, however, an opposing secondary effect that will now be examined. Consider a shell with internal ring frames. In a positive wave, the moment M_θ gives rise to an additional compressive strain in the circumferential direction in the shell. Due to the Poisson effect an axial strain arises in the shell, so that an additional compressive membrane force, ΔN_x, in the axial direction resists this strain. From equilibrium considerations (6.26), the force increment ΔN_x is accompanied by an additional compressive membrane force ΔN_θ in the circumferential direction. This additional compressive force has a radial component that resists radial deformation (Figure 6.7a). In the case of external

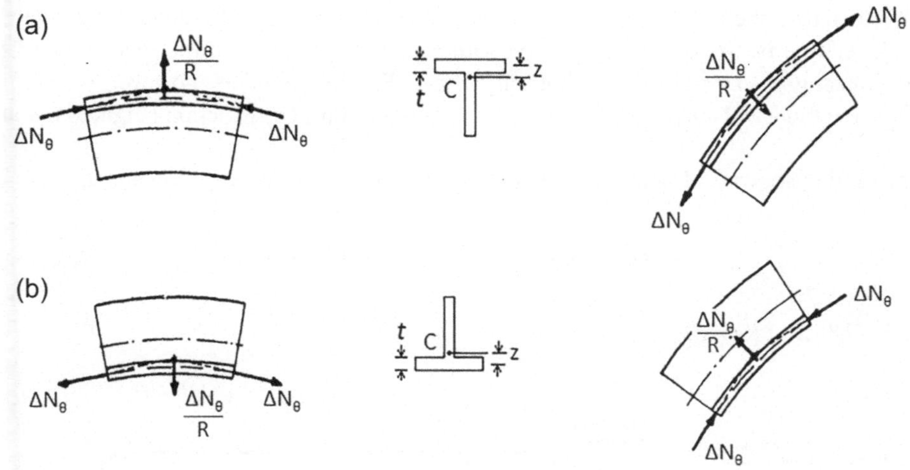

FIGURE 6.7 The secondary eccentricity effect [19]. (a) Internal ring frame. (b) External ring frame.

ring frames, the additional force ΔN_θ is tensile, and therefore, it assists deformation (Figure 6.7b). In a negative wave, the same argument applies, and the additional membrane force ΔN_θ resists the deformation for inside rings, whereas it assists it for external ring frames.

The effects of ring frame eccentricity can therefore be summarised as follows:

1. *Primary Effect*: external ring frames increase the bending stiffness in the circumferential direction more so than internal ring frames
2. *Secondary Effect*: internal ring frames increase the extensional stiffness in the circumferential direction more so than external ring frames.

In the case of long cylinders, N_θ is very small, and the difference in the actual bending stiffness in the circumferential direction for internal and external ring frames, given by (6.27) and (6.28), is also very small. On the contrary, M_θ is relatively large, and therefore, the ΔN_θ produced by it is important. Hence, for long cylinders, the critical load for internal stiffening is larger than that for external stiffening. It should be remembered that this is due to the Poisson effect (ν). If the Poisson effect is neglected ($\nu = 0$), external ring frames yield higher critical loads than internal ring frames, even in the case of long cylinders (Table 6.2).

TABLE 6.2
Effect of Poisson's Ratio on the Eccentricity Effect $I_f/Rt^3 = 5$ $A_f/L_f t = 0.5$ $e/t = \pm 5$

			p_{cr}		
R/t	***L/R***	ν	**+e (internal)**	**−e (external)**	(p_{cr}^-/p_{cr}^+)[a]
250	0.5	0	42.79	68.03	1.590
		0.3	40.88	56.57	1.384
		0.5	34.98	44.32	1.267
2,000	2.0	0	67.51	70.00	1.037
		0.3	69.14	57.69	0.834
		0.5	62.88	45.12	0.717

[a] Ratio of critical buckling pressures for shell with external (p_{cr}^-) and internal (p_{cr}^+) ring frames.

In ring-stiffened shells, N_θ diminishes rapidly with (L/R) while M_θ remains relatively large. Hence, ΔN_θ caused by M_θ is large compared with N_θ for long shells and the secondary effect can dominate, causing *inversion* of the eccentricity effect (Figure 6.9). This, however, does not affect ring-stiffened cylindrical shells used in submarine pressure hull construction (Figures 6.8 and 6.9) (Table 6.3).

For the section shown in Figure 6.8, the distance between successive bulkheads, L_b, is

$$L_b = (42 - 14)(1.140)\text{ m} = 31.92\text{ m}$$

Hence $L/R = 7.88$ and $R/t = 62.12$

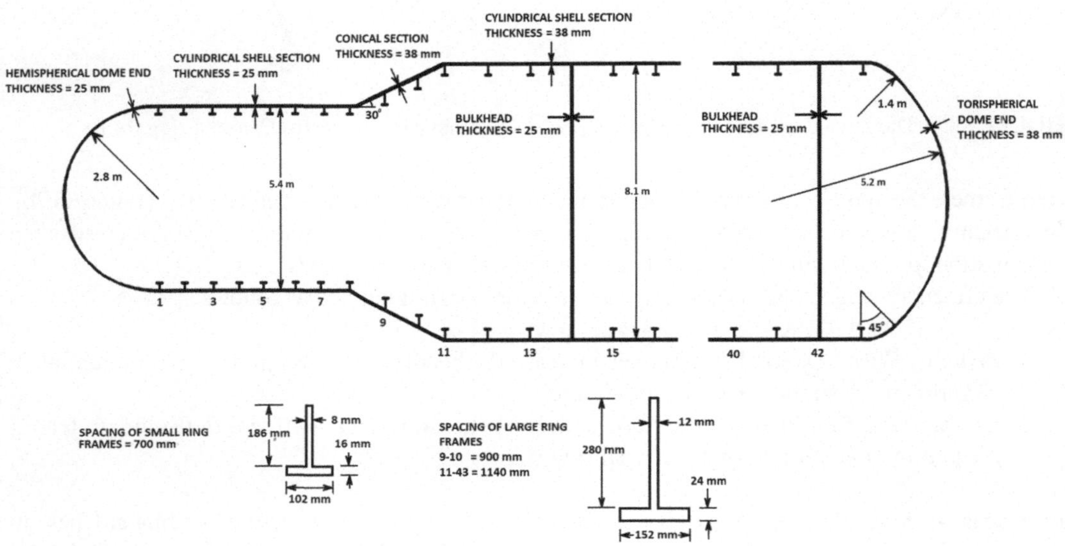

FIGURE 6.8 Submarine section (Kendrick).

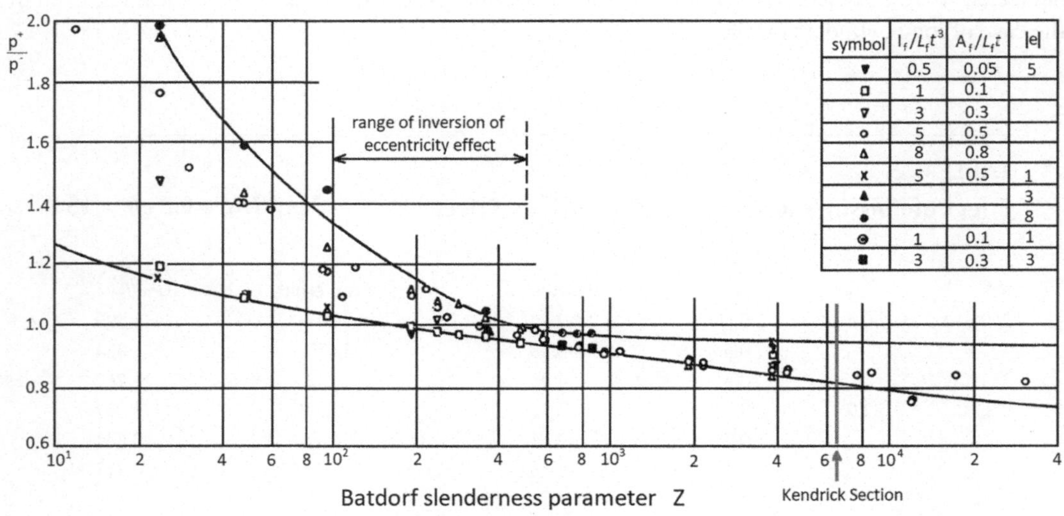

FIGURE 6.9 Eccentricity effect versus Batdorf slenderness parameter (*Z*) [19].

TABLE 6.3
Effect of Eccentricity on Buckling Pressure (Lateral and Hydrostatic)
$I_f/Rt^3 = 5$ $A_f/L_f t = 0.5$ $e/t = \pm 5$

				p_{cr} (Hydrostatic)			p_{cr} (Lateral)		
Case	L/R	R/t	Z	+e (internal)	−e (external)	p_{cr}^-/p_{cr}^+	+e (internal)	−e (external)	
1	0.5	50	11.98	6.23	12.28	1.972	19.46	32.39	1.664
4		100	23.85	15.08	26.66	1.767	31.86	48.59	1.525
8		250	59.62	40.88	56.57	1.384	63.15	79.37	1.257
13		500	119.2	76.80	91.45	1.190	100.48	113.95	1.134
23		1,000	238.5	132.4	140.14	1.058	158.54	162.98	1.028
33		2,000	477.0	217.0	207.78	0.957	242.35	232.01	0.957

Substituting in the expression for the Batdorf parameter, we obtain:

$$Z = \sqrt{1-\nu^2}\,(L_b/R)^2(R/t) = 6{,}316$$

We observe from Figure 6.9 that for this type of cylindrical section, inversion of eccentricity does not arise.

6.3.4 Fabrication Issues

As discussed previously, the form of the weld-induced initial distortions in ring-stiffened cylinders depends on whether the cylindrical shell is internally or externally stiffened (Figure 6.10).

In the case of internal framing, welding is only necessary to ensure that the frame is securely attached to the shell plating and to provide rotational stiffness to the shell–frame intersection in order to avoid lateral-torsional buckling (tripping) of the frame. As such, welding is not essential to the integrity of the pressure hull as the compressive load resulting from the hydrostatic pressure is transmitted to the frame even in the absence of welding. In the case of external frames, the load transfer between a frame and the cylindrical shell involves inducing a tensile stress field in the fillet welds on either side of the intersection of the frame web and the shell plating. Any defect or weld cracking could lead to separation of the shell from the frame resulting in pressure hull collapse [4]. It is therefore seen that from the point of view of fabrication, internal stiffening provides a greater degree of safety.

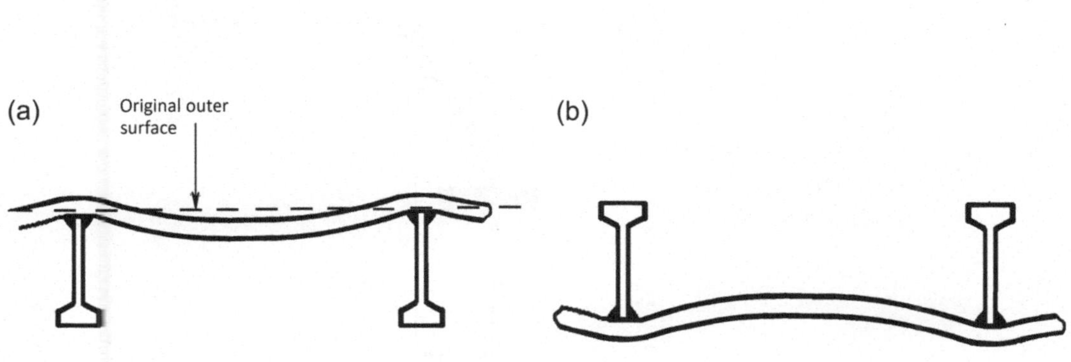

FIGURE 6.10 Weld-induced initial distortions in the shell plating in the axial direction. (a) Internally stiffened. (b) Externally stiffened.

6.4 COLLAPSE MODES OF INTERNAL RING FRAMES

The nomenclature for a typical shell–frame intersection is shown in Figure 6.11.

It was mentioned previously that ring frame failure can result in premature overall collapse between consecutive transverse bulkheads. Failure modes include buckling, which can take a variety of forms, as well as yielding failure. In the case of local buckling, one structural member suffers failure, such as the frame web, shown in Figure 6.12.

Flexural-torsional buckling may also occur, in which case the frame bends and twists about the line of attachment to the shell plating (Figure 6.13).

The external hydrostatic pressure results in compressive and shear stresses in the ring frame, and we can consider the web and flange as plate elements each of which has its particular boundary conditions. Buckling can take the following forms:

- Flange buckling
- Web buckling
- Flexural-torsional buckling about the line of attachment to the shell

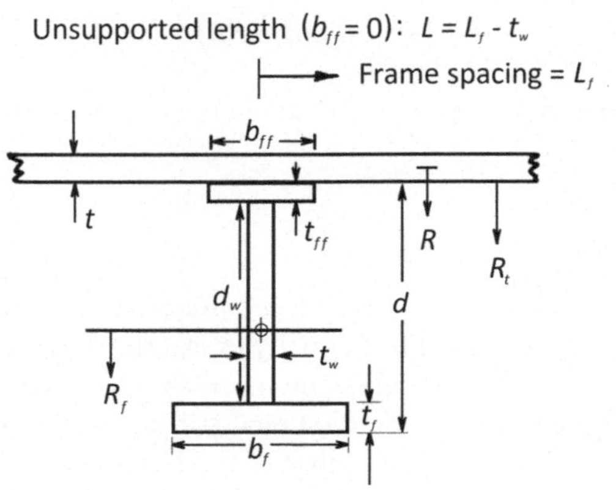

FIGURE 6.11 Nomenclature of ring-stiffened shell section.

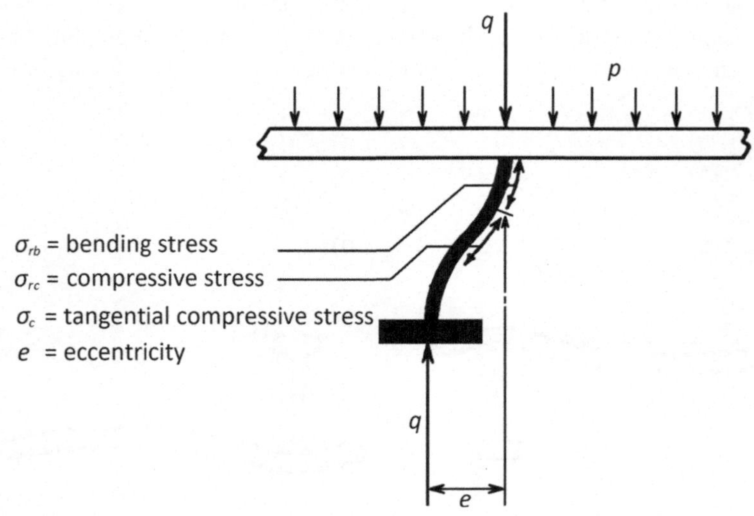

FIGURE 6.12 Buckling of ring frame web.

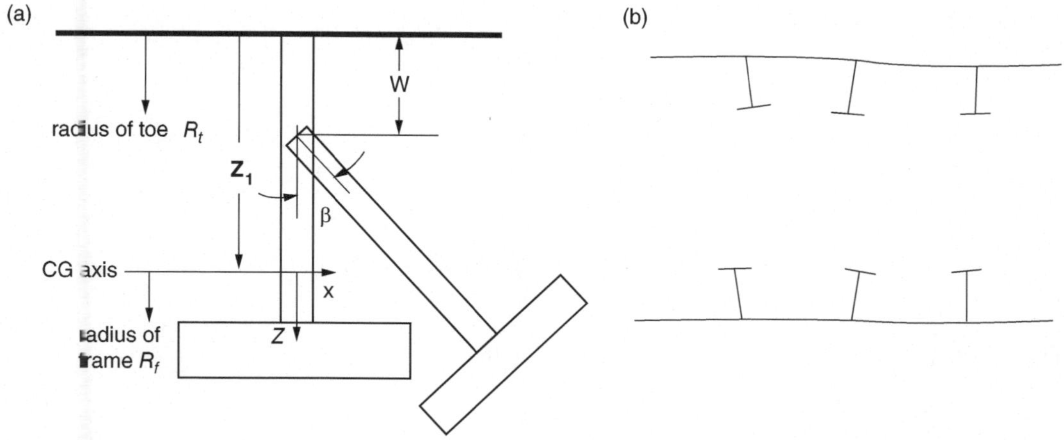

FIGURE 6.13 Flexural-torsional buckling of a ring frame. (a) Geometrical definition. (b) Shell deformation.

The above failure modes do not bring about any change in the form of the cylindrical cross section. This can arise only as a result of local loss of stiffness of the frame. In order for the ring frame to perform efficiently, deep webs and relatively thick flanges are required. In this way, the moment of inertia and thus the stiffness of the frame increase. The result is an increase in frame section overall buckling strength and also avoidance of flange local buckling.

In such an arrangement, however, the web may suffer buckling as it is relatively thin. The critical buckling pressure depends on not only slenderness but also on the degree of fixity to the shell plating. If the section does not have sufficient bending or torsional stiffness, flexural-torsional buckling of the frame may occur.

The situation deteriorates with the use of high-tensile steel. Compressive stresses are then higher than when mild steel is used since scantlings can be reduced. The reduction in thickness results in a corresponding reduction in elastic buckling stress. For the same applied loading, the stresses increase and the buckling strength reduces. This results in a significant reduction in safety margin, thereby necessitating a suitable selection of safety factors.

6.4.1 Ring Frame Flexural-Torsional Buckling

The mean flexural-torsional buckling stress of the ring frame can be calculated assuming that it is simply supported along the line of attachment to the shell plating. This is a conservative assumption because it has been shown that along the line of attachment to the shell plating approaches, the clamped condition prevails, and for this reason, the buckling pressure is significantly higher. Since the frame section does not distort, the strain corresponding to a translation w and a rotation β, shown in Figure 6.13a, is:

$$\begin{aligned}\varepsilon &= \left[\frac{-w+(Z+z_1)(1-\cos\beta)+x\sin\beta}{R_t-(Z+z_1)}\right] \\ &= \frac{1}{R_t}\left[-w+\frac{Z+z}{2}\beta^2+x\beta\right]\end{aligned} \tag{6.29}$$

The internal strain energy is:

$$
\begin{aligned}
U &= \pi R_t E \iint \varepsilon^2 \, dA \\
&= \frac{\pi E}{R_t} \iint \left[w^2 - w(Z + z_l)\beta^2 + x^2\beta^2 \right] dA \\
&= \frac{\pi E}{R_t} (w^2 A_f - w A_f Z \beta^2 + I_f \beta^2) = q R_t^2 w
\end{aligned}
\tag{6.30}
$$

The external work of the applied load q per unit length is:

$$W = \pi R_t q w \tag{6.31}$$

From (6.30) and (6.31), we have:

$$E(w^2 A_f - w A_f Z \beta^2 + I_f \beta^2) = q R_t^2 w \tag{6.32}$$

The solution $\beta=0$, $w=qR_t^2/EA_f$ is of no practical interest, and for there to exist a solution with $\beta \neq 0$, the following conditions should hold:

$$w = qR_t^2 / EA_f \quad \text{and} \quad \sigma = I_f / A_f Z$$

Solving for q and σ:

$$q = \frac{EI_f}{R_t^2 Z} \tag{6.33a}$$

and

$$\sigma_t = \frac{EI_f}{A_f R_t z_1} \tag{6.33b}$$

If we assume that the tripping stress, σ_t, arises in way of the centroidal axis of the section, the flexural-torsional (tripping) buckling pressure is:

$$p_t = p_{fy} \left(\frac{E}{\sigma_0} \right) \left(\frac{R_t}{R_f} \right) \frac{I_f}{A_f R z_1} \tag{6.34}$$

The tripping stress formula (6.33b) is valid in the case of symmetrical sections that are welded to the internal surface of the cylindrical shell. It underestimates the buckling loads of external frames and of solid sections. An important feature of this type of failure is that it is precipitated by a very small degree of out-of-circularity in way of the line of attachment of the ring frame to the cylindrical shell. When buckling occurs, the circumferential and the radial stiffness of the ring frame reduce, and it is thus necessary to increase the relevant safety factor.

This equation is based on the assumption that the web does not distort but remains flat when the section fails. A second assumption is that the web is simply supported along the line of attachment to the cylindrical shell. The latter is a conservative assumption given that the true condition is closer to clamping.

Despite the above, it has been found that this relationship is sufficiently accurate for design purposes unless $b_f/d_w > 0.6$ and $d_w/t_w > 20$ (thin web). The flexural-torsional buckling stress depends, as mentioned above, on the degree of rotational fixity. If the moment that develops along the line of attachment is small and the angle of rotation is large, it is possible for the flexural-torsional buckling stress to be lower than the buckling stress and for this form of failure to become critical. Otherwise, material yielding is the critical failure mode.

Example 6.2

For the T-bar shaped ring frames of Example 6.1, we have:

$$I_f = (3.5)\left(10^6\right)\text{mm}^4$$

$$z_1 = 200 \text{ mm}$$

Calculate the flexural-torsional buckling pressure of the ring frame and assess whether the structure has adequate strength.

Solution

From (6.34), we have:

$$p_t = (5{,}990)\left(\frac{207{,}000}{540}\right)\left(\frac{3{,}800}{3{,}750}\right)\frac{(3.5)10^6}{(4{,}500)(400)(200)}$$

$$= 2.262 \text{ N/mm}^2$$

$$= 0.45 p_0 \left(\text{inadequate strength}\right)$$

The situation becomes more complicated as a result of ring frame *initial tilt,* which reduces the flexural-torsional buckling stress, rendering it the critical failure mode. An initial tilt of the order of the web thickness results in an increase in web stress of up to 25%. For modern submarine structures, the average stress in the ring frame is approximately $(2/3)\sigma_0$ when the vessel is operating at its design depth, and thus, a small initial deformation is not expected to result in the collapse of the section. The consequences of ring frame initial tilt are considered in a later section.

Summarising, the factors that affect ring frame behaviour are:

- Ring frame buckling, especially the polar moment of inertia of the section about the line of attachment with the cylindrical shell
- The web depth d_w in relation to the hull diameter $(d_w/2R)$
- The initial tilt of the ring frame
- The total stress in way of the ring frame flange which is the sum of the residual stresses and the stresses resulting from the applied load
- Whether the ring frames are internal or external. External ring frames retain a greater reserve of strength
- The web thickness in relation to its depth (d_w/t_w)
- The modulus of elasticity of the material (E)

6.4.2 Ring Frame Yielding

Ring frame yielding can result from asymmetrical deformation of the section. In specific regions of the section, stress concentrations arise as a result of welding with other structural members (external ballast tanks, decks, etc.). If the ring frames were perfectly circular, yielding in the frame would not occur before yielding in the cylindrical shell plating because the circumferential stress would be less than in the plating. In the case of ring frames with imperfections in addition to circumferential membrane stress, circumferential bending stresses also arise. These are due to the out-of-circularity and also due to the additional stiffness of the frame flange. The stresses are largest in the frame flange than in the shell.

The number of lobes depends also on the number and distribution of longitudinal weld seams. As a result, it is possible to increase structural strength by using a different distribution of longitudinal welds.

Ring frame analysis can be carried out either by considering the frame and attached shell plating together or by considering the frame in isolation but attached to the cylindrical shell. We shall first consider the frame and shell plating as a unit. The elastic buckling pressure is given by the Bresse equation (4.1). The critical pressure then arises when $n=2$ ($n=1$ has no physical meaning). For $n=2$,

$$p_{2cr} = \frac{3EI_{sf}}{R^3 L_f}$$

In the previous problem, ($R=4{,}000\,\text{mm}$, $t=32\,\text{mm}$, $E=207\,\text{GPa}$), we assume that ring frames are added at a spacing of $L_f = 750\,\text{mm}$, and thus, the moment of inertia I_{sf} is $1.83\times10^8\,\text{mm}^4$. If we compare the effect of ring frames with the section strength, we note that the yield pressure increases from 0.03 to 2.37 MPa. Clearly, the strength of the section increases considerably as a result of the presence of the ring frames.

6.4.2.1 Approximate Calculation of Ring Frame Yield Pressure

To a first order of accuracy, we can estimate the yield pressure on the basis of the following simplifying assumptions:

- Interframe shell bending is ignored.
- The transverse membrane stress is assumed to remain constant over the whole shell surface.
- The cylindrical shell is considered to be perfectly circular.

The force acting on the section in the circumferential direction is then:

$$F = \sigma(A_f + L_f t) \tag{6.35}$$

For equilibrium,

$$2F\theta = 2pL_f\theta t$$

and

$$F = pL_f R \tag{6.36}$$

Hence,

$$p_{fy} \cong \sigma\left(\frac{t}{R}\right)\left[1+\left(\frac{A_f}{L_f t}\right)\right] \tag{6.37}$$

This is seen to be identical to (6.1). In the case of Example 6.1, substituting in (6.37), we obtain:

$$p_{fy} \cong 4.32\left(1+\frac{4{,}500}{24{,}576}\right) = 5.13\ \text{MPa}$$

Expression (6.37) gives a critical pressure that is much higher than the actual one (Example 6.1).

6.4.2.2 Exact Calculation of Ring Frame Yield Pressure

To obtain a more accurate estimate of the pressure to cause yield in the ring frame, we assume that:

- The bulkheads are perfectly circular.
- The initial distortion at the midpoint between successive transverse bulkheads is given by the Fourier series

$$\sum_{n=1}^{\infty} C_n \sin n\theta$$

whereas at other locations, it is given by $\sin(\pi x/L_b)$, where x is the distance to the bulkhead, and L_b is the transverse bulkhead spacing.
- The total stress in the frame flange can be calculated using superposition of the bending and membrane stress components in the circumferential direction.

From (4.85) and (6.34), we calculate the axisymmetric stress in the flange:

$$\sigma_f = \left(\frac{p}{p_{fy}}\right)\sigma_0 \tag{6.38}$$

To this we need to add the bending stress that is induced by out-of-circularity, given by (6.51). The following quadratic equation for the ring frame yield pressure, p_{fyn}, is obtained:

$$\left(\frac{p_{fyn}}{p_{fy}}\right)^2 - \left(\frac{p_{fyn}}{p_{fy}}\right)\left\{\frac{Ee_f}{\sigma_0}\frac{c_n}{R^2}(n^2-1) + \frac{p_n}{p_{fy}} + 1\right\} + \frac{p_n}{p_{fy}} = 0 \tag{6.39}$$

Example 6.3

Calculate the frame yield pressure for the previous example for $n=3, 4, 5$.

Solution

Substituting in (6.39), we obtain:

$$\left(\frac{p_{fyn}}{5.990}\right)^2 - \left(\frac{p_{fyn}}{5.990}\right)\left\{(0.115)(n^2-1) + \frac{p_n}{5.990} + 1\right\} + \frac{p_n}{5.990} = 0$$

Substituting for n and p_n from Example 5.1, we obtain:

$$p_{fy3} = 2.986 \text{ N/mm}^2$$

$$p_{fy4} = 2.773 \text{ N/mm}^2$$

$$p_{fy5} = 2.891 \text{ N/mm}^2$$

We conclude that the critical eigenform is $n=4$, whereas the pressure to cause yielding in the ring frame is only $0.56p_0$.

6.5 GENERAL INSTABILITY OF CYLINDERS FITTED WITH INTERMEDIATE HEAVY FRAMES

In this section, we shall describe the approach developed to predict the strength of a ring-stiffened cylindrical shell fitted with intermediate heavy frames. The original approach to this problem was developed by Kendrick, who applied the Ritz method, described in Sections 5.3 and 5.5. This was modified by Reynolds, who introduced deflections functions in line with test results. The end result was a rigorous approach that was, however, considered not suitable for design purposes, and to this end, Blumenberg, [3], on the basis of test results, proposed a simple formula that is sufficiently accurate and can be used in design.

Early test results indicated that a heavy frame can be effective in increasing the overall collapse strength of a ring-stiffened cylindrical shell so that in principle, they can be used to this end. In this section, we shall give some details of a test program described in [3] that was initiated in order to determine the effect of:

- Cylinder length
- Intermediate heavy frame spacing
- Intermediate heavy frame strength

on the elastic overall collapse pressure of ring-stiffened cylindrical shells. Twenty-four models were tested, divided into four groups (Table 6.4). In each group, only the heavy frame strength was

TABLE 6.4
Test Model Data and Collapse Pressures

Model Group	Model Number	$\frac{I_{FSe}}{I_{fse}}$	Collapse Pressure (psi)	Mode of Failure: N	Mode of Failure: Longitudinal Buckling Pattern
1	1	1.00	240	2	Overall collapse
	2	5.66	299	2	Ditto
	3	12.92	407	2	Ditto
	4	18.75	464	2	Ditto
	5	25.92	555	2	Ditto
	6	49.95	795	2	Ditto
2	7	46.98	777	2	Overall collapse
	8	61.89	913	2	Ditto
	9	67.23	963	2	Ditto
	10	85.33	1,055	4	Between heavy frames
	11	103.71	1,020	4	Ditto
	12	1.00	480	3	Overall collapse
3	13	12.92	785	3	Overall collapse
	14	25.92	948	3	Ditto
	15	49.95	1,048	4	Between heavy frames
	16	1.00	621	3	Overall collapse
	17	5.66	797	3	Ditto
	18	12.92	967	3	Ditto
4	19	18.75	1,063	3	Overall collapse
	20	25.92	1,075	4	Between heavy frames
	21	25.92	758	2	Overall collapse
	22	46.98	1,115	2	Ditto
	23	67.23	1,471	2	Ditto
	24	85.33	1,704	5	Between heavy frames

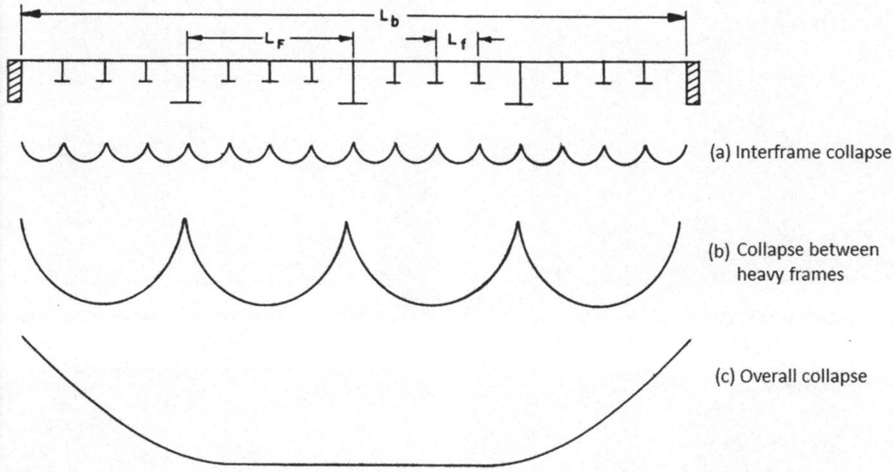

FIGURE 6.14 The three possible collapse modes of a stiffened cylindrical shell fitted with intermediate heavy frames [3].

varied, from that of an ordinary ring frame to that of frame that was considered to be fully effective. A fully effective frame is one that causes failure to occur between, and not including, adjacent heavy frames (collapse mode (b) in Figure 6.14).

To investigate the effect of cylinder length on overall collapse pressure, the models in groups 1–3 were designated $L_b/R=12.37$, 7.41, and 4.93, respectively. The models in group 4 were used to assess the effect of heavy frame spacing on overall collapse strength by comparing with the results from the group 1 models. For the group 4 models, $L_b/R=11.94$, but the distance between the heavy frames (L_F) was 40% smaller.

Model particulars, experimental collapse pressures, and corresponding buckling modes are listed in Table 6.4. A series of photographs (Figure 6.15) illustrates the collapse mode for each individual model. In column 3 of the table:

I_{FSe} = effective moment of inertia of a section consisting of a heavy frame and the effective shell about its centroid

I_{fse} = effective moment of inertia of a section consisting of an ordinary frame and the effective shell about its centroid

A study of the series of photographs that follows reveals the nature of collapse. In certain cases, it appears that failure has arisen between heavy frames; a study of strain measurements revealed, however, that overall collapse took place first, and this was followed by collapse between the heavy frames

The following relationship was found to provide the best estimate of the collapse pressure observed during the tests [3]:

$$p_{cr} = \frac{(I_{Fse} - I_{fse})}{(I'_{Fse} - I_{fse})}(p_F - p_B) + p_B \tag{6.40}$$

where

p_{cr} = critical pressure for elastic overall collapse

p_B = critical pressure for elastic overall collapse of the cylinder with the heavy frames replaced by ordinary frames

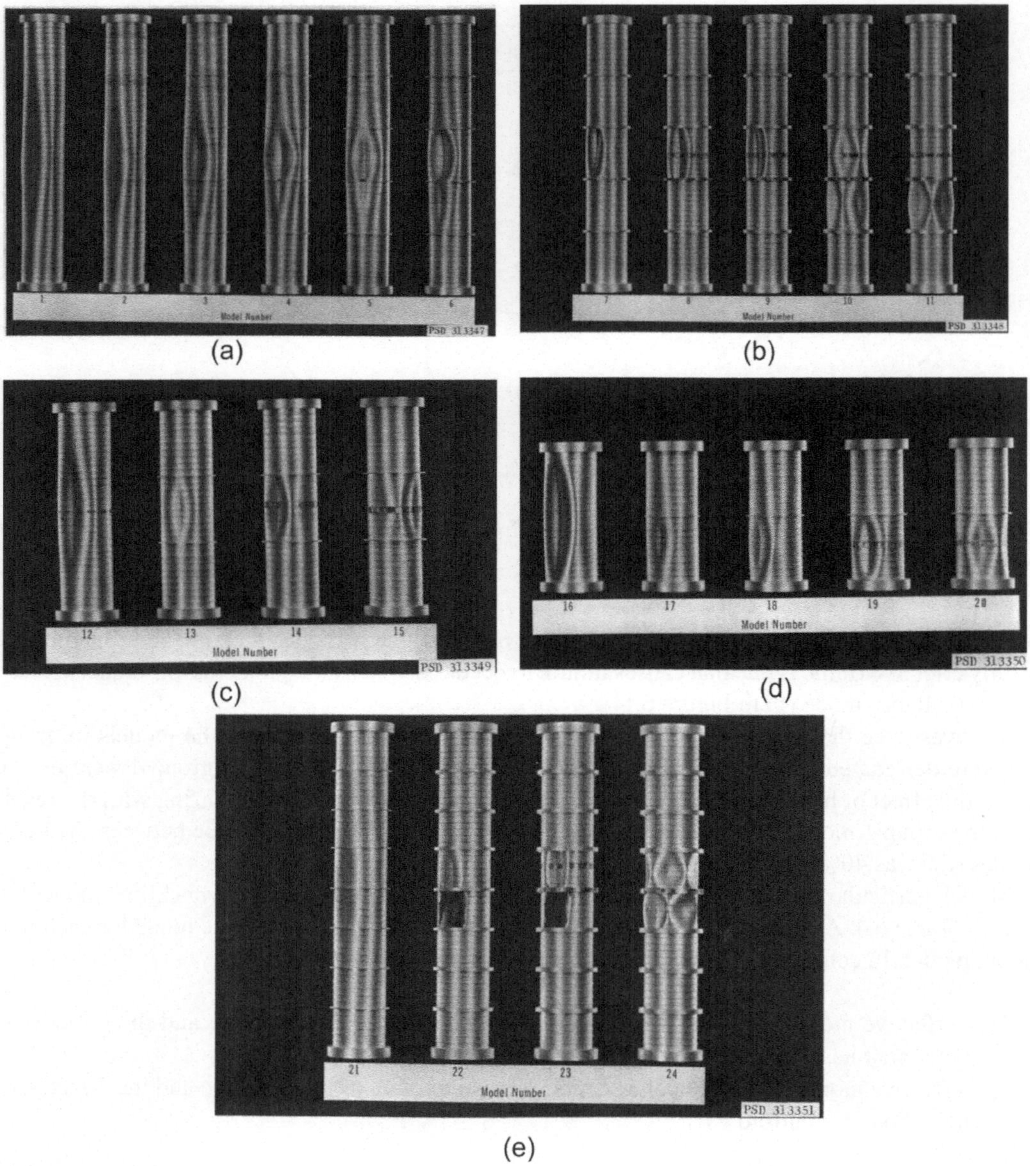

FIGURE 6.15 Test models with intermediate heavy frames following collapse. (a) Models 1–6: Overall collapse. (b) Models 7–8–9: Overall collapse. Models 10–11: Between heavy frames. (c) Models 12–13–14: Overall collapse. Model 15: Between heavy frames. (d) Models 16–17–18–19: Overall collapse. Model 20: Between heavy frames. (e) Models 21–22–23: Overall collapse. Model 24: Between heavy frames.

p_F = critical pressure for elastic overall collapse of the uniformly stiffened cylinder of length equal to the heavy frame spacing

I'_{FSe} = moment of inertia of the minimum strength of a fully effective heavy frame given by:

$$I'_{FSe} = \frac{p_F L_F R_C{}^3}{(n^2-1)E} \tag{6.41}$$

R_C = radius from the axis of the cylinder to the centroid of the effective heavy frame-shell section
n = critical buckling mode for p_B

The values of p_B and p_F can be determined using the graphs included in Figures 6.10 and 6.11. Substituting (6.41) in (6.40) gives:

$$p_{cr} = \frac{(I_{Fse} - I_{fse})(p_F - p_B)}{\dfrac{p_F L_F R_C^{\,3}}{(n^2-1)E} - I_{fse}} + p_B \tag{6.42}$$

This expression has been validated against the test results described previously.

6.6 INITIAL DISTORTIONS AND THEIR EFFECT ON BENDING STRESSES

Real structures differ from perfect ones for which the equations of equilibrium that have been obtained up to this stage are valid. The most important geometrical imperfection in a cylindrical shell is out-of-circularity, whose magnitude increases as the applied loading increases. In real materials, the deformations that take place are relatively small prior to the onset of yielding. If, however, the form of the structure changes at some loading stage (change of buckled shape), then the elastic range is extended.

To determine the permissible out-of-circularity, it is necessary first to assess the effect of applied loading on the deformation of the cylindrical shell. Only after specifying the maximum permitted out-of-circularity can ring frame dimensions be calculated.

A second type of geometrical imperfection is the *initial tilt* of the ring frames, which also results in bending stresses.

Finally, a third type of geometrical imperfection that arises is the *lack of sphericity* of domed ends (hemispherical, torispherical). The strength of these structures reduces considerably in the presence of geometrical imperfections (Chapters 10 and 11).

6.6.1 Out-of-Circularity of Unstiffened Cylindrical Shells

In this instance, the imperfection can be expressed as the following Fourier series [11]:

$$w_0 = \sum_{m=1}^{\infty}\sum_{n=1}^{\infty} c_{0mn} \cos n\theta \sin\left(\frac{m\pi x}{L}\right) \tag{6.43}$$

The shell deflection component is expressed as:

$$w_s = w_{sx} + \sum_{m=1}^{\infty}\sum_{n=1}^{\infty} C_{0mn}\left\{\frac{p}{p_{cmn} - p}\right\}\cos(n\theta)\sin\left(\frac{m\pi x}{L}\right) \tag{6.44}$$

where p_{cmn} is the buckling pressure when m longitudinal buckles and n circumferential lobes occur, and w_{sx} is the axisymmetric component of the deformation. The bending stress in the circumferential direction, $\sigma_{\theta b}$, is then:

$$\sigma_{\theta b} = \frac{Et\nu}{2}\left(\frac{\partial^2 w_{sx}}{\partial x^2}\right) + \left[\frac{Et}{2R^2(1-\nu^2)}\right]\sum_{m=1}^{\infty}\sum_{n=1}^{\infty}\left\{\left[n^2 - 1 + \nu\left(\frac{m\pi R}{L}\right)^2\right] c_{0mn}\left[\frac{p}{p_{cmn} - p}\right]\cos(n\theta)\sin\left(\frac{m\pi x}{L}\right)\right\} \tag{6.45}$$

6.6.2 Out-of-Circularity of Stiffened Cylindrical Shells

In a corresponding manner, the geometrical imperfections of the cylindrical shell between frames are given by:

$$w_0 = \sum_{m=1}^{\infty}\sum_{n=1}^{\infty} c_{0mn}\cos(n\theta)\sin\left(\frac{m\pi x}{L_b}\right) \tag{6.46}$$

If we make use of only the first term of the above expansion, the imperfection is given by:

$$w_0 = c_n\cos(n\theta)\sin\left(\frac{m\pi x}{L_b}\right) \tag{6.47}$$

The deflection is then subjected to a magnification $p/(p_n - p)$ as a result of the beam-column effect. Thus,

$$w = w_0\left(\frac{p_n}{p_n - p}\right) \tag{6.48}$$

where w_0 is given by (6.43).

The change in curvature, $\Delta\kappa$, has components that are due to the change in radius and to the bending of the section:

$$\Delta\kappa = \pm\frac{1}{R^2}\left(w + \frac{d^2w}{d\theta^2}\right) \tag{6.49}$$

Example 6.4

Calculate the bending stresses in the frame flange of the previous example, if the out-of-circularity equals $R/200$ and the external pressure $p = 2.495$ N/mm² (corresponding to a depth of 250 m).

Solution

The critical pressures for $n = 2$–5 have been calculated in Example 6.3. Substituting in (6.51), we obtain:

$n = 2$

Hence, $n^2 - 1 = 3$

$$\sigma_{fb2} = \left(\frac{2.495}{21.475}\right)\left[\frac{(207{,}000)(240)(3)}{(200)(4{,}000)}\right] = 22\ \text{N/mm}^2$$

$$\sigma_{fb3} = 208\ \text{N/mm}^2$$

$$\sigma_{fb4} = 253\ \text{N/mm}^2$$

$$\sigma_{fb5} = 208\ \text{N/mm}^2$$

Comparing these results with those of Example 6.3, we note that the elastic buckling pressure occurs for $n = 3$, whereas the maximum bending stress arises in the flange occurs for $n = 4$ because of the presence of the term $(n^2 - 1)$ in the Bresse equation (4.1).

The mean stress in the ring frame is given by (4.85) and because the relationship between p_f and σ_f is linear, we can use the result of Example 3:

$$\sigma_f = \left(\frac{2.495}{4.34}\right)540 = 310.4 \text{ N/mm}^2$$

The maximum stress that occurs in the frame flange is given by the composite relation:

$$\sigma_{fb} = \sigma_f + \sigma_{fb4} = 310 + 253 = 563 \text{ N/mm}^2 (> \sigma_0)$$

Hence, at a depth of 250 m yielding will arise in the ring frames. They therefore need to be redesigned.

From engineering beam theory,

$$\frac{\sigma_{fb}}{e_f} = E\Delta\kappa \tag{6.50}$$

where e_f is the distance from the flange to the neutral axis of the shell-frame cross section.

Substituting in (6.50) from (6.47–6.49) we obtain:

$$\sigma_{fb} = \frac{p}{p_n - p}\left(\frac{Ee_f}{R^2}\right)c_n(n^2 - 1) \tag{6.51}$$

where:

p_n = overall buckling pressure (m buckles in the longitudinal and n in the circumferential direction)

c_n = out-of-circularity

The above expression gives the maximum bending stress at a mid-length anti-node. It can be used provided the pressure p does not approach the elastic buckling pressure, p_n.

6.6.3 Initial Tilt of Ring Frames

Initial tilt in ring frames gives rise to bending stresses that are magnified by $\sigma_f/(\sigma_t - \sigma_f)$ when external pressure is applied. The effect of initial tilt has been studied using various methods and [21] used numerical integration of the differential equation. A tilt equal to the frame web thickness produces an increase in stress of up to 25%. The nature of the deformation and more specifically the number of lobes affect the increase in bending stress. Thus, for $n = 3$, the additional bending stress is larger than that due to $n = 0$. This is because the elastic buckling stress occurs for $n = 3$. In real structures, tilt may exceed the frame web thickness, and for this reason, care is required during fabrication.

6.7 RESIDUAL STRESSES AND THEIR EFFECT ON LOCAL STRESSES

The presence of residual stresses brings about reductions in the strength and stiffness of metal structures. In the past, they have been of concern as a source of premature cracks and failure by fracture, and in order to ameliorate the situation, various types of thermal and mechanical processes have been used (stress relieving, peening, overloading). As will become apparent, residual stresses are closely linked to geometric imperfections.

In pressure hulls, they are primarily the result of *forming actions* and of *weld shrinkage*. We shall consider each in turn. In order to measure residual stresses, the methods that have been used can be classified as experimental, analytical, and numerical.

Forming actions give rise to permanent plastic behaviour and arise when sections are either hot-rolled or cold-rolled. Following hot rolling, uneven cooling of various parts leads to the formation of tensile and compressive stress fields that result in permanent, plastic deformations. Cold rolling on the other hand involves the cold bending of flat plates to adopt a cylindrical form, the cold

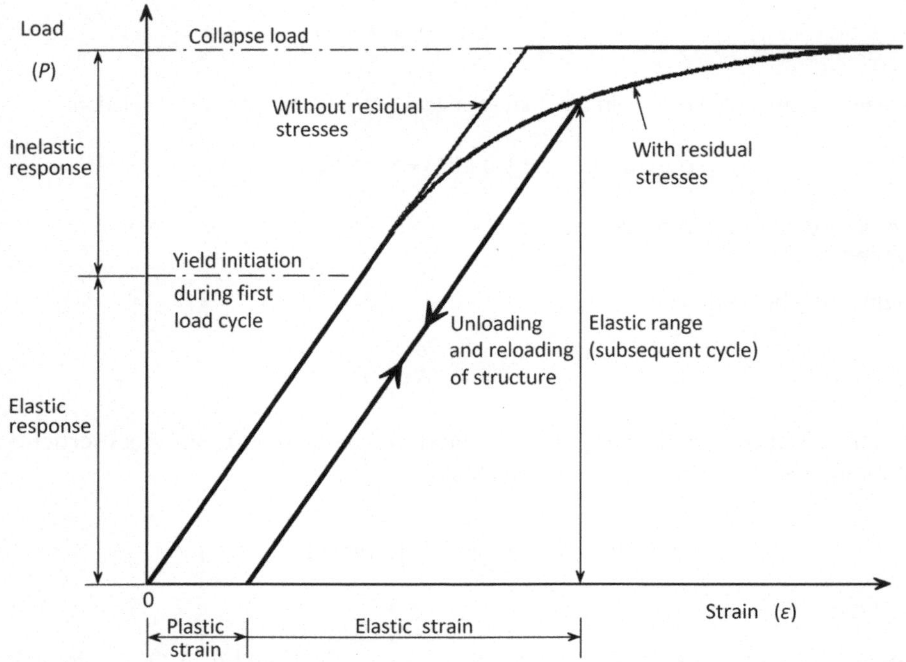

FIGURE 6.16 Effect of residual stresses on pre-collapse behaviour (plating).

bending of stiffeners (ring frames), and cold cambering in order to correct existing curvature. The most important of these forming actions is cold bending, which will be discussed in the next section.

When a structure that has not been stress-relieved is subjected to external loading, in the early stages of loading, the strains are elastic. The upper limit of elastic behaviour is lower than the collapse pressure, and during the first cycle of loading at some point, the strains cease to be elastic. However, if the structure is reloaded, the strains remain elastic until the applied pressure becomes equal to the maximum pressure that was reached during the previous load cycle (Figure 6.16).

6.7.1 Cold-Bending Residual Stresses in the Cylindrical Shell

At low pressures, inelastic behaviour is due to residual stresses that arise primarily as a result of *cold bending* of the cylindrical shell and to a lesser extent because of welding of the ring frames to the cylindrical shell. Cold-bending residual stresses can give rise to yielding at low pressures, but even in their absence, yielding will occur at a pressure that is lower than the collapse pressure (Figure 6.16). They have been found to have a negligible effect on the collapse strength of the cylindrical shell.

The residual stresses that result from cold bending are caused by the imposition of excess bending which results in partial yielding (Figure 6.17a). Following removal of the bending load, the plating springs back to its permanent deformed condition (Figure 6.17b). The residual stresses that result are shown in Figure 6.17c and are obtained by adding the stress distributions due to over-bending and spring-back. It is possible to incorporate work-hardening rules but a simple elasto-plastic model has been found to be sufficiently accurate for submarine steels. Larger errors are likely to result from the lack of knowledge of structural properties of the particular plate rather than from neglect of work hardening (or the Bauschinger effect). A theoretical method of estimating the residual stresses due to cold bending has been proposed by Mitchell and used by Smith and Mackay. This is a finite strip method which successively estimates the net stress in layers for each structural component.

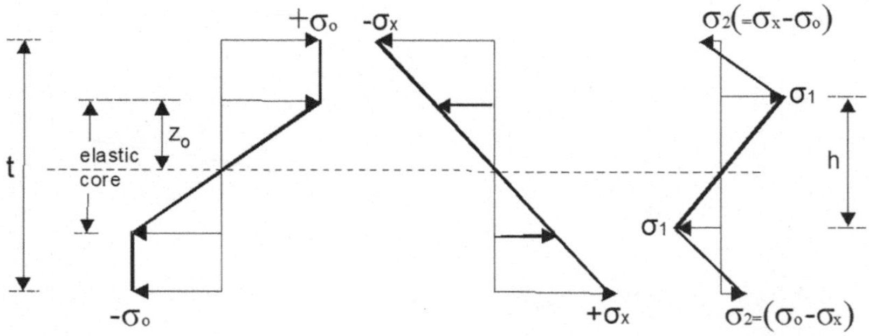

FIGURE 6.17 Induction of cold-bending residual stresses. (a) Over-bending. (b) Spring-back. (c) Resulting distribution.

Summing the net load in each fibre and dividing by the area of the cross section give the total stress in each component.

We shall consider the elasto-plastic model described by Robertson [18].

Following over-bending, the radius of curvature decreases to R_{in}, and after spring-back, it increases to R_{fin}. If the section behaves according to the Love–Kirchhoff hypothesis, (linear strain variation about the plate neutral surface), when bent to a radius R_{in} the strain and stress are:

$$\varepsilon_\theta = z/R_{in} \tag{6.52}$$

$$\begin{aligned} \sigma_\theta &= (E/R_{in})z_0 \quad \text{for} \quad z_0 \le (\sigma_o/E)R_{in} \\ &= \sigma_0 \quad \text{for} \quad z_0 > (\sigma_0/E)R_{in} \end{aligned} \tag{6.53}$$

where $2z_0$ is the depth of the elastic core. Assuming elastic spring-back, when the plate increases to radius R_{fin}, the corresponding circumferential strain and stress become:

$$\varepsilon_\theta = z_0/R$$

$$\begin{aligned} \sigma_\theta &= Ez_0/R \quad &\text{for} \quad z_0 \le (\sigma_0/E)R_{in} \\ &= \sigma_0 - \mathrm{E}z_0\left(1/R_{in} - 1/R_{fin}\right) \quad &\text{for} \quad z_0 \ge (\sigma_0/E)R_{in} \end{aligned} \tag{6.54}$$

Expression (6.54) describes the residual stress distribution that has the characteristic zigzag shape shown in Figure 6.17c.

The radius of curvature R_{in} which must be applied to the plate in order to achieve a final radius R_{fin} can be found from the condition that the bending moment must be zero in the final state:

$$M = 2\int_0^{\sigma_0 R_{in}/E} \frac{Ez^2}{R_{fin}} dz + \int_{\sigma_0 R_{fin}/E}^{t/2} \sigma_0 z - Ez^2\left(\frac{1}{R_{in}} - \frac{1}{R_{fin}}\right) dz = 0 \tag{6.55}$$

where t is the plate thickness.

Kendrick introduced the parameter $Q = (E/\sigma_0)(t/R_{in})$ and showed that

$$\frac{R_{in}}{R_{fin}} = \frac{4 - 3Q^2 + Q^3}{Q^3} \tag{6.56}$$

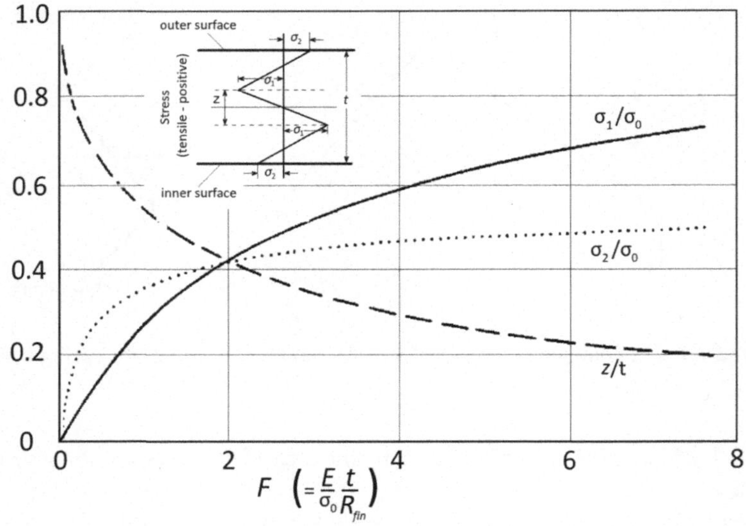

FIGURE 6.18 Effect of material properties and geometry on cold-bending residual stresses in cylindrical shell [12].

The quantities defining the residual stress distribution that results (Figure 6.18 insert) are:

$$\frac{\sigma_1}{\sigma_0} = \frac{R_{in}}{R_{fin}} = \frac{4 - 3Q^2 + Q^3}{Q^3} \tag{6.57a}$$

$$\frac{\sigma_2}{\sigma_0} = \frac{4 - Q^2}{2Q^2} \tag{6.57b}$$

$$\frac{z}{t} = 2Q \tag{6.57c}$$

These three quantities are plotted in Figure 6.18 against the parameter F, where:

$$F = \left(\frac{E}{\sigma_0}\right)\left(\frac{t}{R_{fin}}\right) = Q\left(\frac{R_{in}}{R_{fin}}\right) = \frac{4 - 3Q^2 + Q^3}{Q^2} \tag{6.58}$$

In the case of submarine hulls,[2] F typically has a value of 3, and the circumferential residual stress peaks due to cold bending are about half the yield stress. For equilibrium, the force corresponding to the tensile stresses has to be equal that corresponding to the compressive stresses. In way of the inner surface, the stress σ_2 is tensile, whereas on the outer surface of the shell, it is compressive.

Residual stresses have an important influence on cold-formed plates to the extent that nonlinear response is observed for stresses that are less than $0.4\sigma_0$. Nonetheless, the plastic collapse pressure is not affected by residual stresses. These do, however, affect ring frame strength since the hoop (circumferential) stiffness reduces when subjected to compression. This will be considered in the next section.

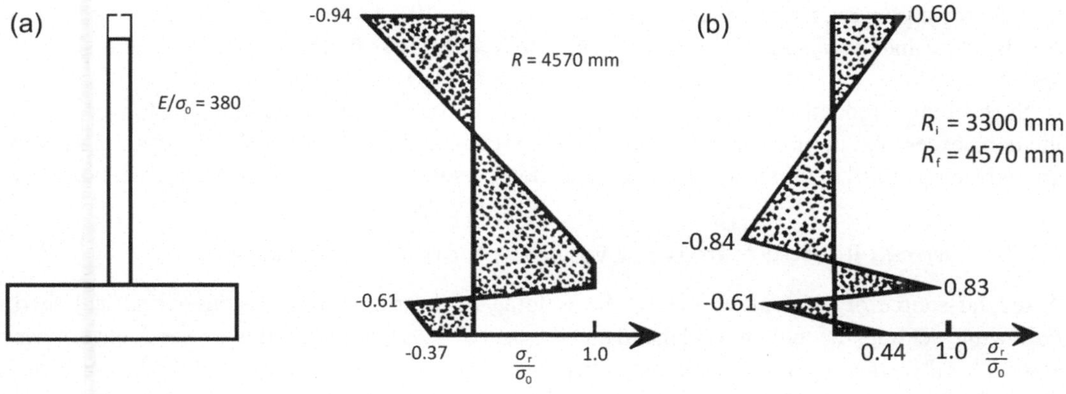

FIGURE 6.19 Cold bending residual stresses in ring frames [8]. (a) Single bending. (b) Double bending.

6.7.2 Cold-Bending Residual Stresses in Ring Frames

As in the case of the shell plating, during the cold bending of ring frames in order to induce circularity prior to welding to the cylindrical shell, residual stresses are formed, and they have a similar pattern to that encountered in plates. In this case, they can cause a significant loss in strength of up to 30% and also a stiffness loss since they are loaded in compression. Yielding can lead to collapse, and for this reason, care needs to be exercised. Figure 6.19 includes residual stress patterns in T-bar ring frames attached to shell plating that result from cold bending.

The effect that cold bending has on the collapse strength of a typical submarine section has been studied theoretically using finite element software (STAGS) by Cresswell and Dow [6].

Figure 6.20 that is based on their paper shows the detrimental effect that cold bending has on collapse strength. In this figure, the collapse load has been plotted against $\sqrt{\varepsilon}$, where ε is an imperfection parameter ($\varepsilon = 100\ w_0/R$). It is evident that as the amplitude of the imperfection increases,

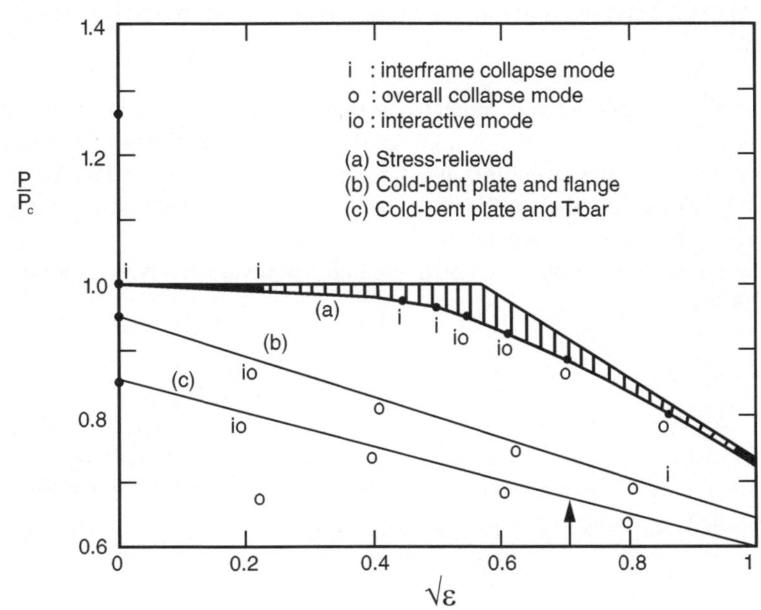

FIGURE 6.20 Effect of cold bending on collapse strength of submarine pressure hull [6].

the accompanying reductions in collapse load range from 10% to 40% depending on the extent of cold bending and the magnitude of the imperfection (w_0/R). These in themselves are serious strength losses.

Such strength reductions can be countered by making use of appropriate safety factors or else by inducing excessive curvature and then adopting the required form. In this way, the residual stresses are reduced. A third option is to use built-up sections, a solution that is not problem-free, however.

6.7.3 Welding Residual Stresses in Cylindrical Shell and Ring Frames

A second source of residual stresses is the welding of the ring frames to the cylindrical shell. Following welding, the molten weld metal and the heat-affected zone cool down to ambient temperature. This drop in temperature is accompanied by a contraction of the material of the heat-affected zone which includes the fillet welds on either side of the ring frame web as well as neighbouring material. The contraction takes place in the circumferential direction of the pressure hull and also in the plane of the cross section of the welds. As a result, a tensile residual stress field is set up. Since the magnitude of the temperature drop is several times greater than that of the yield stress, these shrinkage stresses are equal to the yield stress in tension. The situation is analogous that of welding residual stresses in flat stiffened plates.

For sections that are internally stiffened, the weld shrinkage that takes place consists of:

1. A contraction of the weld material and the heat-affected zone in the circumferential direction, which are then placed in tensile yield. This is balanced by compressive stresses in the plating such that equilibrium in the circumferential direction is maintained (Figure 6.21a).
2. A contraction of the weld material in the cross sections of the fillet welds. As a result of this contraction, the shell plating rotates about the attachment to the stiffener web, thereby introducing a permanent distortion in the inward direction (Figure 6.21b). At the same time, the reaction on the ring frame acts so that the outer surface of the stiffened cylindrical shell is displaced by w_0 in the outward direction. The net total permanent distortion at any point along the shell is then $w(x)$. A consequence of this increase in diameter is to give rise to additional circumferential stresses in the ring frame and cylindrical shell.

The net effect of these stress fields is indicated in Figure 6.21a. The permanent initial distortion $w_p \approx 0.1t$, which indicates the presence of high residual stresses. Yielding of the ring frames may lead to overall collapse of the structure, and for this reason, it is necessary to calculate frame stresses accurately. For externally stiffened cylindrical shells, in the first case, the shell plating stresses are tensile, whereas in the second, they are compressive.

If a sinusoidal type of distortion is assumed, the deflection $w(x)$ at any point along the shell plating is given by [8,18]:

$$w(x) = w_0 - w_p \sin\left(\frac{\pi x}{L}\right) \tag{6.59}$$

where the origin of the x-axis is at the plate–frame intersection, the deflection is positive in the outward direction, and w_0 is not known (Figure 6.21b). The circumferential bending stresses in the shell plating and the ring frame are then, respectively,

$$\sigma_{\theta 2}^{p}(x) = w(x)\left(\frac{E}{R_0}\right) \tag{6.60a}$$

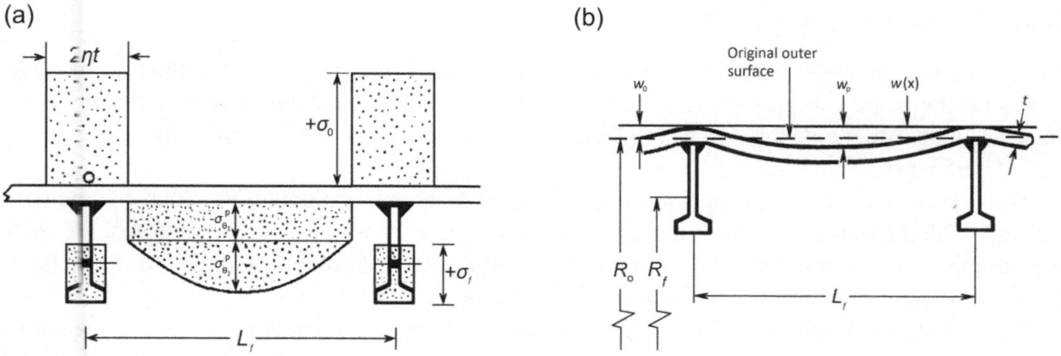

FIGURE 6.21 Welding residual stresses and distortions in a ring-stiffened cylindrical shell. (a) Residual stress distributions in circumferential direction. (b) Shell plate distortions due to shrinkage of cross-section of the fillet welds.

$$\sigma_{\theta 2}^{f} = w_0\left(\frac{E}{R_0}\right) \tag{6.60b}$$

The corresponding loads in the plate and ring frame are then

$$2t\int_0^{L/2} w(x)\left(\frac{E}{R_0}\right)dx \quad \text{and} \quad w_0\left(\frac{E}{R_0}\right)A_f$$

For equilibrium, these must balance each other. Thus,

$$w_0 = \left(\frac{2}{\pi}\right)\frac{w_p}{1+(R_o/R_f)(A_f/tL_f)} \tag{6.61}$$

The circumferential residual stresses in the plate due to longitudinal weld shrinkage are for $0 \le x < \eta t$, equal to $+\sigma_0$. Over the remaining portion of the shell plate,

$$\sigma_{\theta 1}^{p} = -\frac{2\eta t}{L_f - 2\eta t}\sigma_0 \tag{6.62}$$

Transverse shrinkage gives rise to stresses

$$\sigma_{\theta 2}^{p} = \left(\frac{Et}{10R_0}\right)\left[\frac{2}{\pi}\right]\left[1 + A/tL_f\right] - \sin\left(\frac{\pi x}{L}\right) \tag{6.63}$$

These vary in a sinusoidal manner across the plate breadth (Figure 6.21a). In the ring frame, the average stress is given by

$$\sigma_{\theta}^{f} = \left(\frac{Et}{5R_0}\right)\left[\frac{1}{\pi}\left[1 + A/tL_f\right]\right] \tag{6.64}$$

where $A = A_f(R/R_f)^2$. The resulting distributions are shown in Figure 6.21a.

6.8 SUBMARINE ACCIDENTS

In this section, we shall discuss available data concerning submarine accidents, the aim being to assess (a) the quality of pressure hull design, (b) the effect that accidents may have on structural integrity, and (c) methods that can be implemented to introduce improved survivability of the vessel and of personnel.

Data concerning submarine accidents are available in the public domain in books dealing specifically with the subject, a limited number of journal articles as well as internet websites. We shall not attempt to re-analyse these data but will make reference to available studies in order to draw relevant conclusions in accordance with our aims, as stated above.

A study of post-World War 2 accidents provides a statistical breakdown of causes, included in Figure 6.22. It is seen that structural failure is not listed, so that it may be concluded that submarine pressure hulls have been designed in a satisfactory manner, in the sense that structural failure has not occurred at depths equal to or less than the design diving depth, under normal (peacetime) operating conditions. It has to be added that submarines, being weapons of war, may be subjected to loads resulting from underwater explosions that are significantly greater than hydrostatic pressure, the intent being to sink the vessel. The results from studies of structural response to such loads are not available in the public domain. A rational approach to this question would be to introduce a risk-based approach to design by introducing partial safety factors that allow for (a) the probability of occurrence of an underwater explosion and (b) the consequences of such an explosion. However, as discussed in Section 9.2.2, design philosophies differ.

The consequences of loss of structural integrity are dire and lead to sinking of the vessel, so that it is necessary to consider the specific instances in which a submarine has been lost. This is a difficult task because the primary evidence is lost together with the vessel. In certain cases, it may be possible to retrieve information from the seabed and also to reassess communications and tracking data, in an attempt to reconstruct the events leading to the loss. Marine accidents in general are complex events, and accident analysis requires an interdisciplinary approach. Methodologies to analyse accidents concerning surface ships are available in the literature and can be implemented to study individual cases, for example, [1,13]. Methodologies such as these are used by national organisations charged with accident analysis such as the U.S. National Transportation Safety Board and the UK Marine Accident Investigation Branch. The ultimate aim of accident analysis is to draw lessons that may lead to improved design and vessel operation.

It is seen from Figure 6.22 that there are numerous types of events that may or may not lead to loss of a vessel. The most serious of these are flooding, collision, explosion, and fire, all of which have led in one or more instances to the loss of a particular vessel. The question then arises as to what measures need to be taken to avoid losses in such circumstances, given that no questions arise with regard to pressure hull integrity per se.

In the past, the study of serious submarine accidents has led to improved design practice as indeed in the case of surface ships (*RMS Titanic*). In particular, the loss of the *USS Thresher* that took place in 1963 that involved the implosion of the pressure hull when the vessel at one point exceeded its collapse depth led to the introduction of measures to minimise the probability of occurrence of the root cause of this particular type of accident. In the case of the *USS Thresher*, it was considered to be flooding that was initiated by a burst piping system. The measures that were subsequently introduced consisted of the implementation of a quality assurance system (SUBSAFE) to ensure that the hull remains watertight and that a vessel can recover from unanticipated flooding. This has proven to be a success, in the sense that accidents of this type have not occurred since, at least in the US submarine fleet. Indeed, a statistical analysis of accidents has shown that measures introduced in conjunction with technological advances relating to improved tools of analysis and design, have led, over a period of time, to improved safety for both vessels and submariners (Table 6.5).

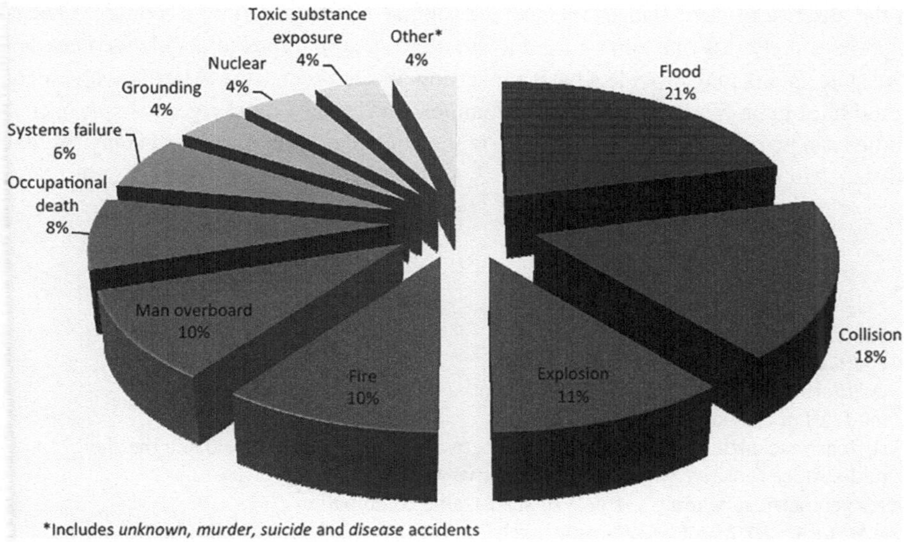

FIGURE 6.22 Relative contribution of types of submarine accidents (1945–2005) [20].

As seen from Figure 6.22, the causes of accidents are not limited to flooding, so that measures need to be taken in the case of the other important root causes of submarine losses. Clearly, because of the variety of accident types and submarine arrangements, a systematic and comprehensive approach is needed. There exist more than one way to go about this.

a. By making use of rules for submarine construction. In the case of surface ships, design codes cater for the eventualities of accidents by the introduction of the Accidental Limit State (ALS) [5,10]. The ALS represents excessive structural damage as a result of accidents such as collisions, grounding, explosion and fire, hazards that are of primary concern in the case of submarines as well. In the case of submarines, it would be necessary to specify appropriate design load scenarios for which the minimum required pressure hull strength would need to be specified in order to ensure structural integrity (avoid implosion). Indeed, Lloyds Register Rules for Naval Ships (surface ships) include what are termed "Military Distinction notations" as well as a series of confidential notations that concern structural performance under wartime loads, such as external and internal air blast and whipping assessment.

TABLE 6.5
Mean values for SUBSUNK and Submarine Fatalities [20]

Parameter	Period	Mean		Correlation (R^2)[a]	Ratio of Means[b]
SUBSUNK	1946–1974	0.92	SUBSUNK/500 available	0.99	3.0
	1975–2005	0.31	subs	0.98	
Submarine Fatalities	1946–1974	38	Fatalities/30,000 submariners	0.98	3.1
	1975–2005	12		0.91	

[a] Correlation (R^2) is a measure of variation in the data. As $R^2 \rightarrow 1$ a regression approaches a perfect fit.
[b] Ratio = $\text{Mean}_{(1946\text{–}1974)}/\text{Mean}_{(1975\text{–}2005)}$.

b. In the absence of such clauses in rules for submarine construction, it would be necessary to prepare scenarios that can be used to simulate specific types of accidents. Descriptions of past accidents may provide a basis for setting up such scenarios. Alternatively, scenarios would have to be set up in a pro-active manner, this being a more rigorous approach. Such studies can be parameterised, so that variations in one or more conditions may reveal critical stages in structural integrity.

NOTES

1 Notation

A coefficient of axial displacement

B coefficient of circumferential displacement

E_f modulus of elasticity of ring frame

e stiffener eccentricity, being the distance between the stiffener centroid and the shell mid-surface (positive when interior, negative when exterior)

$\overline{M}_x, \overline{M}_\theta$ geometrical bending stiffness of shell-frame combination $= M_x/(d^2w/dx^2)$, $M_\theta/(d^2w/d\theta^2)$ respectively

N_{x1}, $N_{\theta 1}$ pre-buckling membrane force resultants

Z Batdorf slenderness parameter $= \sqrt{1-\nu^2}\,(L_b/R)^2(R/t)$

2 For submarine cylindrical pressure hulls $500 > E/\sigma_0 > 250$, $150 > R/t > 50$.

REFERENCES

1. Ayeko, M. (1999). "Integrated Safety Investigation Methodology (ISIM) – A systematic approach to learning from accidents." *RINA International Conference on Learning from Marine Accidents*, London.
2. Baruch, M., Singer, J. (1963). Effect of eccentricity of stiffeners on the general instability of stiffened cylindrical shells under hydrostatic pressure. *J. Mech. Eng. Sci.* Vol. 5, No.1, 23–27.
3. Blumenberg, W.F. (1965). "The effect of intermediate heavy frames on the elastic general-instability strength of ring-stiffened cylinders under external hydrostatic pressure. DTMB Report No 1844.
4. Burcher, R., Rydill, L. (1994). *Concepts in Submarine Design*. Cambridge: Cambridge University Press.
5. Caridis, P.A. (2025). *Global Strength of Ships. Analysis and Design using Mathematical Methods.* Cambridge: Cambridge University Press, 848 p.
6. Cresswell, D.J., Dow, R.S. (1986). The application of nonlinear analysis to ship and submarine structures. In: Smith C.S. and Clarke D.J. (Ed). *Advances in Marine Structures.* Barking, Essex: Elsevier Applied Science Publishers Ltd, 174–200 pp.
7. Donnell, L.H. (1933). "Stability of thin-walled tubes under torsion." NACA Rep. No. 479.
8. Faulkner D. (1977). Effects of residual stresses on the ductile strength of plane welded grillages and of ring stiffened cylinders. *J. Strain Anal.* Vol. 12, No. 2, 130–139.
9. Faulkner D. (1984). "Design of pressure loaded structures". *Informal Lecture Series on Offshore Structure Design*. Dept Naval Arch & Ocean Eng. Univ. of Glasgow.
10. International Association of Classification Societies (IACS). (2022). *Common Structural Rules for Oil Tankers and Bulk Carriers*. London.
11. Kendrick, S. (1985). Ring-stiffened cylinders under external pressure Chap. 3. In: Narayanan R. (Ed). *Shell Structures – Stability and Strength*. London: Elsevier Applied Science Publishers.
12. Kendrick, S.B. (1970). Externally pressurised vessels Chap. 9. In: Gill S.S. (Ed). *The Stress Analysis of Pressure Vessels and Pressure Vessel Components*. Oxford: Pergamon Press, 405–511 pp.
13. Kristiansen, S., Koster, E., Schmidt, W.F., Olofsson, M., Guedes Soares, C., Caridis, P. (1999). "A new methodology for marine casualty analysis accounting for human and organisational factors." *RINA International Conference on Learning from Marine Accidents*, London.
14. Lunchick, M.E. (1963). "Plastic general-instability pressure of ring-stiffened cylindrical shells." DTMB Report No 1587.
15. Lunchick, M.E., Short Jr., R.D. (1957). "Behavior of cylinders with initial shell deflection." *Proc. ASME Applied Mech. Div. Conf.* (also DTMB Report 1150).
16. Pulos, J.G. (1963). "Structural analysis and design considerations for cylindrical pressure hulls." DTMB Rept. No 1639.

17. Pulos, J.G., Salerno, V.L. (1961). "Axisymmetric elastic deformations and stresses in ring-stiffened, perfectly circular cylindrical shell under external hydrostatic pressure." DTMB Rept. No 1497.
18. Robertson, L.M. (1990). "Analytical calculations of fatigue loading of submarine hulls." MRL Tech. Report MRL-TR-90-26, Victoria, Australia.
19. Singer, J., Baruch, M., Harari, O. (1965). "Further remarks on the effect of eccentricity of stiffeners on the general instability of stiffened cylindrical shells." Scientific Rept. No. 2, Technion Res. & Develop. Found. TAE Report No. 42, Israel.
20. Tingle, C. (2009). Submarine accidents. A 60-year statistical assessment. *Prof. Safety*. Vol. 54, 31–39.
21. Vaughan, H. (1963). "The stress distribution in ring frames with initial tilt." NCRE Rept. R 473, Naval. Const. Res. Est. Dunfermline.
22. Wilson, L.B. (1966). The elastic deformation of a circular cylindrical shell supported by equally spaced circular ring frames under uniform external pressure. *Trans. RINA*. Vol. 108, 63–72.

7 Truncated Cones and Major Internal Structures

7.1 TRUNCATED CONES IN SUBMARINE PRESSURE HULLS

The geometry of an unstiffened truncated cone can be defined by the cone semi-vertex angle α at the apex, the radius at one end, and the cone height h or alternatively by the two radii.

Truncated cones are surfaces of revolution as indeed are all elements of pressure hulls of underwater vehicles. The geometrical properties of surfaces of revolution are studied using differential geometry, whereas the structural response of such members is described using thin-walled structures theory, in texts such as [11,17,25,36,39].

Research into cone behaviour has been carried out since the earliest part of the 20th century and continues to this day. It has to be said that much research, both theoretical and experimental, has focused on cones failing in the elastic range. It is only more recently that results concerning inelastic failure, which are of interest in underwater pressure hulls, are forthcoming.

The equations of equilibrium for thin elastic shells obtained by Love [19] and Meissner [23] were applied to cones by Dubois in his doctoral thesis in 1919 [10]. He obtained solutions of two second-order differential equations that were expressed in terms of infinite series and asymptotic formulas. Flügge [11] later solved these using second-order Bessel functions. An alternative solution leading to the same results was obtained by Watts and Burrows [40]. They applied their results to the intersection of a truncated cone with the large end of a truncated complete cone. These solutions, however, are not suited to everyday use, and other investigators have sought alternative approaches [41].

A number of publications have been based on the *equivalent cylinder* approach [7,24,32,43], that was developed by Bauersfeld and Gecklerer [12]. Tokugawa also made use of the equivalent cylinder approach and arbitrarily assumed the meridian of the shell to be a beam simply supported at both ends and loaded by a linearly varying pressure [38]. He proposed that the analysis of a truncated cone be based on that of an equivalent cylinder whose radius is equal to the radius of curvature of the cone at the maximum deflection point. A number of definitions of the radius of curvature have been used. The equivalent cylinder approach is not, however, considered suitable for cones failing by inelastic buckling [2]. In the case of axially compressed cones, the corresponding equivalent cylinders fail at a much higher pressure than the cones they are meant to represent and thus lead to unsafe results.

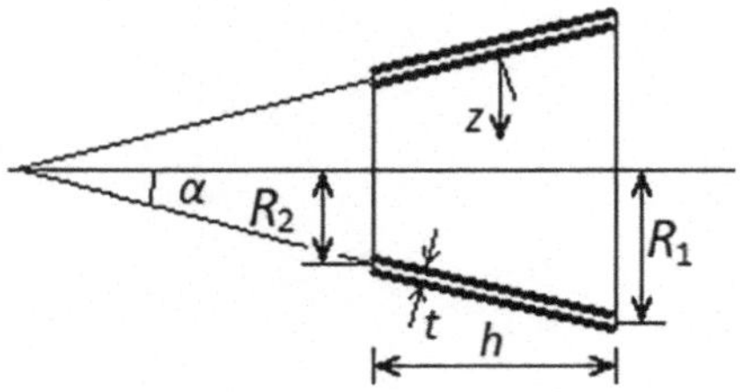

FIGURE 7.1 Geometry of truncated cone [39].

DOI: 10.1201/9781003489153-7

These solutions are treated as being approximate. One other approximate approach was introduced by Hetényi, who treated the cone as a set of tapered longitudinal beams on an elastic foundation [13].

An energy approach was implemented by Hoff [14], who obtained the equations of equilibrium in the axial and circumferential directions. Solution of these equations has, however, proven difficult [25].

7.1.1 Loading of Cones

In the case of a truncated cone that forms part of a submarine pressure hull, the loads consist of:

1. External hydrostatic pressure
2. Axial loading due to compression of the pressure hull end-closures
3. Loads acting at the cone ends required to ensure geometric compatibility with the adjacent structure

The stress resultants that are present at the cone ends consist of shear loads acting in the radial direction and bending moments. These arise as a result of the difference in stiffness of the cone and the adjoining member, which is usually a ring-stiffened cylindrical shell. For cones subjected to uniform lateral pressure (external or internal), we note that because of the change in radius along the length, the bending stiffness will also vary, in contrast to cylindrical shells that maintain a constant radius. As a result, the deflection in the radial direction will not be symmetric about the end supports.

7.1.2 Membrane Theory of Conical Shells

In this section, we shall provide the derivation of the equations of equilibrium of a complete (non-truncated) conical shell subjected to rotationally symmetric loading on the basis of the equations of equilibrium given in Section 3.2. Equations (3.9a–c) have been derived for shells of revolution in general, and in the case of a conical shell, they need to be adapted accordingly. Since in this case, the meridian curvature is zero, we cannot make use of this variable, and we introduce as a substitute variable the distance along a meridian, s (Figure 7.2). The force N_s then replaces N_φ.

$$\text{Hence} \quad \varphi = \text{const} \quad r_\varphi = \infty \quad r_a = x\cot\varphi \quad r_\varphi d\varphi = dx \quad N_\varphi = N_x$$

We then divide equations (3.9a) and (3.9b) by $r_\varphi \cos\varphi$ and (3.9c) by r_a. Substituting the above expressions in these, we obtain:

$$\left.\begin{aligned} &\frac{\partial N_\vartheta}{\partial\vartheta}\frac{1}{\cos\varphi} + \frac{1}{r_\varphi}\frac{\partial}{\partial\varphi}(N_{x\vartheta} r_a \sin\varphi)\frac{1}{\cos\varphi} + N_{\vartheta x} + p_x r_a \tan\varphi = 0 \\ &\frac{1}{r_a}\frac{\partial}{\partial\varphi}(N_\varphi r_a \sin\varphi)\frac{1}{\cos\varphi} + \frac{\partial N_{x\vartheta}}{\partial\vartheta}\frac{1}{\cos\varphi} - N_\vartheta + p_y r_a \tan\varphi = 0 \\ &N_\varphi \frac{r_a}{r_\varphi} + N_\vartheta + p_z r_a = 0 \end{aligned}\right\} \tag{7.1}$$

These equations are applied to the case of the conical shell by applying the required limit progressions to obtain:

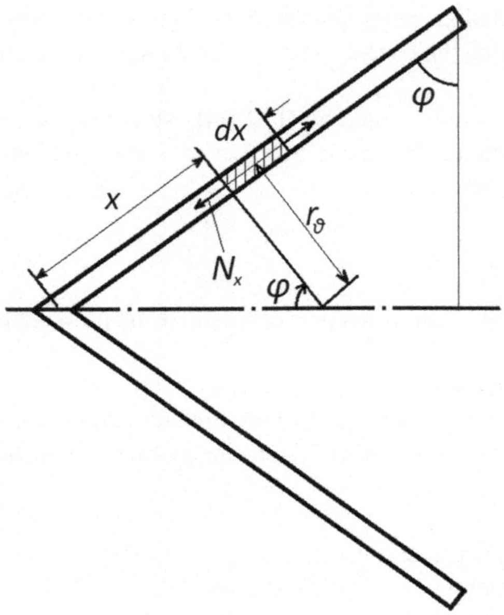

FIGURE 7.2 Nomenclature for the membrane theory of conical shells [27].

$$\left.\begin{aligned}
&\frac{\partial N_\vartheta}{\partial \vartheta}\frac{1}{\cos\varphi}+\frac{\partial}{\partial x}(N_{x\vartheta}x)+N_{x\vartheta}+p_x x=0\\
&\frac{\partial}{\partial s}(N_x x)+\frac{N_{x\vartheta}}{\partial \vartheta}\frac{1}{\cos\varphi}-N_\vartheta+p_y x=0\\
&N_\vartheta+p_z x\cot\varphi=0
\end{aligned}\right\}\tag{7.2}$$

These are the equations of equilibrium of a conical shell in which membrane forces are induced.

7.2 EQUATIONS OF EQUILIBRIUM OF CONICAL SHELLS

The equilibrium of a shell of revolution is expressed in the form of the two second-order differential equations, obtained by Meissner [23], who applied the equations for thin elastic shells derived by Love [19]. These are included in Chapter 3 as equations (3.15).

Dubois applied elastic analysis to these equations for the case of conical shells and obtained a differential equation for the normal displacement of the shell. He obtained an exact solution of the homogeneous form of this equation by making use of special infinite series and asymptotic formulas. This solution has, however, proven difficult to apply, and much effort has been expended in developing more readily applicable solutions, involving in certain cases approximations. In this section, we shall describe an alternative, exact solution that is more amenable to design calculations. The one limitation to this is that it is assumed that the results obtained at one end of the cone do not affect results at the other end. In other words, it applies to what may be termed a *long cone*.

7.2.1 The Dubois Equation

The differential equation obtained by Dubois [10] from the Meissner equations for a conical shell of constant thickness that is subjected to external hydrostatic pressure is:

$$x^2 \frac{d^4w}{dx^4} + 2x \frac{d^3w}{dx^3} - 2 \frac{d^2w}{dx^2} + \frac{12(1-\nu^2)w}{t^2 \tan^2 \alpha} = \frac{9(1-\nu^2)px^2}{Et^3} \quad (7.3)$$

where x = coordinate taken along the generator from the apex (Figure 7.3)
w = displacement normal to the shell surface
ν = Poisson's ratio
t = cone thickness
α = semi-vertex angle
E = modulus of elasticity
p = applied pressure

The stress resultants are

$$M_x = D\left(\frac{d^2w}{dx^2} + \frac{\nu}{x}\frac{dw}{dx}\right)$$

$$M_\theta = D\left(\nu \frac{d^2w}{dx^2} + \frac{1}{x}\frac{dw}{dx}\right)$$

$$N_x = D\tan\alpha\left(\frac{d^3w}{dx^3} + \frac{1}{x}\frac{d^2w}{dx^2} - \frac{1}{x^2}\frac{dw}{dx}\right) - \frac{px\tan\alpha}{2} \quad (7.4)$$

$$N_\theta = \frac{d}{dx}(xN_x)$$

$$Q_x = D\left(\frac{d^3w}{dx^3} + \frac{1}{x}\frac{d^2w}{dx^2} - \frac{1}{x^2}\frac{dw}{dx}\right)$$

where
D = flexural rigidity
M_x = bending moment in a meridional plane
M_θ = bending moment in circumferential direction

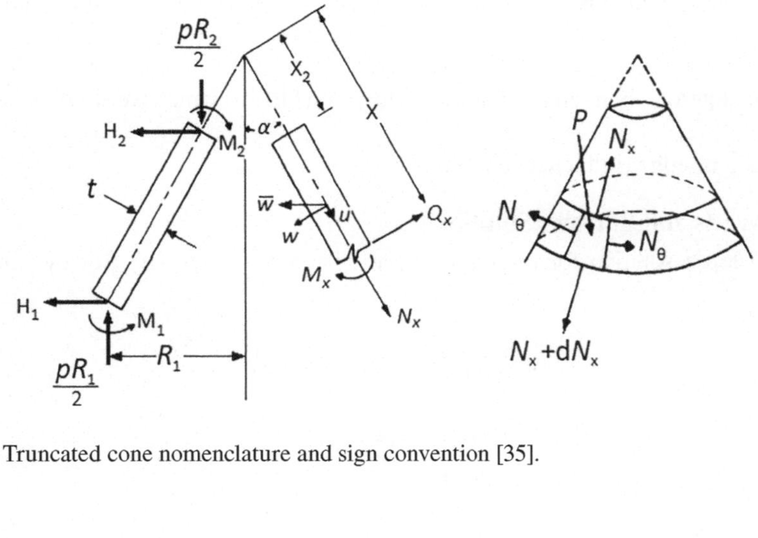

FIGURE 7.3 Truncated cone nomenclature and sign convention [35].

N_x = stress resultant along the meridian
N_θ = stress resultant in the circumferential direction
Q_x = shear force transverse to shell edge

7.2.2 Approximate Solutions for Edge Bending Response of Conical Shells

As mentioned previously, the Dubois equation for conical sections encounters considerable difficulties in its practical application, especially when a conical section is attached to a shell of revolution that has a different geometry, such as a cylinder or a sphere. For this reason, efforts were made early on to introduce approximations that enable the extraction of solutions without an inordinate amount of effort. We shall discuss two approximations that have found practical use in submarine pressure hull analysis and design. Differences with the exact solution have been quantified and have been found to be acceptably low.

The first approximation involves introducing the assumption that the edges of the cone are far apart enough for the deformations and stresses that arise at one end as a result of interaction with an adjoining member for them not to affect those at the other end. On this basis, Taylor and Wenk [35] developed an exact solution of the Dubois equation.

The second approximation concerns problems that include intersections of shells of revolution in general and was introduced by Geckeler [12]. This approximation has since found widespread application in pressure vessel design and is based on the realisation that the edge shear force and moments at an intersection are highly localised and attenuate rapidly. As a result, it is permissible for the lower-order derivatives in the equation of equilibrium of a shell of revolution to be omitted. Furthermore, and for the same reason, it is acceptable to assume that the radius of the sections on either side of an intersection is constant. The accuracy and limitations of the Geckeler approximation are discussed in [44].

When issues concerning the design of truncated cones that form part of submarine pressure hulls arose in the early 1950s in the United States, it was not known whether it is permissible or not to introduce the Geckeler approximation. To provide an answer to this question, two studies were carried out. In the first, it was confirmed that the second-order term of the Dubois equation can be omitted, thus leading to a simpler solution [42]. In the second study, the omission of the third-order term as well was considered [30]. Results obtained using the Raetz and Pulos solution have been compared with those of the exact solution obtained by Taylor and Wenk.

7.2.3 The Taylor–Wenk Solution for Conical Shells

In this section, we shall present the exact solution of Dubois' differential equation obtained by Taylor and Wenk [35], which has acted as a benchmark for subsequent solutions. Their solution is based on:

1. An earlier solution obtained by Watts and Burrows, [40], to which we shall make reference to
2. The method of edge (influence) coefficients

7.2.3.1 The Watts and Burrows Solution

In order to develop a solution for the equilibrium of a cone, Watts and Burrows introduced the notation:

$$L(\ )=x\frac{d^2}{dx^2}(\ )+\frac{d}{dx}(\ )-\frac{(\ )}{x} \tag{7.5}$$

where

L = Meissner's operator for a conical shell
p = external pressure
$V = Q_x x \tan\alpha$
$W = \dfrac{dw}{dx}$
$D = \dfrac{Et^3}{12(1-\nu^2)}$

and obtained the equations of equilibrium for a conical shell in the following form:

$$L(V) + EtW\cot\alpha = 1.5px\tan\alpha \tag{7.6a}$$

$$L(W) - \frac{V\cot\alpha}{D} = 0 \tag{7.6b}$$

These two second-order equations can be expressed as a single fourth-order equation:

$$LL(W) + \lambda^4 W = \frac{1.5U^4 px}{Et^3} \tag{7.7}$$

where

$$\lambda^4 = \frac{12(1-\nu^2)}{t^2\tan^2\alpha} \tag{7.8a}$$

$$U^4 = 12(1-\nu^2) \tag{7.8b}$$

Integrating (7.7) leads to the Dubois equation (7.3). The constant term appearing in (7.3) represents translation parallel to the axis (rigid-body motion) and hence can be neglected without affecting the elastic analysis. The solution to the cone problem was obtained by Watts and Burrows using the two second-order equations (7.6). The stress resultants and displacements derived from their solution are:

$$N_x = \frac{1}{x}\left(C_1 ber_2\xi + C_2 bei_2\xi + C_3 ker_2\xi + C_4 kei_2\xi\right) - \frac{px\tan\alpha}{2}$$

$$Q_x = \frac{\cot\alpha}{x}\left(C_1 ber_2\xi + C_2 bei_2\xi + C_3 ker_2\xi + C_4 kei_2\xi\right)$$

$$\begin{aligned} M_x = \frac{t}{2U^2 x}\Big[&C_1\left(\xi bei_2'\xi + 2\nu bei_2\xi\right) - C_2\left(\xi ber_2'\xi + 2\nu ber_2\xi\right) \\ &+ C_3\left(\xi kei_2'\xi + 2\nu kei_2\xi\right) - C_4\left(\lambda ker_2'\xi + 2\nu ker_2\xi\right)\Big] + \frac{pt^2\tan^2\alpha}{8(1-\nu)} \end{aligned} \tag{7.9}$$

$$\chi = \frac{dw}{dx} = \frac{U^2}{Et^2}\left(C_1 bei_2\xi - C_2 ber_2\xi + C_3 kei_2\xi - C_4 ker_2\xi\right) + \frac{3px\tan^2\alpha}{2Et}$$

$$\bar{w} = -\frac{\sin\alpha}{Et}\left[C_1\left(\frac{\xi}{2}ber_2' - \nu ber_2\xi\right) + C_2\left(\frac{\xi}{2}bei_2' - \nu bei_2\xi\right)\right.$$

$$\left. + C_3\left(\frac{\xi}{2}\ker_2' - \nu\ker_2\xi\right) + C_4\left(\frac{\xi}{2}kei_2' - \nu kei_2\xi\right)\right] + \frac{px^2\sin\alpha\tan\alpha}{Et}\left(1 - \frac{\nu}{2}\right)$$

Where:

C_1, C_2, C_3, and C_4 are constants of integration determined from the boundary conditions, given below:

$$C_1 = \left[Q_1x_1t\tan\alpha\left(\xi_1 ber_2'\xi_1 + 2\nu ber_2\xi_1\right) + 2M_1U^2x_1bei_2\xi_1\right]\left[t(C + 2\nu G)\right]^{-1}$$

$$C_2 = \left[Q_1x_1t\tan\alpha\left(\xi_1 bei_2'\xi_1 + 2\nu bei_2\xi_1\right) - 2M_1U^2x_1ber_2\xi_1\right]\left[t(C + 2\nu G)\right]^{-1}$$

$$C_3 = \left[Q_2x_2t\tan\alpha\left(\xi_2\ker_2'\xi_2 + 2\nu\ker_2\xi_2\right) + 2M_2U^2x_2kei_2\xi_2\right]\left[t(C + 2\nu G)\right]^{-1}$$

$$C_4 = \left[Q_2x_2t\tan\alpha\left(\xi_2 kei_2'\xi_2 + 2\nu kei_2\xi_2\right) - 2M_2U^2x_2\ker_2\xi_2\right]\left[t(C + 2\nu G)\right]^{-1}$$

ber_2, bei_2 are real and imaginary second-order Bessel functions (Appendix 4)

ker_2, kei_2 are real and imaginary second-order Kelvin functions (Appendix 4)

$$\xi = 2\lambda\sqrt{x}$$

The primes denote differentiation with respect to ξ

7.2.3.2 Junctions of Dissimilar Shells: The Method of Edge Coefficients

Figure 3.5 includes the types of shells of revolution that are encountered in submarine pressure hulls. It is seen that a variety of forms are used, each of which has its own strength and stiffness characteristics as well as the pattern of deformation adopted prior to and following failure. When these are welded to each other such that a watertight envelope is produced, geometric compatibility is required at the junction of dissimilar forms. To ensure geometric compatibility when the pressure hull is subjected to hydrostatic pressure, a system of forces is induced in way of each junction.

Additionally, the deformations of the members on either side of a junction will differ from those in the isolated case. If the external loading is axisymmetric (external hydrostatic pressure), the deformation pattern will vary in the axial direction alone.

At the junctions, both membrane and bending stresses arise, and these need to be accounted for in design. These stresses can be determined by enforcing conditions of compatibility of deformations and equilibrium of forces and moments. The approach followed to introduce these into a solution is the *method of edge* (*or influence*) *coefficients* [31].

In the case of an intersection of a small diameter cylinder with a truncated cone, the application of external pressure results in the deformation of the connection in the inwards direction; the external fibres are then in compression and the internal fibres in tension (Figure 7.4). The opposite effect arises in the case of a junction with a large diameter cylinder. When axial loading is applied in addition to hydrostatic pressure, this also has to be taken account of. Figure 7.4 shows the membrane stresses that arise in an isolated truncated cone. These are the meridional stress, σ_x, and the circumferential stress, σ_θ.

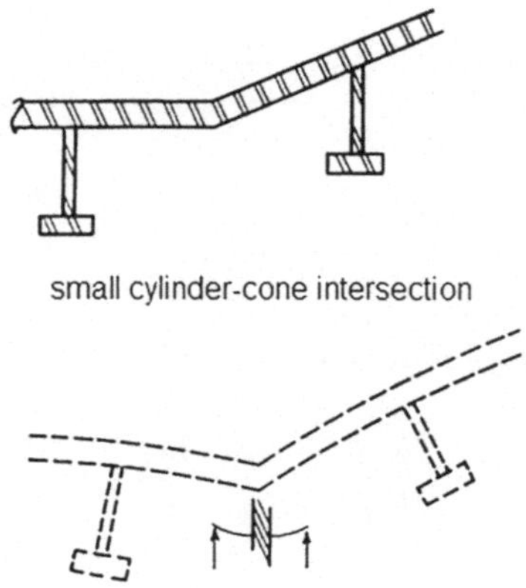

FIGURE 7.4 Deformation of small cylinder–cone intersection due to external pressure.

Figure 7.5 shows the edge forces and moments arising at the ends of shells of revolution that intersect. These are resolved into components parallel to and perpendicular to the axis of revolution. In this case, the axial component of the pressure is equal to $pR/2$, where R is the local radius of the shell, measured from the axis of revolution to the mid-thickness. For each of the two shells shown, the stresses are functions of x, a meridional coordinate, the edge loads H, M where H is the shear force component acting normal to the axis of revolution, M is the bending moment, and p is the applied pressure:

$$\sigma_x = \sigma_x\left(x, H_A, M_A, H_B, M_B, p\right) \tag{7.10a}$$

$$\sigma_\theta = \sigma_\theta\left(x, H_A, M_A, H_B, M_B, p\right) \tag{7.10b}$$

The edge forces and moments (H, M) are determined by applying geometrical compatibility that involves introducing the deflection $\overline{w}$ normal to the axis of revolution and the angle of rotation of a shell element θ in the meridional plane. These quantities are expressed in terms of functions of (H, M) as follows [31]:

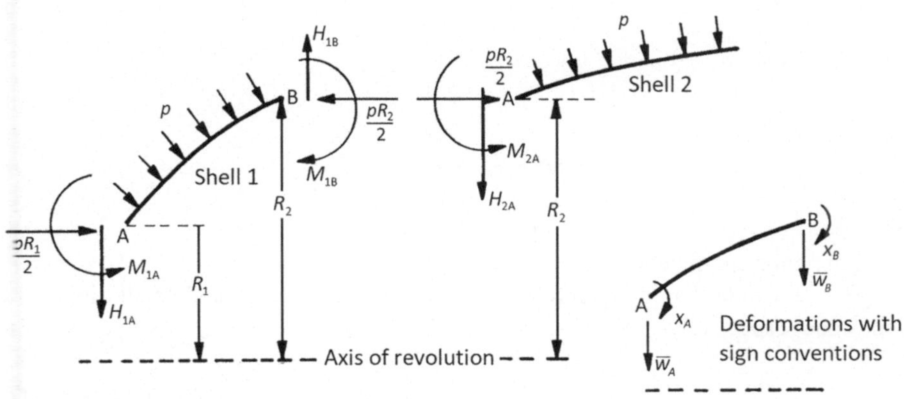

FIGURE 7.5 Loads and deformations in way of shell intersections [31].

$$\bar{w}_{1B} = \frac{\partial \bar{w}_{1B}}{\partial H_{1B}} H_{1B} + \frac{\partial \bar{w}_{1B}}{\partial M_{1B}} M_{1B} + \frac{\partial \bar{w}_{1A}}{\partial H_{1A}} H_{1A} + \frac{\partial \bar{w}_{1B}}{\partial H_{1A}} M_{1A} + \frac{\partial w_{1B}}{\partial p} p \tag{7.11a}$$

$$\theta_{1B} = \frac{\partial \chi_{1B}}{\partial H_{1B}} H_{1B} + \frac{\partial \theta_{1B}}{\partial M_{1B}} M_{1B} + \frac{\partial \theta_{1B}}{\partial H_{1A}} H_{1A} + \frac{\partial \theta_{1B}}{\partial H_{1A}} M_{1A} + \frac{\partial \theta_{1B}}{\partial p} p \tag{7.11b}$$

where $\bar{w}_{1B}$ and θ_{1B} are the deflection and rotation of shell 1 at B (Figure 7.5) and the quantities

$$\frac{\partial \bar{w}_{1B}}{\partial H_{1B}}, \frac{\partial \bar{w}_{1B}}{\partial M_{1B}}, \frac{\partial \bar{w}_{1A}}{\partial H_{1A}}$$

are the edge coefficients.

For geometrical compatibility at point B, we have:

$$\bar{w}_{1B} = \bar{w}_{2A} \tag{7.12a}$$

$$\theta_{1B} = \theta_{2A} \tag{7.12b}$$

For equilibrium at point B:

$$H_{1B} = H_{2A} \tag{7.13a}$$

$$M_{1B} = M_{2A} \tag{7.13b}$$

These equations are used to determine the unknown forces and moments that can then be substituted in (7.10) to evaluate stresses.

7.2.3.3 The Exact Solution of Taylor and Wenk

At an intersection, it is necessary to develop relationships between the edge loads and deflections for all members. These are expressed in terms of the edge coefficients as follows:

$$\theta_1 = a_1 M_1 + b_1 H_1 + c_1 p \tag{7.14a}$$

$$\bar{w}_1 = d_1 M_1 + e_1 H_1 + f_1 p \tag{7.14b}$$

$$\theta_2 = a_2 M_2 + b_2 H_2 + c_2 p \tag{7.14c}$$

$$\bar{w}_2 = d_2 M_2 + e_2 H_2 + f_2 p \tag{7.14d}$$

where

a_1, b_1, etc are edge coefficients

Subscripts 1, 2 denote that the quantities are evaluated at the large and small diameter ends of the cone respectively

The shear force component H normal to the shell axis and the shear force Q at an edge are related by:

$$Q_x = H_x \cos\alpha \pm \frac{pR}{2}\sin\alpha \tag{7.15}$$

The plus sign is used for the large diameter end, whereas the minus sign is used for the small diameter end of a truncated cone.

Explicit expressions for the edge coefficients a_1, b_1, etc., are obtained by using three loading conditions at the edge (M_x, H_x, p). These are represented by setting

$$\begin{aligned} M_1 &= (M_1,\ 0,\ 0) \\ H_1 &= (0, H_1,\ 0) \\ p &= (0,\ 0, p) \end{aligned} \tag{7.16}$$

Three corresponding sets of values of C_1 and C_2 are found. These are then substituted in (7.9) to obtain the following expressions for the coefficients:

$$\begin{aligned}
a_1 &= \frac{\theta_1^{\,M}}{M_1} = \frac{U^3}{E}\sqrt{\frac{2R_1}{t^5\cos\alpha}}.\Omega_1 \\
b_1 &= \frac{\theta_1^{\,\mathrm{H}}}{\mathrm{H}_1} = -\frac{U^2}{E}\frac{R_1}{t^2}.\Omega_2 \\
c_1 &= \frac{\theta_1^{\,p}}{p} = -\frac{U^2R_1^{\,2}\tan\alpha}{2Et^2}\Omega_2 + \frac{3R_1\tan\alpha}{2Et\cos\alpha} - \frac{3(1+\nu)\sqrt{R_1}\sin^2 a}{\sqrt{2}UE\sqrt{t}\cos^5\alpha}\Omega_1 \\
d_1 &= \frac{\overline{w_1}^{\,M}}{M_1} = \frac{U^2R_1}{Et^2}.\Omega_2 \\
e_1 &= \frac{\overline{w_1}^{\,H}}{H_1} = -\frac{U}{E}\sqrt{\frac{2R_1^{\,3}\cos\alpha}{t^3}}.\Omega_3 \\
f_1 &= \frac{\overline{w_1}^{\,p}}{p} = -\frac{U}{E}\sqrt{\frac{R_1^{\,5}\sin^2\alpha}{2t^3\cos\alpha}}.\Omega_3 + \frac{(1-\nu/2)R_1^{\,2}}{Et\cos\alpha} - \frac{U^2R_1\tan^2\alpha}{8E(1-\nu)}\Omega_2 \\
a_2 &= \frac{\theta_2^{\,M}}{M_2} = -\frac{U^3}{E}\sqrt{\frac{2R_2}{t^5\cos\alpha}}.\Omega_4 \\
b_2 &= \frac{\theta_2^{\,\mathrm{H}}}{\mathrm{H}_2} = -\frac{U^2}{E}\frac{R_2}{t^2}.\Omega_5 \\
c_2 &= \frac{\theta_2^{\,p}}{p} = -\frac{U^2R_2^{\,2}\tan\alpha}{2Et^2}\Omega_5 + \frac{3R_2\tan\alpha}{2Et\cos\alpha} + \frac{3(1+\nu)\sqrt{R_2}\sin^2 a}{UE\sqrt{2t}\cos^5\alpha}\Omega_4 \\
d_2 &= \frac{\overline{w_2}^{\,M}}{M_2} = \frac{U^2R_2}{Et^2}.\Omega_5 \\
e_2 &= \frac{\overline{w_2}^{\,H}}{H_2} = -\frac{U}{E}\sqrt{\frac{2R_2^{\,3}\cos\alpha}{t^3}}.\Omega_6 \\
f_2 &= \frac{\overline{w_2}^{\,p}}{p} = \frac{U}{E}\sqrt{\frac{R_2^{\,5}\sin^2\alpha}{2t^3\cos\alpha}}.\Omega_6 - \frac{(1-\nu/2)R_2^{\,2}}{Et\cos\alpha} - \frac{U^2R_2\tan^2\alpha}{8E(1-\nu)}\Omega_5
\end{aligned} \tag{7.17}$$

where

$$\Omega_1 = \frac{\xi_1 G}{\sqrt{2}(C + 2\nu G)}$$

$$\Omega_2 = \frac{-A}{C + 2\nu G}$$

$$\Omega_3 = \frac{\xi_1 B - \dfrac{4\nu^2 G}{\xi_1}}{\sqrt{2}(C + 2\nu G)}$$

$$\Omega_4 = \frac{-\xi_2 G_K}{\sqrt{2}(C_K + 2\nu G_K)} \tag{7.18}$$

$$\Omega_5 = \frac{-A_K}{C_K + 2\nu G_K}$$

$$\Omega_6 = \frac{\xi_2 B_K - \dfrac{4\nu^2 G_K}{\xi_2}}{\sqrt{2}(C + 2\nu G)}$$

and

$$A = \xi_1 \left(ber_2'\xi_1 bei_2\xi_1 - bei_2'\xi_1 ber_2\xi_1 \right)$$

$$B = \left(ber_2'\xi_1 \right)^2 + \left(bei_2'\xi_1 \right)^2$$

$$C = \xi_1 \left(ber_2\xi_1 ber_2'\xi_1 + bei_2\xi_1 bei_2'\xi_1 \right)$$

$$G = \left(ber_2\xi_1 \right)^2 + \left(bei_2\xi_1 \right)^2$$

$$A_K = \xi_2 \left(ker_2'\xi_2 kei_2\xi_2 - kei_2'\xi_2 ker_2'\xi_2 \right) \tag{7.19}$$

$$B_K = \left(ker_2'\xi_2 \right)^2 + \left(kei_2'\xi_2 \right)^2$$

$$C_K = \xi_2 \left(ker_2'\xi_2 ker_2'\xi_2 + kei_2\xi_2 kei_2'\xi_2 \right)$$

$$G_K = \left(ker_2'\xi_2 \right)^2 + \left(kei_2\xi_2 \right)^2$$

The special Ω functions included in (7.17) are plotted in Figures 7.6–7.8. Use has been made of series expressions such that accuracy up to six significant figures is ensured. For values of ξ exceeding the range included the expressions given in the figures may be used. It is seen that for $\xi > 10$, the functions approach unity. These functions are a measure of the inaccuracy of the approximate solutions, and it is noted that the maximum discrepancy of the approximate solution (for $\xi = 10$) is of the order of 10.5%. Their value lies in providing an exact solution and also in avoiding the calculation of the Bessel functions.

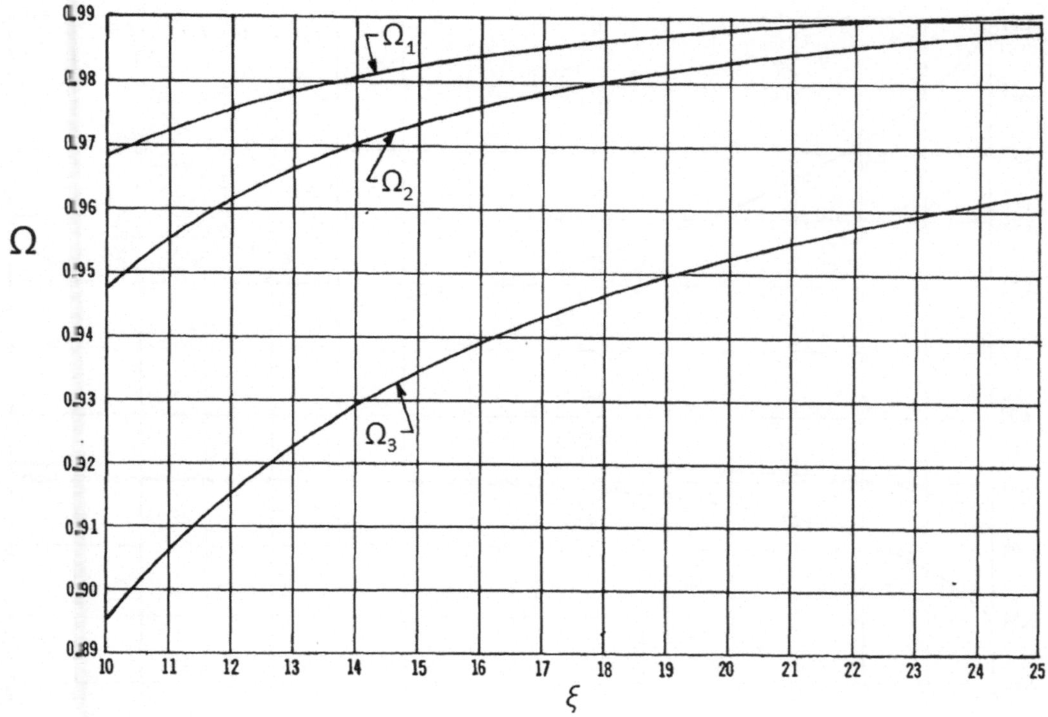

FIGURE 7.6 Omega functions used for edge rotation and displacement coefficients at a large end of a truncated cone [35].

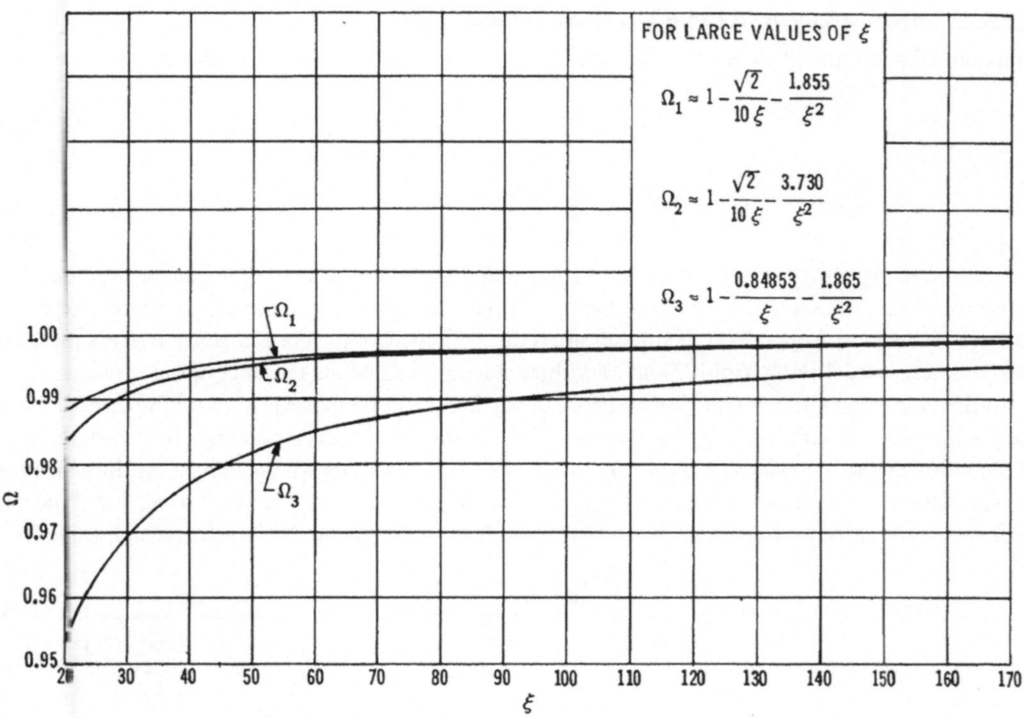

FIGURE 7.7 Omega functions used for edge rotation and displacement coefficients at a small end of a truncated cone [35].

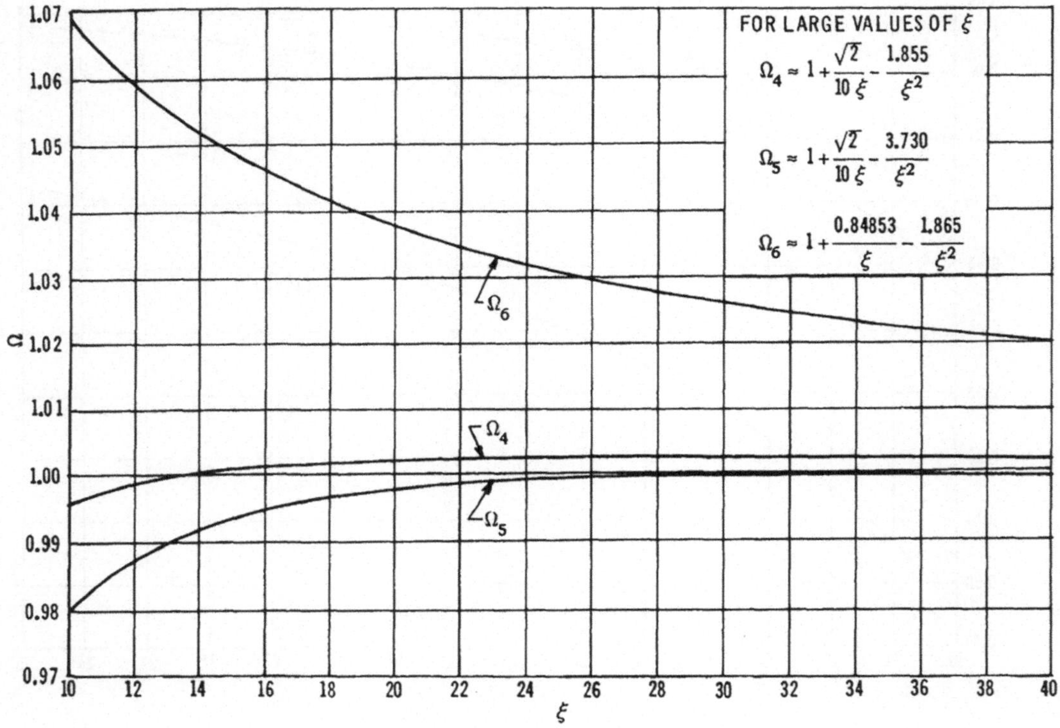

FIGURE 7.8 Omega functions used for edge rotation and displacement coefficients at a large end of a truncated cone [35].

7.2.3.4 Stress Analysis in Cylinder–Cone Intersection

In a conical shell, the meridional and circumferential stresses (σ_x, σ_θ) are given by:

$$\sigma_x = \frac{N_x}{t} \pm \frac{6M_x}{t^2} \tag{7.20a}$$

$$\sigma_\theta = \frac{N_\theta}{t} \frac{6M_\theta}{t^2} \tag{7.20b}$$

It is therefore necessary to determine N_x, N_θ, M_x, and M_θ, and to this end, (7.9) can be used using the constants of integration (C_1, C_2, etc.) for the appropriate boundary conditions. Expressions for C_1, C_2, etc., are included in Appendix 3. If stresses are required in way of the cone edges, the stress resultants can be determined directly from H and M without having to calculate the integration constants.

In the case of a cylinder–cone intersection, the cylinder can be treated as a limiting case of a cone section for which $\alpha = 0^0$. If the approximate solution is considered accurate enough, based on the values of the Ω functions, the values of H, M can be calculated using the approximate relations in which case the Ω functions are set equal to unity. The approximate forms of c_1 and f_1 are then used, and if the shell thickness is the same for both members, $t_1 = t_2$, $b_{11} - b_{12} = d_{11} - d_{12}$. At the large cone end, we have

$$M_1 = -\frac{c_{11}p}{a_{11}+a_{12}} = \frac{\sqrt{R_1 t}\tan\alpha p}{\sqrt{8}\sqrt[4]{12(1-v^2)}\left[1+\sqrt{\cos\alpha}\right]}\left[R_1\sqrt{\cos\alpha_1} - \frac{3t}{\sqrt{12(1-v^2)\cos\alpha}}\right] \tag{7.21a}$$

$$H_1 = \frac{(-f_{11}+f_{12})}{e_{11}+e_{12}}p = -\frac{R_1\sin\alpha}{2(\cos\alpha+\sqrt{\cos\alpha})}p + \frac{(1-v/2)\sqrt{R_1 t}}{\sqrt{2}\sqrt[4]{12(1-v^2)}}\left(\frac{1-\sqrt{\cos\alpha}}{\cos\alpha}\right)p \tag{7.21b}$$

7.2.4 The Wenk and Taylor Approximate Solution

In an effort to simplify the solution of the Dubois equation, Wenk and Taylor solved the equation in the following form [42]:

$$x^2\frac{d^4w}{dx^4}+2x\frac{d^3w}{dx^3}+\frac{12(1-v^2)w}{t^2\tan^2\alpha}=\frac{9(1-v^2)px^2}{Et^3} \tag{7.22}$$

Omitting the second-order term enables the use of known functions rather than special or complex functions to obtain a solution. It is, however, necessary to assess the effect of this omission. This was performed by comparing the amplitude of the omitted term with the amplitude of each of those retained. The following results were obtained, based on the solution of (7.22):

$$\frac{\left|-2\dfrac{\partial^2 w}{\partial x^2}\right|_{max}}{\left|2x\dfrac{\partial^3 w}{\partial x^3}\right|_{max}}=\frac{2}{\xi-2} \tag{7.23a}$$

For $\xi > 69$, the error is limited to 3%. Also,

$$\left|2x\frac{\partial^3 w}{\partial x^3}\right|_{max} << \left|x^2\frac{\partial^4 w}{\partial x^4}\right|_{max} \tag{7.23b}$$

It is seen that the omitted second-order term is always has a smaller magnitude than any of the remaining terms.

7.2.5 The Geckeler Approximation for Conical Shells

The homogeneous form of the Dubois equation is:

$$x^2\frac{d^4w}{dx^4}+2x\frac{d^3w}{dx^3}-2\frac{d^2w}{dx^2}+\frac{12(1-v^2)w}{t^2\tan^2\alpha}=0 \tag{7.24}$$

In accordance with the Geckeler approximation, the second- and third-order terms may be omitted, leading to:

$$x^2\frac{d^4w}{dx^4}+\frac{12(1-v^2)w}{t^2\tan^2\alpha}=0 \tag{7.25}$$

If the origin is taken at one edge of the cone rather than at the apex, the above differential equation assumes the following form:

$$\frac{d^4w}{dx^4}+4\beta^4x=0 \tag{7.26}$$

$$\text{where}\quad \beta^4=\frac{3(1-v^2)\cos\alpha}{t^2R^2} \tag{7.27}$$

and R is a constant average radius. The form of (7.26) is identical to that of an unstiffened cylinder subjected to overall bending[1], and for this reason, it is referred to as the *equivalent cylinder approximation.* As the cone approaches the geometry of a cylinder ($\alpha \to 0^0$), equations (7.26) and (7.27) reduce to those of a cylinder. Equation (7.26) may be termed the Geckeler equivalent for any shell of revolution, regardless of its shape. The definitions of w, x, and β depend on the type of shell being considered.

The solution of (7.26) assumes the following form:

$$w = e^{-\beta x}(c_1\cos\beta x + c_2\sin\beta x) + e^{\beta x}(c_3\cos\beta x + c_4\sin\beta x) \tag{7.28}$$

The boundary conditions are

$$w \to 0 \quad \text{at } x \to \infty \tag{7.29a}$$

$$M_x = M_i \quad \text{at } x = 0 \tag{7.29b}$$

$$Q_x = Q_i \quad \text{at } x = 0 \tag{7.29c}$$

where M_i and Q_i are the edge bending moment and shear force, respectively. From (7.29a), we have

$$c_3 = c_4 = 0 \tag{7.30}$$

From (7.29b) and (7.29c)

$$c_1 = \frac{1}{2D\beta^3}(-Q_i + \beta M_i) \tag{7.31a}$$

and

$$c_2 = -\frac{M_i}{2D\beta^2} \tag{7.31b}$$

Substituting in (7.28), we obtain

$$w = \frac{e^{-\beta x}}{2D\beta^3}\left[(-Q_i + \beta M_i)\cos\beta x - \beta M_i \sin\beta x\right] \tag{7.32}$$

The stresses induced in the shell are given by (7.20a) and (7.20b) where the upper sign denotes the outer fibre and the lower sign the inner fibre. The corresponding strains are obtained using Hooke´s law:

$$\varepsilon_x = \frac{1}{E}(\sigma_x - \nu\sigma_\theta) = \frac{1}{Et}\left(N_x - \nu N_\theta \pm \frac{6M_x}{t} \mp \nu\frac{6M_\theta}{t}\right) \tag{7.33a}$$

$$\varepsilon_\theta = \frac{1}{E}(\sigma_\theta - \nu\sigma_x) = \frac{1}{Et}\left(N_\theta - \nu N_x \pm \frac{6M_\theta}{t} \mp \nu\frac{6M_x}{t}\right) \tag{7.33b}$$

The stress and moment resultants are given by:

$$
\begin{aligned}
M_x &= D\left(\frac{d^2w}{dx^2}+\frac{v}{x}\frac{dw}{dx}\right)=D\left(\frac{d^2w}{dx^2}-\frac{v\sin\alpha}{R}\frac{dw}{dx}\right)\\
M_\theta &= D\left(v\frac{d^2w}{dx^2}+\frac{1}{x}\frac{dw}{dx}\right)=D\left(v\frac{d^2w}{dx^2}-\frac{\sin\alpha}{R}\frac{dw}{dx}\right)\\
N_x &= D\tan\alpha\left(\frac{d^3w}{dx^3}+\frac{1}{x}\frac{d^2w}{dx^2}-\frac{1}{x^2}\frac{dw}{dx}\right)-\frac{px\tan\alpha}{2}=Q_x\tan\alpha-\frac{pR}{2\cos\alpha}\\
N_\theta &= \frac{d}{dx}(xN_x)=-\frac{d}{dx}\left(\frac{R}{\sin\alpha}Q_x\tan\alpha\right)-\frac{pR}{\cos a}\\
&= D\tan\alpha\left(\frac{R}{\sin\alpha}\frac{d^4w}{dx^4}-2\frac{d^3w}{dx^3}-\frac{\sin\alpha}{R}\frac{d^2w}{dx^2}-\frac{\sin^2\alpha}{R^2}\frac{dw}{dx}\right)-\frac{pR}{\cos\alpha}\\
Q_x &= D\left(\frac{d^3w}{dx^3}+\frac{1}{x}\frac{d^2w}{dx^2}-\frac{1}{x^2}\frac{dw}{dx}\right)=-D\left(\frac{d^3w}{dx^3}-\frac{\sin\alpha}{R}\frac{d^2w}{dx^2}-\frac{\sin^2\alpha}{R^2}\frac{dw}{dx}\right)
\end{aligned}
\tag{7.34}
$$

7.3 TESTS ON TRUNCATED CONICAL SECTIONS

A large number of tests on conical sections have been carried out since the middle of the 20th century [4,9, 15]. These cover a wide range of structural configurations, types of load action as well as boundary conditions. In this section, we shall focus on a small number of series of tests whose results are available in the open literature that have been performed in order to provide guidance in submarine pressure hull design. The tests discussed have been selected in order to provide an overall understanding of the effect that various parameters have on the buckling response of truncated conical sections, both unstiffened and stiffened. Results from the earlier tests (Krenzke, Raetz) ought to be treated as providing a qualitative understanding of response and also as benchmarks for modern theoretical approaches. All test results can be used to provide a benchmark for finite element analysis, for the development of design codes as well as interactive design formulas.

In the cases in which unstiffened cones have been considered, the results are not directly applicable to submarine pressure hull design, but the insight provided leads to an improved understanding of interframe behaviour of stiffened conical shells in the inelastic range, especially in the case of combined loading [2,33]. Comparisons with analytical formulations, finite element analysis, and design codes have also been reported and are discussed. An interactive design equation for ring-stiffened conical sections is presented and its derivation explained [8,9].

The series of tests discussed are listed in Table 7.1.

7.3.1 Effect of Stiffeners at Cone–Cylinder Junctions

The earliest reported tests were carried out in the United States at the David Taylor Model Basin on truncated conical shells attached to cylindrical sections at the two ends [18]. The series consisted of six tests that were performed in order to assess the effect of placing stiffeners at the junctions on the collapse pressure of the conical section. In order to provide a basis for comparison, three of the models had circumferential reinforcing rings at both cone–cylinder junctions, whereas the other three did not (Figure 7.9). The shell thickness differed for each model.

Strain gauges were attached to the models in way of the cone–cylinder junctions (Figure 7.11a), and longitudinal and circumferential strains were recorded throughout the loading range in each case. Models of different thickness were tested so that the effect of thickness on collapse pressure could also be assessed.

TABLE 7.1
Experiments on Truncated Conical Sections for Submarine Pressure Hull Design

Researcher	Aim of Test	Description of Models Tested
Krenzke [18]	1. Determine collapse pressures of cone sections without stiffeners at the junctions 2. Determine the effect of stiffeners at the junctions of the cone section with the adjoining cylinders and the mode of collapse and collapse pressure. 3. Compare test collapse pressures with expressions for elastic buckling	Six unstiffened cones attached to cylindrical shells at ends with and without stiffeners at cone–cylinder junctions
Raetz [29]	1. Extend the previous studies of both the elastic behaviour and the collapse strength of conical shells to other geometries. 2. Assess effect of cone angle, shell thickness, and cone–cylinder reinforcement 3. Compare observed collapse pressures with analytical predictions	Unstiffened cones attached to cylindrical shells at ends with and without stiffeners at cone–cylinder junctions under external hydrostatic pressure
Cho et al: [9]	1. Predict failure mode allowing for imperfections 2. Predict failure mode using FEA 3. Compare FEA with design codes 4. Assess interaction formula for failure prediction	Ring-stiffened conical shells under external hydrostatic pressure
Błachut [3]	1. Predict failure load 2. Assess code predictions	Unstiffened conical shells under external pressure, axial compression and combined loading
Błachut & Ifayefunmi [4]	1. Validate FEA against single load tests. 2. Predict elasto-plastic interactive buckling using FEA.	Four unstiffened conical shells under external hydrostatic pressure (2) and axial compression (2)

FIGURE 7.9 Models before testing [18]. (a) Model without stiffeners and (b) model with stiffeners.

These tests were carried out to determine:

1. The actual collapse pressures of conical sections without stiffeners at the junctions
2. The effect of stiffeners at the junctions of the conical section with the adjoining cylinders on the mode of collapse and collapse pressure.

In the case of the conical sections fitted with ring stiffeners at the junctions, the collapse pressure appears to be associated with that pressure at which yielding first occurs at points other than at the junction, so that the junction is not the weakest position.

Figure 7.10 shows that the collapse pressure is increased by placing stiffeners at the junctions. The effect becomes more pronounced with increases in cone thickness. For a full-scale submarine pressure hull having a shell thickness that exceeds 25 mm (1 inch), the effect is seen to be particularly important, and heavy ring-stiffeners are required.

Figure 7.11a is a gauge location diagram for Model 5 (one half of the section shown), whereas Figure 7.11b is a pressure–strain plot for gauge 110. Using the pressure–strain plots obtained for each gauge, the *strain sensitivity factors*, being the ratio of the strain to the pressure in the linear elastic range, are obtained. These are plotted in Figure 7.11c. This is a representative plot of strain sensitivity factors (in effect in-plane flexibility) at different positions for the models fitted with stiffeners (Model No. 2, 4, and 6).

The difference in flexibility across the stiffener cross section is a measure of the effect of the presence of the stiffener. Thus, the difference reduces with increasing cone thickness. From Figure 7.12, it is seen that there was appreciable rotation of the stiffeners at both junctions. At the small diameter end, the rotation was towards the cone, thus increasing the stress in the cone, whereas at the large diameter end, rotation was towards the cylinder. Furthermore, it was found that excessive rotation of the stiffener at the junction decreases the collapse pressure. Therefore, it is imperative to design

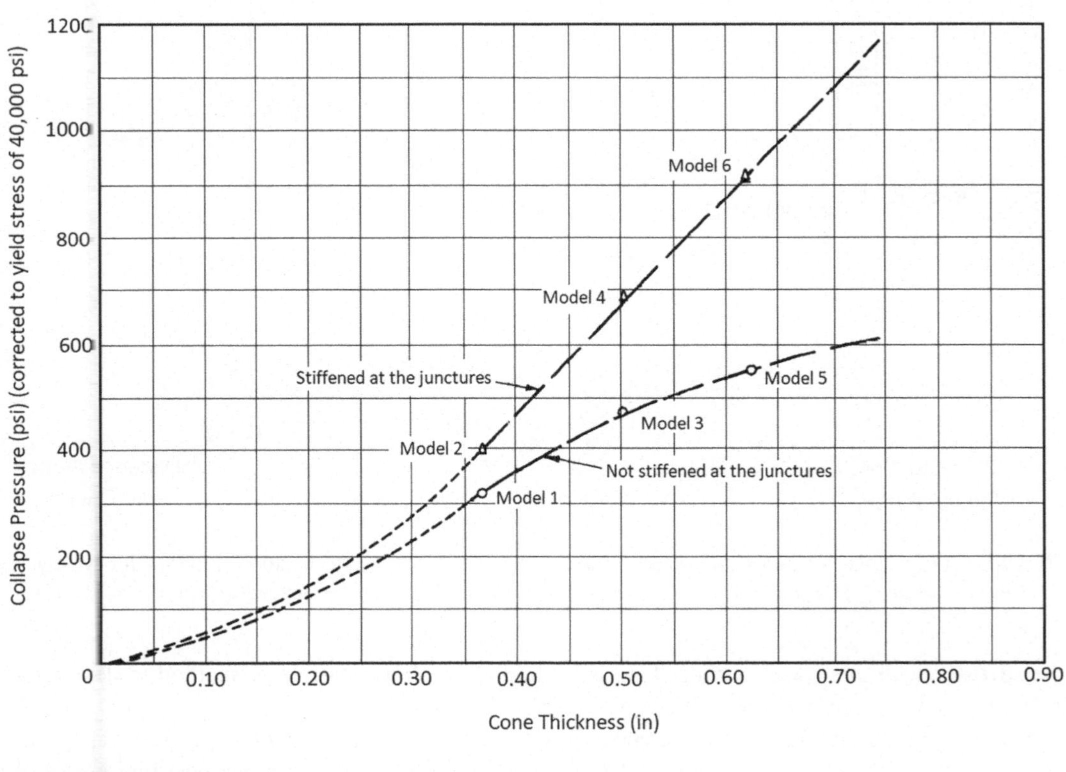

FIGURE 7.10 Collapse pressure versus cone thickness [18].

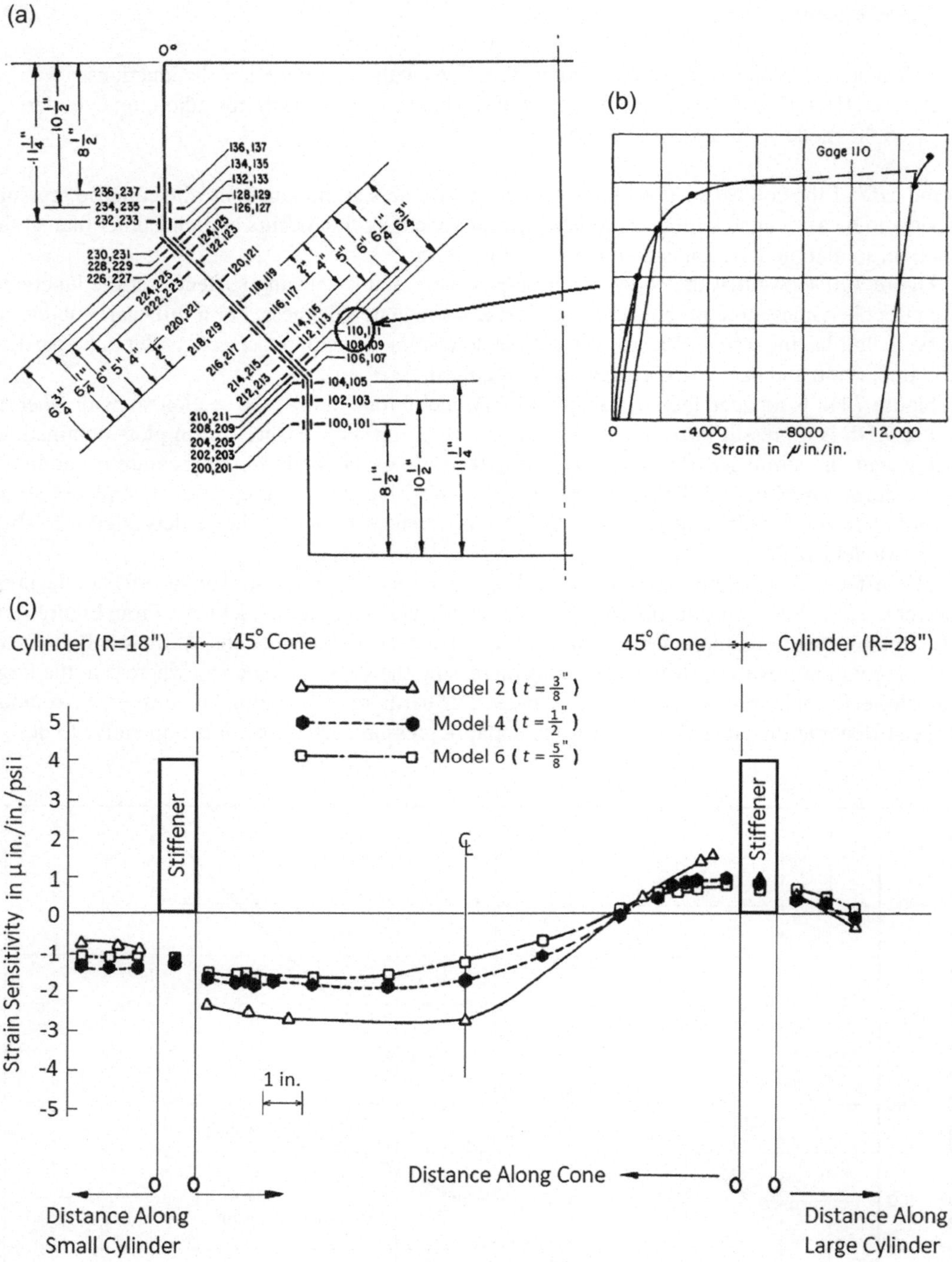

FIGURE 7.11 Gauge locations and strain measurements [18]. (a) Gauge location diagram for Model 5. (b) Pressure-strain plot for gauge 110. (c) Average circumferential strain in models with stiffeners at junctions.

the stiffener such that rotation under load is limited as much as possible. A method to design the stiffener to ensure zero rotation is included in [18].

It was observed that for all models, yielding occurred at the large junction at pressures appreciably below the collapse pressure. For models without stiffeners at the junctions, yielding next took

FIGURE 7.12 Models following collapse [18]. (a) Models without stiffeners. (b) Models with stiffeners.

place at the small junction; for models with stiffeners, yielding next took place away from the small junction at about one-third to one-half of the distance to the large end.

7.3.2 Effect of Cone Angle and Presence of Stiffeners at Cone–Cylinder Junctions

In this series of tests performed at the David Taylor Model Basin, the conical shells had larger length-to-thickness and diameter-to-thickness ratios than those of the earlier series and were intended to fail in the asymmetric buckling range [29]. The purpose of these tests was to extend the previous studies of both the elastic behaviour and the collapse strength to other geometries. The following parameters were varied to determine their effects on strain distributions and collapse pressures:

- Cone angle
- Shell thickness
- Cone–cylinder reinforcement

Figure 7.13 includes a schematic diagram of Model 5 with details of the arrangement of the ring stiffeners at the ends of the conical section.

Following the procedure adopted in the previous series of tests, strain sensitivity coefficients were plotted for each model. Figure 7.14 shows the measured values as well as the theoretical strains predicted using the method described in [30] for Model 1. This method was originally devised to be used for cones with small vertex angles, but it was concluded from this series of tests that it can be used for a larger range of vertex angles. A comparison of results obtained for models with and without ring stiffeners at the junctions showed that in the former case, the circumferential strains are reduced at the large cone–cylinder junctions. It was also found that both circumferential and longitudinal strains are reduced in way of the small cone–cylinder junctions in the presence of a ring stiffener.

The longitudinal strains in the conical shells at the large junctions were little affected because the ring stiffeners rotated in the meridional planes. This reinforces the conclusion reached in the

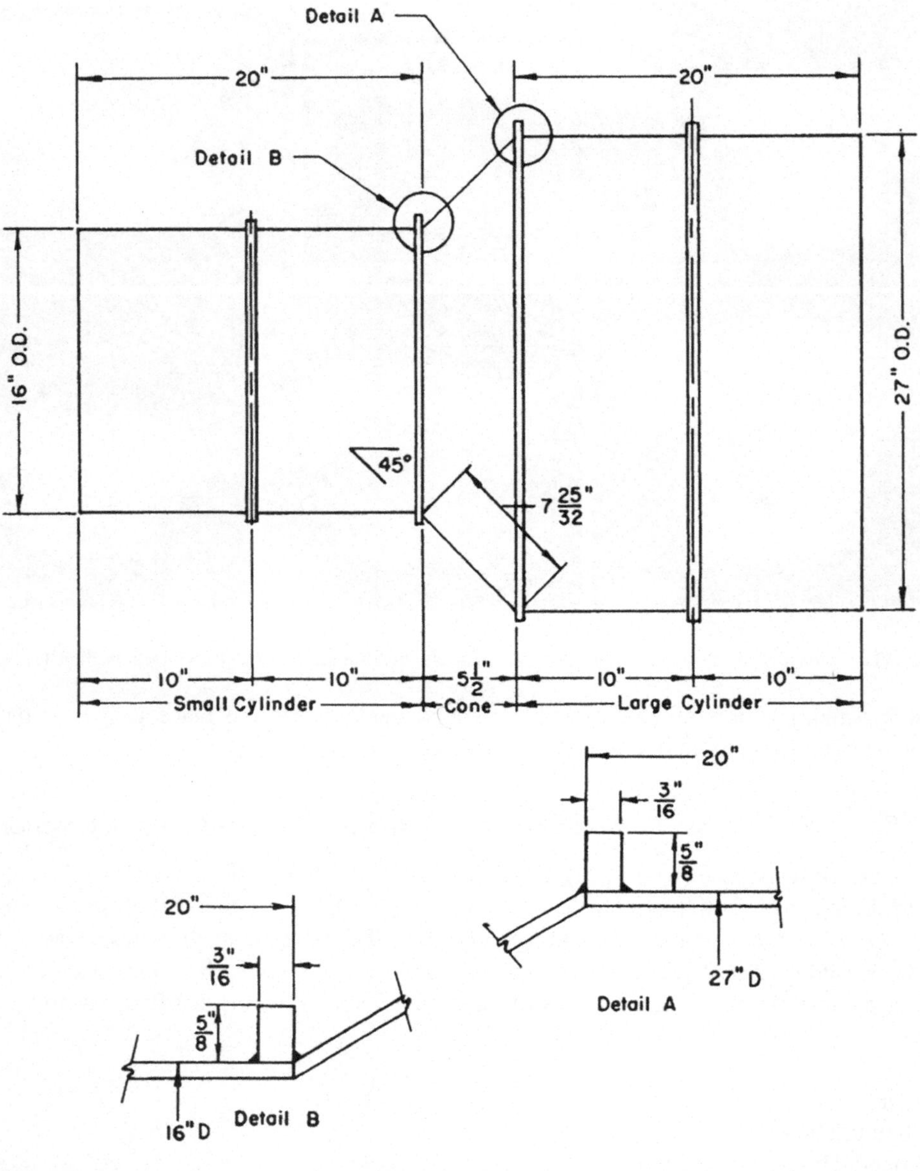

FIGURE 7.13 Schematic diagram of Model 5 [29].

previous study that ring stiffeners at the junction need to be designed such that rotation is eliminated. In the region of the conical shells far from the junction, both longitudinal and circumferential strains were little affected by the presence of the ring stiffeners at the ends. It has been shown theoretically [30] that the effect of a ring stiffener disappears at a distance along the shell of approximately $3.0/\beta$.

Models 3, 4, and 5 following collapse are shown in Figure 7.15. Collapse data from the tests as well as predicted values are included in Table 7.2. The values obtained using the theoretical approaches correspond to elastic buckling of simply supported conical shells without stiffening rings at the ends. Comparing the test values with the theoretical predictions, it is seen that failure occurred in the inelastic range, with the cone not being able to attain the elastic buckling stress predicted by the theoretical approaches. It is of interest at this point to compare the results of the

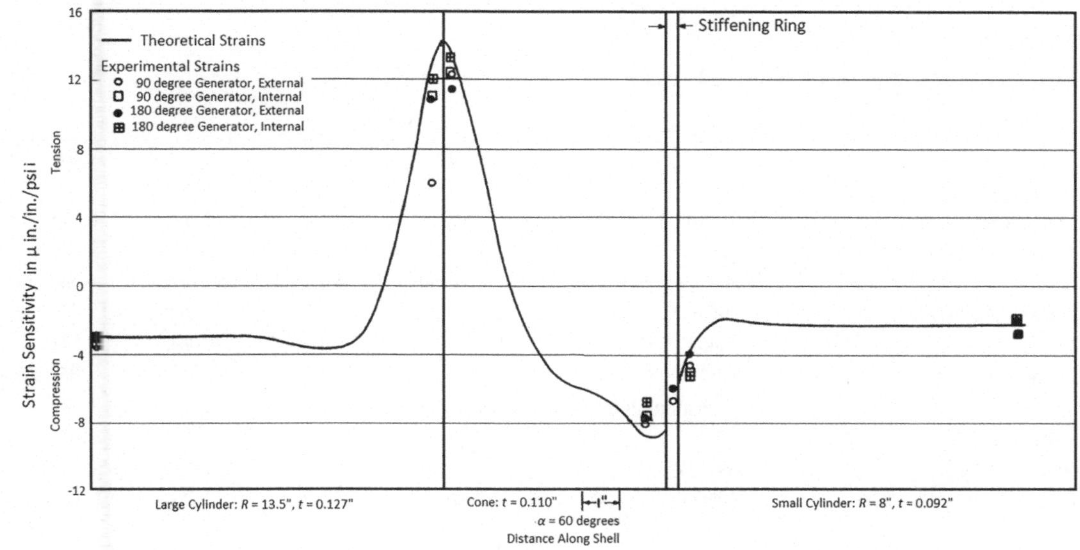

FIGURE 7.14 Theoretical and experimental circumferential strain sensitivity in Model 1 [29].

FIGURE 7.15 Models 3, 4 and 5 following collapse [29].

theoretical approaches. The differences between them are attributed to the selection of deflection functions in each case.

Figure 7.16 shows the exact positions at which failure occurred in each model. It is seen that in all cases with the exception of Model 6, failure arises with the formation of one of more lobes in the vicinity of the longitudinal weld seam(s). In the case of Model 5 (retest) in which longitudinal stiffeners were fitted, failure occurs adjacent to the outer longitudinal stiffener, away from the weld seam.

TABLE 7.2
Model Particulars and Collapse Pressures

	Model					
	1	**2**	**3**	**4**	**5**	**6**
Cone angle (deg.)	60	60	30	45	45	45
Cone thickness (in)	0.110	0.110	0.094	0.106	0.106	0.095
Yield point of core material (psi)	42,300	42,300	65,100	66,000	66,000	63,400
Location of stiffening rings	Small end	Both ends	Both ends	None	Both ends	None
Length of cone (in)	6.36	6.36	11.00	7.78	7.78	7.78
βl [a]	5.453	5.453	13.427	8.081	8.081	8.581
Collapse pressure (psi)	97	115	105/120 [b]	163	135/148	135
	Calculated Collapse Pressure (psi)					
Tokugawa [37]	237	237	195	334	334	105
Niordson [26]	399(12) [c]	399(12)	146(10)	290(11)	290(11)	220(11)
Bunich, Paily & Piskovitina [5]	202(8)	202(8)	161(8)	237(8)	237(8)	179(9)

[a] βl is the argument of the trigonometric expressions used to describe the deflection of the conical shell under pressure where $\beta = \sqrt[4]{\frac{3(1-\nu^2)}{(Rt)^2}}$ and l = length along the shell between junctions.

[b] The first pressure is the pressure at which the model failed in the vicinity of a suspicious weld. The second was obtained during a retest after the first lobe had been straightened out and the cone had been reinforced with longitudinal stiffeners.

[c] The numbers in parentheses denote the number of lobes in buckling.

7.3.3 Elasto-Plastic Failure of Unstiffened Cones under Uniaxial Loading

In this series of tests, reported in [4], two specimens of machined unstiffened conical shells with flanges were subjected to hydrostatic pressure (models C1 and C2) and two to axial compression (models C3 and C4). The test results were used as a benchmark for finite element modelling of the tests using ABAQUS software. This was used to simulate response under combined hydrostatic pressure and axial compression.

The models prior to and following collapse are shown in Figure 7.17, whereas results from the use of finite element analysis for the two types of loading are included in Figure 7.18.

Table 7.3 (p.243) includes quantitative results for the experiments and the numerical analysis, followed by a comparison of the two.

It is seen that in general, the FE predictions of the collapse pressures overestimate the experimental values by about 5%. These results are based on average wall thickness, average geometry, and on using engineering stress–strain modelling of the material. It is expected that if detailed data on initial geometrical imperfections (initial distortions and wall thickness) as well as more accurate data on material strength properties had been introduced, the numerical predictions would have been much closer to the experimental values.

In the case of the axially compressed cones, the numerical predictions of the collapse force exceeded the experimental values by 13% for C3 and 30% for C4. This is considered to be excessive, and this discrepancy is attributed on the one hand to the need for more detailed data as discussed above and on the other to the need to measure residual stresses in the test models, since these were not stress-relieved prior to testing.

7.3.4 Elasto-Plastic Failure of Unstiffened Cones under Combined Loading

In this series of tests, aspects of inelastic buckling behaviour of unstiffened cones were studied [3].

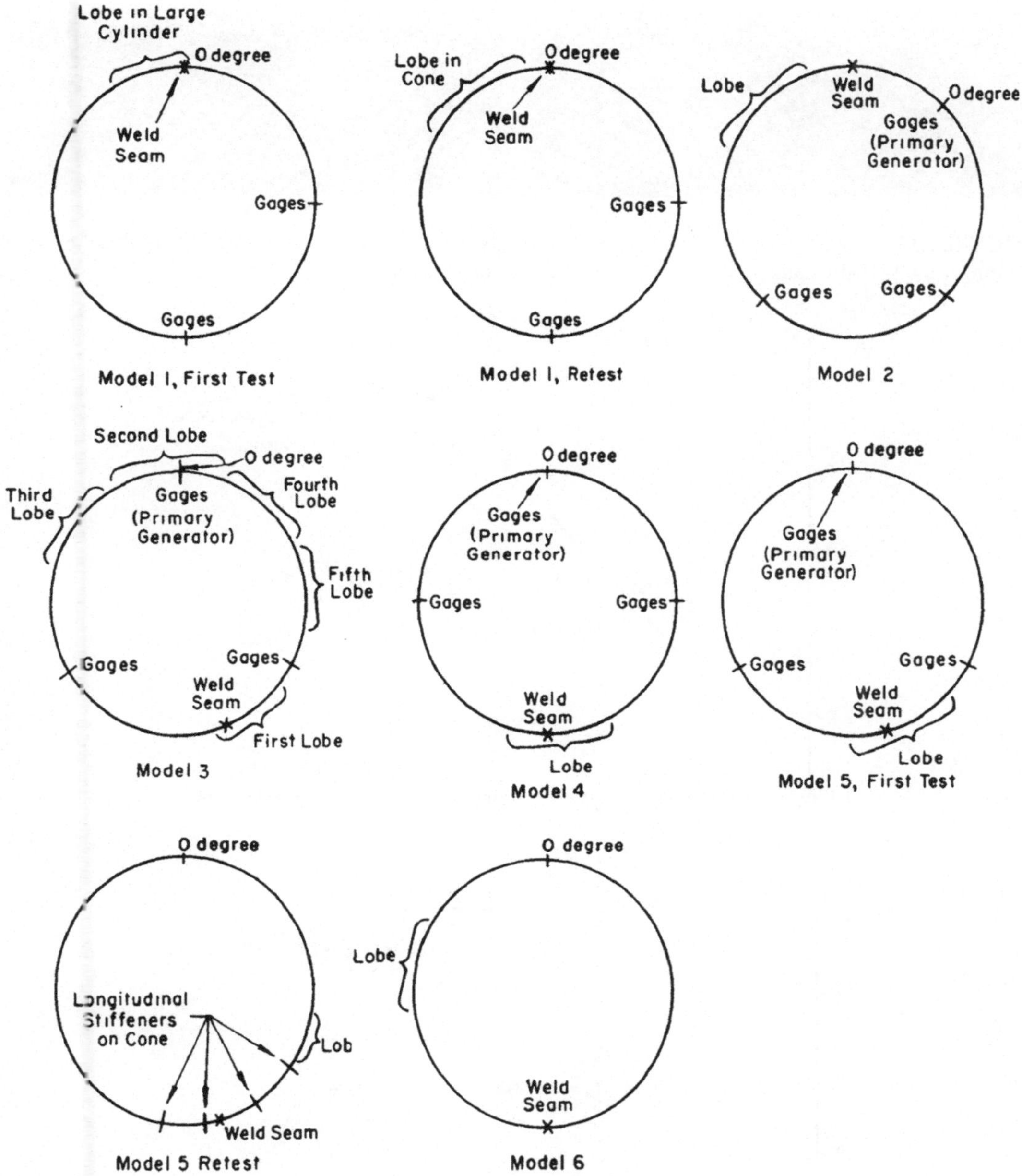

FIGURE 7.16 Regions of failure of all models [29].

Firstly, the applicability of the equivalent cylinder concept, originally defined for elastic behaviour, was reassessed in the case of elasto-plastic buckling of cones subjected to external hydrostatic pressure and axial compression acting independently. It was found that in the case of cones failing in the elasto-plastic range it is not possible to make use of the equivalent cylinder concept.

Secondly, the elasto-plastic buckling of cones subjected to combined loading was studied. A series of tests was carried out, and these were modelled numerically and results compared. The tests were conducted on cones subjected to axial compression, external pressure, and combined loading. Each specimen was clamped in way of the larger radius. Particulars are listed below.

FIGURE 7.17 Test specimen used and qualitative results for the two types of loading [4]. (a) Model prior to testing, (b) C1- Hydrostatic pressure, and (c) C4- Axial Compression.

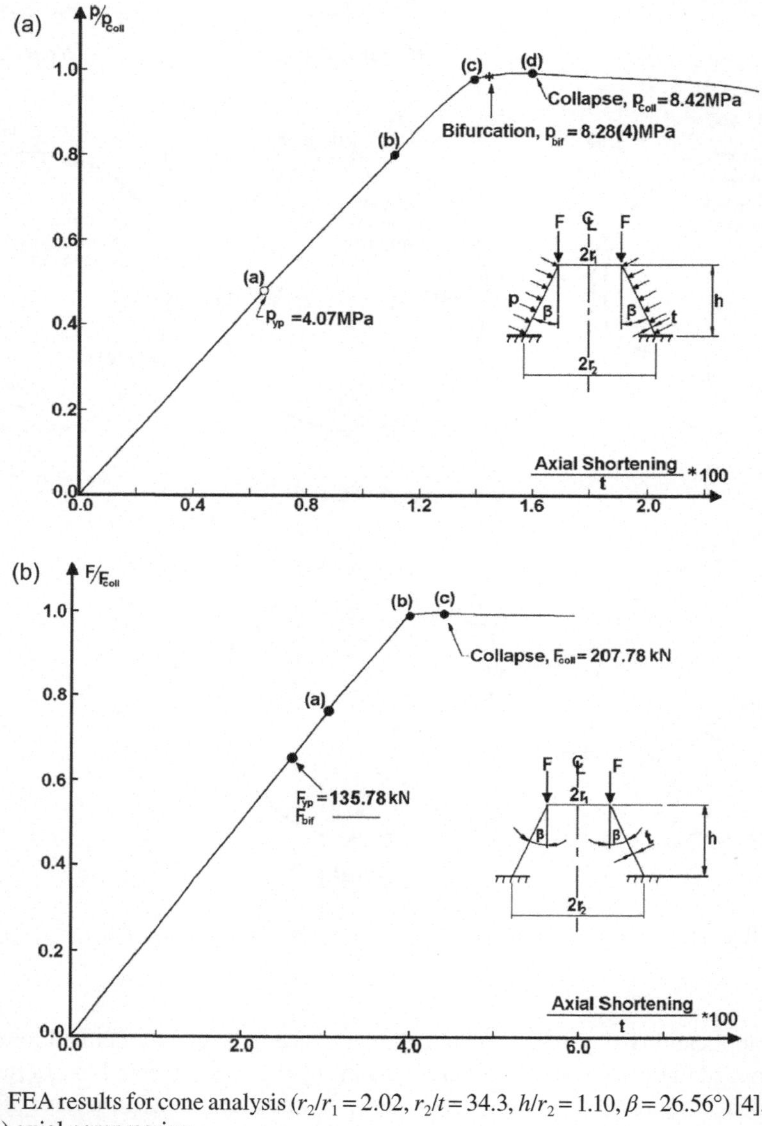

FIGURE 7.18 FEA results for cone analysis ($r_2/r_1 = 2.02$, $r_2/t = 34.3$, $h/r_2 = 1.10$, $\beta = 26.56°$) [4]. (a) Hydrostatic pressure and (b) axial compression.

$$(r_2 / r_1) = 2.02 \quad \beta = 26.561° \quad (h / r_2) = 1.01$$

$$E = 210.49 \text{ GPa} \quad \sigma_y = 230.6 \text{ MPa} \quad \nu = 0.281$$

TABLE 7.3
Comparison of Experimental and Numerical Results (A: ABAQUS, B: BOSOR5)

		Experimental				Numerical				Expt./Numerical			
		C1	C2	C3	C4	C1	C2	C3	C4	C1	C2	C3	C4
A	p (MPa)	8.0(4)[a]	8.0(4)	–	–	8.36	8.40	–	–	0.96	0.95	–	–
	F_{coll} (kN)	–	–	238.98	279.8	–	–	211.95	215.3	–	–	1.13	1.30
B	p_{bif} (MPa)	8.0(4)	8.0(4)	–	–	8.1(7)	8.2(7)	–	–	0.99	0.98	–	–
	F_{coll} (kN)	–	–	238.98	279.8	–	–	210.46	215.7	–	–	1.14	1.30

[a] Numbers in brackets denote number of circumferential waves in the eigenshape or the experiment.

TABLE 7.4
Experimental and Numerical Results for Unstiffened Cones Subjected to Axial Compression And Lateral Pressure

			Numerical		*Experimental/Numerical*	
		Experimental	ABAQUS-S8R	BOSOR5	ABAQUS	BOSOR5
Cone	*Load Type*	*F*(kN), *P* (MPa)	*F*(kN), *P* (MPa)	*F*(kN), *P* (MPa)	*F*(kN), *P* (MPa)	*F*(kN), *P* (MPa)
CS2	Lateral	0.0, 4.62(7)[a]	0.0, 3.54	0.0, 4.13(8)	-, 1.31	-, 1.12
CS5	pressure	0.0, 4.62(7)	0.0, 4.42	0.0, 4.10(8)	-, 1.13	-, 1.15
CS3	Axial compression	232.0, 0.0	231.0, 0.0	228.84, 0.0	1.00, -	1.01, -
CS4		230.0, 0.0	218.82, 0.0	214.27, 0.0	1.05, -	1.07, -
CS1		65.81, 4.76(7)	54.58, 3.99	56.40, 4.08(8)	1.21, 1.19	1.17, 1.17
CS6	Combined loading	152.2, 4.14(6)	143.82, 3.91	143.03, 3.89(7)	1.07, 1.07	1.06. 1.06
CS7		126.6, 4.48(6)	112.26, 3.98	112.93, 4.0(8)	1.13, 1.13	1.12, 1.12
CS8		189.1, 3.28	172.78, 2.96	175.93, 3.0(6)	1.10, 1.11	1.08, 1.09
CS9		237.1, 2.35	218.0, 2.16	210.31, 2.08(6)	1.09, 1.09	1.13, 1.13
CS10		231.9, 1.12	223.59, 1.06	214.34, 1.04(5)	1.04, 1.06	1.08. 1.08

[a] Numbers in brackets denote the number of lobes observed/predicted.

A summary of results and a comparison of collapse values measured during the tests with the numerical predictions using ABAQUS and BOSOR5 are included in Table 7.4.

Figure 7.19 shows the spread of plastic straining at the force required to produce axial collapse (Figure 7.19a) and at the pressure to cause collapse (Figure 7.19b). It is seen that in the case of axial compression, failure is observed in way of the smaller radius, which is where the stresses reach their maximum value throughout the loading procedure. In the case of external pressure, inelastic buckling occurs, and a plastic hinge forms at the collapse pressure at the clamped end, where a reaction force and a bending moment act.

When a cone is subjected to combined external pressure and an independent axial compressive force, the response depends on the relative magnitudes of the pressure acting and the force. An axial force that is acting on a cone causes the shell plating to move outwards, whereas external pressure loading results in the opposite. Hence, when both loads act concurrently, in principle, the buckling strength is not significantly affected. Comparing results for cones with cylindrical shells under combined loading, it is seen that in the case of a cone, the axial force does not lower the buckling pressure to the same extent that it does in the case of a cylindrical shell.

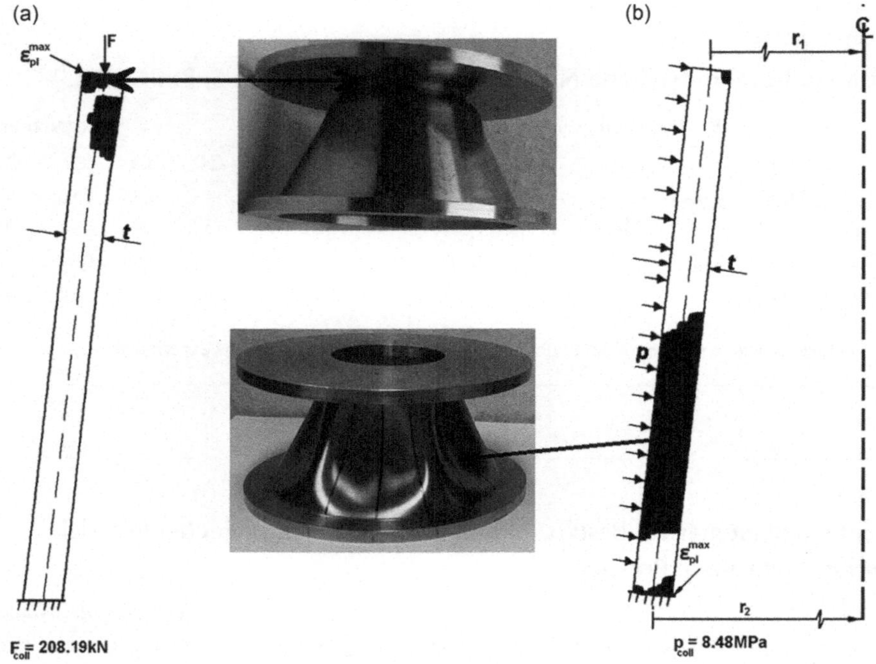

FIGURE 7.19 Plastic strain distributions at collapse in cones subjected to (a) axial compression and (b) external pressure [3].

When two different types of loading are applied concurrently, results are presented in a *combined loading* plot. In this, both loads are normalised with respect to the corresponding values at failure ($F/F^0_{coll}, P/P^0_{coll}$). Figure 7.20a presents results for a cone with $r_2/t = 75.0$. It is seen that in addition to the failure curve (collapse envelope), a region defining the initiation of yielding is outlined so that the shaded region that is shown is that in which post-yield straining takes place. The two sub-regions A and B specify the axial force-dominated response and the pressure-dominated response, respectively. Analogous behaviour is exhibited in the case of cones of different geometries (Figure 7.20b).

Table 7.4 summarises the failure loads measured during testing as well as numerical predictions using ABAQUS and BOSOR5 software. It can be seen that when the axial compressive load $P/P_{col} < 0.66$ (cone CS6), behaviour is dominated by the lateral pressure failure mode that consists of a series of lobes. As the value of P/P_{col} is increased, the failure mode changes with the number of lobes decreasing. When the lateral pressure load is equal to zero, asymmetric buckling is observed (CS4). When failure arises as a result of combined loading, the failure mode differs from either of the uniaxial loading cases. We observe features of both lateral pressure failure (lobes) and asymmetric buckling (axial compression) (CS8, CS9 and CS10 shown in Figure 7.21).

The collapse envelope shown in Figure 7.20 was obtained using finite element analysis assuming that the material is elastic perfectly plastic and average values for the cone geometry. It can be seen that for "pressure dominated" failures, the experimental data points closely follow the interactive diagram, whereas in the case of axial pressure dominated failure, the collapse curve underpredicts the observed collapse strength.

It is seen that the observed collapse values exceed those predicted using finite element analysis. It is seen from columns 6 and 7 (experimental/predicted collapse) that the largest discrepancy arises in the case of external pressure (cones CS2 and CS5). In this case, the bias varies from 1.13 to 1.31 for ABAQUS and from 1.12 to 1.15 for BOSOR5.

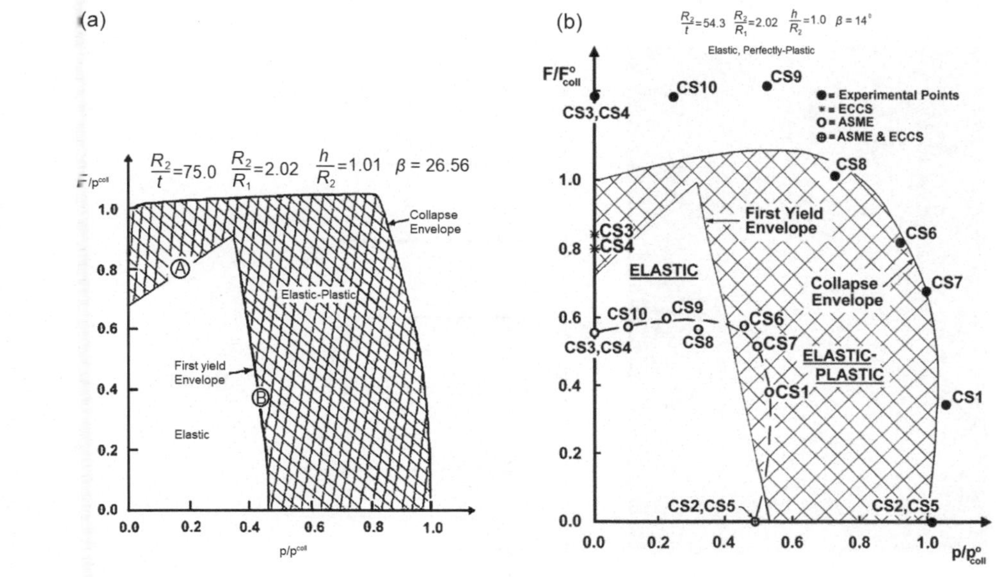

FIGURE 7.20 Combined loading plots for cones loaded in axial compression and lateral pressure [3]. (a) Features of a typical combined loading plot. (b) Test data and ASME, ECCS design curves.

FIGURE 7.21 Failure modes observed under combined loading [3].

With regard to the codes considered (ASME and ECCS), the following two observations can be made:

i. Under a single incremental loading, there is a large margin of safety against buckling along the whole interactive domain
ii. Code recommendations for some loading combinations fall into the elastic–plastic domain.

7.3.5 Collapse Strength Prediction of Ring-Stiffened Cones Using FEA

More recently, a series of tests on welded ring-stiffened steel conical shells subjected to external hydrostatic pressure was carried out in Korea [9]. These tests were performed in order to:

1. Examine the different failure modes and determine the ultimate strength in each case
2. Act as a benchmark for finite element studies and a proposed design equation
3. Compare results obtained using the proposed design equation with those obtained using existing design codes

Four test models were fabricated, following standard shipyard practice and scantlings (Figure 7.22). The ring-stiffeners used were externally welded flatbars instead of internal T-bars. A characteristic

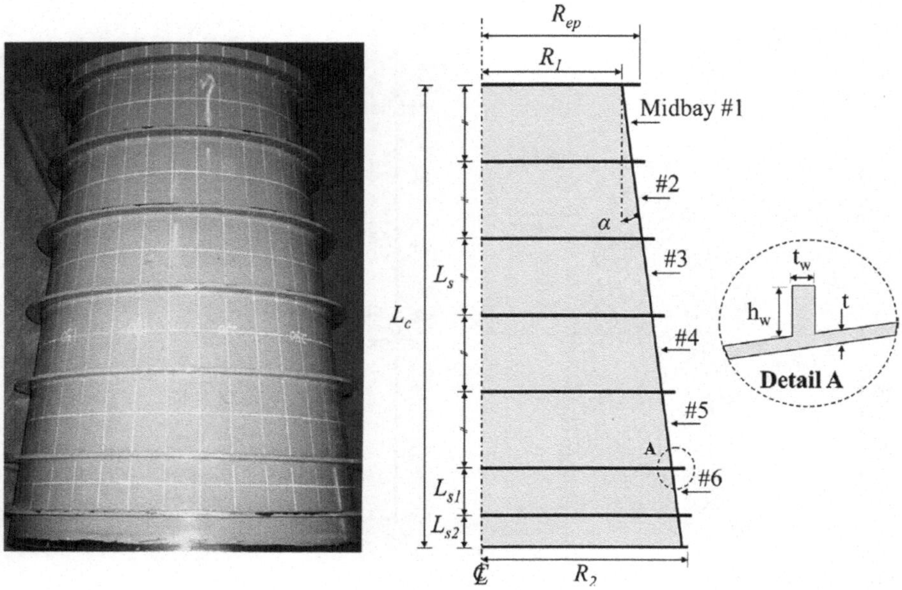

FIGURE 7.22 Test model of ring stiffened cone and nomenclature [9].

property of conical sections is that the stiffness reduces with increasing radius. To compensate for this, an additional ring-stiffener was added adjacent to the large diameter end, thereby dividing the stiffener spacing into L_{s1} and L_{s2} (Figure 7.22).

Details concerning the models are included in Table 7.5. Of these, models RSC-1, RSC-2, and RSC-3 were designed to fail in a local mode, whereas model RSC-4 was designed to fail in an overall mode (weaker ring-stiffeners).

Initial distortions can result in significant reductions in ultimate strength. An initial distortion having an amplitude of 0.4t results in a reduction in collapse pressure of about 40%. It is thus necessary to allow for them, as indeed for the presence of welding residual stresses. For present purposes, the initial distortions were measured using the chord gauge method, described in Section 4.2.2. Due to the small scale of the specimens, the initial distortions were found to be much larger than those measured in a full-scale structure. Thickness measurements were also obtained using the ultrasonic gauge method. Figure 7.23 includes measurements of initial distortions and shell thickness for model RSC-1.

TABLE 7.5
Model Dimensions (as Built) [9]

Model	Shell				Stiffener			End Plate	
	Cone Angle (α) (deg)	Overall Length (L_c) (mm)	Thickness (t)	Small and Large Radius R_1, R_2)	Height and Thickness h_w, t_w	Main and Extra Spacing (L_s, L_{s1}, L_{s2})	Number of Stiffeners	Radius (R_{ep}) (mm)	Thickness (t_{ep})
RSC-1	7.7	725	2.40	202, 300	30 × 5.81	120,75,50	6	232	20
RSC-2	7.7	625	2.33	214, 299	25 × 3.86	100,75,50	6	244	20
RSC-3	13.6	725	2.38	125, 299	30 × 5.86	120,75,50	6	155	20
RSC-4	12.7	620	2.40	160, 300	15 × 3.11	60,50,30	10	190	20

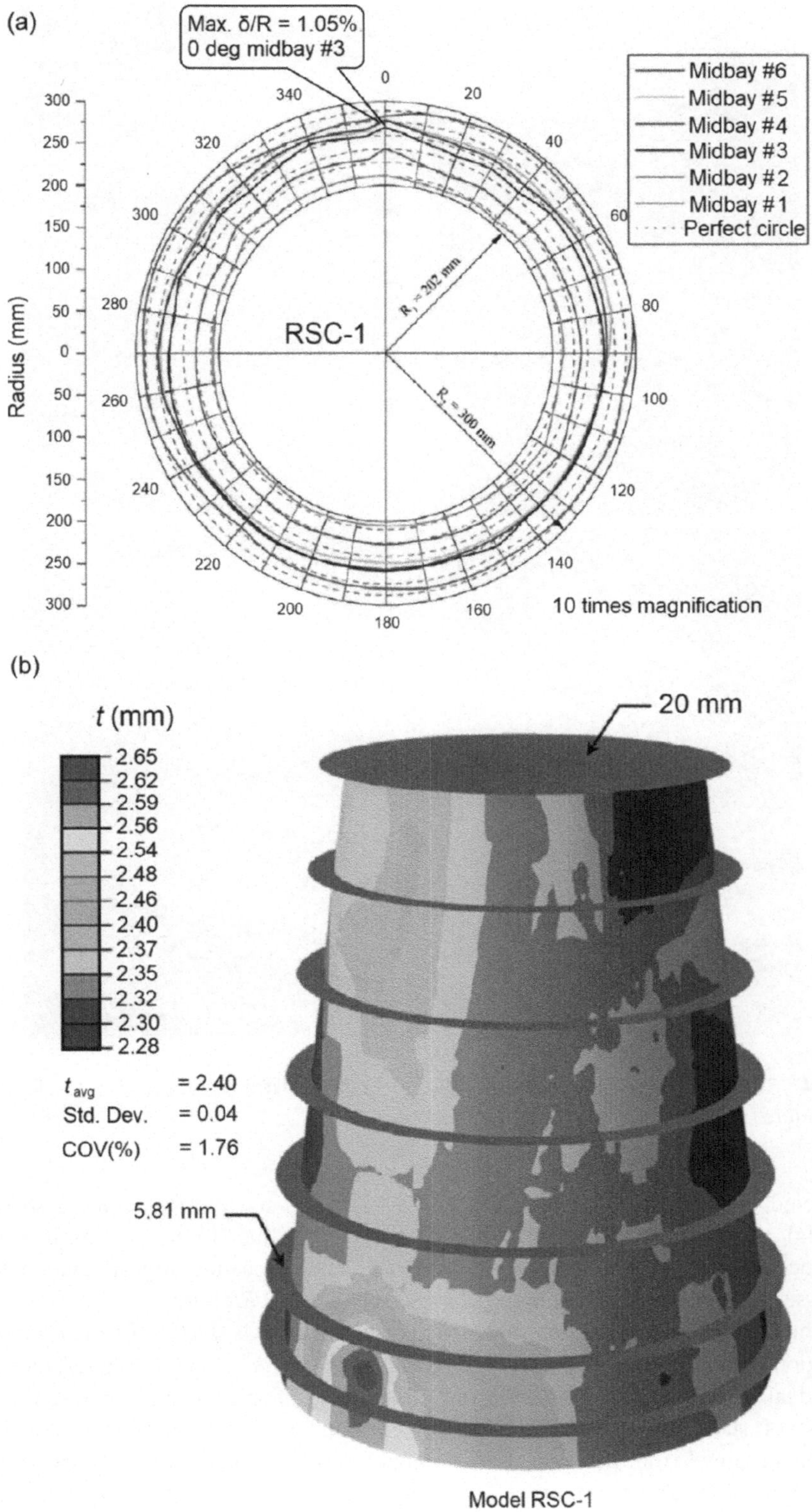

FIGURE 7.23 Initial imperfections for Model RSC-1.[9]. (a) Initial distortions and (b) cone thickness.

FIGURE 7.24 Test models following collapse [9]. (a) Local interframe failure (RSC-1, RSC-2, RSC-3). (b) Overall failure (RSC-4).

As expected, Models RSC-1, RSC-2, and RSC-3 failed by interframe local shell buckling (Figure 7.24a), whereas Model RSC-4 failed by overall buckling (Figure 7.24b). It was observed that failure occurs in bay 5, which is the weakest one, since it has the largest value of R/t. Failure then progresses to adjacent bays that have progressively smaller R/t ratios.

Finite element modelling of the tests was carried out using the ABAQUS software package. Data on initial distortions and thickness were introduced using the measured data for each test model, whereas welding residual stresses were introduced using the Faulkner formula. The resulting distribution of residual stresses is shown in Figure 7.25a. The assumed boundary conditions are shown in Figure 7.25b.

The results obtained from the finite element analysis were compared to the test results in both qualitative (Figure 7.26) and quantitative (Figure 7.27) terms. For the representative results included in Figures 7.26 and 7.27, it is seen that close agreement was observed at both a qualitative and a quantitative level.

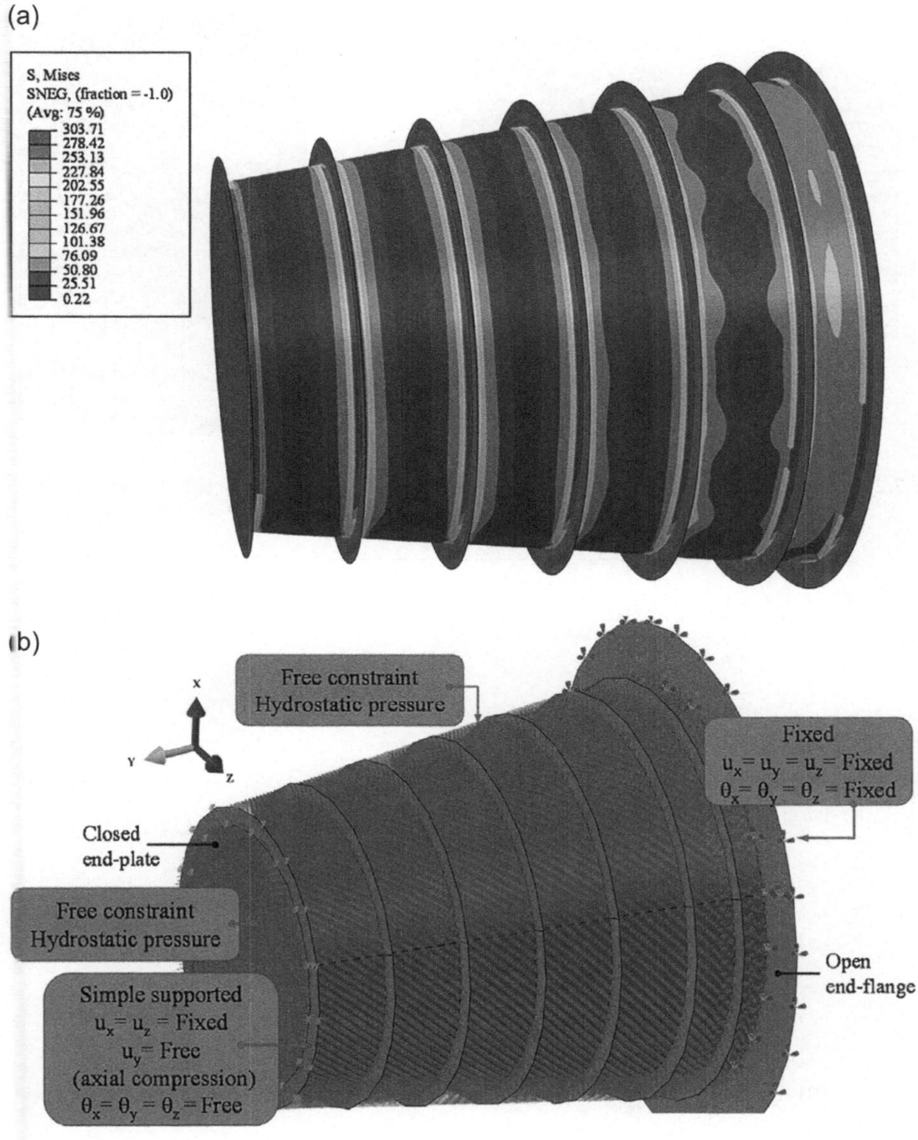

FIGURE 7.25 FE model of ring-stiffened cylinder [9]. (a) Welding residual stress distribution. (b) Boundary conditions.

Table 7.6 provides a comparison of test results with collapse predictions using finite element analysis, several design codes (ABS, DNVGL, and PD5500), and a proposed interaction equation [9].

7.4 DESIGN OF RING-STIFFENED CONICAL SECTIONS

The design of cones used in submarine pressure hulls necessitates the combined use of design codes, design formulas, and where necessary numerical analysis, as will become clear in this section. The design problem concerns the failure modes of ring-stiffened cones attached to stiffened cylindrical shells that are subjected to external hydrostatic pressure. As in the case of cylindrical shells, internal ring frames are attached to truncated cones. These should be normal to the cone

FIGURE 7.26 Qualitative comparison of finite element model with test results (Model RSC-1) [9].

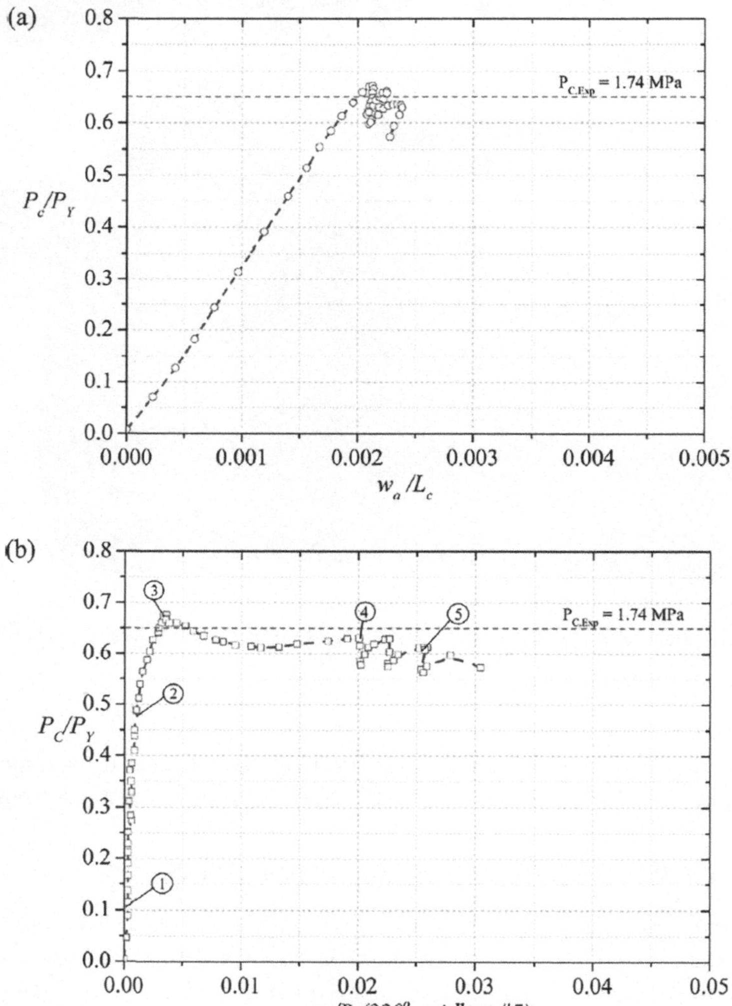

FIGURE 7.27 Quantitative comparison of ultimate strength capacity versus. (a) Axial compression and (b) radial compression [9].

TABLE 7.6
Comparison of Test Results with Predictions from Various Sources

		Bias ($P_{c,exp}/P_{red}$), X_m				
Model	$P_{c,exp}$	**ABS**	**DNV/GL**	**PD5500**	**Cho et al:**	**FEA**
RSC-1	1.74	1.06	0.91	0.90	0.92	0.96
RSC-2	2.06	1.21	0.98	1.03	1.06	0.97
RSC-3	1.96	1.14	0.89	0.96	1.00	1.03
RSC-4	2.50	1.15	0.83	0.92	1.00	0.95
Mean		1.14	0.91	0.95	0.99	0.98
COV(%)		5.39	6.50	5.96	6.03	3.80

axis since buckling strength is improved and bending stresses are reduced. Furthermore, as seen in Section 7.3.1, it is necessary to attach a heavy ring frame in way of the cone–small cylinder junction.

The failure modes that need to be considered are:

- Elastic shell buckling
- Interframe shell yielding
- Interframe collapse
- Overall buckling

In an ideal world, it would be desirable for the cone thickness to be tapered so that the bending stiffness is constant along the length of a meridian. This is not possible in practice so if the larger diameter end is used as a basis for design, an inherently safe structure will result. However, the European standard makes provision, in addition to a cone with constant shell thickness and stiffener dimensions, for the use of a stepped thickness and varying stiffener pitch (Figure 7.28).

The equivalent cylinder approach can be used to calculate (a) the elastic buckling pressure and (b) the pressure to cause interframe yielding in the circumferential direction. For elastic buckling, the von Mises formula can be used (2.34), whereas for shell yielding, (4.8) can be used, both suitably modified to account for cone taper. As mentioned earlier, it has been found that for inelastic buckling of unstiffened cones the equivalent cylinder approach cannot be used [2]. Clearly, since the diameter of each bay changes, it is necessary to allow for the geometry in each case. The mean dimensions of the ring frames can then be used.

An alternative and less conservative approach would be to perform a finite element analysis of each bay.

The pressure to cause overall buckling can be determined using the Bryant formula (4.29), whereas the loss of strength due to out-of-circularity (OOC) can be calculated using the approach outlined for ring-stiffened cylinders in Section 5.5.

7.4.1 Use of Rules and Codes

The primary aim of rules and codes is to ensure structural adequacy for the duration of the operating life of a structure. In doing so, the procedures used do not require the use of sophisticated numerical tools. The major disadvantage of this approach is that a structure is generally overdesigned. It may then become necessary to resort to the use of tools such as finite element or finite difference procedures that have been validated for use for a specific problem. In [16], general criteria for the design of ring-stiffened cylindrical shells, stiffened cones as well as dome ends are given.

With regard to cone design, it has been shown that the EN 13445-3 design curve for cylinders and cones (Figure 9.10) gives a pressure margin of 1.5 that is applicable to cones as well as

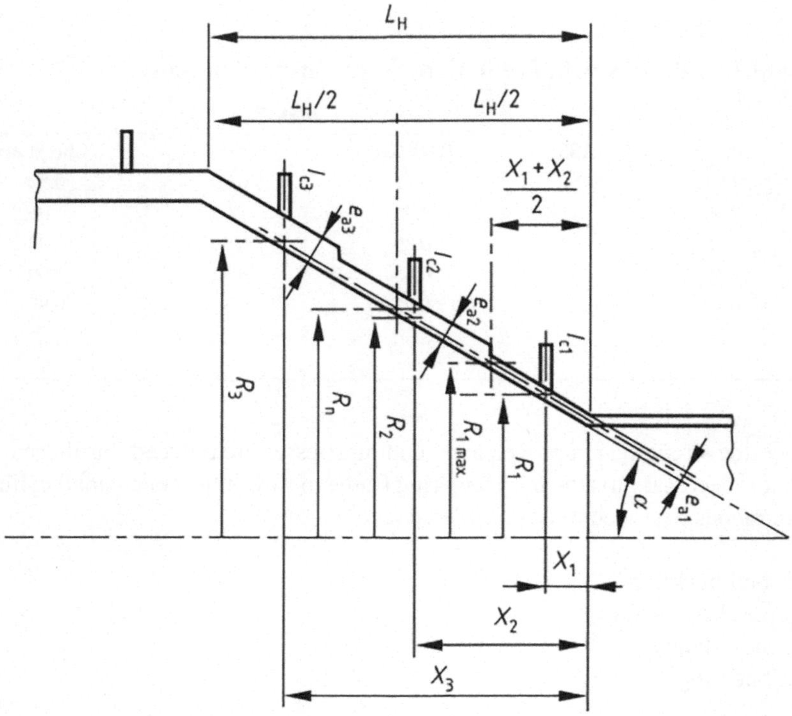

FIGURE 7.28 Stepped thickness and varying stiffener pitch in truncated cone (EN 13445-3).

cylinder–cone intersections. In order to use this curve in cone design, it is necessary to redefine the non-dimensional parameters used as ordinate and abscissa as follows:

$$K = \frac{p_e}{p_y} \tag{7.35a}$$

$$p_{\text{allowable}} = \Delta p_y \tag{7.35b}$$

where p_e = lowest interstiffener buckling pressure (buckling program required)

p_y = pressure at which maximum circumferential mean stress reaches the yield stress (axisymmetric analysis program required)

To validate the use of this curve for use in cone design, a comparison of the experimental collapse pressure with the predictions of the curve was carried out. Results are summarised in Table 7.7.

Table 7.8 includes the most important design variables that need to be evaluated in accordance with selected design codes.

In designing ring-stiffeners welded to conical shells, it is necessary to check the following:

1. Maximum stress in the stiffener
2. Pressure to cause stiffener tripping
3. Pressure to result in local buckling of webs and flanges
4. Stiffener flexural inertia

TABLE 7.7
Validation of Curve 1 in Figure 9.10 for Cone Design

Experiment Number	1	2	3	4	5	6
Yield stress (N/mm²)	360	600	380	410	310	340
Collapse end of cone	Large	Large	Large	Small	Large	Small
p_e	7.2	5.3	3.8	11.2	10.4	15.1
p_y	3.7	4.6	2.7	4.7	3.2	4.3
p_e/p_y	1.9	1.1	1.4	2.4	3.3	3.8
p_c predicted	3.28	3.24	2.11	4.32	3.14	4.28
p_c experiment	3.38	4.22	2.41	4.12	3.24	4.07

TABLE 7.8
Design Variables Required in Ring-Stiffened Cone Design According to Selected Codes

Code	Variables Required	Theoretical Basis
ABS	Inter-stiffener strength of the shell for each bay	Use of design curve (p_c/p_m)
	Pressure to cause yielding in each stiffener	Formula(s)
	Overall buckling pressure	Modified Bryant
DNV-GL	Inelastic interframe collapse pressure	Salerno & Pulos
	Axisymmetric yield pressure	Formula
	Asymmetric buckling pressure	Formula
	Overall buckling pressure	Modified Bryant
EN13445	Interframe shell collapse pressure	Use of design curve (p_m/p_y)
(PD5500)	Stiffener yielding pressure	Formula(s)
	Overall buckling pressure	Bryant

Distinction is made between ordinary and heavy stiffeners and also between internal and externally welded stiffeners.

In way of the boundaries, the junction rotates and the cone angle increases. In the case of the cone–small diameter cylinder, a "punching-in" effect arises. To reduce the magnitude of the resulting high circumferential stresses, a heavy ring frame is needed in way of the junction or else the thickness can be increased (Figure 9.4c). In the case of the cone–large diameter cylinder junction, the opposite effect occurs ("punching-out") and high longitudinal stresses result. In this case, the use of a heavy ring frame is ineffective because although rotation can be eliminated high bending stresses result. The best practice is either to thicken the shell or else introduce a gradual change of slope using a moulded transition [16].

The design of a cone–small cylinder intersection can in general be carried out using design codes. The European code makes provision of the reinforcement of the cone–small cylinder intersection by requiring an increase in thickness and/or the addition of local stiffening in order to ensure that the maximum local circumferential stress does not exceed acceptable limits. The recommended procedure involves calculating the maximum circumferential stress in the cylinder and the maximum circumferential stress in way of the junction in the absence of a heavy ring frame. A stress analysis technique is required for the latter calculation.

7.4.2 Interaction Equation for the Ultimate Strength of Ring-Stiffened Cones

Using as a basis the series of tests on ring-stiffened truncated cones presented in Section 7.3.5, comparisons of the predicted ultimate strength were carried out and are summarised in Table 7.9 [8].

In addition to the three design codes (ABS, DNV-GL, and PD5500), the collapse pressure was calculated using (a) finite element analysis and (b) the following interactive equation, given below [8,9]:

$$\left[\frac{P_C}{\rho_L P_m}+\frac{P_C}{\rho_T P_t}+\frac{P_C}{\rho_{OA} P_n}\right]^2+\left[\frac{P_C}{P_Y}\right]^2=1 \tag{7.36}$$

Details concerning the variables included in (7.36) and its development are included in Section A4.2. From Table 7.9, it is confirmed that the finite element analysis for which accurate data concerning initial distortions, thickness, and residual stresses have been provided is the most accurate of all approaches, having a coefficient of variation (COV) of 3.8%. It is also seen that (7.36) is much more accurate than the design code formulations. The presence of imperfections is allowed for in the design equation through the use of knockdown factors whose values have been obtained using regression analysis of test data. Details are included in Section A4.2.

7.5 BULKHEADS

7.5.1 The Different Roles of Submarine Bulkheads

Bulkheads fitted to submarines assume multiple roles, analogous but not identical to those on board surface ships. They are used to:

- Contribute to pressure hull strength and retention of form
- Subdivide the pressure hull into watertight compartments for operational and other reasons
- Contribute to personnel safety by forming part of an escape system
- Provide support to major internal structural members (decks) and act as tank boundaries

With regard to pressure hull strength, bulkheads contribute to the overall collapse strength of a compartment by limiting its unsupported length, which should not exceed twice the pressure hull diameter ($L_c/D < 2$). A compartment may be designed to remain watertight for a variety of reasons (protection of personnel, of vital equipment and the prevention of the spreading of damage resulting from fire, flooding, or the action of underwater explosions). Bulkheads may also be designed to form part of an integrated escape system for personnel. This last design requirement requires special considerations.

TABLE 7.9
Comparison of Test Results with Predictions from Various Sources

		Bias ($P_{c,exp}/P_{red}$), X_m				
Model	$P_{c,\,exp}$	**ABS**	**DNV/GL**	**PD5500**	**Cho et al:**	**FEA**
RSC-1	1.74	1.06	0.91	0.90	0.92	0.96
RSC-2	2.06	1.21	0.98	1.03	1.06	0.97
RSC-3	1.96	1.14	0.89	0.96	1.00	1.03
RSC-4	2.50	1.15	0.83	0.92	1.00	0.95
Mean		1.14	0.91	0.95	0.99	0.98
COV(%)		5.39	6.50	5.96	6.03	3.80

The inclusion of a bulkhead in a submarine structure imposes increased size requirements for the pressure hull. It is expected that the vessel's length increases by 1.5–2.0 m resulting from the space required and additional weights involved. If, however, a bulkhead is designed to withstand the lateral pressure that acts at the deep diving depth, the increase in length of the submarine and that of the cost will be of the order of 4%. The additional costs are thus related to:

- Increased vessel size
- Bulkhead construction
- Increased life-support systems installed in adjacent compartments

7.5.2 Types of Bulkheads and Structural Types

From the point of view of structural design, two types of bulkheads are used in submarines.

The first type concerns those that are designed to withstand the external hydrostatic pressure transmitted to them by the pressure hull, termed *lightweight* or *subdivision* bulkheads. These are not meant to withstand significant lateral pressure such as that which will arise when a submarine is flooded when submerged. However, they are expected to withstand limited lateral pressure such as that arising in the case of local flooding when the submarine is surfaced and also to prevent the propagation of damage such as fire from one compartment to an adjacent one (Figure 7.29).

The second type concerns bulkheads that are designed to support the pressure hull and also to withstand large lateral pressure (*safety* or *pressure* bulkheads). These act as holding or rescue bulkheads, preventing the flooding of a compartment and facilitating the escape of personnel. The design pressure in this case may be assumed to correspond to continental shelf depth (200–300 m).

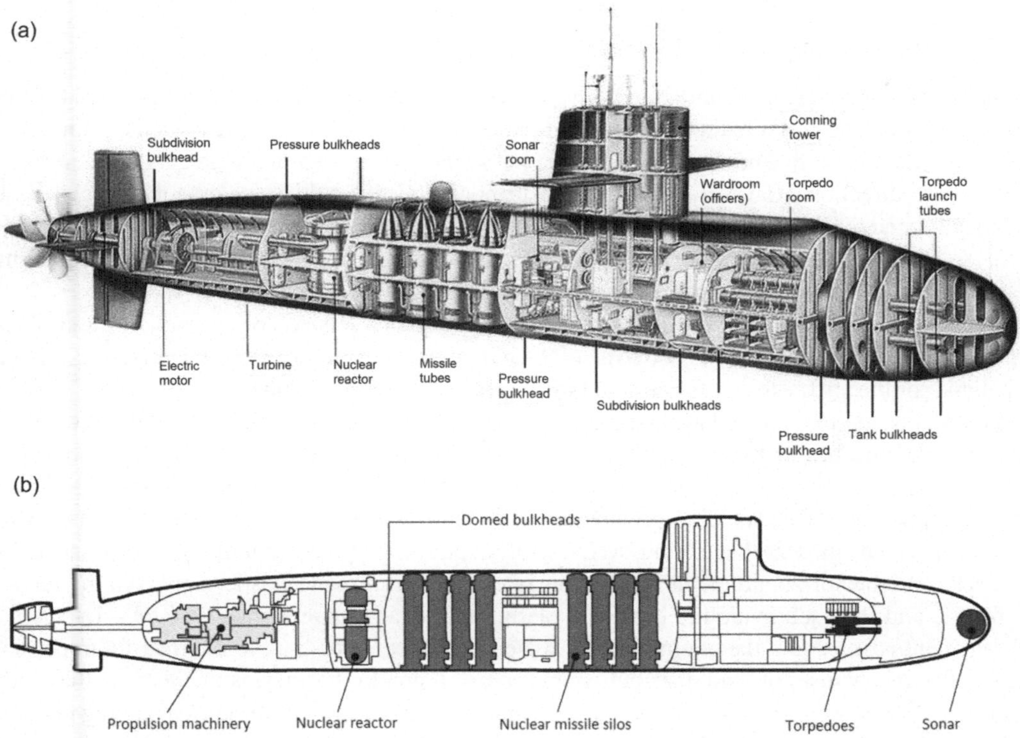

FIGURE 7.29 Types of bulkheads fitted on nuclear-powered submarines. (a) Flat stiffened bulkheads and (b) Domed bulkheads (*Le Triomphant*) [45].

It is possible to increase the design pressure, but this entails a weight penalty. One other role of safety bulkheads is to contain possible damage arising in the compartment housing the nuclear reactor in which case the boundary bulkheads ought to be able to limit the spreading of fission products.

Mention should also be made of *tank* bulkheads that act as boundaries of various types of tanks positioned within the pressure hull (high-speed diving tanks, fuel and ballast tanks). These are required to withstand the same pressure as the pressure hull and in addition a certain pressure increase when blowing the tanks using compressed air at the deep diving depth. Tanks forming part of the exostructure are not subjected to external hydrostatic pressure so that they have to be designed for the local pressure loading that acts.

Three different structural types are used for submarine bulkheads:

- Flat, orthogonally stiffened structure
- Longitudinal cylindrical bulkheads
- Spherical cap

When lateral pressure loading arises in the case of compartment flooding, the most efficient type from the point of view of structural design is the spherical cap, because membrane stresses are induced, leading to a structure that is much lighter than orthogonally stiffened plating. For this reason, it is commonly used for personnel access openings. If, however, the bulkhead forms part of an escape system and is connected to a rescue sphere, it is not possible to use the spherical cap arrangement because the supporting ring would be interrupted in way of the rescue sphere [1].

The advantage of flat orthogonally stiffened bulkheads is the ability to introduce openings and multiple penetrations.

Cylindrical bulkheads form the longitudinal shell plating of saddle and bottom ballast tanks (Figure 7.30b).

7.5.3 Stiffening System and Supporting Structure

In the case of lightweight flat bulkheads, the primary loading to which they are subjected to is the external hydrostatic pressure that acts in the in-plane direction so that they are loaded in radial compression. The most efficient means of avoiding buckling is to arrange for stiffeners to be arranged in the radial direction. However, this creates insurmountable difficulties at a practical level because bulkheads need to be fitted with openings for personnel access as well as equipment (piping, cables, etc.). As a result, all bulkheads are fitted with an orthogonal stiffening system, with access openings being limited in size to the maximum possible extent (Figures 7.31 and 7.32).

Advantage is taken of the presence of the main deck and of a second deck as well, if fitted. These are treated as fixed boundaries to which a bulkhead is attached. Bulkhead–deck boundaries are thus beneficial for the stiffness and strength of both types of structural members.

The deck(s) are thus treated as the primary stiffening system of the bulkhead. In order to ensure that the structure has sufficient buckling strength, additional vertical (secondary) and horizontal stiffening is used (Figure 7.33). It is important to consider the arrangement of vertical stiffeners given the presence of decks. Two possible arrangements can be used, the first one involving the use of continuous girders that extend over the full depth of the pressure hull and are connected to the decks and the horizontal members. The horizontal beams provide elastic support to the vertical stiffeners, and they subdivide the bulkhead plating into near-square elements ($a / b \cong 1$). Such an arrangement ensures that the plating and its stiffening system cooperate to the maximum possible extent. The mechanism of load distribution and transmission to the various members is then indeterminate. For this reason, all stiffeners are designed as indeterminate structures, having clamped ends with a constant cross section along their length. In the case of continuous girders, the clamping effect of the pressure hull is small, and a considerable part of the girder load is transferred to

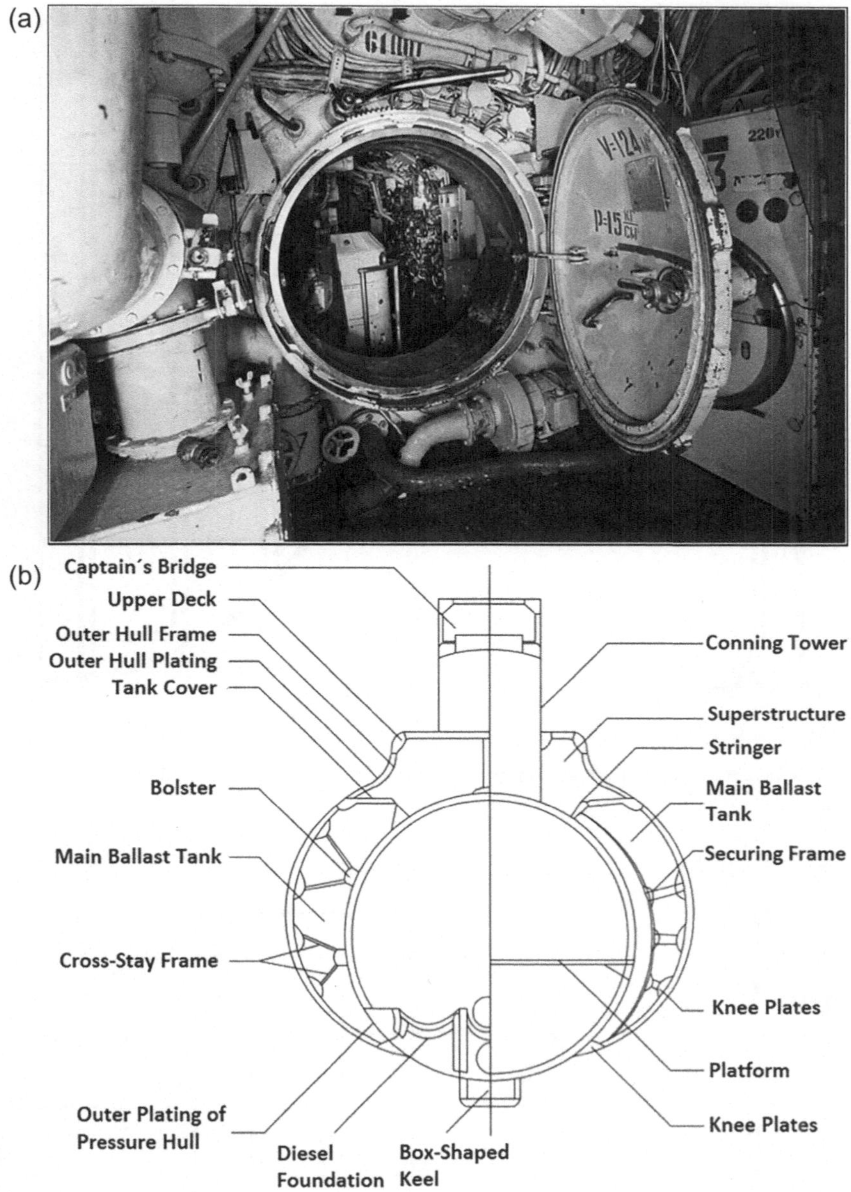

FIGURE 7.30 (a) Personnel access opening in the form of a spherical cap. (b) Saddle tank arrangement.

the deck. At the girder–deck connection, maximum bending moments and shear forces arise. This can be avoided if the girders are not aligned above and below the deck, as shown in Figure 7.32 [1].

With regard to the bulkhead supporting structure, a number of alternative arrangements are possible (Figure 7.33). In the case of large submarines, the bulkhead–twin deck connection is common and the most effective one, whereas on smaller vessels, bulkheads are supported by and provide support to the single deck. In the absence of decks, deep horizontal girders are used. A third alternative arises when a bulkhead acts as a boundary of a tank or a cofferdam in which case internal stiffening provides sufficient support.

At a local level, it is necessary to pay attention to detail structural design in way of bulkhead-pressure hull attachments. The reason for this is primarily because the significantly higher stiffness of the bulkhead compared to a ring frame, the shell plating and adjacent ring frame bend and twist

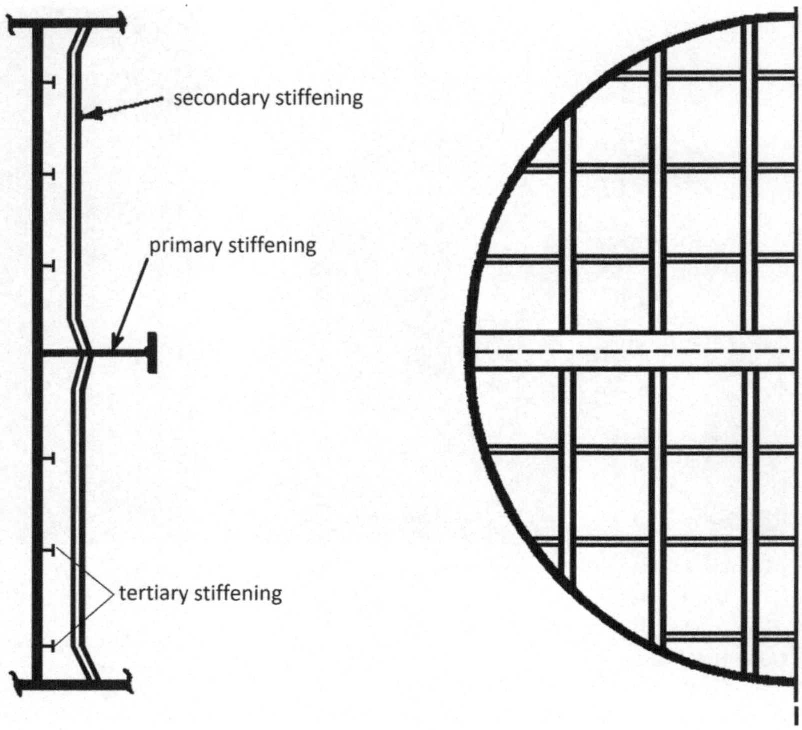

FIGURE 7.31 Classes of stiffening members on a transverse bulkhead fitted with continuous vertical members [21].

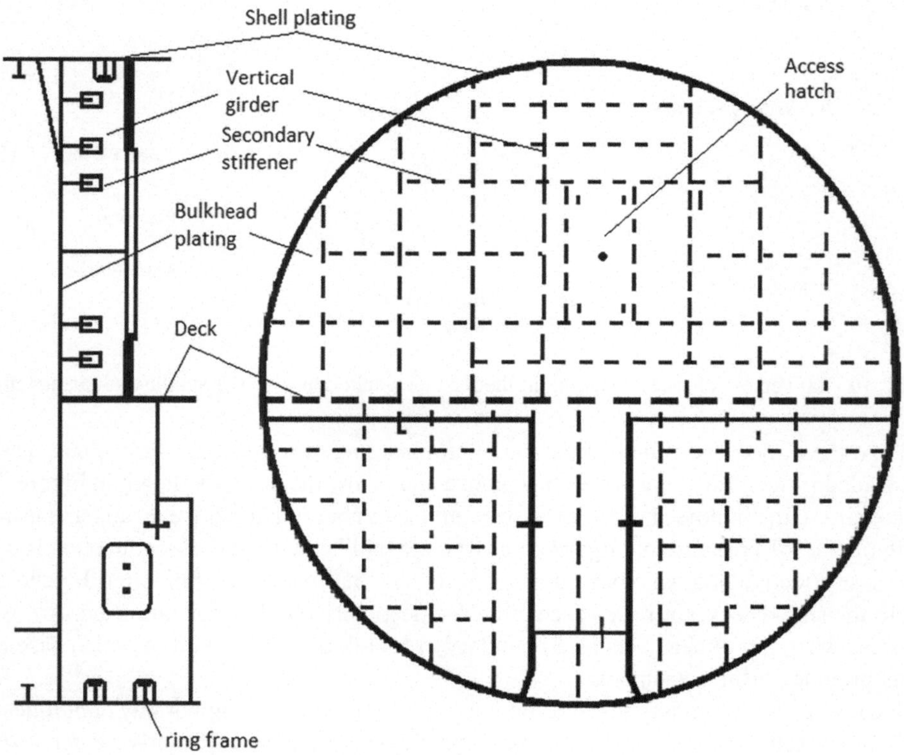

FIGURE 7.32 Primary and secondary stiffening members of a flat bulkhead [1].

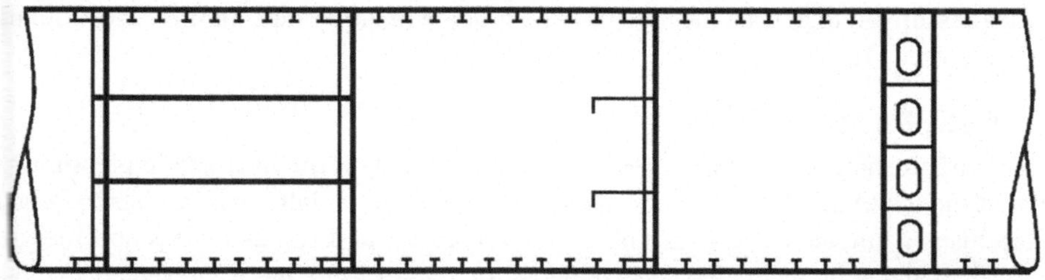

FIGURE 7.33 Alternative arrangements of bulkhead supporting structure [22].

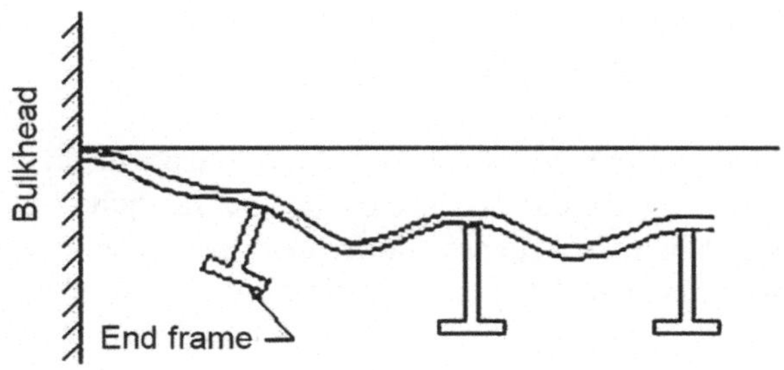

FIGURE 7.34 Bending and twisting of the shell plating and ring frame in way of bulkhead-shell connection [20].

about the connection to the bulkhead (Figure 7.34). It is also important to design out stress concentrations that can act as fatigue initiation sites.

7.5.4 Design Philosophy for Submarine Bulkheads

7.5.4.1 Elastic Design

A flat stiffened bulkhead has to be able to withstand the normal loading that it is subjected to for the duration of its operating life. This loading consists essentially of the external hydrostatic pressure that varies from zero at the surface to that which corresponds to the design diving depth. It is necessary to ensure that no permanent deformations are induced each time the hydrostatic pressure reduces to zero as the vessel ascends to the surface. To ensure this, it is necessary to follow an *elastic design* procedure, that is, stresses have to be less than the yield stress by an appropriate margin. In doing so, it is necessary to allow for the different deformations resulting from bending and shear loads when the equivalent (Mises) stress reaches yield (for high-strength steels, the 0.2% proof stress). At this point, the shear stress equals $\sigma_{0.2} / \sqrt{3}$ so that the deformations due to shear are significantly samller than those arising when a plastic hinge arises due to bending action. The partial safety factor allowing for shear failure can then be taken equal to 1.0, whereas the stress in stiffener flanges is limited to the yield stress (0.2% proof stress). Strain hardening in high-strength steels provides an additional margin.

There arise circumstances in which a transverse bulkhead is subjected to extraordinary loads. One example arises when a submarine exceeds its deep diving depth and approaches the pressure hull collapse depth (overdiving). In this case, the induced stresses in the plating are not permitted to exceed the material yield stress, whereas those in the joints are not permitted to exceed $0.8\sigma_0$ (Russian Navy) [34]. In the German Navy, a pressure bulkhead is required to withstand the external hydrostatic pressure up to at least the collapse depth of the pressure hull. In this case, a small safety factor of 1.1 is used [1].

7.5.4.2 Plastic Design

In different circumstances, a compartment may be flooded with the hydrostatic pressure rising to equal that at the depth at which the submarine finds itself. In the latter case, it has to be acknowledged that this is a unique event in the life of the structure, but it is nonetheless necessary to safeguard personnel and equipment. To succeed in this, ingress of water has to prevented so that failure in the form of a fracture cannot be permitted to arise. Permanent distortions may, however, be permitted because otherwise design requirements would impose an unduly severe weight penalty. The path that has to be followed then is that of *plastic design* with regard to the extraordinary loads that are not expected to arise more than once in the lifetime of a submarine.

7.5.5 Design Loads

A flat stiffened bulkhead is subjected to (Figure 7.35):

1. Radial pressure whose magnitude corresponds to the collapse depth of the pressure hull
2. Lateral pressure whose magnitude depends on whether the bulkhead is a pressure bulkhead or a lightweight bulkhead. In the former case, permanent set is permitted at the design pressure, whereas in the latter case, elastic design is carried out.

The lateral pressure loading results in bending of the vertical and horizontal members. This induces high shear loading in the webs since the unsupported span is small (low span-to-web-depth ratio). The vertical stiffeners are thus designed as beams supported by decks or girders. The beam section consists of the beam section and the effective bulkhead plating. When elastic design is used, such as in the case of tank bulkheads beam bending theory in combination with plate bending theory is applied. In the case of emergency conditions, plastic design is implemented, and in this case, permanent set is allowed.

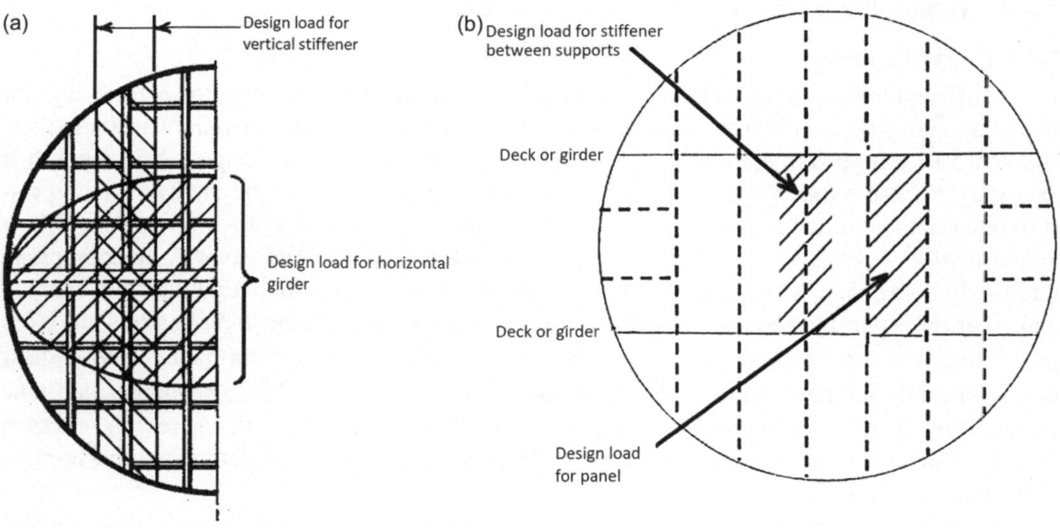

FIGURE 7.35 Design load distributions for bulkhead members. (a) Horizontal girder (deck) and vertical stiffener. (b) Stiffener intercostal and plating.

Nonlinear finite element analysis can be used to assess the ultimate strength of the bulkhead and leads to a lighter structure. When preparing a finite element model, it is necessary to include the supporting structure as well (cylindrical shell plating and decks/girder). A 2.5D model is thus required.

7.6 DECKS AND TANKS

In submarines, the primary role of decks is the support of equipment, and they are also used to provide living spaces for personnel [6]. In the case of lower decks, they can also function as tank tops. In that sense, their role differs from that of decks in surface ships in which they contribute in a major way to the hull girder bending strength. Figure 7.29a and b show the extent and arrangement of decks inside the pressure hull of a nuclear submarine. Decks are not continuous throughout the length of the pressure hull and terminate in way of particular bulkheads. They generally extend over the entire breadth of the pressure hull and can be either welded to it (*fixed decks*) or *resiliently mounted* (Figure 7.36).

Fixed decks provide support to the pressure hull along its entire length as well as to the bulkheads at their extremities. The effect the presence of a fixed deck has on pressure hull response is shown in Figures 8.19b (overall collapse) and 8.20b (interframe buckling mode). It is seen to act as a rigid boundary, thereby bringing about changes to the buckling modes and deflections of the pressure hull. Equipment fitted to a fixed deck needs to be fitted with shock mounts. A fixed deck has to be designed to be subjected to compressive loading transmitted to it by the pressure hull, although its presence is not allowed for in pressure hull design. If compressive loading is not allowed for it is possible that buckling may arise when the submarine dives beyond a critical depth [5] (Figure 7.37).

One other disadvantage of fixed decks is the fact that they transmit vibration arising from the operation of equipment to the sea environment, thereby reducing the stealth qualities of a submarine. This is an important reason for which alternative arrangements have been sought, with the *resiliently mounted* deck contributing to a solution of this problem (Figure 7.36b). In this case, the entire deck is mounted on shock-absorbing equipment along the connection with the pressure hull and the deck does not then act as a line of support for the pressure hull. The equipment fitted to the deck is then shock-protected. From the point of view of production, resiliently mounted decks offer significant advantages compared to fixed decks. Each deck module can be prefabricated with equipment installed prior to being introduced to the pressure hull. This results in significant economies, in accordance with the 1–3–8 rule [22]:

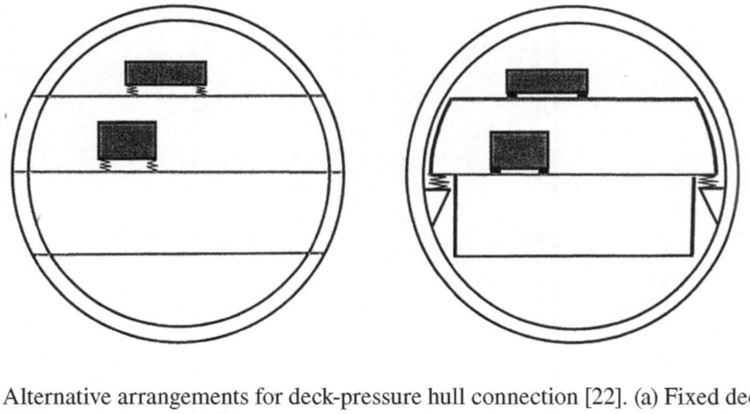

FIGURE 7.36 Alternative arrangements for deck-pressure hull connection [22]. (a) Fixed deck and (b) resiliently mounted deck.

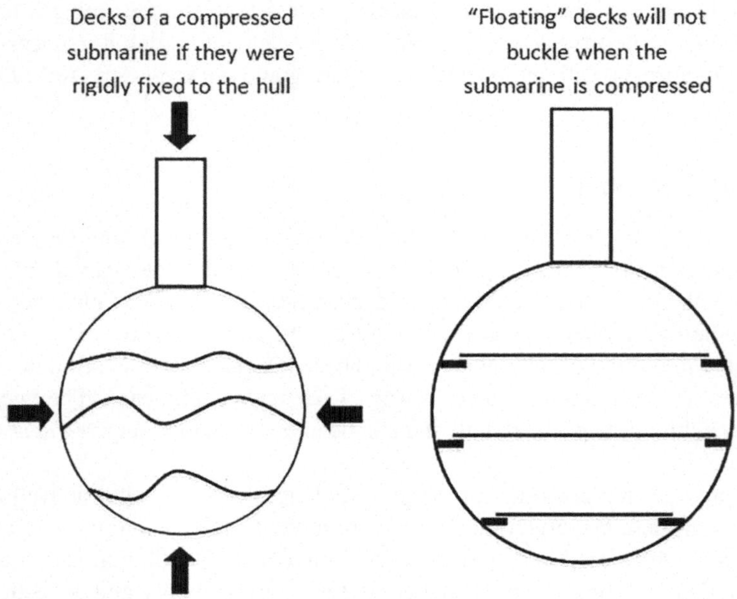

FIGURE 7.37 Effect of compressive loading of decks due to external hydrostatic pressure.

1 = Cost of work undertaken on module inside the shop
3 = Cost of work undertaken with open access to hull units
8 = Cost of work undertaken within the pressure hull

Submarine decks are stiffened using an orthogonal system as shown in Figure 7.38. They are designed to withstand local lateral loads arising from equipment as well as internal tank pressures. As such, they are not expected to provide support to the pressure hull. In general, a uniformly distributed load is assumed, except for the case of particularly heavy items. As seen in Figure 7.38,

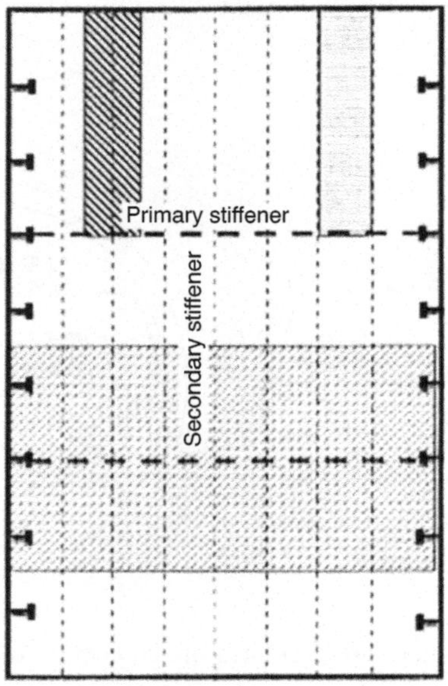

FIGURE 7.38 Deck stiffening arrangement and the area supported by individual members [22].

decks are supported by deep beams that span the breadth of the pressure hull as well as lighter, more closely spaced longitudinal stiffeners. The deep beams are spaced such that they form a ring with the pressure hull ring frames, whereas the longitudinal stiffeners are aligned with the vertical girders attached to the bulkheads (lower side).

Beam bending theory can be used over the region located some distance from the transverse bulkheads. A stiffener is thus assumed to act in unison with an effective breadth of plating. An important question arising in submarine deck design is the choice of boundary conditions and the loads transmitted there. In the case of fixed decks, allowance has to be made for the in-plane compression resulting from the external hydrostatic pressure acting on the pressure hull. This question does not arise in the case of resiliently mounted decks. In way of the deck–bulkhead boundaries, allowance has to be made for the shear loads transmitted by the bulkhead to the deck. To avoid failure, the deck thickness should be increased over a suitable distance from the intersection with the bulkhead. For a deck installed at mid-height, a distance of $2D/3$, where D is the diameter of the pressure hull, and a deck plate thickness of $1.2–1.5t$, where t is the shell thickness, is recommended.

With regard to tanks located within the pressure hull, generally below the (lower) deck, these can be designed according to beam theory. The loads consist of the hydrostatic pressure acting internally. Fatigue issues may need to be considered in the case of cyclically loaded tanks. Main ballast tanks (Figure 7.30b) on the other hand play an important role, and their structural integrity is of paramount importance when a submarine is surfacing or undergoing diving manoeuvres. Under normal operating conditions, they are expected to withstand routine blowing, whereas in an emergency, they are required to sustain sudden blow pressures without suffering damage. The estimation of design pressures and the development of design equations and related curves for main ballast tanks are included in [28].

NOTE

1 Equation (276) on p. 468 [35].

REFERENCES

1. Abels, F., Niessen, E. (1986). "The pressure-tight bulkhead in the submarine". *RINA International Symposium*, Naval Submarines.
2. Błachut, J. (2011). On elastic–plastic buckling of cones. *Thin-Walled Struct.* Vol. 49, 45–52.
3. Błachut, J. (2012). Interactive plastic buckling of cones subjected to axial compression and external pressure. *Ocean Eng.* Vol. 48, 10–16.
4. Błachut, J., Ifayefunmi, O. (2010). Plastic buckling of conical shells. *J. Offsh. Mech. Arc. Eng.* Vol. 132, 041401.
5. Bunich, L.M., et al. (2016). "Stability of a truncated conical shell under the action of uniform external pressure." Res. & Adv. Develop. Div., Avco Manufacturing Corp Rept. (Transl. from Russian by Zarymovych M.I., Columbia Univ.).
6. Burcher, R., Rydill, L. (2014). *Concepts in Submarine Design.* Cambridge: Cambridge University Press 289 p.
7. Burrows, W.R. (1950). "Pressure Vessel Heads under Internal Pressure. Part 1: General Considerations."
8. Cho, S.-R., Muttaqie, T., Do, Q.-T., So, H.-Y., Sohn, J.-M. (2018). Ultimate strength formulation considering failure mode interactions of ring-stiffened cylinders subjected to hydrostatic pressure. *Ocean Eng.* Vol 161, 242–256.
9. Cho, S.-R., Muttaqie, T., Do, Q.-T., Park, S.-H., Kim, S.-M., So, H.-Y., Sohn, J.-M. (2019). Experimental study on ultimate strength of steel-welded ring-stiffened conical shell under external hydrostatic pressure. *Mar. Struct.* Vol. 67, 102634.
10. Dubois, F. (1917). "Über die Festigkeit der Kegelschale (On the strength of the conical shell)." Doctoral Thesis, Eidgenössische Technische Hochschule, Zürich.
11. Flügge, W. (1934). *Statik und Dynamik der Schalen.* Berlin: Julius Springer.
12. Geckeler, J. (1926). "Über die Festigkeit Achsensymmetrischer Schalen" (On the Strength of Axisymmetric Shells)." *Forschunsarbeiten auf dem Gebiete des Ingenieurwesens*, Berlin No. 274 1–52.

13. Hetényi, M. (1946). *Beams on Elastic Foundation: Theory with Applications in the Fields of Civil and Mechanical Engineering.* Ann Arbor: The University of Michigan Press, 256 p.
14. Hoff, N.J. (1955). Thin circular conical shells under arbitrary loads. *J. App. Mech.* Vol. 22, 557–562.
15. Ifayefunmi, O. (2014). A survey of buckling of conical shells subjected to axial compression and external pressure. *J. Eng. Sci. & Tech. Rev.* Vol. 7, No. 2, 182–189.
16. Kendrick, S. (1982). Design for external pressure using general criteria. *Int. J. Mech. Sci.* Vol. 24, No. 4, 209–218.
17. Kraus, H. (1967). *Thin Elastic Shells: An Introduction to the Theoretical Foundations and the Analysis of Their Static and Dynamic Behavior.* New York: John Wiley & Sons, 471 p.
18. Krenzke, M.A. (1959). "Hydrostatic tests of conical reducers between cylinders with and without stiffeners at the cone-cylinder junctures." DTMB Report 1187, Washington, DC.
19. Love, A.E.H. (1927). *A Treatise on the Mathematical Theory of Elasticity*, 4th Ed. New York: Dover, 672 p.
20. Lunchick, M.E. (1960). "Plasticity research on submarine pressure hulls conducted at the david taylor model basin." *Proceedings of the Second Symposium on Naval Structural Mechanics,* Held at Brown University, Rhode Island, 347–368.
21. MacNaught, D.F. (1967). Strength of ships Chap. IV. In: Comstock, J.P. (Ed). *Principles of Naval Architecture* (revised), New York: SNAME.
22. McCormick, J. (2009). *An Overview of Submarine Design. Lect. 1 in Design of Submarine Structures.* Glasgow: ASRANet Course.
23. Meissner, E. (1915). Über Elastizität und Festigkeit dünner Schalen" (On the elasticity and strength of thin shells). *Vierteljahrschrift der Naturforschenden Gesellschaft*, Vol. 60, 23–47.
24. Moir, C.M. (1933–36). Direct and bending stresses in cylindrical and conical thin walls. *Glasgow Royal Techn. College J.* Vol. 3, 142.
25. Nash, W.A. (1995). *Hydrostatically Loaded Structures: The Structural Mechanics, Analysis and Design of Powered Submersibles.* New York: Elsevier Science, 181 p.
26. Niordson, F.I.N. (1947). Buckling of conical shells subjected to uniform external lateral pressure. *Trans. Roy. Inst. Tech.* Vol. 29, 119.
27. Pflüger, A. (1961). *Elementary Statics of Shells*, 2nd Ed. (Transl. Galatay E.) New York: McGraw-Hill Book Co, 140 p.
28. Pohler, C.H., Bernent, A.A., Wilson, D.S., Skinner, W.A. (2020). Submarine main ballast tanks: Theory and methods for refined structural design. *Mar. Technol. Sname.* Vol. 6, No. 02, 171–196.
29. Raetz, R.V. (1960). "An experimental investigation of the strength of small-scale conical reducer sections between cylindrical shells under external hydrostatic pressure." DTMB Report 1397, Washington, DC.
30. Raetz, R.V., Pulos, J.G. (1958). "A procedure for computing stresses in a conical shell near ring stiffeners or reinforced intersections." David Model Basin Report 1015.
31. Reynolds, T.E. (1966). "A summary of submarine structural research. Part 1. Conventional hull configurations. Chap. 5: Stresses in pressurized shells of revolution." DTMB Report 2264, Washington, DC.
32. Rhys, C.O. (1951). "Geckeler approximation applied to stresses near the juncture of small cylinder and reducer." PVRC Report No 12.
33. Ross, C.T.F., Andriosopoulos, G., Little, A.P.F. (2008). Plastic general instability of ring stiffened conical shells under external pressure. *Appl. Mechan. Mater.* Vol. 13–14, 213–223.
34. Shimanskiy, Y.A. (1964). "Structural mechanics of submarines. Part 1 practical methods and examples of calculations for the hull strength of submarines." (Transl. *DTMB Science Translation Service* Translation 305-1).
35. Taylor, C.E., Wenk Jr., E. (1956). "Analysis of stresses in the conical elements of shell structures." *Proceeding of Second U.S. National Congress of Applied Mechanics.* (also DTMB Report 981).
36. Timoshenko, S.P., Woinowsky-Krieger, S. (1959). *Theory of Plates and Shells*, 2nd Ed. Tokyo: McGraw-Hill Kogakusha, 580 p.
37. Tokugawa, T. (1932). "Experiments on the elastic stability of a thin-walled cone under uniform normal pressure on all sides, and an approximate method for computing its collapsing pressure." *Congr. Appl. Mech. Leag. Shipb. Assoc. (Zosen Kyokai) Misc. Publ.* Vol. 125, 151–169.
38. Tokugawa, T. (1940). "An approximate method for calculating the collapse pressure of thin, cylindrical, conical and spherical shells under uniform pressure." *Zosen Kyokai* (Shipb. Assoc.) *Meeting.*
39. Vinson, J.R. (1993). *The Behavior of Shells Composed of Isotropic and Composite Materials.* Dordrecht: Kluwer Academic Publishers 416 p.

40. Watts, G.W., Burrows, W.R. (1949). The basic elastic theory of vessel heads under internal pressure. *J App. Mech.* Vol. 71, 55–73.
41. Watts, G.W., Lang, H.A. (1952). Stresses in a pressure vessel with a conical head. *Trans ASME*, Vol. 74, No. 3, 315–326.
42. Wenk Jr., E., Taylor, C.E. (1953). "Analysis of stresses at the reinforced intersection of conical and cylindrical shells." DTMB Report 826.
43. Wetterstrom, E. (1947). "Discontinuity stresses in pressure vessels". *Master's Thesis*, Purdue University.
44. Zingoni, A., Pavlović, M. (1993). A note on the accuracy of the geckeler approximation. *Eng. Comp.* Vol. 10, 369–379.
45. https://philippeberiou.fr/SM_SNLE-NG-caract.php?r=1&Largeur=1920&Hauteur=1080&wf=1647&hf=899

8 Finite Element Analysis of Submarine Pressure Hulls

8.1 NUMERICAL PROCEDURES FOR UNDERWATER STRUCTURES

The use of numerical procedures in the analysis and design of underwater structures dates from the 1970s when the finite difference-based program BOSOR (*B*uckling-*O*f-*S*hells-*O*f-*R*evolution) was used in the analysis of submarine pressure hulls. The BOSOR series of programs was developed in the United States [3] originally for space structures. Subsequent to and in addition to, finite element software has been used in the analysis and design of pressure hulls.

As seen in Chapter 6, the response of pressure hulls involves elasto-plastic behaviour with collapse occurring in the inelastic range. In order to predict the collapse load, it thus becomes necessary to make use of large deflection analysis and to allow for the spread of plasticity. It is therefore necessary to make use of software that can model nonlinear response to calculate the collapse load.

The post-collapse response of underwater structures is unique in that unloading does not take place in the manner experienced in structures exposed to atmospheric conditions but involves a sudden *implosion* that occurs in the space of a few milliseconds and involves deep plastic collapse and crushing of the structure beyond recognition (Figure 8.1). Clearly, it is necessary to use purpose-oriented numerical tools to study such a type of response. In addition to hydrostatic pressure loading, underwater structures and submarines in particular may be subjected to the effects of underwater explosions at some point in their operating life. Such explosions may be due to underwater bombs, torpedoes, or mines located on the seabed. The design of such structures thus necessitates the assessment of strength against these types of loading. As in the case of post-collapse response, such a study needs to focus on nonlinear dynamic response. Before considering structural response, it is necessary, however, to model the time and space distribution of the explosive loading in the absence of the structure [4,5]. The study of structural response is a fluid–structure interaction problem, and in addition to determining the effect an explosion has on the solid boundary, it is necessary to also

FIGURE 8.1 *HMS Stoic* following implosion collapse.

DOI: 10.1201/9781003489153-8

allow for hydroelastic effects. These affect the response of both the structure and also that of the surrounding fluid. It is thus a complex task to model, given that hydrostatic pressure is also present. A number of studies are now appearing in the literature, such as [10,13,17]. Post-collapse response is being considered theoretically, experimentally as well as using numerical procedures. In this chapter we shall restrict ourselves to the use of numerical methods aiming at the prediction of the collapse load due to hydrostatic pressure alone. The use of finite element procedures in submarine pressure hull design is discussed in Chapter 9.

In the case of submarine structures, benefits accrued from the use of numerical methods and in particular the finite element method, in general terms are [6]:

a. Rendering the design process more accurate
b. Identifying non-conservativeness or redundancy
c. Validating existing design methods in situations not open to experimental investigation
d. Solving post-design and post-construction problems such as the assessment of structural damage and the granting of concessions on material and geometrical tolerances

The modelling used has to be able to represent the failure modes in ring-stiffened cylindrical shells as well as those of end closures that may be torispherical, hemispherical, or other (Figure 8.2a). These have been described in other chapters. The importance of initial geometrical distortions and residual stresses due to cold bending or welding has been underlined so that these need to be accurately represented in any numerical model.

Cylindrical shells are designed to collapse by interframe elasto-plastic buckling, and this may be followed by general instability. Some interaction between the two failure modes may arise. In this, Finite Element Analysis (FEA) can be helpful since this is a nonlinear problem (Figure 8.2b). It has been found that the collapse load may be predicted sufficiently accurately from the pre-collapse regime, obviating the need for a post-collapse analysis.

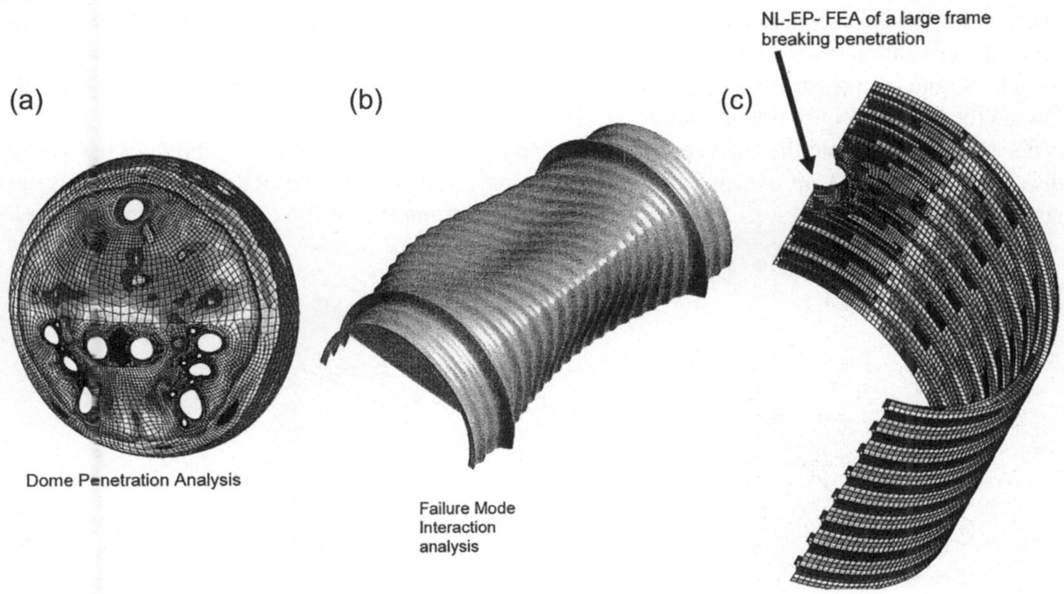

FIGURE 8.2 Uses of FEA in submarine pressure hull analysis and design [12]. (a) Dome penetration analysis. (b) FMIA. (c) NLFEA of penetration.

With regard to mesh requirements, in the case of a full model representation, fairly simple flat shell elements can be used; these lead to computational economy since only moderate rotations are allowed [6].

With regard to element type, in most cases, curved isoparametric is used although 3D shell and flat plate elements are also used.

With regard to the type of analysis, linear elastic FEA can be used in fatigue studies to determine stresses. However, in the case of buckling and collapse problems, it is necessary to use nonlinear elasto-plastic FEA [12].

Nonlinear elasto-plastic FEA requires care when specifying boundary conditions, the introduction of imperfections (initial distortions and residual stresses), the selection of appropriate elements, and mesh density. As in the case of surface ship structures, good practice involves the use of two numerical models. The first concerns the use of a coarse mesh to represent a major portion of the pressure hull, excluding local details such as penetrations. The second is used to model the structural detail of concern at a local level. The use of the first model assists in determining the boundary conditions to be used in the second model.

8.2 NONLINEAR ANALYSIS OF UNDERWATER STRUCTURES

Nonlinear finite element analysis is described in detail in many textbooks, (e.g. [2], [15]), and we shall restrict our discussion to aspects that are important to the study of underwater structures. As mentioned above, the two types of nonlinearity that arise are geometrical and material nonlinearity. Geometrical nonlinearity may be due to large deflections, buckling, post-buckling change in buckling mode as well as large strains. This type of nonlinearity may thus involve either small or large strains and is treated accordingly. Material nonlinearity on the other hand is due to yielding that may or may not be followed by work-hardening. In this case, the relationships between stresses and strains are nonlinear. The study of underwater structures generally involves the use of material as well as small-strain geometrical nonlinearities. The differences between a linear and a nonlinear analysis are illustrated in Figure 8.3.

It is seen that in a linear analysis **KU** = **F,** that is, the displacement vector **U** is a linear function of the load vector **F**, **K** the stiffness matrix, being constant. The difference between the two types of analysis arises because in nonlinear analysis, the stiffness matrix **K** changes throughout the loading range as a result of geometrical and material nonlinearities. The term *tangent stiffness matrix* is used to denote the fact that the properties of **K** change continuously throughout the load range, following the initiation of nonlinear behaviour. In order to obtain the true load–deflection response, it is necessary to trace out the correct load–deflection curve. A large number of techniques have been developed to this end. In nonlinear finite element procedures that have been used in underwater structures applications as well as other applications involving thin-walled structures, the approach

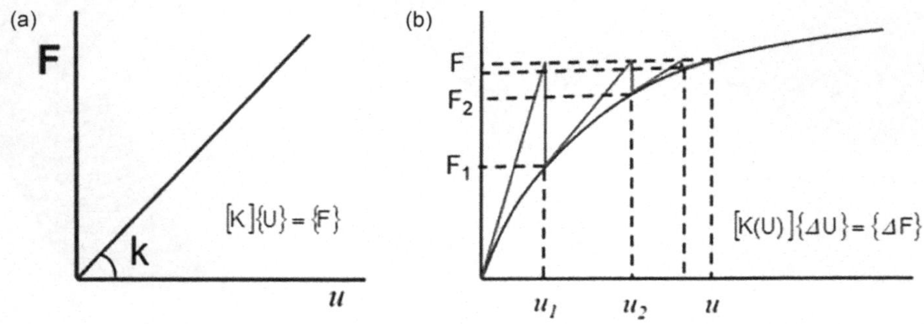

FIGURE 8.3 Linear and nonlinear analysis. (a) Linear analysis. (b) Nonlinear analysis.

involves the combined use of the *modified Newton–Raphson* method with *arc-length* methods. We shall discuss the basis of these in the following sections. In Figure 8.3:

K = global tangent stiffness matrix
U = global displacement vector
ΔU = global incremental displacement vector
F = global load vector
ΔF = global incremental load vector

The matrix equation of equilibrium for the structure is rearranged in the form $\mathbf{U} = \mathbf{K}^{-1}\mathbf{F}$, and the vectors **U** and **F** are replaced by finite increments $\Delta\mathbf{U}_n$ and $\Delta\mathbf{F}$, respectively, so that $\Delta\mathbf{U} = \mathbf{K}^{-1}\Delta\mathbf{F}$. The load–deflection relationship is thus expressed in terms of increments in load and deflection. We note that the increment in the force vector, **ΔF**, is maintained constant. The load–deflection history is thus described by the sequence of increments **ΔF**, $\Delta\mathbf{U}_n$, where $n = 1, 2, \ldots$, defining the states:

$$\mathbf{F}_{n+1} = \mathbf{F}_n + \Delta\mathbf{F} \tag{8.1a}$$

$$\mathbf{U}_{n+1} = \mathbf{U}_n + \Delta\mathbf{U}_{n+1} \tag{8.1b}$$

where $\Delta\mathbf{U}_{n+1}$ is initially not known. We can implement these expressions to propose an *incremental method* that makes use of the current tangent stiffness matrix, $\mathbf{K}_n$. If use is made of $\mathbf{K}_n$ at the beginning of the n+1-th increment, we have:

$$\Delta\mathbf{U}_{n+1} = \mathbf{K}_n^{-1}\Delta\mathbf{F} \quad n = 1, 2, \ldots \tag{8.2}$$

The result is shown in Figure 8.4. It is seen that as loading is increased, the solution based on (8.2) diverges from the true path, and the structure is predicted to be much stiffer than it is in reality. It is thus necessary to introduce some form of correction in order to approximate the true load–deflection curve. This can be done by the use of a suitable *successive approximation scheme* at each load level. The method used to perform these approximations in such types of problems is described in the next section.

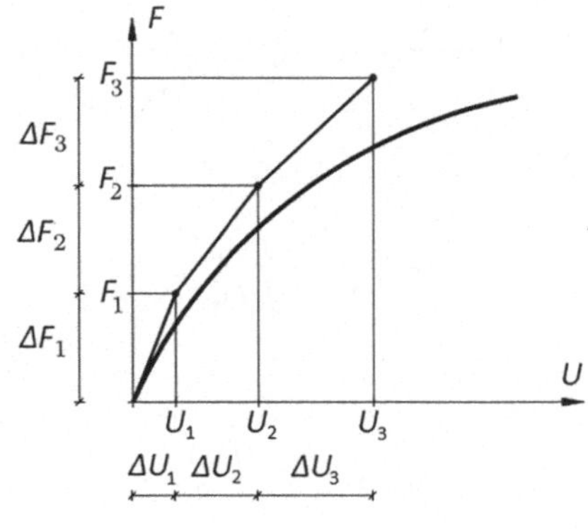

FIGURE 8.4 Explicit incremental method.

8.3 THE NEWTON–RAPHSON METHOD IN NONLINEAR FINITE ELEMENT ANALYSIS

Before discussing the use of the Newton–Raphson method in nonlinear finite element analysis, we shall give a brief description of the basis of the method, which was devised in order to obtain solutions to algebraic equations using a numerical approach. It is based on *linear approximation* and, if used properly, is efficient in determining the roots of an equation that may be difficult to solve in closed form.

We start from the definition of the derivative of a function $f(x)$, given its values at x_0 and $x_0 + h$ where h is small so that $x_0 + h$ lies in the neighbourhood of x_0:

$$f'(x_0) \rightarrow \frac{f(x_0+h)-f(x_0)}{h} \tag{8.3}$$

as $h \rightarrow 0$. If, however, h retains a constant value,

$$f'(x_0) \cong \frac{f(x_0+h)-f(x_0)}{h} \tag{8.4}$$

Rearranging,

$$f(x_0+h) \cong f(x_0)+hf'(x_0)$$

We let r be the exact solution of the equation so that

$$r = x_0 + h \tag{8.5}$$

The value of h is then the difference between the approximate and the exact solution. Hence,

$$f(x_0+h) = f(r) = 0 \cong f(x_0)+hf'(x_0) \tag{8.6}$$

Therefore,

$$h \cong -\frac{f(x_0)}{f'(x_0)} \tag{8.7}$$

Hence,

$$r = x_0 + h \cong x_0 - \frac{f(x_0)}{f'(x_0)}$$

The new estimate x_1 for r is then

$$x_1 \cong x_0 - \frac{f(x_0)}{f'(x_0)} \tag{8.8}$$

In general, therefore,

$$x_{i+1} \cong x_i - \frac{f(x_i)}{f'(x_i)} \tag{8.9}$$

It is seen that it necessary to iterate until the value of x_{i+1} is sufficiently close to r, the solution of the equation. It therefore becomes necessary, following the evaluation of each approximant to introduce an accuracy check. Accuracy of the solution is specified in terms of a required number of decimals.

The method is successful in reaching a solution of an algebraic equation by successive linear approximations provided the interval h is such that the neighbourhood solution $f(x_0 + h)$ is close to the exact solution of the equation. The method will break down if the solution coincides with an inflection point ($f'(x) = 0$).

We shall now consider the application of the Newton–Raphson method to nonlinear finite element analysis. The first issue that arises is the proximity of the approximate solution to the exact solution. If this is not the case, the numerical solution will diverge, and this becomes immediately apparent. To avoid this from occurring, it is necessary to select a sufficiently small load increment $\mathbf{\Delta F}$.

At equilibrium, the external force vector $\mathbf{F}_n$ and the vector of internal stress resultants $\mathbf{G}_n$ are equal, that is,

$$\mathbf{F}_n - \mathbf{G}_n = \mathbf{0} \tag{8.10}$$

The first step following the application of a load increment $\mathbf{\Delta F}$ involves calculating the difference, or residual force $\mathbf{R}_{n+1}(\mathbf{U}_n, \mathbf{F}_{n+1})$ between the resulting external load $\mathbf{F}_{n+1}$ and the internal stress resultant vector $\mathbf{G}_{n+1}$, from:

$$\mathbf{R}_{n+1}(\mathbf{U}_{n+1}, \mathbf{F}_{n+1}) = \mathbf{F}_{n+1} - \mathbf{G}_{n+1} \tag{8.11}$$

At equilibrium $\mathbf{R(U,F)} = \mathbf{0}$. Once, however, the load is increased by $\mathbf{\Delta F}$ equilibrium is no longer satisfied since the deflection vector $\mathbf{G}_n$ corresponds to the previous equilibrium state. It is therefore necessary to determine an approximate value of the increment in the deflection vector, $\mathbf{\Delta U}_{n+1}$, and to iterate in order to obtain the value of $\mathbf{U}_{n+1}$ for which $\mathbf{R}_{n+1} = 0$.

Since the solution for the nth load increment is known, we have:

$$\mathbf{G}_{n+1} = \mathbf{G}_n + \Delta\mathbf{G} \tag{8.12}$$

where $\mathbf{\Delta G}$ is the change in the stress resultant vector corresponding to the change in stresses and displacements from n to $n + 1$. This vector can be approximated using the tangent stiffness matrix that was obtained for the n-th increment, $\mathbf{K}_n$:

$$\Delta\mathbf{G} \cong \mathbf{K}_n \Delta\mathbf{U} \tag{8.13}$$

where $\mathbf{\Delta U}$ is the vector of incremental displacements and

$$\mathbf{K}_n = \frac{\partial \Delta\mathbf{G}_n}{\partial \Delta\mathbf{U}} \tag{8.14}$$

Substituting (8.12) and (8.13) in (8.10), we obtain:

$$\mathbf{K}_n \Delta\mathbf{U} = \mathbf{F}_{n+1} - \mathbf{G}_n \tag{8.15}$$

We solve for $\mathbf{\Delta U}$ following which we can obtain an approximation for the displacement vector $\mathbf{U}_{n+1}$ from:

$$\mathbf{U}_{n+1} \cong \mathbf{U}_n + \Delta\mathbf{U} \tag{8.16}$$

The important stages in this procedure consist of the determination of $\mathbf{\Delta G}$ and $\mathbf{K}_n$. Since we have introduced an approximate calculation of $\mathbf{\Delta G}$ through (8.13), errors are introduced and issues of numerical stability that depend on increment size may arise. It therefore becomes necessary to iterate until the residual vector $\mathbf{R}_i$ is less than the tolerable limit. The following two steps thus constitute necessary components of the algorithm:

1. Perform a check to ensure that equilibrium is satisfied to within the desired degree of accuracy
2. If equilibrium is not satisfied, modify the state of deformation of the body

At each load increment (n), iterations (i) are carried out to ensure that (8.10) is satisfied within acceptable levels of accuracy. The algorithm consists of (8.15) and (8.16) in the following form:

$$\mathbf{K}_{n+1}^{i-1}\Delta\mathbf{U}^i = \mathbf{F}_{n+1} - \mathbf{G}_{n+1}^{i-1} \tag{8.17a}$$

$$\mathbf{U}_{n+1}^{i} \cong \mathbf{U}_{n+1}^{i-1} + \Delta\mathbf{U}^i \tag{8.17b}$$

An important observation from (8.17a) is that the tangent stiffness matrix has to be re-evaluated in each iterative step, which in the case of large structures is computationally expensive. In order to avoid having to perform this calculation each time, the Newton–Raphson algorithm is modified so that $\mathbf{K}$ is calculated only at certain stages. In what is termed the *modified Newton–Raphson method*, $\mathbf{K}$ is calculated only at the beginning of each load step, in which case (8.17a) assumes the following form:

$$\mathbf{K}_{n+1}\Delta\mathbf{U}^i = \mathbf{F}_{n+1} - \mathbf{G}_{n+1}^{i-1} \tag{8.18}$$

The number of iterations necessary for satisfactory convergence increases in this case, but not having to assemble the stiffness matrix, each iteration renders the procedure more efficient from a computational perspective. In addition to the modified Newton–Raphson method, there exist a number of other methods that are used in numerical procedures (secant method, etc.).

8.4 THE ARC-LENGTH METHOD

As mentioned in the previous section, in all Newton-based methods, the load increment $\mathbf{\Delta F}$ is maintained constant. This, however, encounters difficulties and is inefficient especially when the change in the stiffness matrix from one increment to the other is large and also when the load–deflection curve obtains a negative gradient due to unloading in the post-collapse range (Figure 8.5). This situation arises in the study of underwater structures such as cylindrical shells following inelastic buckling. In the case of spherical shell structures subjected to hydrostatic pressure, following buckling, snap-through occurs (Figure 8.5). It is thus not possible to trace out the true load–deflection path using load control. In the case of snap-back displacement control cannot be used.

Of the methods developed to deal with this the most successful is the *arc-length method* that is applied to the equilibrium curve in combined load–displacement space [14]. The method was devised by Riks [22] and was adapted for use in finite element procedures by Crisfield [7,8]. Subsequent developments are reviewed in [16,20]. The key point of the method is that the prescribed load increment is treated as a first estimate, and it changes value in successive iterations.

Equation (8.11) thus becomes:

$$\mathbf{R}_{n+1}(\mathbf{U}_{n+1},\mathbf{F}_{n+1}) = \lambda\mathbf{F}_{n+1} - \mathbf{G}_{n+1} \tag{8.19}$$

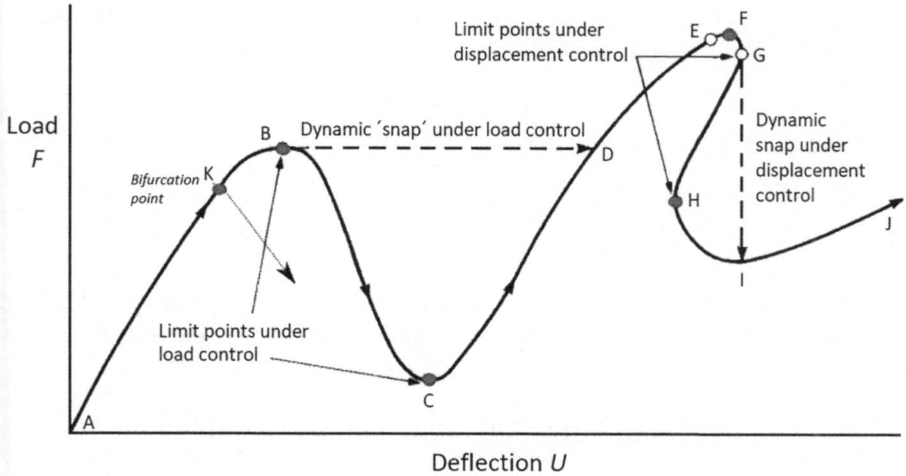

FIGURE 8.5 Snap-through and snap-back in post-collapse response of structures [8].

where λ is the load factor. Solutions of equation (8.19) are obtained as previously, by using an incremental approach in which convergence is achieved using an iterative technique at each load increment.

In the arc-length method, both the displacement vector and the load factor at the current load step, $\mathbf{U}_{n+1}$ and λ_{n+1}, respectively, are computed as increments, $\mathbf{\Delta U}$ and $\Delta\lambda$, from their respective values at the previously converged load step, $\mathbf{U}_n$ and λ_n, as

$$\mathbf{U}_{n+1} = \mathbf{U}_n + \Delta\mathbf{U} \tag{8.20a}$$

$$\lambda_{n+1} = \lambda_n + \Delta\lambda \tag{8.20b}$$

where the subscripts $n + 1$ and n respectively denote the current and the previously converged load step. Using (8.20a) and (8.20b), the residual vector for the current load step can be written as:

$$\mathbf{R}(\mathbf{U}_{n+1}, \lambda_{n+1}) = \lambda_{n+1}\mathbf{F}_{n+1} - \mathbf{G}(\mathbf{U}_{n+1}) = 0 \tag{8.21}$$

The system of nonlinear equations in equation (8.21) is solved for the displacement $\mathbf{U}_{n+1}$ and also for the loading parameter λ_{n+1}. This approach increases the number of degrees of freedom (DOF) by 1, making this an under-determined system of equations. An additional equation is therefore introduced, called the *arc-length* or *constraint* equation. This equation relates the predicted displacement to the predicted load and in the case of a single DOF structure has the following form:

$$\Delta U^2 + (\lambda\Delta F)^2 = \Delta s^2 \tag{8.22}$$

where Δs is the "arc length". For a structure with a large number of degrees of freedom,

$$\Delta\mathbf{U}^T\Delta\mathbf{U} + \psi\Delta\lambda^2\mathbf{F}^T\mathbf{F} = \Delta s^2 \tag{8.23}$$

where Δs is the arc-length parameter, and ψ is a scalar parameter. For $\psi = 0$, the cylindrical arc-length method is recovered, and for $\psi = 1$, the method becomes the spherical arc-length method. Equations (8.20a) and (8.20b) constitute a system of equations for increments $\Delta\mathbf{U}$ and $\Delta\lambda$ that is solved using the (modified) Newton–Raphson scheme for a selected value of Δs.

Starting with an initial guess ($\Delta\mathbf{U}^1$, $\Delta\lambda^1$), the displacement increment $\Delta\mathbf{U}$ and the load increment $\Delta\lambda$ are computed by iterative updates as:

$$\Delta\mathbf{U}^{i+1} = \Delta\mathbf{U}^i + \Delta\mathbf{U} \tag{8.24a}$$

$$\Delta\lambda^{i+1} = \Delta\lambda^i + \delta\lambda \tag{8.24b}$$

where i is the iteration counter that obtains values $i = 1, 2, 3, \ldots, i_{max}$, and i_{max} is the maximum number of iterations. Using equation (8.24a and 8.24b) $\mathbf{U}_{n+1}$ and λ_{n+1} at the current iteration $i + 1$ can be written as:

$$\mathbf{U}_{n+1}^{i+1} = \mathbf{U}_n + \Delta\mathbf{U}^{i+1} = \mathbf{U}_n + \Delta\mathbf{U}^i + \delta\mathbf{U} = \mathbf{U}_{n+1}^i + \delta\mathbf{U} \tag{8.25a}$$

$$\lambda_{n+1}^{i+1} = \lambda_n + \Delta\lambda^{i+1} = \lambda_n + \Delta\lambda^i + \delta\lambda = \lambda_{n+1}^i + \delta\lambda \tag{8.25b}$$

for $i = 1, 2, 3, \ldots i_{\max}$. By applying the Newton–Raphson scheme to solve (8.21) and (8.23), we get the following matrix system:

$$\begin{bmatrix} \mathbf{K}(\mathbf{U}_{n+1}^i) & -\mathbf{F} \\ a^T & b \end{bmatrix} \begin{Bmatrix} \delta\mathbf{U} \\ \delta\lambda \end{Bmatrix} = -\begin{Bmatrix} \mathbf{R}(\mathbf{U}_{n+1}^i, \lambda_{n+1}^i) \\ A(\Delta\mathbf{U}^i, \Delta\lambda^i) \end{Bmatrix} \tag{8.26}$$

where

$$\mathbf{K}(\mathbf{U}_{n+1}^i) = \left.\frac{\partial\mathbf{G}}{\partial\mathbf{U}}\right|_{\mathbf{U}_{n+1}^i} \tag{8.27a}$$

$$a = 2\left[\Delta U^i\right] \tag{8.27b}$$

$$b = 2\psi\left[\Delta\lambda^i\right]\left[\mathbf{F}^T\mathbf{F}\right] \tag{8.27c}$$

$$A(\Delta\mathbf{U}^i, \Delta\lambda^i) = \left[\Delta\mathbf{U}^i\right]^T\left[\Delta\mathbf{U}^i\right] + \psi\left[\Delta\lambda^i\right]^2\mathbf{F}^T\mathbf{F} - [\Delta s]^2 \tag{8.27d}$$

The matrix system in (8.26) is solved for ($\delta\mathbf{U}$, $\delta\lambda$) until a predefined convergence criterion is satisfied. An iterative process is assumed to have converged when[1]:

$$\|\mathbf{R}_{n+1}\| < \varepsilon_R R_{ref} \quad (\text{out} - \text{of} - \text{balance convergence}) \tag{8.28a}$$

and/or

$$\|\mathbf{U}_{n+1}\| < \varepsilon_u U_{ref} \,(\text{DOF increment convergence}) \tag{8.28b}$$

where:

$\mathbf{R}_{n+1}$ is the residual force vector given by (8.11)
$\|\mathbf{R}_{n+1}\|$ is a vector norm of the residual force vector. This is a scalar measure of its magnitude
ε_R, ε_U are tolerances
R_{ref}, U_{ref} are reference values

In general, only out-of-balance convergence is checked, and to this end, the Euclidean ($L2$) norm is used. This is given by:

$$\|\mathbf{R}_{n+1}\|_2 = \sqrt{\sum R_i^2} \tag{8.29}$$

The evolution of solution increments in the arc-length method is illustrated in Figure 8.6.

8.5 SUBMARINE PRESSURE HULL ANALYSIS

Having described the most important features of finite element procedures for the analysis of structures in the inelastic range, we shall discuss the implementation and modelling of submarine pressure hulls.

Pressure hull analysis can be carried out in one or more of the following ways:

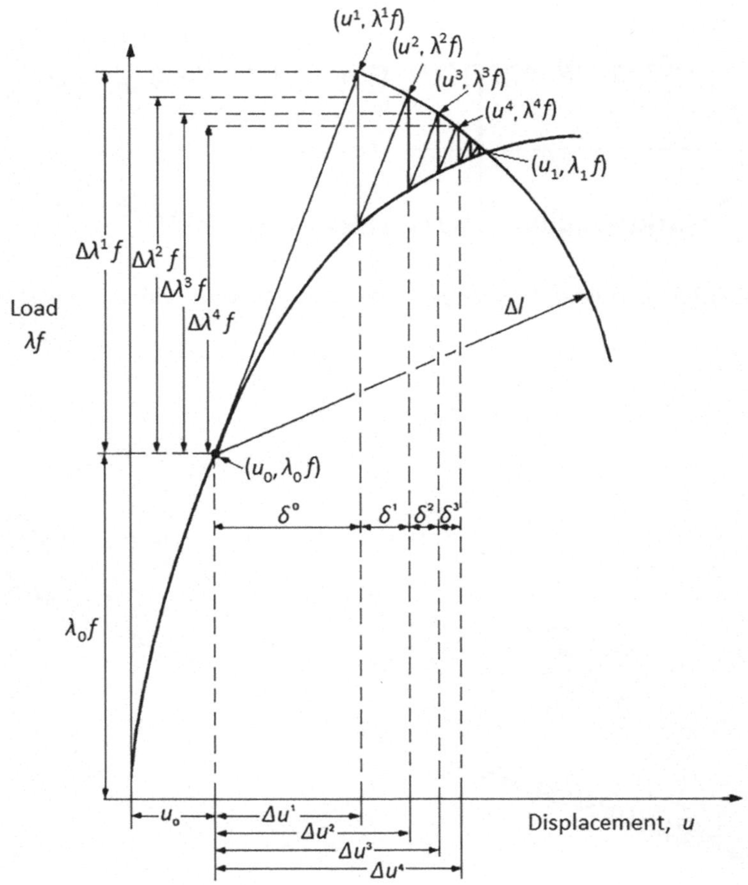

FIGURE 8.6 The arc-length method applied to a structure with one degree of freedom. (Adapted from [7]).

1. By making use of the equations of pressure hull strength and collapse, as described in Chapters 4–6.
2. By developing and processing a finite element (or finite difference) model of a representative portion of the pressure hull or by modelling the entire pressure hull, ignoring appendages, penetrations, internal structure,[2] and cutouts, as shown in Figure 8.7.
3. By developing and processing a finite element model of the entire hull, including appendages, internal structure, penetrations, and cutouts (Figure 8.8).
4. By performing laboratory tests on scale models

The choice of method depends on the needs of the study and the tools available. Experience has shown that the use of results from laboratory tests has proven invaluable in order to ensure sufficient strength of pressure hulls, because it is only by performing tests that real conditions can be represented accurately. In particular, initial imperfections that include out-of-circularity of the shell and residual stresses in both the shell and ring-stiffeners need to be allowed for because they have a detrimental effect on strength. Mathematical formulas cannot allow for this to the extent necessary, and the combined use of laboratory tests and analytical expressions has formed the basis of rules for pressure vessel design [9]. With regard to finite element procedures, their accuracy also needs to be assessed by comparing results with laboratory tests carried out on scale models. The results of a study performed to this end will be described in the next section.

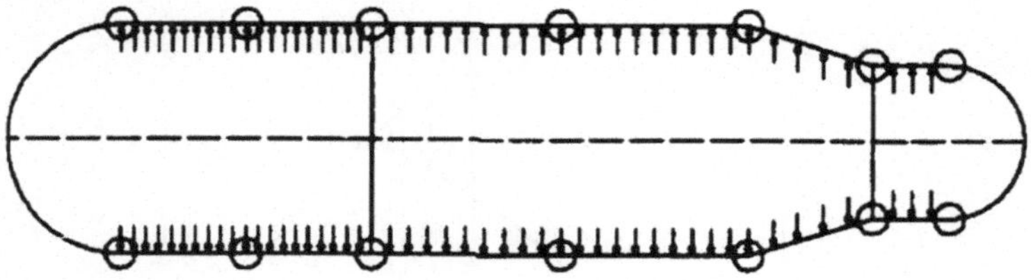

FIGURE 8.7 Model of pressure hull consisting of cylindrical sections, conical sections, and end closures [21].

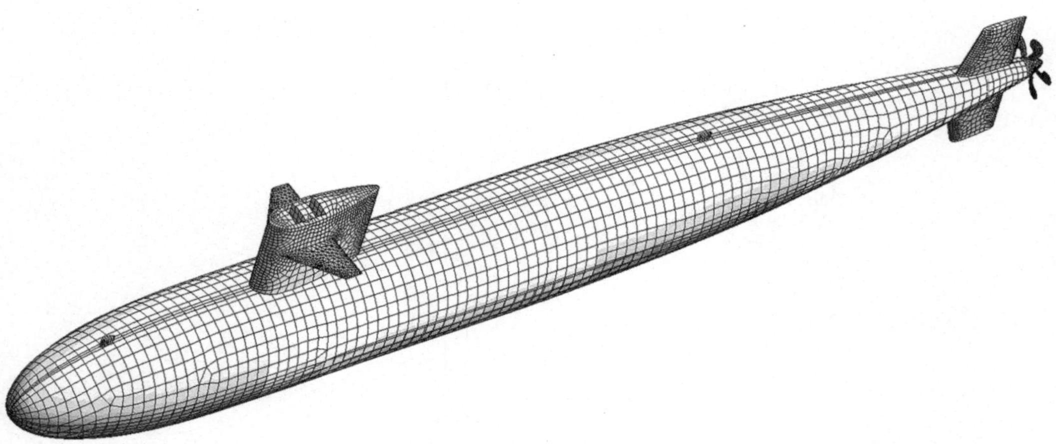

FIGURE 8.8 Finite element model of entire submarine.

8.6 DETERMINATION OF COLLAPSE PRESSURE USING FINITE ELEMENT ANALYSIS

When compared to analytical methods, FE procedures can model a real structure to the level of structural detail required, including defects and imperfections, so that in principle, it ought to be possible to obtain an exact prediction of pressure hull collapse provided element selection, discretisation, and solver accuracy are optimised. The aim of the work described in this section was to develop a benchmark that can be used in assessing FE software that is used in the analysis and design of submarine pressure hulls, and it was carried out at the UK naval research facilities [11].

This enables numerical modelling standards to be set. A number of tests on scale models were performed, following which finite element studies were carried out, and the results compared with the tests results as well as those of tests carried out at earlier stages.

The models were fabricated such that the initial imperfections present were of the same type as those that arise in real pressure hulls as a result of the erection process. They consisted of cylindrical ring-stiffened shells, corresponding to a representative portion of the pressure hull to which end closures were attached. No attempt was made to represent the entire submarine steel structure, and thus, conic sections, appendages, and internal structure were omitted, the aim of the study being to assess finite element procedures in predicting the collapse load of the cylindrical ring-stiffened pressure hull. Figure 8.9 shows a model that failed by interframe yielding.

The purpose of the study was to assess the performance of the numerical procedure with respect to the major modes of failure, and this necessitated the testing of models that fail distinctly in each of these: interframe buckling, interframe yielding, overall collapse, and interactive overall with local failure. Additionally, the effects of geometrical imperfections, both local and overall, material properties, and material models and the effect of residual stresses were studied. To this end, models of four different lengths (25, 28 36, and 40 frames) were tested.

Initial geometrical imperfections regarding out-of-circularity and frame tilt were measured and recorded. Residual stresses due to cold bending and welding were calculated using the finite element procedure, because analytical methods do not allow for material hardening and Poisson effects. Welding residual stresses result in interframe bending of the shell, and this was allowed for directly through the use of local initial deflection measurements.

The numerical analysis involved the use of PATRAN to represent initial geometrical distortions and subsequently the structural models. Analysis was carried out using ABAQUS procedures, and second-order thin shell elements were used with a density of six or eight elements between frames (Figure 8.10).[3] The following two types of boundary conditions were considered:

FIGURE 8.9 Scale model failing by interframe yielding.

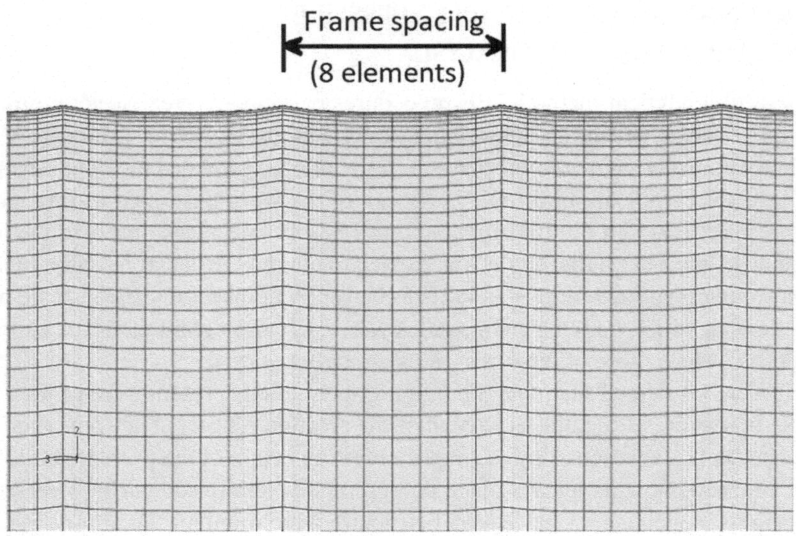

FIGURE 8.10 FE mesh for cylindrical shell.

- Type 1: Effectively fully clamped, with one end free to move axially as a plane
- Type 2: Modified to relax the axial suppression, allowing the ends of the cylinders to warp

The short models (25 and 28 frames) were tested to simulate interframe collapse, (buckling and yielding). Figure 8.11 shows the test specimen after failure due to interframe buckling and the corresponding numerical model. Figure 9.8c (p.309) shows a test specimen whose failure involved interactive collapse (overall failure and interframe). Test and numerical results and comparisons for the short models are summarised in Tables 8.1 and 8.2.

Figure 8.12 includes details concerning the effect of imperfections on collapse pressure. The two out-of-circularities (OOC) considered differed in that the latter contained details of additional significant modes (up to $n = 9$).

With regard to residual stresses, they resulted in a drop of interframe collapse pressure of about 3%. Residual plasticity and the Bauschinger effect brought about collapse strength reduction of 0.5%. Material anisotropy was not modelled numerically using the software used.

The figures in brackets included in Table 8.2 refer to models for which the minimum rather than the average thickness was used. Details on scantlings are not included in [11] so that it is not possible to draw conclusions on the different failure modes observed, although it may be inferred that differences in scantlings were responsible for the differences in failure modes.

The experimental and the (unmodified) analytic collapse pressures are plotted in nondimensional form in Figure 8.13. The pressures p_{m1} and p_{c5} are defined by equations (3.81) and (5.8), respectively.

Results for the longer models cylinders (36 and 40 frames) are included in Table 8.3.

In the case of fabricated frames, the residual stresses resulted in a 1% reduction in collapse strength, whereas in the case of cold-bent frames, the reduction was much larger (10%–15%).

A general conclusion is that it is necessary to provide accurate descriptions of shape, scantlings, residual stress, and material properties for an FE model to be able to produce a reliable prediction of the collapse pressure of a submarine. In this case, the predicted value lies in the range ±6%, providing the collapse mode is correctly identified. If interframe collapse arises, the analysis has to be repeated using the minimum measured plate thickness.

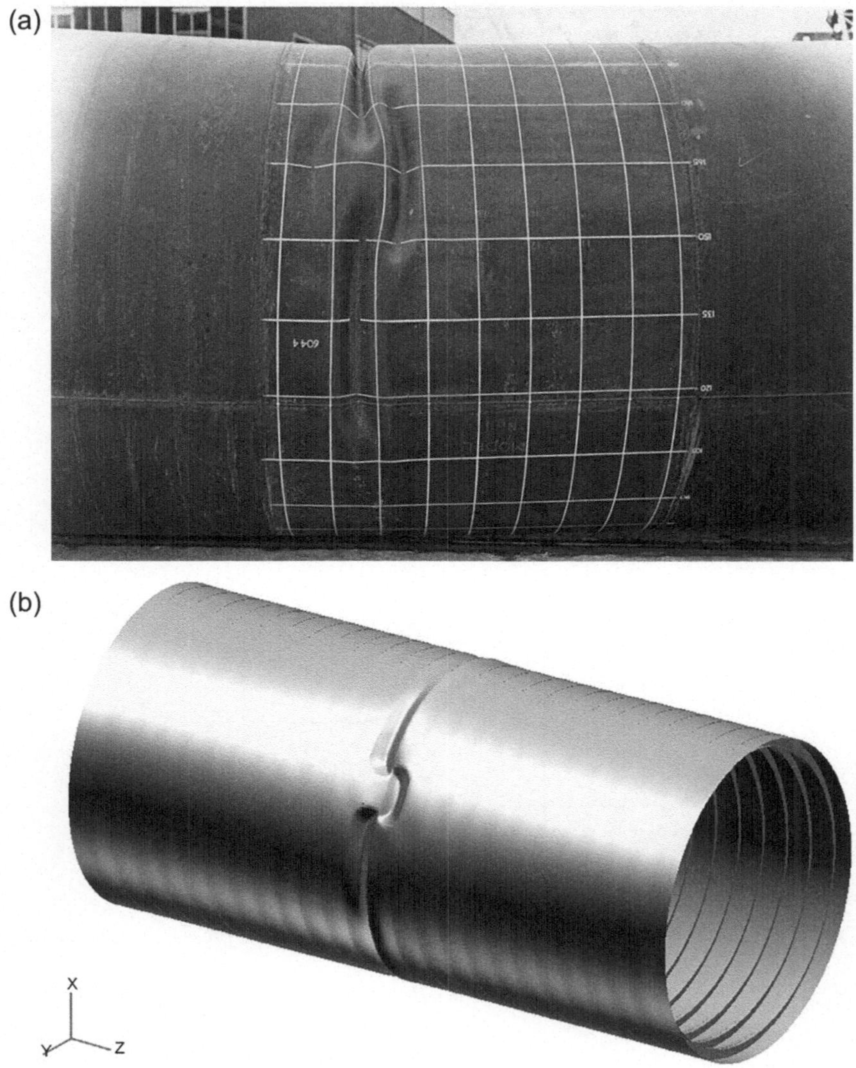

FIGURE 8.11 Interframe buckling collapse of a 25-frame model. (a) Test specimen. (b) Finite element model.

TABLE 8.1
Predicted/Experimental Collapse Pressures for the Two 25-Frame Models

	Type 1 BCs	Type 2 BCs
Model 1 ($n = 3$)	1.007	0.997
Model 6 ($n = 2$)	1.200	1.062

TABLE 8.2
Failure Mode and Predicted Collapse/Experimental Collapse Pressures for Short Models

Model	Mode	PFE/P_{expt}
11	Buckle	1.14 (1.006)
13	Buckle	1.002
15	Yield	1.098 (1.059)
17	Buckle	1.062 (1.019)
18	Yield	1.074 (1.054)
19	Yield	0.963 (0.920)
20	Overall	0.999

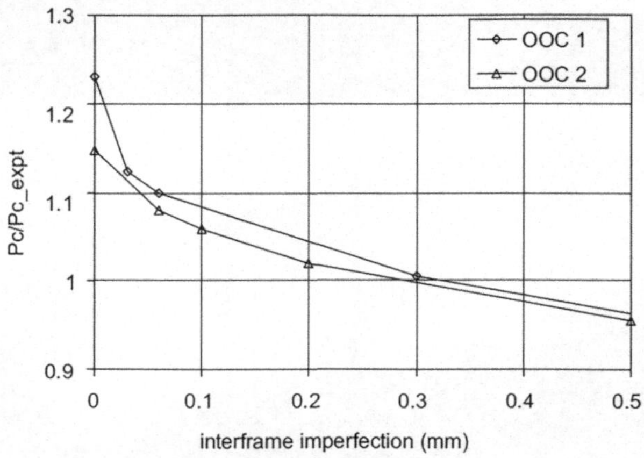

FIGURE 8.12 Effect of interframe geometrical imperfection on collapse pressure of short models.

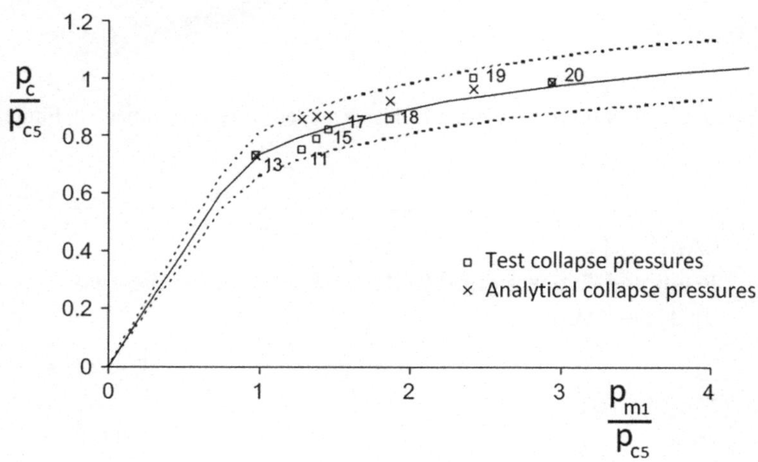

FIGURE 8.13 Analytical and test collapse pressures nondimensionalised with respect to interframe yield pressure.

TABLE 8.3
Failure Modes and Predicted/Test Collapse Pressures for Long Models

Model	Mode	p_{fe}/p_{expt}
25 frame	Interframe	1.017
28 frame	Interframe	1.015
36 frame	Overall ($n = 3$)	1.051
40 frame	Overall ($n = 2$)	0.986

8.7 EFFECT OF SHELL THICKNESS, INITIAL IMPERFECTIONS, AND HEAT TREATMENT ON CYLINDRICAL PRESSURE HULL COLLAPSE STRENGTH

The results from a series of studies carried out to investigate the effect of a number of parameters on the collapse strength and the collapse mode of ring-stiffened cylindrical shells used in naval submarines will be described in this section [6]. The following parameters were considered:

1. Shell geometrical imperfections, expressed in terms of the imperfection parameter ε, where:

$$\varepsilon = 100\left(\frac{w_0}{R}\right) \tag{8.30}$$

   where w_0 is the $n = 2$ shape amplitude, and R is the cylindrical shell radius to mid-thickness. Also, small amplitude geometrical imperfections (w_i) at the interframe level, where:

$$w_i = 0.05\%R \tag{8.31}$$

2. Shell plate thickness, t. Three values of thickness were considered: a typical value, t, $1.25t$, and $t/1.2$.
3. Heat treatment. Three cases were considered: Stress relieved, cold-bent plate and flange, and cold-bent plate and T-ring stiffener
4. Residual stresses
5. Combined effect of residual stresses and overall imperfections

Results are given in Figures 8.14 and 8.15 in which the collapse modes observed in the numerical analysis are indicated by the letters "i" (interframe), "o" (overall collapse), and "io" (interactive collapse involving both modes). Nonlinear elasto-plastic analyses were performed using the STAGS software. In order to reduce computational time, a single ring-stiffener with attached shell plating was modelled. Repeating symmetry boundary conditions were introduced, and ring frame dimensions were increased slightly in comparison to actual scantlings in order to ensure the relative values of overall and interframe collapse that arise in finite-length pressure hulls. The results are plotted in terms of the collapse pressure p, non-dimensionalised with respect to the collapse pressure of the pressure hull with typical scantlings, versus $\sqrt{\varepsilon}$ for values in the range 0.0–1.0, where ε is the imperfection parameter; this was done in order to make clear that the dependence of overall collapse pressure on the $n = 2$ mode is less than linear.

In the case of the thick shell ($1.25t$), the loss of strength is quasi-linear for "large" imperfections. The strengthening effect due to the thicker shell plating results in an increase in collapse pressure of nearly 30%, in the absence of imperfections. In this case, only does interframe collapse arise.

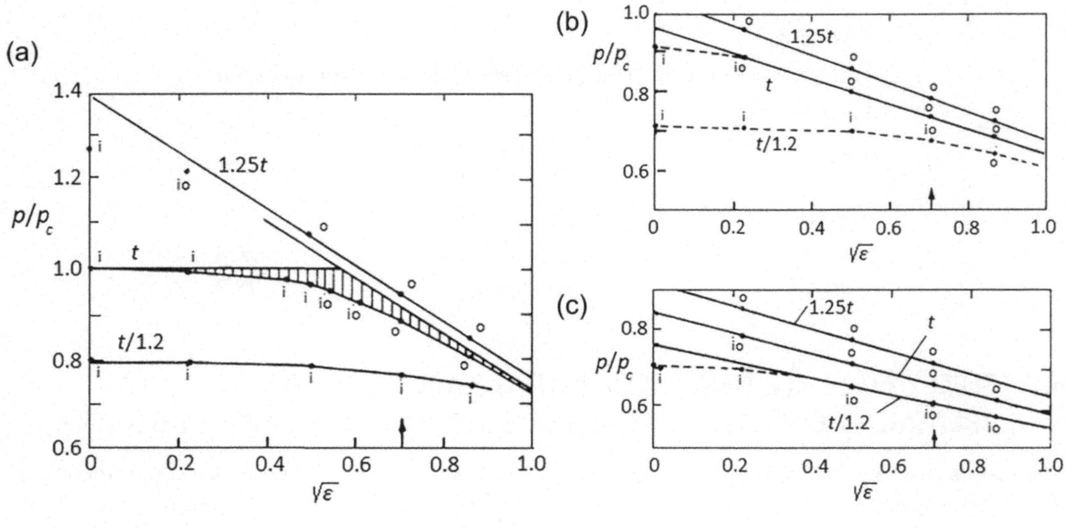

FIGURE 8.14 Ring-stiffened cylinder collapse pressure against $n = 2$ shape amplitude for three shell thicknesses (t = typical for submarine design) and three material idealisations. (a) Stress relieved. (b) Cold-bent plate and flange. (c) cold-bent plate and T-ring stiffener.

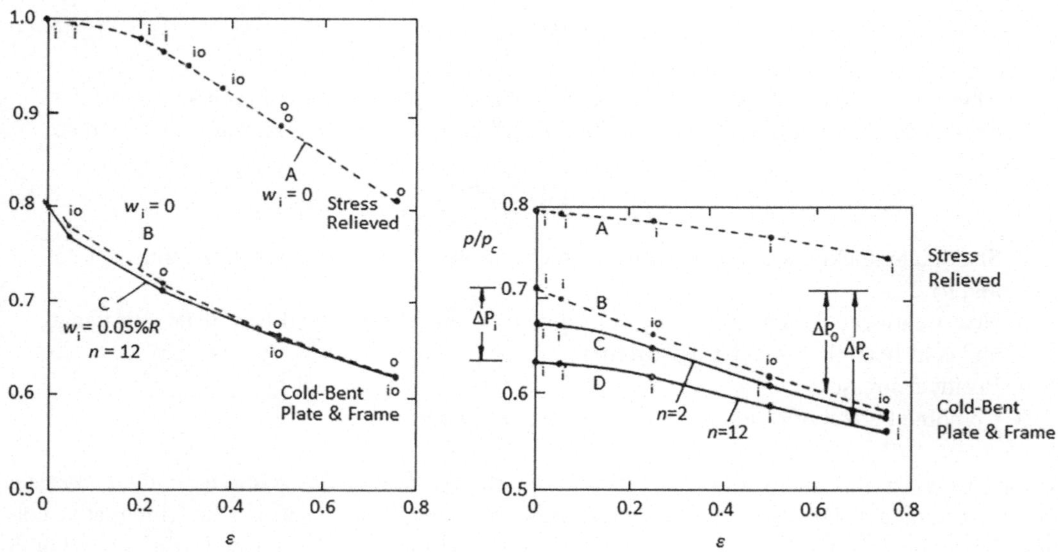

FIGURE 8.15 Combined effect of residual stresses, overall shape imperfection amplitude ($n = 2$) and interframe imperfections w_i on cylindrical shell collapse pressure[4]. (a) Typical shell thickness. (b) Thin shell (A, B: $w_i = 0$, C, D: $w_i = 0.05\%$R).

For small imperfections, interactive collapse is observed, and this changes to overall collapse for medium and large imperfections ($1.0 > \sqrt{\varepsilon} > 0.5$).

In the case of the typical shell thickness (t), it is seen that in the range $0.4 > \sqrt{\varepsilon} > 0.0$, the presence of imperfections has little detrimental effect on the collapse pressure. For this range of imperfections, the failure mode is interframe collapse. Increases in imperfections beyond this point bring about reductions in collapse strength. The collapse mode changes to interactive and for larger imperfections to overall collapse. For $\sqrt{\varepsilon} = 1.0$, the loss of strength equals approximately 25%. The interaction effects account for a maximum loss of strength of approximately 5%.

In the case of the lower shell thickness, (t/1.2), the collapse pressure reduces by 20% for imperfections $0.5 > \sqrt{\varepsilon}$. Thereafter, the collapse pressure reduces further, whereas the collapse mode is interframe throughout.

Figures 8.14b and c show the combined effect of cold-bending residual stresses and the overall $n = 2$ mode. Effective stress–strain curves were obtained and were used in the analysis, in order to model the presence of residual stresses. In Figure 8.14b, the shell and web were bent separately and the web of the ring-stiffener was initially stress-free. In Figure 8.14c, the ring-stiffener was bent as a unit. In the case of the cold-bent plate and flange, the benefit of an increased thickness is not as evident as in the case of the stress-relieved section. In this case, overall collapse arises, and there is a quasi-linear reduction in strength with increasing imperfection, as in the previous case. In the case of the typical shell thickness (t), in contrast to the previous case, there is a quasi-linear loss of strength with increasing imperfections ε. The collapse mode is overall collapse throughout the range of imperfections considered. In the case of the section with reduced shell thickness, the response is similar to the previous case: a reduction in strength of nearly 30% that is unaffected by imperfections in the range $0.5 > \sqrt{\varepsilon} > 0.0$, whereas for larger imperfections, there is a corresponding loss of strength. The collapse mode changes from interframe to interactive to overall collapse and for $\sqrt{\varepsilon} = 1.0$ the loss of strength approaches 40%.

It is seen that for all three shell thicknesses considered, there is a loss of strength compared to the previous cases. This assumes a quasi-linear from in all cases and the collapse mode for the thick and normal thickness is interactive for the lower range of imperfections, whereas for the larger imperfections, it changes to overall collapse. In the case of the reduced shell thickness, the collapse mode is initially interframe, and for $\sqrt{\varepsilon} > 0.4$, it becomes interactive. The loss of strength is 30%, 40%, and 50% respectively in the three cases considered.

Figures 8.15a and b illustrate the effect of the presence of, in addition to cylindrical shell imperfections, that of interframe small amplitude imperfections. If these are present, interframe collapse can occur at collapse pressures that are much less than in the case of the geometrically perfect structure. As seen in Figure 8.15a (typical shell thickness) in the case of cold-bent frames, this effect is small, whereas in Figure 8.15b (thin shell), comparing curves A and B with C and D that this effect may be significant. Curves C and D are obtained from the following formula:

$$\Delta p_c^{\ 2} = \Delta p_0^{\ 2} + \Delta p_i^{\ 2} \tag{8.32}$$

Where

Δp_c = drop in collapse pressure for combined overall/local imperfections
Δp_0 = drop in collapse pressure due to overall imperfections ($n = 2$)
Δp_i = drop in collapse pressure due to local (interframe) imperfections

8.8 EFFECT OF INTERNAL STRUCTURE ON PRESSURE HULL STRENGTH

In this section, we shall describe the results of a study [21] carried out to investigate the effect of the presence of internal structural members and penetrations on pressure hull collapse strength. Internal members include decks and tanks (Figure 8.16), and their presence, as well as that of penetrations and possible dents that may result from contact damage, leads to non-axisymmetric behaviour that needs to be understood and quantified.

8.8.1 Finite Element Modelling

An axisymmetric cylindrical section of the pressure hull was modelled free of secondary structure (Figure 8.17a), and to this were added in stages individual elements of the secondary structure so that the effect of each of these could be assessed. The final model included the pressure hull and all major secondary structure members (Figure 8.17b).

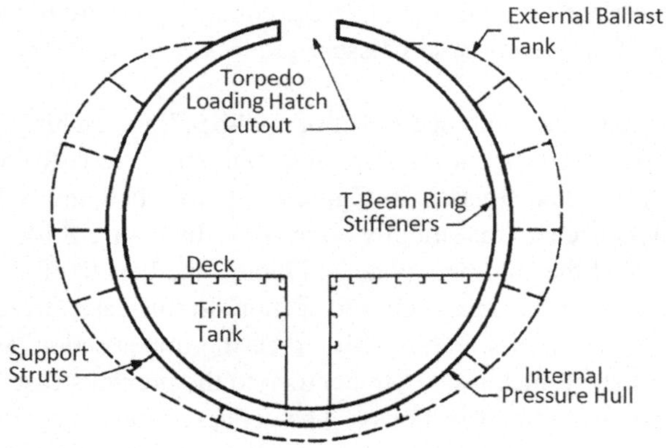

FIGURE 8.16 Pressure hull cross section showing secondary structure [21].

As seen in Figure 8.17a, the model extended over six frames and represented one half of the cylindrical section. This was discretised using eight elements in the circumferential direction and two elements between frames and can thus be treated as a coarse model suitable for representing the overall one half-sine wave buckling mode. The shell plating was modelled using eight-noded isoparametric shell elements, and the ring-stiffeners were modelled using compatible, three-noded isoparametric eccentric beam elements. The model was assumed to be supported by rigid bulkheads at the two ends. In the case of Model 3 (Figure 8.17b), the trim tank consisted of a deck on top and a vertical bulkhead on its side. The deck and bulkhead stiffeners were modelled by smearing them into equivalent plating.

In the case of Model 2 that was used to represent the effect of an indent, this was given a sinusoidal shape with a maximum depth equal to the depth of the ring stiffener and a half-period equal to the frame spacing.

In order to model interframe collapse, it was necessary to increase the mesh density to four elements across the frame spacing and 30 elements around the half-circumference (Figure 8.18a).

8.8.2 Analysis of Results (Overall and Interframe Buckling, Stresses, and Collapse Loads)

The results obtained are summarised in Table 8.4. These relate to buckling, stress, and collapse loads for all models. The results are nondimensionalised with respect to the critical loads for overall buckling (Bryant) (equation 3.11), yielding according to the von Mises yield criterion (5.5) and collapse [18].[5] The percentage comparisons are made with respect to the axisymmetric pressure hull data included in column 1.

8.8.2.1 Overall Buckling

Overall buckling load values for the non-axisymmetric three-dimensional finite element models including secondary structure are included in line 1 of Table 8.4. The form of the lowest overall modes for each model is shown in Figure 8.19a–g.

Comparing Figure 8.19a and b, it is seen that in the case of the deck and trim tank model, the longitudinal half-sine wave and four circumferential waves of the axisymmetric model are retained. The presence of the deck connection results in an increased stiffness of the lower portion of the section, and this is evident by the reduced deflections in this region. The increase in section stiffness resulted in a 6% increase in the overall buckling load.

The effect of the presence of the torpedo hatch cutout is shown in Figure 8.19c. It is seen that in way of the cutout, deflections increase significantly. The longitudinal half-sine wave shape is

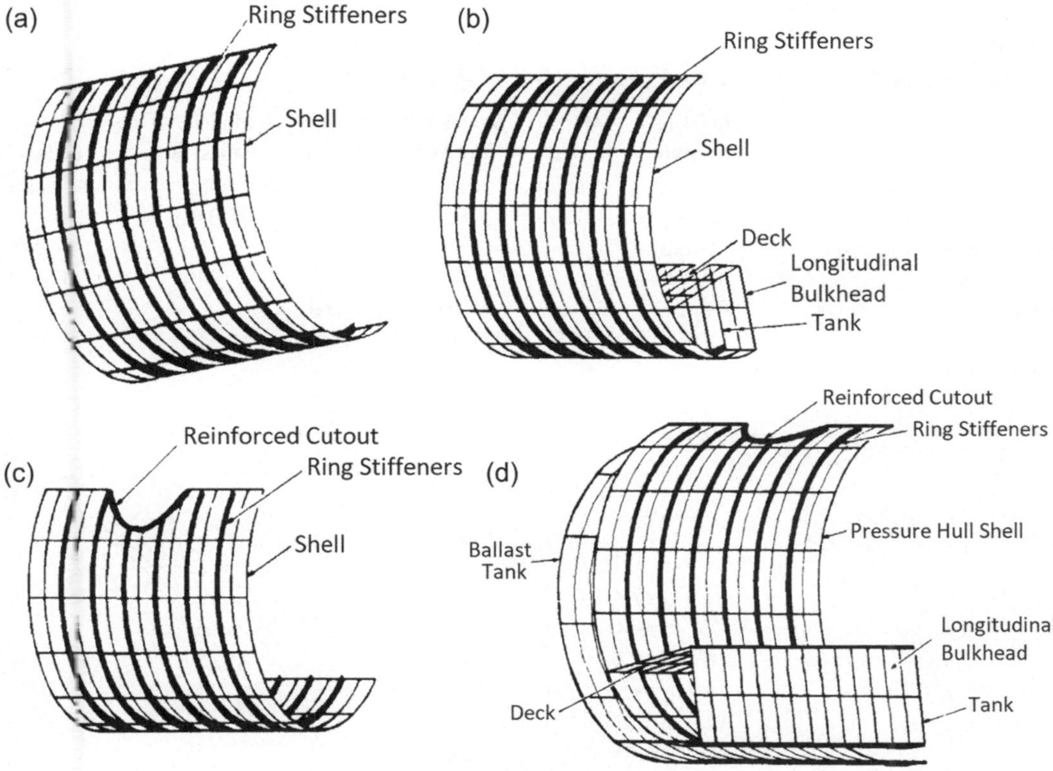

FIGURE 8.17 Selected finite element models used for overall buckling collapse studies [21]. (a) Axisymmetric model (Model No 1), (b) Model 3: Model 1 with the trim tank, (c) Model 4 – Model 1 with a symmetric, (d) Full submarine structure (Model No 6) reinforced torpedo loading hatch cutout.

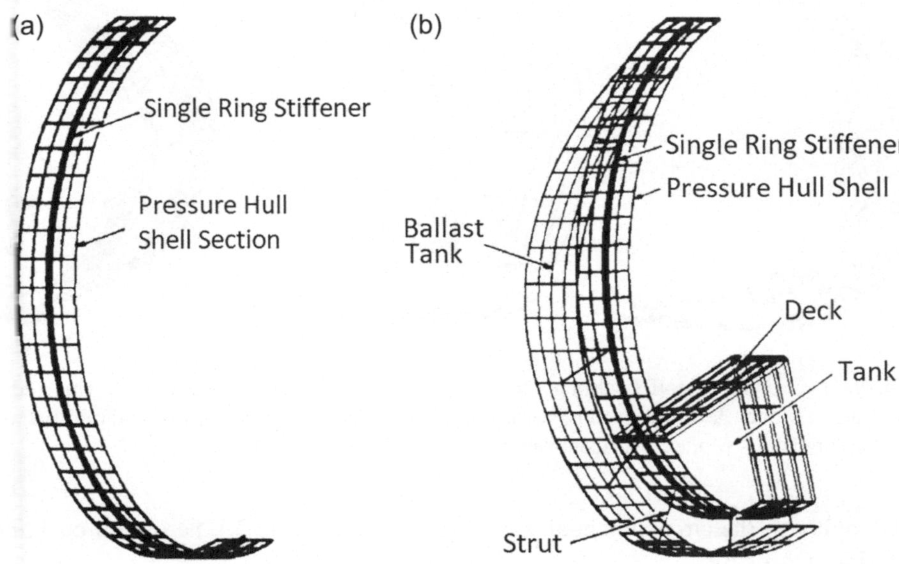

FIGURE 8.18 Selected finite element models used for interframe collapse studies [21]. (a) Refined grid model for interframe bucking. (b) Model 10 – The interframe model incorporating the pressure hull, deck, trim tank, and external ballast tank.

TABLE 8.4
Effect of Secondary Structure on Pressure Hull Response [21]

	Model 1+7 Axisymm.	Model 2 Dent	Model 3+8 Trim Tank	Model 4 Cutout	Models5–9 Ballast Tank	Model 6–10 Full Sub
Overall Buckling Load	0.93 (4)	0.72 (4)	0.99 (4)	0.88 (5)	0.96 (5)[6]	0.98 (4)
	0%	−22%	6%	−5%	34%	5%
Interframe Buckling Load	1.09 (12)		1.5 (12)		1.46 (13)	1.48 (13)
	0%		37%		34%	36%
Maximum Circumfer. Stress	0.95	4.2	1.18	2.60	1.06	2.11
	0%	342%	24%	173%	11%	122%
Plastic Collapse Stress (ADINA)	1.03	0.76	1.03	0.81	1.03	
	0%	−26%	0%	−21%	0%	

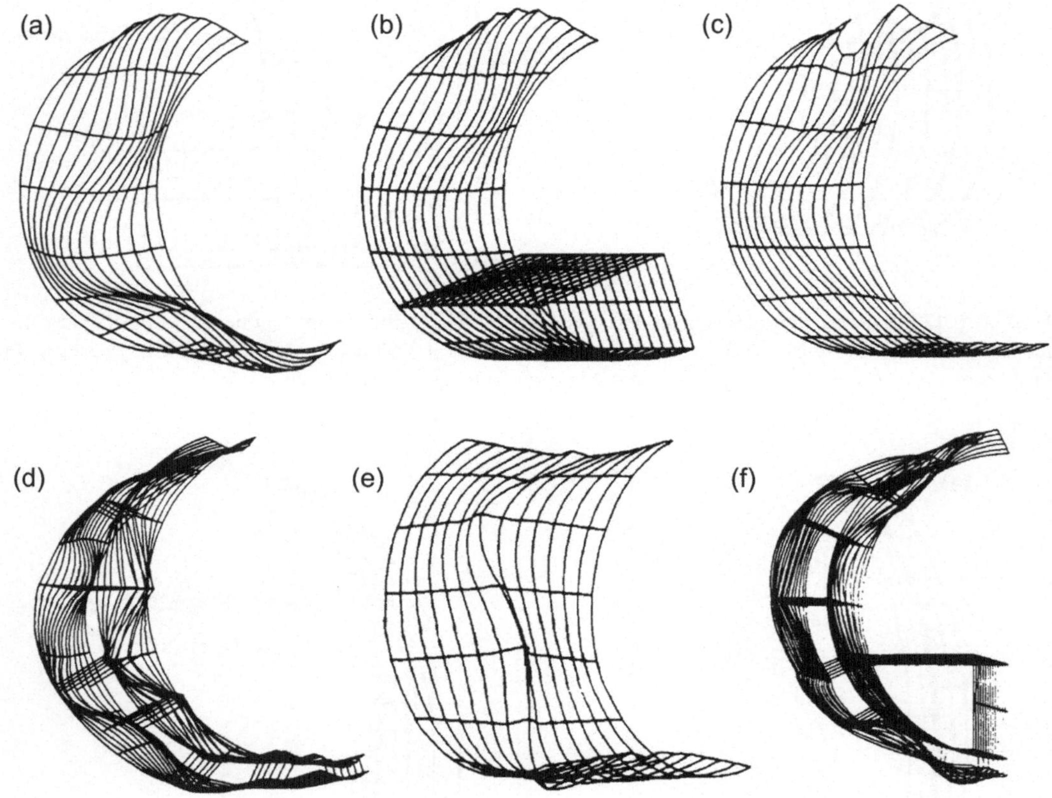

FIGURE 8.19 Minimum overall buckling modes [21]. (a) Axisymmetric – Model 1, (b) Addition of Deck-Trim Tank – Model 3, (c) Cutout in Pressure Hull – Model 4, (d) Addition of external ballast tank, (e) Dented Pressure Hull – Model 2, (f) Complete Submarine Section – Model 6.

maintained, although the circumferential wave number increases to 5. The overall buckling load reduced by 5% in this case.

The effect of adding the external ballast tank to the pressure hull is shown in Figure 8.19d. As shown in Figure 8.16, the external ballast tank is attached to the pressure by struts. Since the ballast tank is not subjected to external hydrostatic pressure, this is applied only to the pressure hull.

Compression stresses are transferred to the ballast tank structure via the struts, and the stiffening effect results in an increase in the overall buckling load of 3%.

In the case of the model with the indent (Figure 8.19e), it is seen that this influences the mode shape, and also the deflections are much larger in the indented region. The longitudinal half-sine wave and four circumferential waves are maintained. The presence of the indent resulted in a reduction of the overall buckling load of 22%, which is significant.

The overall buckling mode shape of the complete submarine section model (Model 6) is shown in Figure 8.19f. It is observed that the number of waveforms in the circumferential direction remains equal to four, and the magnitudes of the deflections are largest in way of the cutout. Comparing with the axisymmetric pressure hull, it is seen that the overall buckling load is 5% higher, as a result of the stiffening effect of the trim tank and the external ballast tank.

8.8.2.2 Interframe Buckling

The interframe buckling loads for the non-axisymmetric three-dimensional finite element models are included in line 2 of Table 8.4. The buckling mode shapes are included in Figure 8.19a–d.

In the case of Model 8 (deck and trim tank model) shown in Figure 8.18b, the stiffening effect of the trim tank compared to the axisymmetric model is clear in the lower part of the section. The number of waveforms in the circumferential direction (12) does not change. The presence of the deck and trim tank leads to an increase in buckling load of the pressure hull-trim tank assembly of 37%, which is significant.

Figure 8.20d shows the mode shape for the pressure hull to which the external ballast tank is added. The mode shape is similar to that of the axisymmetric pressure hull (Figure 8.20a), but in this case, the number of circumferential waveforms increases to 13 from 12. The unpressurised ballast tank adds stiffness, and its presence results in a 34% increase in the buckling load of the pressure hull.

The composite effects of the presence of the secondary structure on the buckling mode are seen in Figure 8.20d. The stiffness is increased, as evidenced by the reduced deflections, and the buckling load increases by 36%. The number of circumferential waveforms increases to 13, presumably due to the struts that act as nodal points. In this model, the effect of the cutout is not included.

In the longitudinal direction, the buckling mode consisted of a half-sine wave in all cases.

8.8.2.3 Stresses

Figure 8.21 shows the stress distributions in the axisymmetric model and the complete submarine section. In these, the magnitudes of the stresses are nondimensionalised with respect to the hoop stress at midbay in the axisymmetric model. Comparing the two, it is seen that the presence of the

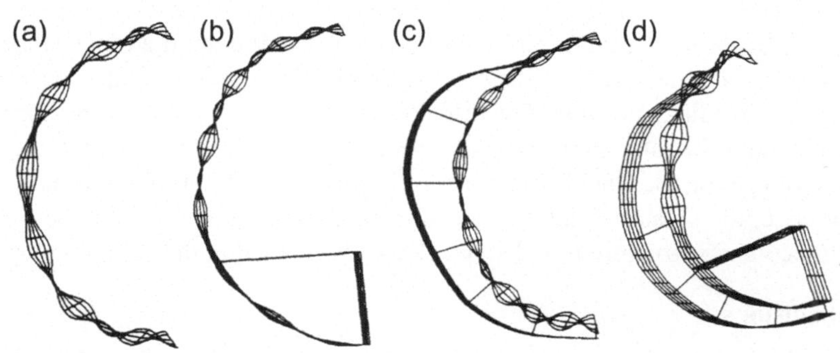

FIGURE 8.20 Minimum interframe buckling mode shapes [21]. (a) Axisymmetric, (b) Deck-Trim Tank with, (c) Ballast Tank with, (d) Complete Section – Model 1 Pressure Hull – Model 8 Pressure Hull – Model 9 - Model 10

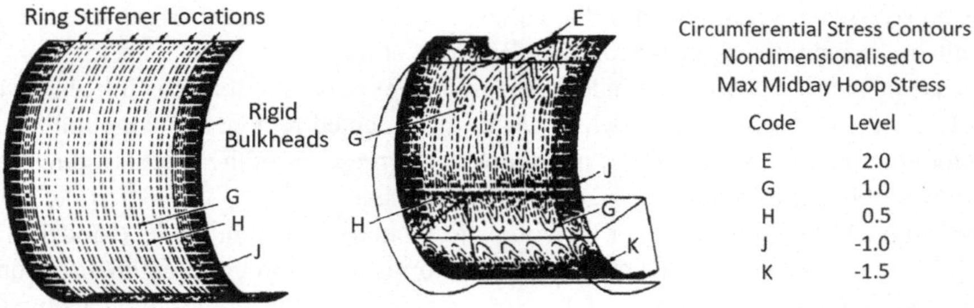

FIGURE 8.21 Pressure hull stress contour patterns for models 1 and 6 [21]. (a) Axisymmetric (Model 1). (b) Complete Submarine Section.

secondary structure results in significant changes to the stress patterns and magnitudes. These were calculated using analytical expressions (4.83).

Comparing Figure 8.21b with 8.21a, it is seen that the stress distribution is no longer symmetric. The largest divergence occurs in way of the cutout where the maximum stress was 122% greater than the maximum midbay hoop stress in the axisymmetric model.

In the case of Model 3, which includes the deck and trim tank, it was seen that the deck connection acts as a hard boundary for the shell. The compressive stress on the outer shell surface reduces as a result of bending arising at the connection. A short distance from the deck increases, and the midbay circumferential stress increases by 24% in way of the outer shell surface.

In the case of Model 4 that included the cutout, there was a significant increase in maximum stress of 173% compared to the Model 1, despite the presence of a large reinforcing stiffener in way of the cutout. This was highly localised in way of the cutout.

In the case of Model 5 that included the external ballast tank, the resulting stress pattern was similar to that observed with the trim tank model. The ballast tank acted as a stiffening member that restrains deformations of the pressure hull. This led to an increase in local stresses of 11% in way of the supporting struts.

In the case of Model 2 that was indented, a considerable increase in bending stresses of 342% compared to the circumferential midbay stress in the axisymmetric model. The detrimental effect that an indent may have is clear although the indent modelled was larger than expected under normal operating conditions.

8.8.2.4 Collapse Strength

The plastic collapse loads of the various models including secondary structure are included in line 4 of Table 8.4; these were obtained using ADINA software. In the case of the axisymmetric hull inelastic collapse did not involve any large deflection behaviour prior to failure. The response was thus linear elastic followed by the spread of yielding until a three-hinge mechanism is formed. Following yielding strain hardening was observed, although collapse occurred at a load slightly above that for initial yielding. For this geometry, it would therefore be sufficient to calculate the load to cause initial yielding to predict collapse. In other cases, with more complicated (non-axisymmetric) geometries, nonlinear analysis is necessary. Where defects such as the indent and the cutout arise, the collapse load is reduced in comparison to the axisymmetric case. In the cases in which secondary structure is added, no change was noted in the collapse pressure.

8.8.2.5 Conclusions

From the preceding comments, it is seen that the presence of secondary structure results in significant changes to buckling loads mode shapes and stress patterns. The defects introduced resulted in lower buckling loads, increased stresses, and changes in mode shapes. The secondary structure added stiffness to the pressure hull and brought about increases in buckling loads, in particular the interframe buckling load. This ought to be accounted for in pressure hull design given that interframe buckling is a primary failure mode. At a local level stresses in way of the connections to the pressure hull increased. Given the observations made earlier concerning the effect of cutouts and penetrations, it is necessary to design the structure at a local level with much accuracy in order to compensate for the presence of the cutouts, as in a typical pressure there are a large number of these.

8.9 BUCKLING OF CYLINDRICAL PRESSURE HULLS WITH OPENINGS

In this section, we shall consider the strength of a cylindrical pressure hull that forms part of the structure of a tourist submarine whose purpose is to transport passengers who wish to gain visual contact with the underwater environment [24]. Tourist submarines are being produced in increasing numbers in recent years in order to cater for this particular sector of the leisure market. Such vessels have large openings in the pressure hull for viewing purposes. These are generally made of acrylic, and it becomes necessary to consider the structural integrity problem that arises with respect to the performance capability of the pressure hull. Issues regarding the geometry and size of the openings, the choice of internal or external stiffening systems, and the ultimate strength of the pressure hull require answers.

The pressure hull generally consists of a ring-framed steel cylindrical shell with hemispherical or conical end closures (Figure 8.22). Compared to naval submarines, the major difference that arises, apart from the presence of the openings in the pressure hull is in the size of the vessel, which is significantly smaller (less than 20 metres in this particular case).

In the present study, two alternative prototypes were considered, whose dimensions are included in Table 8.5. The difference in the prototypes lay in the choice of framing systems (internal or external). The buckling strength and ultimate strength studies for the prototype submarines were validated against tests carried out on a small-scale model (Figure 8.23). It was concluded that the numerical analysis gave accurate results in comparison to the test results.

Before discussing the results of the numerical analysis, we shall mention briefly some of the observations made following the test. The scale model is shown in Figure 8.23a. Hydrostatic pressure is increased gradually, simulating quasi-static conditions up to the collapse pressure of 0.67 MPa, corresponding to a depth of approximately 670 m. Collapse is indicated by a sudden discharge

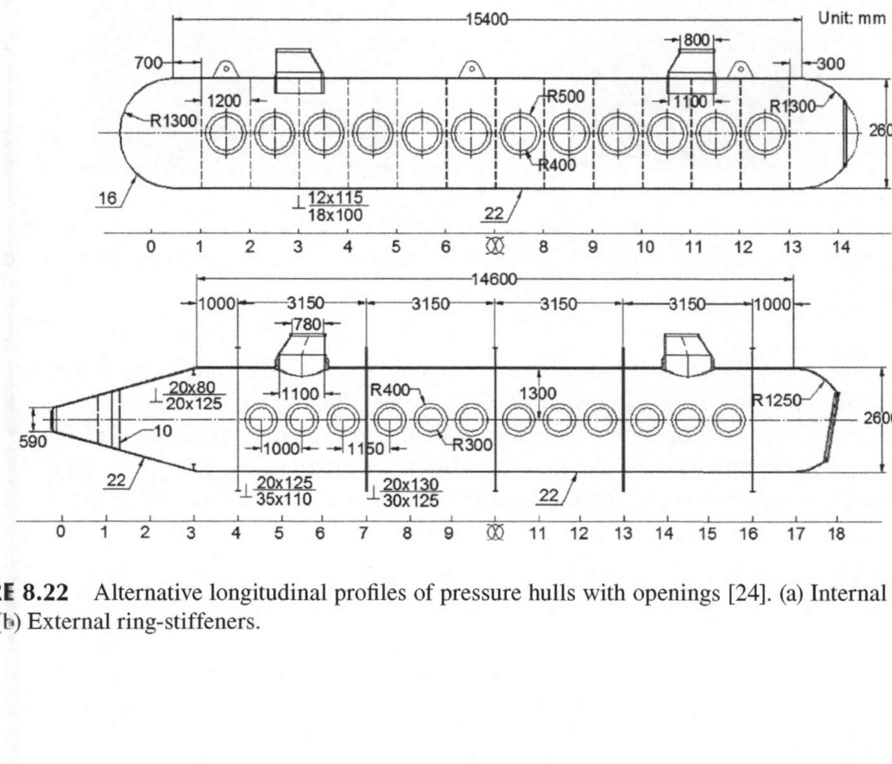

FIGURE 8.22 Alternative longitudinal profiles of pressure hulls with openings [24]. (a) Internal ring-stiffeners. (b) External ring-stiffeners.

TABLE 8.5
Dimensions of Alternative Prototype Tourist Submarines

Dimensions (mm)	Prototype A (Internal Framing)	Prototype B (External Framing)
Overall length	17,800	19,000
Length of cylindrical shell	15,400	14,600
Diameter of cylindrical shell	2,600	2,600
Frame spacing	1,200	3,150
Ring stiffener scantlings	T 115 × 100 × 12 × 18	T 125 × 110 × 2 × 35
Pressure hull thickness	22	22
Hemispherical shell thickness	16	22

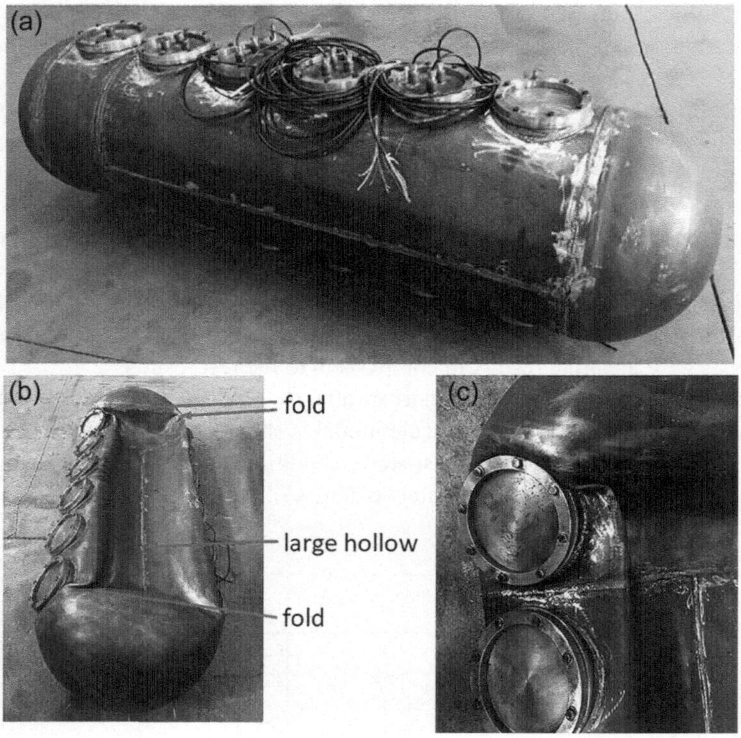

FIGURE 8.23 Scale model subjected to hydrostatic pressure testing to collapse used for validation of the numerical analysis [24]. (a) Prior to testing. (b) Following implosion collapse. (c) Openings (severe deformation side).

of load pressure accompanied by a loud noise (implosion). As seen in Figure 8.23b, collapse involves severe deformation of the cylindrical shell all along its length up to the hemispherical end closures. The deformation consisted of two hollows of unequal depth that extended the full length of the cylindrical section. Figure 8.23c shows that the reinforced openings did not suffer visually observable damage so that the reinforcements are seen to have provided sufficient compensatory strength for the openings.

8.9.1 Finite Element Analysis of the Scale Model

The finite element analysis of the scale model was carried out using the ABAQUS software package and the entire structure was modelled (Figure 8.24). The Riks–Crisfield arc-length algorithm, described in Section 8.4, was used to cater for material nonlinearities before and at the collapse load. Large displacements and finite strain procedures were used.

For the cylindrical pressure hull type S4R shell elements were used, whereas for the reinforced openings solid elements of type C3D8R were used. The S4R element is a four-noded doubly curved finite strain shell element, whereas the C3D8R is an eight-noded solid element. Both elements use reduced integration.

A convergence study was carried out in which element sizes of 25, 20, 15, and 10 mm were used to predict the collapse strength of the scale model. Details concerning the discretisation and the results of the convergence study are included in Table 8.6.

Initial distortions were measured and introduced in the analysis, but residual stresses were not. The initial OOC was 1.0068 with a maximum deviation of 4.5 mm. The boundary conditions used in way of the midship section (plane of symmetry) nodes are indicated in Figure 8.24b. The loading is applied as a uniform pressure to the outer shell of the scale model. It is seen from Table 8.6 that the FE analysis predicts the collapse load accurately, with accuracy increasing as the mesh size is reduced. The predicted post-collapse response of the scale model is shown in Figure 8.25. It is seen that, as occurred during the test, the cylindrical shell collapsed asymmetrically. Furthermore, the numerical analysis correctly predicted that the reinforced openings did not suffer collapse to the same extent as the cylindrical shell.

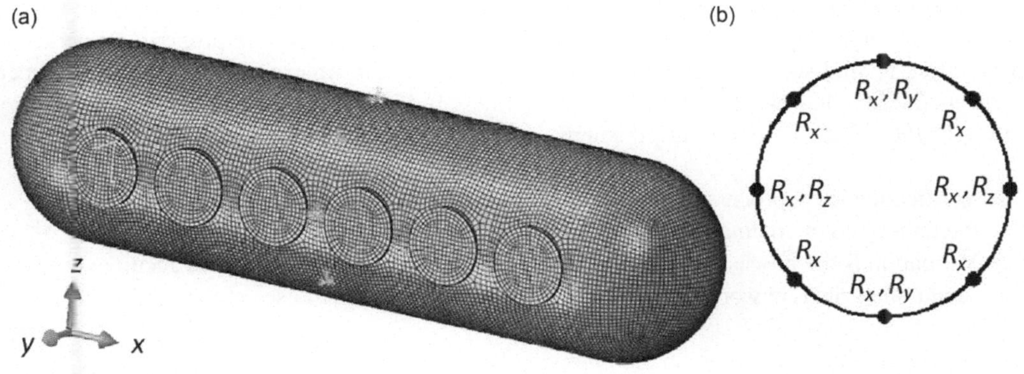

FIGURE 8.24 Finite element model of pressure hull [24]. (a) Mesh. (b) Boundary conditions.

TABLE 8.6
Discretisation of Pressure Hull Structure (Test Collapse Pressure = 0.67 MPa)

Mesh Size (mm)	Number of Nodes	Number of Elements	Collapse Pressure (MPa)	Error (%)
25	13,236	10,568	0.72	+8.5
20	20,270	16,642	0.70	+4.5
15	36,920	31,276	0.65	−3.0
10	86,343	73,716	0.65	−3.0

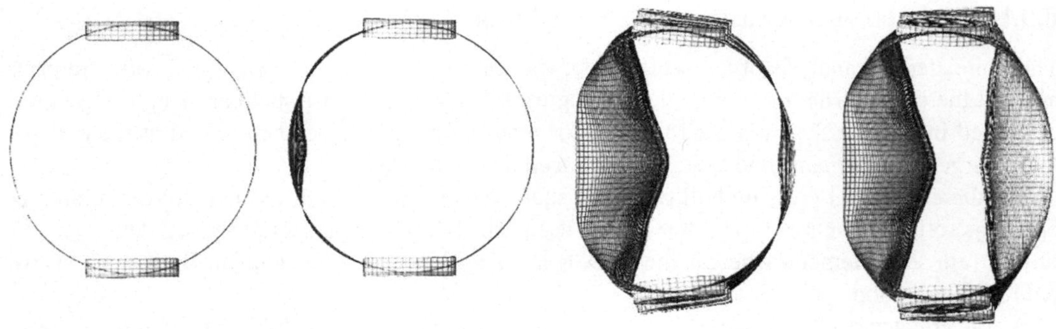

FIGURE 8.25 Post-buckling deformation of cylindrical shell [24]. (a) Collapse pressure. (b) Plastic buckling. (c) Plastic buckling. (d) Final deformation of one side of the other side.

8.10 ENSURING ACCURACY AND RELIABILITY OF NUMERICAL ANALYSIS RESULTS

The selection and use of numerical analysis software to model and solve specific problems raise the question of accuracy and reliability of the results obtained, especially for problems that involve nonlinear behaviour, such as the collapse of submarine pressure hulls. If these questions are not answered, results can be used only to make qualitative recommendations, such as that a particular load does or does not have an effect on structural performance [21].

It therefore becomes necessary to ensure firstly that the software is capable of providing solutions for the range of problems needed and secondly that the software is properly implemented when setting up and solving specific problems. To this end, the American Society of Mechanical Engineers (ASME) has proposed a series of guidelines [1]. In these guidelines, the terms *validation* and *verification* (V&V) are introduced where:

a. Verification is the process of determining that a computational model accurately represents the underlying mathematical model and its solution [1].
b. Validation is the process of determining the degree to which a model is an accurate representation of the real world from the perspective of the intended uses of the model.

The ASME guidelines are based on the following key principles [1]:

1. Verification (addressing programming errors and estimating numerical errors) must precede validation (assessing a model's predictive capability by comparing calculations with experiments).
2. The need for validation experiments and the associated accuracy requirements for computational model predictions are based on the intended use of the model and should be established as part of V&V activities.
3. Validation of a complex system should be pursued in a hierarchical fashion from the component level to the system level.
4. Validation is specific to a particular computational model for a particular intended use.
5. Simulation results and experimental data must have an assessment of uncertainty to be meaningful

In this section, we shall describe in more detail what validation and verification are and how they are carried out, following which we shall discuss their application to the selection and use of finite element procedures in the analysis and design of submarine pressure hulls.

8.11 MODEL BUILDING. VERIFICATION AND VALIDATION

Any problem in structural mechanics involves the use of models that proceed from the general level, at which a *conceptual model* is set up, to a *mathematical model* and finally, a *computational model* [23].

Initially, it is necessary to identify in general terms the physical system to be modelled. A problem statement is formulated, such as "it is required to calculate the elastic buckling pressure of a cylindrical pressure hull subjected to hydrostatic pressure" where our interest lies in the response of the particular structural configuration.

The conceptual model thus comprises of the assumptions and descriptions of the physical processes that represent the solid mechanics behaviour of interest. In the case of a submarine pressure hull, it would in all probability be "a hydrostatically loaded ring-stiffened thin shell of constant diameter made of high-strength steel having initial imperfections that include shell and frame distortions and also residual stresses".

Having specified the conceptual model, the mathematical model is defined. This involves selection of the mathematical equations, boundary and possibly initial conditions, and the material response model. For a submarine pressure hull, equation (3.56) may, for example, be used to calculate the elastic buckling pressure for a ring-stiffened cylindrical shell. A single ring-stiffener with attached shell plating may be analysed, in which case the boundary conditions require symmetry along the edges of the shell plating. In order to model material response, it is necessary to address the properties of high-strength steel, and this will generally require elastic-rigid plastic or work-hardening behaviour.

This is followed by the computational model that involves the numerical implementation of the mathematical model through the use of the software procedure in question. This necessitates the development of a discretised model, the introduction of physical parameters, and the selection of convergence criteria.

The three models are related to each other as shown in Figure 8.26.

Once the steps of the procedure outlined in Figure 8.26 have been followed through, the computational model can be executed and results obtained. These results can then be compared to experimental data that is treated as a benchmark. If the numerical results are not in agreement with the latter, two alternative paths can be followed:

1. Modify the form of the model at either the conceptual or the mathematical level or,
2. Calibrate the model parameters

An example of a change in the mathematical model would to add the effect of axial compression of the ring-stiffened shell. Changes introduced in the conceptual model may be much more complicated, such as allowing for the presence of heavy stiffeners at a specified number of frame spacing. This will necessitate allowing for the strength and stiffness of the additional stiffener and will have repercussions with regard to the computational model. An example of a change in parameters would relate to the magnitude and/or distribution of the initial geometrical distortions of the cylindrical shell.

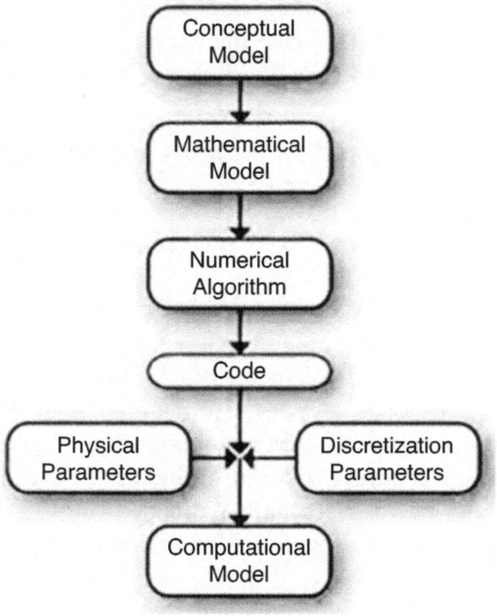

FIGURE 8.26 From conceptual model to computational model [1].

Let us now consider how the development of a computational model described up to this point is linked to the V&V procedure.

Verification includes:

1. *Code Verification*: This involves ensuring that the mathematical model and solution algorithms are working correctly and is a task done primarily by code developers. Users, although they do not have access to the software code, are a source of verification data through the provision of feedback to developers. Code verification involves comparing numerical output to selected benchmark solutions that may be analytical as well as experimental.
2. *Calculation Verification*: This involves checks that the discrete solution of the mathematical model is accurate.

The validation step that follows addresses the question of the adequacy of the conceptual and mathematical models developed by the user. This is done by comparing the results of the model with validation experiments. If the comparisons are satisfactory, the model is considered to have been validated for its intended use. This implies that the conceptual and the mathematical models are acceptable.

The manner in which validation and verification interact with the development of the computational model is illustrated in the flowchart in Figure 8.28. It is seen that physical testing is a necessary and essential part of the process, as it acts as the benchmark for the conceptual and the mathematical models. It is thus necessary to ensure cooperation between the modellers and the experimentalists because the two groups will have different views of the conceptual model since the mathematical and the physical representations of the conceptual model differ. An example is a clamped boundary that is easy to represent in mathematical terms but which in physical terms does not exist. The "Preliminary Calculations" step in Figure 8.27 that links the computational model with the experimental design involves discussions necessary to resolve differences between the modellers and the experimentalists with respect to the conceptual model. Another point that emerges from Figure 8.27 is that the theoretical, that is, the mathematical and computational modelling proceeds in parallel

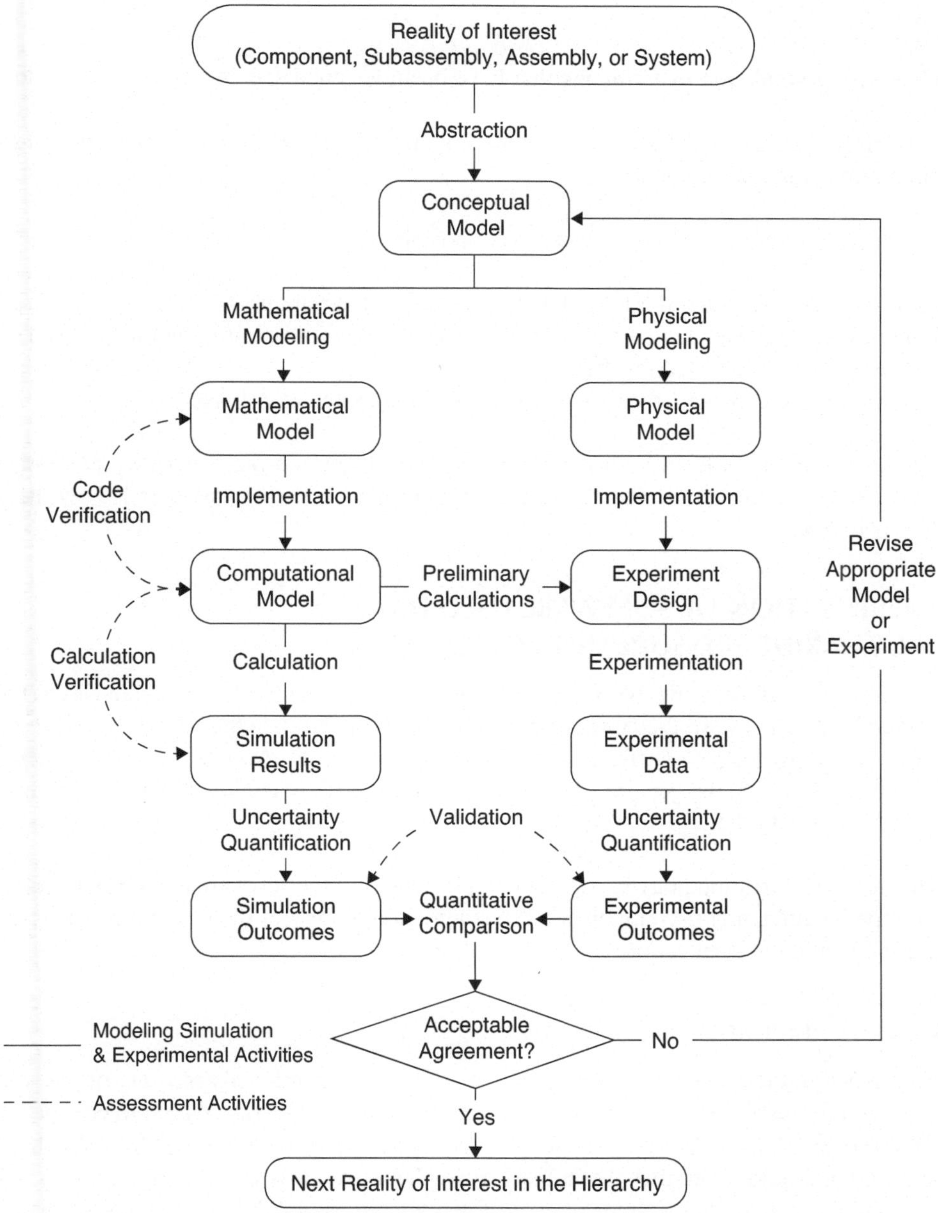

FIGURE 8.27 Validation and verification and the development of the computational model [1].

with the physical modelling (tests) and only interacts at the level of the Preliminary Calculations. The test results should be made known to the modellers after the computational model has been developed. This approach enhances the predictive ability of the computational model as otherwise it would be adjusted during development so as to agree with the test results. Inevitably, tests that are repeated produce somewhat different results. This introduces a degree of uncertainty that has to be allowed for using *uncertainty quantification*. This involves techniques that are used to quantify the effect of parameter variations on simulation outcomes.

Validation involves:

1. Conducting validation experiments that have been designed specifically for the model in question
2. Performing an accuracy assessment, that is, quantifying how well the experimental and simulation outcomes compare

When conducting a validation experiment, it is important to:

- Repeat experiments in order to establish experimental variation.
- Record measurements for other than the important system response quantities in order to validate the data recorded for these.
- Classify errors as being either random (precision) or systematic (bias).

Once results are obtained from both the computational model and the experimental work, these are compared. The method followed has to be outlined in a V&V plan that details how the entire procedure is followed.

8.12 VERIFICATION OF SOFTWARE USED FOR SUBMARINE PRESSURE HULL COLLAPSE

In this section, we shall describe the verification studies carried out by the Canadian Navy (DREA) and reported in [18]. These studies were performed in order to validate the software used to predict the collapse pressure for ring-stiffened cylindrical pressure hulls. A series of FE analyses were performed using ANSYS, the general FE software package used, in conjunction with in-house software, in order to verify that:

1. The numerical methodology is correctly implemented in the software (code verification)
2. The spatial and temporal refinement of the validation models of the test specimens is adequate (calculation verification).

8.12.1 Code Verification

It was mentioned in the previous section that code verification is primarily the duty of the software developers. In this case, code verification was carried out because it was desired to conduct a thorough V&V process that focuses on the specific structural arrangement and loading. Furthermore, code verification would show that the numerical implementation is correct for simple problems, thereby reducing the potential sources of error if at a later stage validation work shows discrepancies in results obtained for more complex problems.

Code verification was carried out by comparing results obtained from numerical analysis with classical analytical solutions for the following problems:

1. Elastic buckling of infinitely long cylindrical shells
2. Elastic buckling of ring-stiffened cylinders
3. Elasto-plastic collapse of a ring-frame using a finite difference solution

The complexity of the solution of these problems is in increasing order. In the first case, the purpose was to study basic linear-elastic shell theory that forms the building block for nonlinear elasto-plastic procedures used to model real problems. In the second case, geometric complexity is increased since a ring-stiffener is included in the elastic buckling analysis. In the third case, nonlinear geometrical and material behaviour was introduced so that the collapse pressure can be

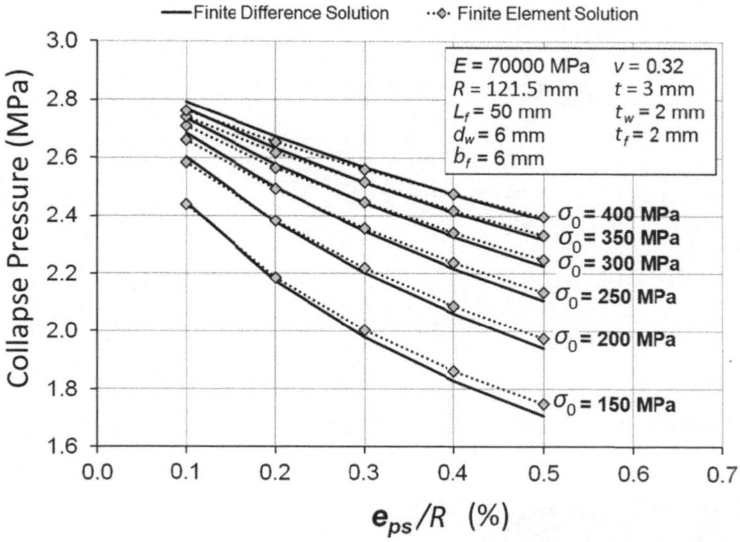

FIGURE 8.28 Verification study for elasto-plastic collapse of ring-stiffened cylinder [19].

predicted. Results obtained from the first and second studies indicated that the software performed satisfactory [18]. In the remainder of this section, we shall discuss the results presented for the third study.

In this case, the benchmark solution was obtained by using Kendrick's finite difference program used to predict the overall collapse pressure of a ring-stiffened cylinder. In this, the model consists of a ring-stiffener and the attached cylindrical shell plating. The cross section is divided into a series of strips in order to be able to introduce residual stresses due to cold rolling. The progression of yielding through the cross section is thus also allowed for. The geometry and material properties of the section considered are included in Figure 8.28.

In the FE analysis, shell elements were used although the FD analysis is based on beam theory. A mesh convergence study led to the use of 72 elements for the 90° arc modelled.

A series of parametric studies was carried out in which the yield stress and the magnitude of the OOC, e_{ps}, were varied to generate 30 cases. In all cases, the OOC was in the critical $n = 2$ mode. Material behaviour was assumed to be elastic-rigid plastic in the FD (one-dimensional stress field). In the FE model, an isotropic von Mises yield surface was used. Isotropic hardening was assumed, thereby ignoring the Bauschinger effect.

Results are included in Figure 8.28 in which the predicted collapse pressure is plotted against OOC for a range of yield stresses (400 MPa > σ_0 > 150 MPa). It is observed that the FD and FE solutions follow the same trends, that is, a reduction in collapse pressure with increasing OOC and reducing yield stress. The differences that are due to OOC arise because in the case of the FE model, a bilinear stress field arises in comparison with the uniaxial stress field of the FD model. The RMS error (difference) between the two models was less that 1%, which is considered small, given the difference in structural theories used. Of the two, the FE model is the most accurate.

8.12.2 Calculation Verification

In this case, calculation verification was carried out to obtain data on the satisfactory levels of:

1. Mesh density
2. Load increment size
3. Convergence tolerances for the iterative solution during each load increment

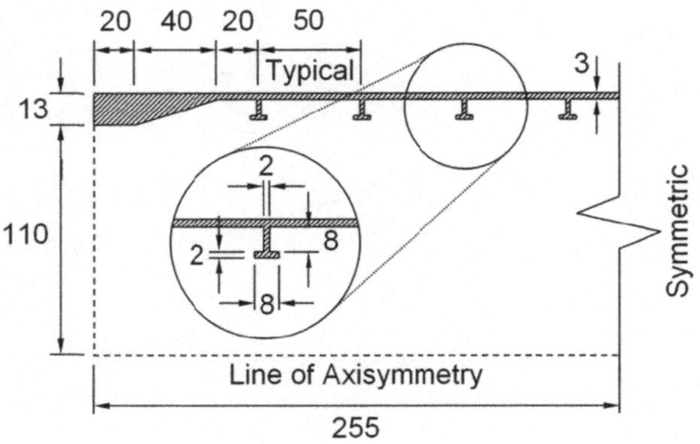

FIGURE 8.29 Geometry of specimen L510-No 18 [19].

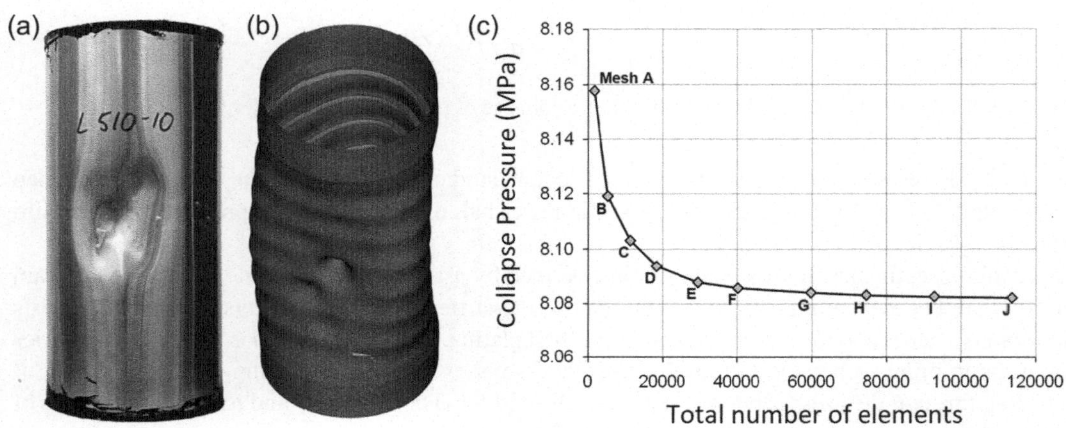

FIGURE 8.30 Specimen L510-No 18 and convergence study. (a) Model following collapse. (b) Finite element model. (c) Results of convergence study.

This was performed for each test specimen, and we shall give results obtained in the case of specimen L510-No18, whose geometry is shown in Figure 8.29.

In this case, the specimen was made of aluminium 6082-T6 with a yield stress of 305 MPa, and the shell and stiffeners were proportioned such that overall collapse took place. An OOC in the critical $m = 1$, $n = 3$ mode at an amplitude of 0.4% was induced. Figure 8.30 shows the actual test specimen following collapse as well as the FE model. In the latter case, the collapse pressure of 8.08 MPa exceeded the test value by 8.7%.

Results from the mesh convergence study are shown in Figure 8.30c in which the predicted collapse pressure is plotted against the total number of elements. For this series of runs, the initial load increment was set at 0.25 MPa and the convergence tolerance at 0.005. On the basis of these results, Mesh H was used in the analysis of specimen L510-No18 as well as other similar specimens. For this mesh, 288 elements were used around the entire circumference of the cylinder, 19 elements between frames, and 4 elements on the stiffener web and flange.

With regard to load increment size, a study was carried out to examine the sensitivity of the predicted collapse pressure to load increment size. In each case, the initial load increment size was specified, with the arc-length method and automatic time-stepping controlling the subsequent load path. In this study, the converged Mesh H and a convergence tolerance of 0.005 were used. The cases considered and corresponding results are included in Table 8.7.

TABLE 8.7
Results from Load Increment Study for L510-No 18

	Load Increment Size (MPa)			
Model	Initial	Last Increment before Collapse	Load Increments to Collapse	Collapse Pressure (MPa)
L1	2.0000	0.0165	6	8.0772
L2	1.0000	0.0128	11	8.0755
L3	0.5000	0.0204	21	8.0825
L4	0.2500	0.0067	42	0.0831
L5	0.1250	0.0006	85	8.0841
L6	0.0625	0.0001	170	8.0844

It is seen that the size of the last load increment decreases as the initial load increment is reduced. This is due to the use of the arc-length method that enforces a constant length in load–displacement space. It is seen from the last column that results are largely insensitive to load increment size. The load increment size of 0.25 MPa was considered to be converged, since the predicted collapse pressure was within 0.1% of the value for an analysis with twice the increment size. A general conclusion was that the initial load increment should be approximately 3% of the expected collapse pressure, provided an arc-length approach is used.

With regard to solution convergence tolerances, the criterion used was (8.27a) where the L2 vector norm was used (8.28). For Mesh H and an initial load increment of 0.25 MPa, a range of convergence tolerances was used ($0.0005 < \varepsilon_R < 0.02$). It was found that the solution was insensitive to the magnitude of the convergence tolerance parameter, that is, it did not affect the predicted collapse pressure. The default tolerance of 0.005 was thus considered adequate for further V&V studies.

NOTES

1 The convergence criteria given are used in ANSYS procedures and are applied in the convergence verification study described in Section 8.12.
2 The transverse bulkheads are included in order to correctly simulate overall collapse.
3 To accurately model interframe buckling considerably more elements are required in both the circumferential and longitudinal directions than for the overall half sine wave mode [12].
4 Collapse is deemed to have occurred when a plastic hinge forms at midbay. When this arises, a three-hinge collapse mechanism is formed with plastic hinges having also been formed in way of the lines of attachment of the shell plating to the frames [18].
5 Results are normalised to the stress-relieved typical shell thickness case for $w_0=0$.
6 The ballast tank buckled at a load 14% lower than that of the pressure hull.

REFERENCES

1. ASME. (2006). *ASME V&V 10–2006: Guide for Verification and Validation in Computational Solid Mechanics*. New York: American Society of Mechanical Engineers (ASME).
2. Bathe, K.-J. (1996). *Finite Element Procedures*. Upper Saddle River, NJ: Prentice-Hall Inc, 1037 p.
3. Bushnell, D. (1976). BOSOR5-program for buckling of elastic-plastic complex shells of revolution including large deflections and creep. *Comp. & Struct.* Vol. 6, 221–239.
4. Cole, R.H. (1965). *Underwater Explosions*. New York: Dover Publications.
5. Costanzo, F.A. (2010). Underwater explosion phenomena and shock physics. *Proceeding of the 28th International Modal Analysis Conference (IMAC-XXVIII)*, Jacksonville, FL.
6. Cresswell, D.J., Dow, R.S. (1986). The application of nonlinear analysis to ship and submarine structures. In: Smith C.S. and Clarke D.J. (Eds). *Advances in Marine Structures*. Barking, Essex: Elsevier Applied Science Publishers Ltd, 174–200.

7. Crisfield, M.A. (1981). A fast incremental/iterative solution procedure that handles "Snap-Through". *Comp. & Struct.* Vol. 13, 55–62.
8. Crisfield, M.A. (1986). Snap-through and snap-back response in concrete structures and the dangers of underintegration. *Int. J. Num. Meth. Eng.* Vol. 22, No. 3, 751–768.
9. European Committee for Standardisation. (2009). *European Standard EN 13445:3 Unfired Fusion Welded Pressure Vessels. Part 3: Design.*
10. Gish, L.A. (2013). "Analytic and numerical study of underwater implosion". *Ph.D. Thesis* Massachusetts Institute of Technology, Cambridge, MA.
11. Graham, D. (2007). Predicting the collapse of externally pressurized ring-stiffened cylinders using finite element analysis. *Mar. Struct.* Vol. 20, 202–218.
12. Graham, D. (2009). "Finite element analysis". Lecture 6–8 ASRANet *Design of Submarine Structures* Online Course.
13. Gupta, S., Parameswaran, V., Sutton, M.A., Shukla, A. (2014). Study of dynamic underwater implosion mechanics using digital image correlation. *Proc. R. Soc. A.* Vol. 470, 20140576.
14. Kadapa, C. (2021). A simple extrapolated predictor for overcoming the starting and tracking issues in the arc-length method for nonlinear structural mechanics. *Eng. Struct.* Vol. 234, No. 1, 111755.
15. Krenk, S. (2009). *Non-Linear Modeling and Analysis of Solids and Structures.* Cambridge: Cambridge University Press, 345 p.
16. Lafontaine, N.M., Wang, X.R., Huang, K.F., Yuan, M.W., Oñate, E. (2018) "On the arc length method: Combining ideas and implementations aspects". Research Report N° PI399, Int. Cen. Num. Meth. Eng. (CIMNE). Available at: https://www.scipedia.com/public/Lafontaine_et_al_2013b (Accessed date February 15, 2024).
17. Lee, Y.-J., Hsu, C.-H., Huang, C.-H. (2008). Pressure hull analysis under shock loading. *Shock Vib.* Vol. 15, 19–32.
18. Lunchick, M. E. (1959). "Yield failure of stiffened cylinders under hydrostatic pressure," David Taylor Model Basin Report 1291, Washington, DC.
19. MacKay, J.R. (2011). "Verification of numerical models for pressure hull collapse predictions." *Technical Memorandum* DRDC Atlantic TM 2011-281.
20. Memon, B.-A., Su, X.-Z. (2004). Arc-length technique for nonlinear finite element analysis. *J. Zhejiang Univ.* Vol. 5, No. 5, 618–628.
21. Pegg, N.G. (1987). "Effects of secondary structure on the stress and stability of submarine pressure hulls." Canadian Navy DREA Report 87/102.
22. Ricks, E. (1979). An incremental approach to the solution of snapping and buckling problems. *Int. J. Solids & Struct.* Vol. 15, 524–551.
23. Schwer, L. (2009). An overview of the ASME V&V-10 guide for verification and validation in computational solid mechanics. *20th International Conference on Structural Mechanics in Reactor Technology (SMiRT 20),* Espoo, Finland, SMiRT 20-Division 3, Paper 2010.
24. Xu, S.-X., Lin, J.-P., Liu, B., Garbatov, Y., Wang, Y., Guedes Soares, C. (2020). Experimental and numerical buckling analysis of cylindrical pressure hulls with multi-circular openings. *Ocean Eng.* Vol. 214, 107689.

9 Design of Submarine Pressure Hulls

9.1 OPERATIONAL REQUIREMENTS

In Chapter 6, Section 6.1, we described the failure modes of submarine pressure hulls. This understanding is a prerequisite to the *design* of the structure of these vessels, a topic that we shall now address.

Submarines are *weight-critical* structures. That is, the limiting factor for a given enclosed volume that operates at a particular depth is the weight of the pressure hull, and the deeper a vessel is expected to operate, the larger the weight of the pressure hull. Since the weight of the pressure hull accounts for a significant proportion of the total weight of the vessel (approximately 25%), it is important to optimise the design from the point of view of pressure hull weight. The *weight-to-displacement (buoyancy) ratio* can be used as a measure of hull efficiency. For an unstiffened cylindrical shell that is loaded only in the lateral direction, the weight per unit length is $W = 2\rho g\pi Rt$, and the displacement per unit length is $\Delta = \rho_w g\pi R^2$. Hence,

$$\frac{W}{\Delta} = 2\left(\frac{\rho}{\rho_w}\right)\frac{t}{R} \tag{9.1}$$

where ρ, ρ_w are the densities of the material used and of seawater. The corresponding expression for a sphere is:

$$\frac{W}{\Delta} = \frac{3}{2}\left(\frac{\rho}{\rho_w}\right)\frac{t}{R} \tag{9.2}$$

Apart from weight, one other limiting factor to operating depth is the strength of the structure in way of penetrations in the pressure hull, such as propeller shafts, periscopes. As the depth increases, it becomes increasingly difficult to ensure absolute watertightness.

As the diameter of the pressure hull is increased, the strength reduces because of the reduction in curvature, so that for a given deck area capability, it is necessary to increase the thickness or else use a different material. There are cases, however, in which the diameter cannot be reduced beyond a certain point for operational reasons such as when essential equipment has to be fitted.

For naval submarines, the desired *operating depth* is determined by such qualities as "undetectability", which involves being able to avoid detection and tracking devices. These may be located within the ocean (other submarines), the surface (warships), or elsewhere. In order to avoid the simplest of all types of detection, which is visual, it is necessary for the vessel to be completely immersed. This necessitates that it can withstand a hydrostatic pressure of a few atmospheres. If a submarine is to operate below the surface, it has to be at a much greater depth in order to avoid collision with the largest of ships, so that a depth of 50 m is required. World War II submarines operated at a depth of up to 150 m, whereas operating depths have since increased and range from 300 m for a conventional diesel-powered vessel to possibly 1,000 m (for the Soviet Alfa SSN with a titanium pressure hull). Apart from detection avoidance, one other reason for depth capability arises from the fact that because modern submarines operate underwater at high speeds, they need ample margins for manoeuvring. Thus, if the control systems malfunction or a navigational error arises, a rapid descent may result that could have catastrophic consequences.

DOI: 10.1201/9781003489153-9

Conventional, non-nuclear submarines need to surface at regular intervals so that operating at depths that are greater than necessary prolongs the rise and descent to and from the surface. One other reason not to operate at depths that are greater than necessary is in order to avoid structural and material problems such as fatigue.

In addition to hydrostatic pressure loading, a submarine is subjected to a number of other types of loads at a global and a local level. At a global level, the structure is subjected to longitudinal bending and torsional loads as any other surface ship, although in the case of rough weather, the submarine dives. When under way, dynamic and flutter loads act on the bow and stern planes. Hydrodynamic loads also act on the superstructure and the rudders, especially when the vessel is turning. There are instances when the structure is subjected to intentional impact loads (docking, tug action, breaking through ice) or unintentional ones, such as minor contact damage whilst berthing. These act on the outer shell and only indirectly affect the pressure hull.

In this chapter, we shall discuss issues relating the pressure hull design, the emphasis being on the design of ring-stiffened cylindrical shells. The design of other pressure hull forms and major internal structures is covered in Chapters 7 and 12.

9.2 PRESSURE HULL FORM

9.2.1 Main Hull

The diving capability of a submerged hull depends on its particular failure modes, which in turn depend on the properties of the material used as well as the structural configuration. A uniform hydrostatic pressure is best withstood by membrane action, and thus, a spherical cross section is the optimum. For this reason, axisymmetric bodies of revolution whose form ranges from a sphere to that of an elongated section are in general use. The designer has to select the appropriate form for the particular application, whether this may be a large nuclear submarine or a submersible used for deep sea exploration. A large variety of configurations has been used, especially in the case of submersibles (Chapter 1).

Submarines constructed during the first half of the 20th century consisted of a pressure hull enclosed within a hull form reminiscent of a surface vessel (Figure 9.1). During the second half of the 20th century, hydrodynamic criteria came to be applied when defining the external form of submarines. The first submarine designed and built on this basis was the experimental vessel *USS Albacore* (Figure 9.1).

If mission requirements include the need for high speed, it is clear from Table 9.1 that a cylindrical section (or at least an axisymmetric section with a high L/D ratio) is to be preferred. Minimum resistance is experienced when $7 > L/D > 6$ with a hydrodynamic form in the shape of a *teardrop*, whose diameter is larger towards the forward end and which tapers off gradually towards the aft end. Spherical vessels on the other hand have the most favourable W/Δ ratio which makes them suited for deep diving, especially as high speed is not a prerequisite.

We therefore see that in the case of naval submarines, competing requirements arise, as hydrodynamic efficiency points towards the need for an elongated axisymmetric form (such as an oblate spheroid)[1] whereas structural efficiency is maximised when the pressure hull is spherical. The modern submarine is thus a compromise, and its present diving capabilities are the result of advances in structural analysis and design, materials performance, and fabrication technology. These have permitted the construction of vessels that operate at depths that are significantly greater than previously whilst maintaining the required hydrodynamic form.

9.2.2 Single or Double Hull Submarine?

The debate as to whether single or double hull submarine designs are preferable has a long history, and even at this point in time views are divided (Figure 9.2). During the 19th century, with the

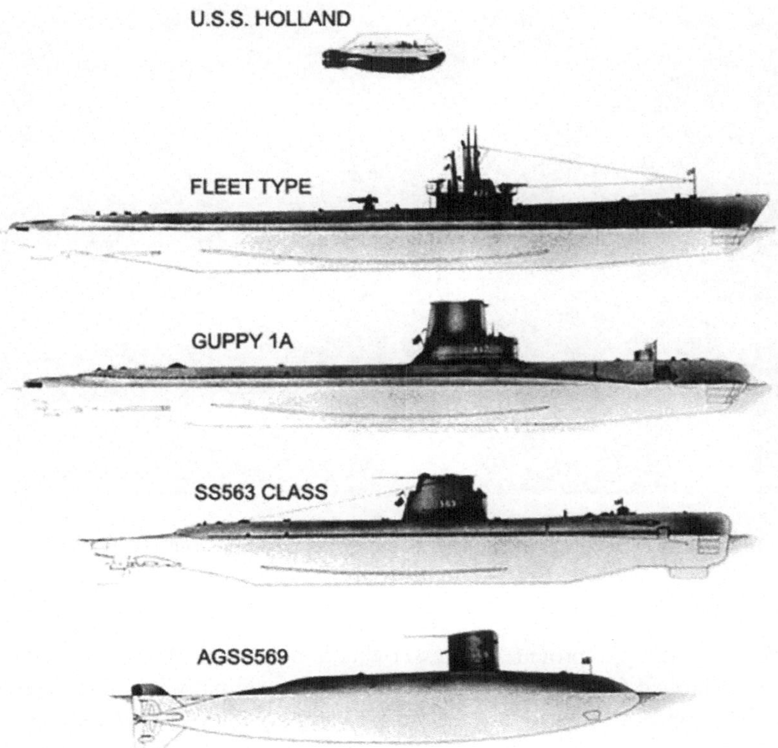

FIGURE 9.1 Evolution of outer hull form (the *USS Albacore* is *AGSS569*) [28].

TABLE 9.1
Advantages and Disadvantages of Pressure Hull Forms [5]

		Advantages		Disadvantages
Sphere	1	*Most favourable W/Δ ratio*	1	Difficult interior arrangements
	2	Through hull penetrations easily made	2	Large hydrodynamic drag
	3	Stress analysis more accurate and less complex	3	Poor manoeuvrability and docking
			4	Difficult to manufacture (sensitive to initial imperfections)
Ellipse	1	*Favourable W/Δ ratio*	1	Fabrication expensive
	2	More efficient internal arrangements	2	Structural analysis difficult
	3	Through hull penetrations easily made		
Cylinder	1	Fabrication easiest	1	*Least efficient W/Δ ratio*
	2	Most efficient interior arrangements	2	Internal stiffeners required
	3	Low hydrodynamic drag	3	Structural analysis difficult for through-hull penetrations

exception of the design proposed by Monturiol (Chapter 1), all submarine pressure hulls were of the single hull type. The design that made an impact and set the standard for many years was the double hull design introduced by Laubeuf in the *Le Narval* in 1897 (Figure 1.21).

FIGURE 9.2 Alternative hull forms in modern submarines [26]. (a) Single hull design. (b) Double hull design.

This was done primarily to provide the vessel with a more seaworthy hull when on the surface. Other advantages claimed for the double hull design when originally introduced are mentioned in Section 1.3.3. This approach was followed by most countries until the 1960s when the US Navy decided to revert to the single hull design and has subsequently been followed by Western countries. In several cases, a hybrid form is used, combining both single and double hull construction. The double hull has continued to be used in the Russian Federation and in China to this day. It is unfortunate that direct primary sources describing Russian design philosophy are not available, and discussion has to rely on Western understanding.

The design philosophies underlying the two approaches differ in the interpretation of the term *invulnerability* [14]. This term has a dual interpretation:

1. Immune to attack
2. Incapable of being damaged

The Western interpretation that is used in submarine design is the former, whereas Russian and Chinese design philosophy is based on the latter, this being is a more stringent criterion. From the US perspective, it is sufficient for a submarine to be undetectable for it to be immune to attack. It is necessary to also introduce safety and reliability, so that the vessel is invulnerable to accidental damage as well. The three pillars of submarine design are therefore stealth, safety, and reliability. One important reason for which the double hull design was abandoned is that a larger hull is needed, with increased drag and powering requirements. In the United Kingdom, one other reason given is the increased cost of construction.

For submarines constructed using HY-80 steel, the required submerged displacement for a range of depths is shown in Figure 9.3. Curves are included for both single and double hull designs, and it is clear that a single hull design is preferable. For the single hull design, the change in the rate of weight increase that is noted for depths exceeding 270 m indicates that this is a maximum operating depth that makes sense to aim for when using this type of steel.

In the second interpretation, it is not sufficient to reduce the probability of attack by stealth. Submarine design has to allow for the possibility of attack, in which case the design has to cater for this. For Russian designers, the ability to withstand combat and accident damage and to restore

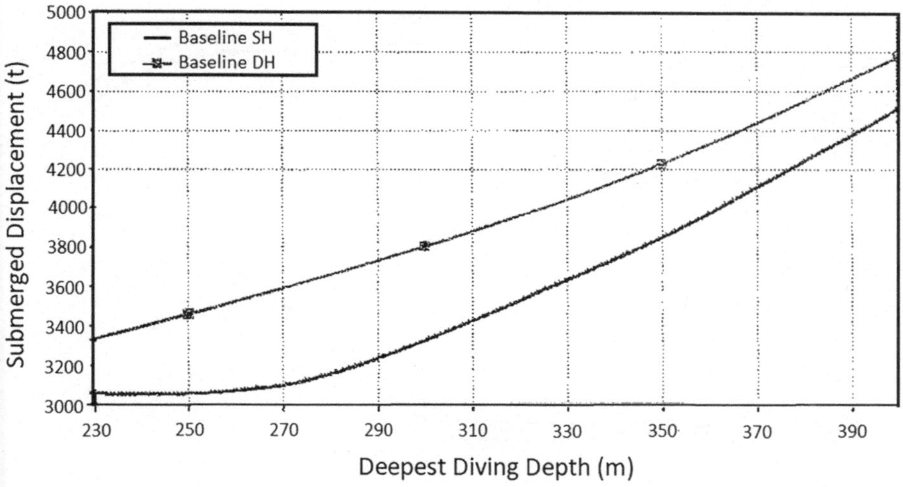

FIGURE 9.3 Deep diving depth versus submerged displacement.

combat capability is essential and must be provided throughout the operating life of the vessel. Combat survivability has to be ensured at two levels:

1. The ability of the structure to survive the initial effect of enemy weapons and
2. The ability to limit the effects of secondary damage resulting from fire and flooding for single and double hull designs [3]

The arguments made by advocates of the double hull design are:

- Increased survivability
- Increased useful volume within the pressure hull
- Reduced costs
- Signature reduction

Increased survivability arises because the outer shell and exostructures provide a degree of protection for the pressure hull against the detonation of weapons within the vicinity of the vessel, thus reducing the impact of explosion. It is possible to introduce armour in the inter-hull space, thereby increasing the probability of survival in case of attack. Should a submarine be damaged as a result of enemy action, the ballast tanks will permit the vessel to remain neutrally buoyant and maintain trim.

Increased useful internal volume results because the ring-stiffeners can be welded externally to the pressure hull. Furthermore, specific items such as mechanical equipment and weapons can be located in the inter-hull space. Space savings are thereby achieved within the pressure hull.

Cost reduction can result from the use of the outer shell plating, which is much thinner than that of the inner hull, to obtain hydrodynamic form.

Signature reduction can arise as a result of the use of degaussing coils as well as measures to reduce noise in the inter-hull spaces. Furthermore, boundary layer control schemes can be introduced, the aim being to enhance the military capability of the vessel.

Finally, the inter-hull space can be used to retrofit a vessel throughout its operating life.

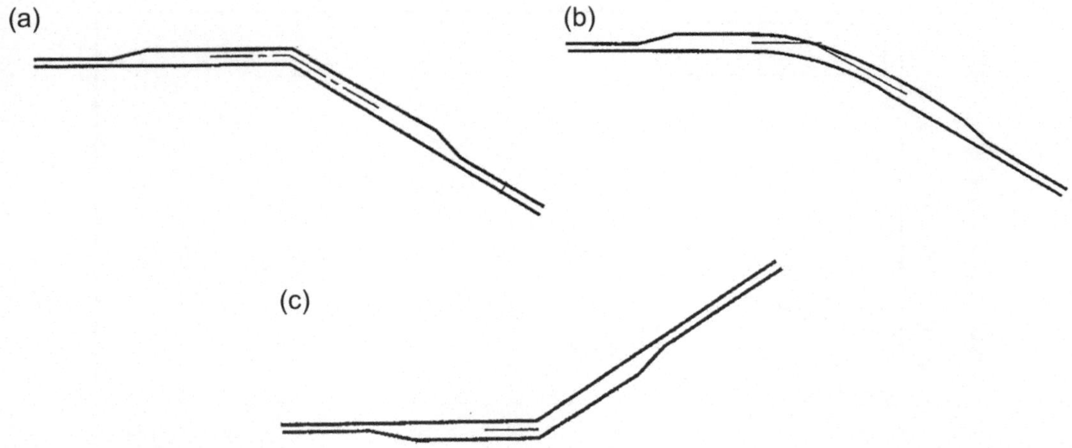

FIGURE 9.4 Cylinder-cone intersections. (a) Intersection with knuckle. (b) Intersection without knuckle. (c) Intersection with knuckle small diameter cylinder end.

9.2.3 Pressure Hull Stiffening, Connections, and End Closures

In the previous chapters, we saw that in order for the cylindrical shell to develop its full strength, it needs to be stiffened. The stiffening takes the form of ring frames with no axial stiffening being required, since the longitudinal stresses are half those in the circumferential direction and also the stiffness and strength of the end closures are much greater than those of the cylindrical pressure hull. Since it is possible to weld the ring frames internally or externally, it is necessary to compare the two alternatives. From the point of view of design, external stiffeners are preferred since there is no loss of internal volume. This choice is available when the external structure can accommodate the external stiffening system. When external stiffening is used, the pressure acts so as to separate the cylindrical shell from the stiffeners, potentially leading to cracking at the toe weld. Furthermore, external stiffeners and their welds are exposed to the sea environment and will thus suffer corrosion and fatigue much more so than internal ones. On the other hand, external frames are more stable than internal stiffeners because the only possible buckling modes are non-symmetric, and these involve higher energy levels. The webs of external stiffeners are in tension, thereby ensuring improved stability. If, however, a single hull design is used, internal ring frames have to be fitted.

The pressure hull is made up of a number of interconnected cylindrical sections that have different diameters for reasons unrelated to structural considerations, such as for the installation of external structures or *exostructures*, for example, external ballast tanks. The most efficient way of attaching cylindrical shell sections whose diameter differs to each other is by use of *truncated cones.*

The intersections shown in Figure 9.4 are in accordance with the current European pressure vessel code, and it is seen that in all cases, an increase in thickness is mandated.

As seen in Figure 9.5, the end closures of submarine pressure hulls can assume a variety of forms.

The purpose of inserts (Figure 9.6) is to eliminate stress concentrations at a junction due to the presence of welds and also because the external hydrostatic pressure loading gives rise to high shear stresses in the cylindrical shell on either side of a bulkhead.

Penetrations of the pressure hull are introduced to provide access hatches and also for the operation of hydraulic, mechanical, and electrical equipment (propellers, thrusters, dive planes and rudders, manipulators, vents). Figure 9.7 shows one of several recommended arrangements that are included in the European Standard BS EN 13445.

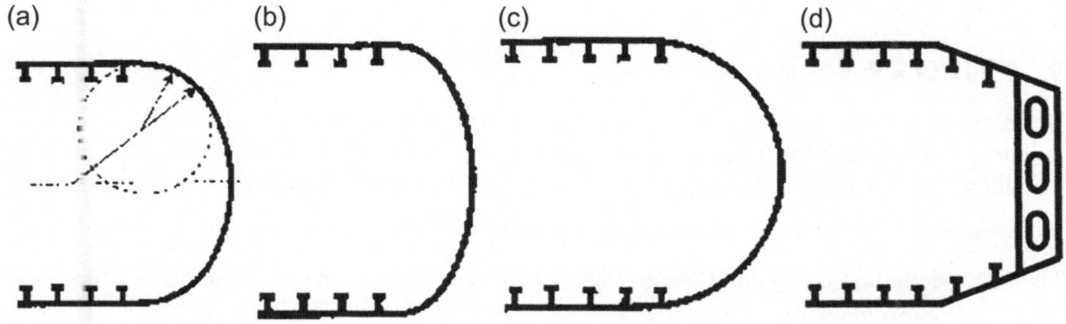

FIGURE 9.5 Usual end closures [22]. (a) Torisphere, (b) ellipse, (c) hemisphere, and (d) flat end.

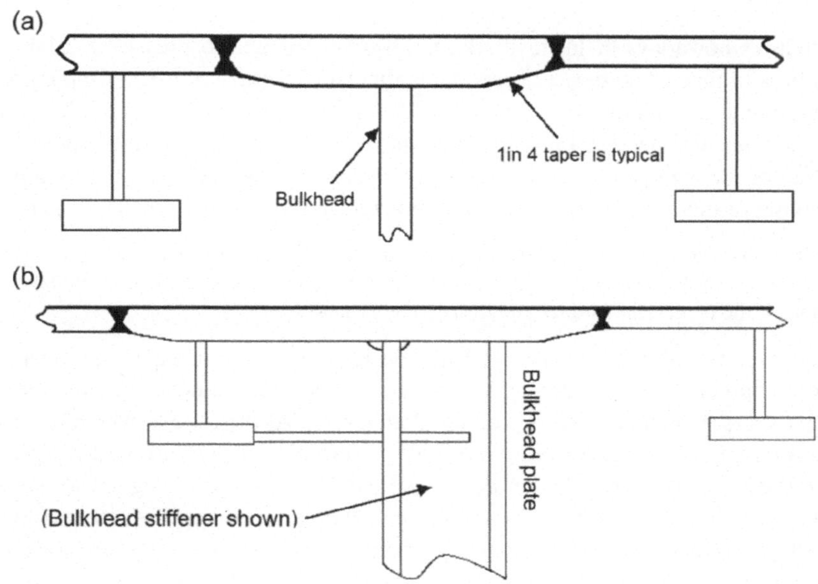

FIGURE 9.6 Pressure hull inserts [22]. (a) Thick insert in way of bulkhead. (b) As (a) but combined with frame.

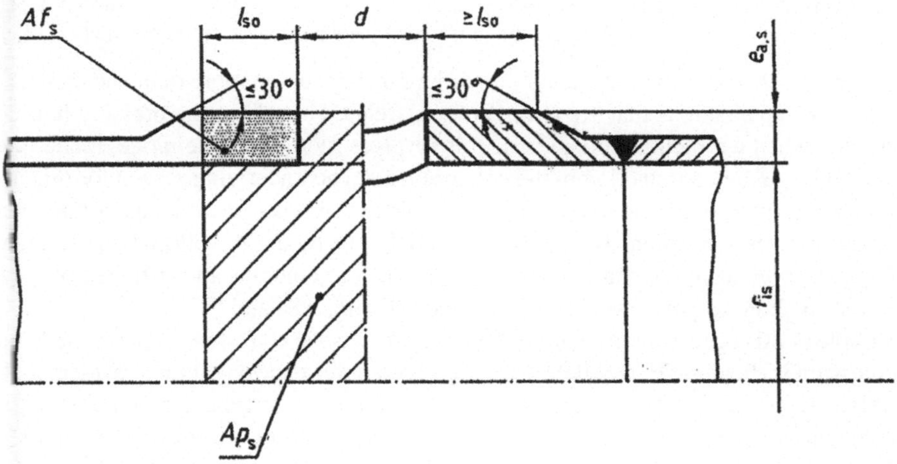

FIGURE 9.7 Arrangement for isolated opening in cylindrical shell with increased shell thickness.

9.3 DESIGN

9.3.1 Design Philosophy and Criteria

Before we discuss the various approaches that have been followed over time in submarine pressure hull design, it is necessary to clarify certain key concepts. In doing so, we may turn to the following definition of *design philosophy* that was proposed in 1962 by a warship and submarine designer [15]:

> Philosophy of design is the broad framework of principles to be used in consummating a design.

The above definition is a general one so that it serves our purpose in general terms. He then proposed the following definition for *design procedure*:

> Design procedure is the detailed summary of steps to be followed including the order in which they are to be taken and the specific mathematical expressions to be used in determining the dimensions of structural members.

Design procedures nowadays include, in addition to mathematical expressions, a large variety of tools from a broad range of numerical procedures that can be initiated using simple programs that are run on hand-held calculators to sophisticated codes. Furthermore, structural design procedures for ships and submarines need to take into account requirements of extraneous organisations that may or may not be incorporated in available procedures. A more specific, but at the same time more general definition of marine structural design that is relevant to present-day needs is given below:

> Marine structural design is the process of selecting form, material and geometrical parameters so that a waterborne vessel fulfils structural loading, commercial or other mission goals in accordance with the requirements of the owner, the builder and the regulatory authorities.

Thinking and practice with regard to pressure hull design have continued to evolve over many years, and in order to appreciate current trends, it is helpful to consider the developments that have taken place in recent years, which have witnessed rapid advances. At the time at which the above quotations were written, a debate concerning the design philosophy of pressure hull design was taking place. One of the key issues was whether the pressure hull ought to be designed so that at collapse, failure would occur in all possible modes or not. At first glance, it appears attractive to decide on an operating depth, select a safety margin, and then ensure that the scantlings are such that for each individual failure mode, the safety margin is adhered to. This would mean that failure would occur simultaneously for all modes, thus ensuring the most efficient use and distribution of material. However, at the time, it was felt that

> … because existing theoretical analyses which have been fully corroborated by experiment do not include effects such as plasticity and imperfection, its adoption for the present was discouraged. The alternate concept of "forcing" the failure to be in the shell yield mode was recommended… [15].

Meanwhile, research was being conducted into failure that occurs in more than one mode, although this was done for applications that are not in any way related to submarine pressure hull design. It was found that when a structure collapses, if this involves more than one mode, mode interaction takes place and results in a reduction in the collapse load. The underlying reason for this is that the structure becomes very imperfection-sensitive. We see that when simultaneous failure occurs, the collapse load is lower than when collapse occurs in only one mode. A well-known example of *failure mode interaction* occurs in compressed members such as columns and stiffened plates that can fail in an overall (Euler type) as well as a local mode (Figure 9.8a and b).

An example of this occurring in a ring-stiffened cylindrical shell test model subjected to uniform external pressure is shown in Figure 9.8c. It appears then in retrospect that the prevalent feeling was correct although for reasons that were not fully understood at the time, since it was not appreciated that concurrent failure modes lead to a loss of strength. In the case of the example given in Figure 9.8c, which concerns tests performed at ARE Dunfermline for the UK Navy, the loss of strength amounted to 13%, which is by no means a negligible amount.

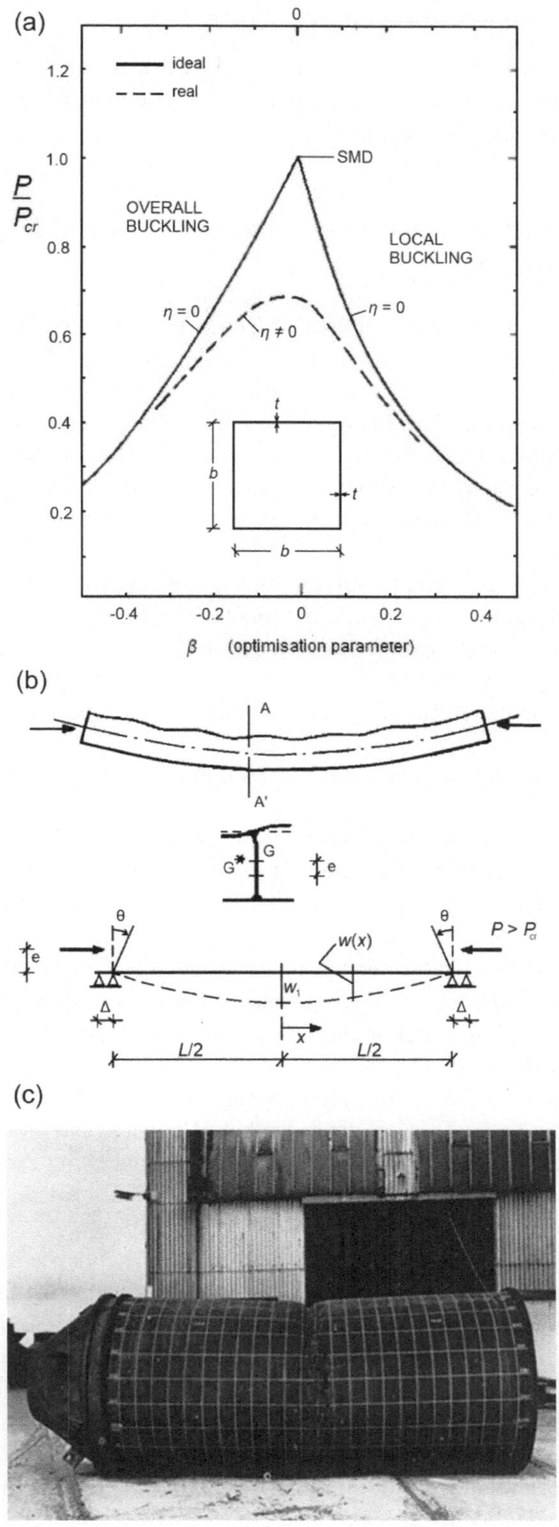

FIGURE 9.8 Failure mode interaction. (a) Interactive buckling in a box column [24]. (b) FMI in a compressed column section. (c) Failure mode interaction (interframe failure and overall collapse) in a submarine pressure hull test model (UK Navy) [13].

9.3.2 Design of Ring-Stiffened Cylindrical Shell

It therefore appears necessary for the designer to select a critical failure mode that will occur at a stress that is lower than that of the other modes. In the case of submarine pressure hulls, the possible failure modes of the stiffened cylindrical shell are:

- Overall buckling
- Interframe collapse (elastic buckling or yielding)
- Ring frame tripping

In selecting the critical failure mode, a number of factors that may be other than purely structural have to be considered, such as (a) the consequences of failure, (b) the effort involved in determining the failure mode with sufficient accuracy, and (c) the cost, in material terms, of avoiding failure in each case. The prime mode of collapse is then that which has the greatest influence on structural weight and thereby has a bearing on the cost of the vessel. Once this is identified, it is allocated an adequate but not excessive margin of safety, whereas other modes of collapse are allocated higher safety margins against elastic buckling (Kendrick). In this way, it is ensured that failure for each mode occurs at different loads.

In the case of a stiffened cylindrical shell, the interframe collapse load can be calculated with a high degree of accuracy, whereas for overall buckling and ring frame collapse, much more effort is required. Furthermore, the weight gain from an accurate determination of the frame tripping load would be significantly less than that of the shell plating. For these reasons, the approach that has been followed and implemented in British and subsequently in European pressure vessel codes is to:

a. Assume a cylindrical shell imperfection of 0.5%R (maximum permitted out-of-circularity) in the critical buckling mode
b. Select interframe buckling as the primary failure mode
c. Set an appropriate safety factor for failure of the unsupported shell
d. Set much higher safety factors against failure for ring frame tripping and overall buckling

By proceeding in this way, the collapse sequence can be predicted; in actual fact, this will be a type of *progressive collapse*, commencing with elasto-plastic buckling of the cylindrical shell. This will be followed by ring frame failure which will precipitate overall buckling of the stiffened cylindrical shell between transverse bulkheads.

By selecting a particular safety factor, the expected *collapse depth* becomes correspondingly greater than the *operating depth*. This margin has to be sufficient to cater for the emergency circumstances cited earlier (accidental or deliberate overdiving), without being excessive. A severe weight penalty on the vessel, which as discussed in Section 9.1 is weight-critical, is thereby avoided. We shall return to this question later on.

It was mentioned previously that the calculation of ring frame collapse requires much effort. For this reason, it was decided that rather than pursue this, to set a larger safety factor against ring frame collapse; this results in increased scantlings of the stiffener cross section. In this way, ring frame collapse is avoided and does not have to be calculated accurately since it is preceded by interframe buckling. An important benefit that stems from the increased strength of the ring frames is the additional support to the ring-stiffened cylindrical shell which translates into an increase of the overall buckling strength.

This then is the approach that is followed in the design of the pressure hull of naval submarines that may be termed *classical* or *traditional*. The seeds for this approach were sown in the early 1960s, and the method reached its fruition with the work of Kendrick and his co-workers in the United Kingdom during the decade that followed [17], [19]. This work led to the drafting of the European pressure vessel code that is in use today.

Subsequent to and concurrently with these developments, a number of other avenues have been explored.

- Firstly, numerical procedures have and are being used, and it has been shown that it is possible to reduce the weight to displacement ratio or conversely, to increase the operating depth for a given W/Δ ratio by using rational procedures. The use of numerical procedures has continued to evolve.
- Optimisation of designs based on the classical approach has been performed.
- Tests have continued and failure mode interaction in ring-stiffened pressure hulls has been studied.
- Reliability-based procedures including optimised design have been used in a number of instances. Recently, more elaborate approaches have been explored involving more the use of more than one safety factor (*partial safety factor approach*).

We shall consider the use of numerical methods in Section 9.4, but we need to return to the previous discussion.

9.3.2.1 Design for Interframe Collapse

The next step then is to determine interframe collapse. In Chapter 4, we saw that yielding failure occurs first in the cylindrical shell in the circumferential direction midway between ring frames. Furthermore, we saw in Chapter 3 that elastic buckling of an unstiffened cylinder occurs at a pressure that is calculated using the von Mises equation,[2] which shows that both the elastic buckling pressure and the buckling mode are dependent on L_f/R. If the length reduces, the number of circumferential lobes and the elastic buckling pressure p_m will both increase. In all cases, however, elastic buckling occurs at a pressure that is lower than that necessary to cause yielding in the circumferential direction, p_{c5}. The situation is analogous to the behaviour of flat plates in compression in which the elasto-plastic buckling load is always less than the squash load. Ring frames effectively reduce the length so that although they are not as rigid as transverse bulkheads, the von Mises relation is equally applicable since from the point of view of the shell plating, the frame spacing can be treated as the distance between rigid end supports.

A large number of test data on cylindrical shell buckling have been collected and plotted as shown in Figure 9.9. The geometries of these test data lie in the range $5.9 < R/t < 250$ and $0.04 < L/R < 50$, and the out-of-circularity does not exceed 0.05%R. It has to be noted that such a level of out-of-circularity (OOC) is enough to reduce the buckling strength by up to 30%. In order to reduce the scatter, the data have been non-dimensionalised with respect to the shell yielding pressure for each individual case, p_{c5}. In the chart,

p_c = experimental collapse pressure
p_{c5} = pressure at which the maximum circumferential mean stress reaches the yield stress as per (6.7)
p_r = calculated lower bound collapse pressure
p_y = pressure at which mean circumferential stress in a cylindrical or conical shell midway between stiffeners, or in a spherical shell, reaches yield point
p_m = lowest elastic (von Mises) interframe buckling pressure

We note that for "long" cylinders, ($p_m/p_{c5} < 1$), the collapse pressure is significantly lower than the yield pressure, and in this region of the diagram, elastic buckling effects dominate. For intermediate values of p_m/p_{c5}, collapse is affected by elasto-plastic buckling, whereas for large values of p_m/p_{c5} (>2.5), yielding dominates and the strength approaches the theoretical maximum. Practical geometries lie in the intermediate range ($p_m/p_{c5} > 1.5$) (Figure 9.9).

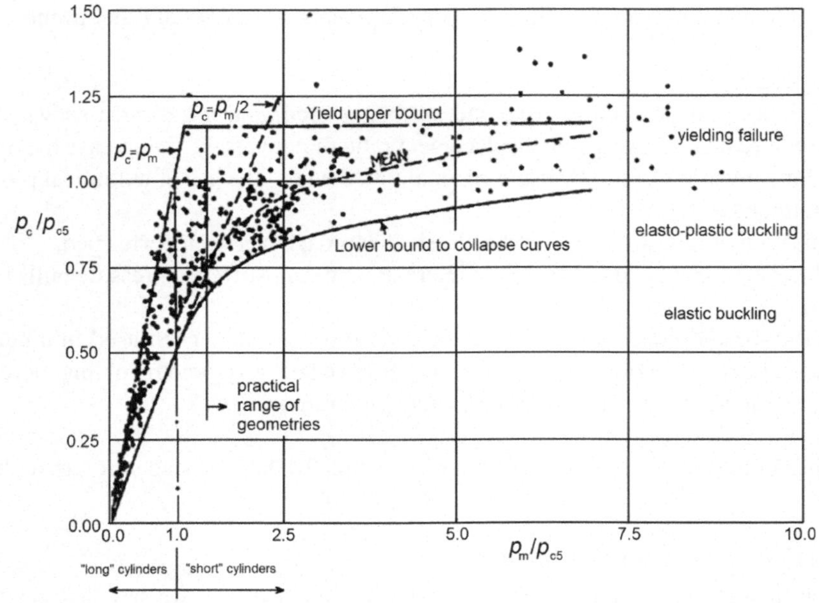

FIGURE 9.9 Test data for interframe collapse of cylindrical shell [9].

From the data that has been plotted, mean and lower bound curves can be plotted. The lower bound curve can be treated as a *design curve* that is defined by the following relationship:

$$Fp_d \leq p_{ck} \tag{9.3}$$

where p_{ck} is the collapse pressure that is defined by the (characteristic) mean curve, p_d is the *design pressure,* and F is a *safety factor.* In deciding on an appropriate safety factor, it is important to bear in mind sources of uncertainty. These may be objective or subjective. Objective uncertainties are operation-dependent and are therefore related to external loading, which is the actual depth experienced. Other objective uncertainties are related to structural response, such as material properties and fabrication imperfections. Subjective uncertainties are due to the approximattions that are inherent in structural calculations (model uncertainty). It is also necessary to bear in mind in-service factors such as deterioration of the condition of the structure with time (corrosion, fatigue) as well as wartime operational loads such as underwater explosions. The designer then has to decide how these uncertainties ought to be allowed for. In the latest European standard, the design curve coincides with the lower bound to the test data (Figure 9.10). The scatter of the results is seen to vary for different geometries, and for $p_m = p_{c5}$, it is seen to be maximum. This is understandable given the increased interaction between elastic buckling and yielding. It has been found that initial imperfections, these being welding residual stresses and out-of-circularity, do not affect interframe collapse. The test data include imperfections that are representative of full-scale data, and thus, no further treatment is necessary. If anything, weld size is comparatively larger in the models so that the effect of scaling would result in more acute effects than in a real structure. There is therefore an inherent degree of pessimism in the design curve with regard to fabrication-related imperfections. One important attribute of the design curve as proposed is that it can be used for non-cylindrical sections as well, such as cones and dome ends. The justification for this lies in the form and solution of the differential equations that describe truncated conical sections.

The question of a suitable design curve is addressed once again in Section 9.5.

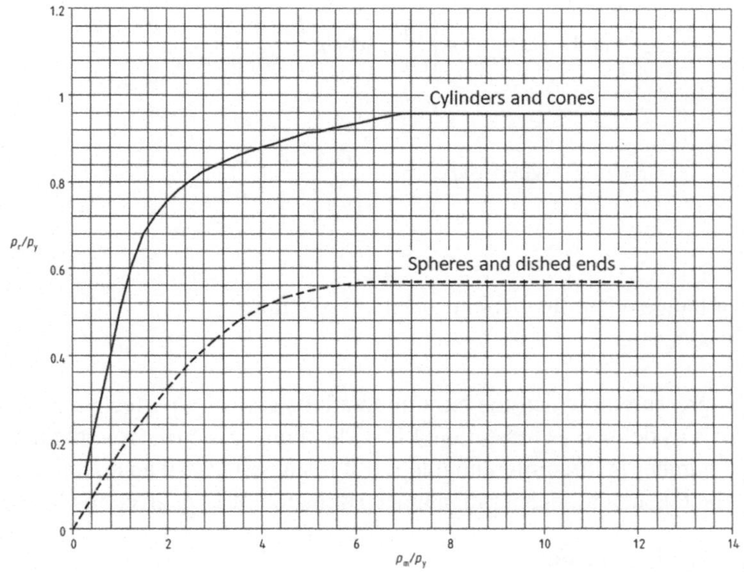

FIGURE 9.10 BS EN 13445-3:2009 design curves for cylindrical and spherical shells [8].

9.3.2.2 Design for Overall Collapse (General Instability)

Overall collapse may occur in the case of widely spaced bulkheads and weak ring frames. Weak frames may fail by yielding or lateral tripping, thereby ceasing to provide support to the cylindrical shell which then collapses between consecutive bulkheads in a two-noded form. In contrast to inter-frame collapse, initial imperfections play an important role and have to be accounted for. Due to the lack of relevant test data, it is necessary either to perform calculations using a numerical procedure or to introduce an appropriate safety factor. The approach that has been followed in European standards has been to require that:

a. The maximum out-of-circularity occurs in the worst possible mode and is distributed as a half-sine wave between consecutive transverse bulkheads
b. The maximum stress in the ring frame, σ_{tf}, is less than the yield stress, $(\sigma_f)_{max} < \sigma_0$, at a pressure that exceeds the design pressure and which is given by $p = Fp_d$ where

$F = 1.8$ for hot-formed or fabricated ring frames
$F = 2.0$ for cold bent frames

The difference in safety factors is due to the presence of residual stresses in cold-bent frames; these reduce strength at the onset of yield (Section 6.7). Standards that use a partial safety factor approach are less conservative. The frame cross section is specified to be such that the ring frame collapse pressure is much closer to the interframe collapse pressure, and a safety factor is introduced to allow for failure mode interaction.

The frame is loaded in compression as a result of the external hydrostatic pressure and in bending as a result of out-of-circularity. The maximum stress in the frame, σ_{tf}, will arise in the extreme flange fibres. It can then be calculated in a manner analogous to that of beams with initial distortions that are loaded in end compression, by adding compressive and bending stresses σ_c, σ_b:

$$\left(\sigma_f\right)_{max} = \sigma_c + \sigma_b \leq \sigma_0 \tag{9.4}$$

$$\text{and} \quad (\sigma_f)_{max} = \left(\frac{p}{p_{of}}\right)\sigma_0 + \frac{Ez_f\delta_n(n^2-1)p}{R^2(p_n-p)} \tag{9.5}$$

$$\text{where} \quad p_{of} = \frac{\sigma_0 t}{1-\left(\frac{\nu}{2}\right)-\gamma}\left[\frac{R_f}{R^2}\right] \tag{9.6}$$

z_f = distance of flange outer fibre from neutral axis of frame
$\delta_n \le 0.005R$

In (9.6), p_{of} is the pressure at which the circumferential stress of a perfect ring frame reaches the yield stress, and γ is as defined as in Section 4.1.5. The flange distance from the neutral axis varies with position and number of circumferential lobes. It is suggested that an effective width $L_e = 0.75L$ or $L_e = 1.56\sqrt{(Rt)}$ be taken, although a rational determination of L_e is given in Chapter 4 (4.43).

9.3.2.3 Design of Ring Frames

Equation (9.4) ensures that the ring frame will be of sufficient strength when considered in its entirety. However, it is necessary to check the strength of each component of the frame section to ensure local stability against buckling. This may take the form of lateral-torsional buckling of the entire ring frame cross-section or buckling of individual component members. In order to avoid lateral-torsional buckling (sideways tripping), stiffeners have a symmetrical cross section, and in proposing a failure load, it has been found sufficient to assume that there is no rotational restraint along the line of attachment to the cylindrical shell (pinned). Based on this assumption, the design condition is then

$$\sigma_t \ge \sigma_0 \tag{9.7}$$

where[3]

$$\sigma_t = \frac{EI_f}{A_s R z_1} \tag{9.8}$$

The assumption of a pinned condition is conservative hence design condition (9.7). To increase σ_t, I_f is increased, and this is most effectively done by increasing the flange width.

Buckling of section members is avoided by implementing the condition

$$\sigma_{cr} \ge 1.5\sigma_0 \tag{9.9}$$

where σ_{cr} is the elastic buckling stress of the web and of the flange. Condition (9.9) is satisfied by proportioning the section so that:

$$\frac{d_w}{t_w} \le 20 \quad \text{for webs of flanged stiffeners} \tag{9.10a}$$

$$\frac{d_w}{t_w} \le 16 \quad \text{for webs of flat} - \text{bar stiffeners} \tag{9.10b}$$

$$\frac{d_w}{t_w} \le 10 \quad \text{for webs of flat bar stiffeners free to rotate} \tag{9.10c}$$

$$\frac{b_f}{t_f} \leq 5 \quad \text{for flanges} \tag{9.10d}$$

assuming $\sigma_0 \leq 414$ MPa. For $\sigma_0 > 414$ MPa, these proportions are reduced by $\sqrt{414/\sigma_0}$.

9.3.2.4 Preliminary Design of Ring-Stiffened Cylindrical Shell

The iterative procedure that is described in this section was proposed in [9] and is considered suitable for a preliminary selection of the structural layout and the scantlings of the cylindrical shell and ring frames. At a later stage, this can be performed using more accurate procedures such as finite element software. The notation used is included in Figure 9.11.

9.3.2.4.1 Determination of Shell Thickness, t, in Compliance with (6.3)

$$Fp_d \leq p_{ck} \tag{9.11}$$

where:

p_d = design pressure (Figure 9.9)
p_{ck} = collapse pressure (lower bound of test data shown in Figure 9.9)
F = safety factor (=1.50)

In (9.11), the proposed safety factor F that has been used in the British Standard BS 5500 code is equal to 1.5 with respect to the lower bound of the collapse test data included in Figure 9.9. This safety factor can be used provided the out-of-circularity does not exceed 0.005R. Other typical values of safety factors are 2.5 for interframe elastic bifurcation buckling and 3.5 for overall elastic bifurcation buckling.

For cold-formed ring frames (that are characterised by corresponding residual stresses), the maximum permissible stress that arises at the design pressure has to be less than or equal to one half the yield stress of the material, so that the safety factor is equal to 2.0.

In the case of annealed frames, the maximum permissible stress may be equal to or less than the yield stress when the pressure is 1.8 times the design pressure.

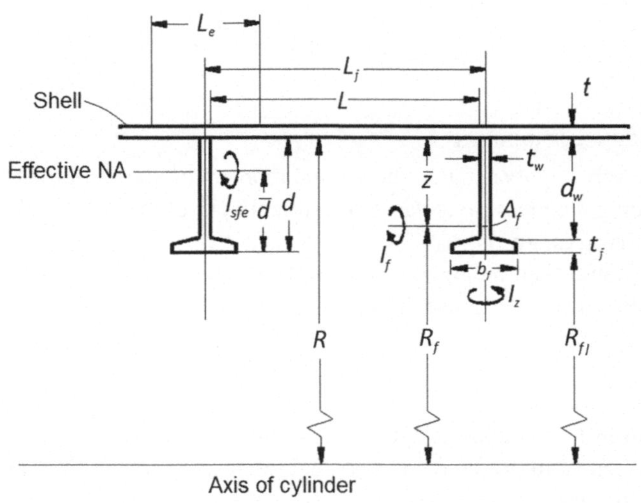

FIGURE 9.11 Geometry of ring-stiffened cylinder.

These safety factors allow for the interaction of local yield failure with overall pressure hull collapse and also the reduction in strength due to residual stresses.

If in Figure 9.9 $2.0 < p_m/p_{c5} < 3.0$, it follows that $p_c/p_{c5} \cong 0.9$. Hence,

$$p_{c5} \cong 1.25 p_c \tag{9.12}$$

where p_{c5}[4] is the pressure to cause yield in the shell plating at midbay given by (3.2). The following relations also hold:

$$\gamma = \frac{A(1-v/2)}{\left(A + bh + 2Nh/\alpha\right)} \qquad \alpha = 1.285/\sqrt{Rt}$$

If we assume that in

$$p_{c5} = \left[\frac{\sigma_o t}{R}\right]\frac{1}{1+\gamma G}$$

the second fraction is equal to 1.2, we obtain

$$p_{c5} = 1.2 p_0 \tag{9.13}$$

$$\text{Hence,} \quad p_{c5} \cong 1.25 p_c \tag{9.14}$$

Solving for the shell thickness, t, we obtain:

$$t = 1.5R\left(\frac{p_d}{\sigma_0}\right) \tag{9.15}$$

This relationship can be used to obtain a first estimate for the shell thickness.

9.3.2.4.2 Frame Spacing

Making use of the effective width of shell plating in bending, L_e, the frame spacing, L_f, is given by:

$$L_f = 1.85\sqrt{Rt} \tag{9.16}$$

9.3.2.4.3 Ring-Stiffener Scantlings

We select the frame web height such that the second moment of area I_c of the frame corresponds to a pressure p_n (this being the pressure to cause overall collapse of the stiffened cylindrical shell with n circumferential halfwaves, given by (5.29).

The pressure p_n should then be equal to $4.5p_d$. If only the first (Bresse) term of (9.10) is used, $p_n = 3.6p_d$. In the latter case, we can use the following expression:

$$I_{sfe} \cong p_d R^3 L_f / E \tag{9.17}$$

This expression is used to calculate the frame scantlings. The ratio of the web height, d_w, to its thickness, t_w, is obtained from (9.10a–d). The second moment of area of the frame together with the effective width of shell plating, I_{sfe}, is given by:

$$I_{sfe} = 0.05d^4 \tag{9.18}$$

The stiffener depth, d, is then calculated from:

$$d \approx 2.02\sqrt[4]{I_{sfe}} \tag{9.19}$$

The flange thickness, t_f is calculated from:

$$t_f = 1.5t_w \tag{9.20}$$

The flange breadth, b_f, using (9.10d), is given by:

$$b_f = \left(1 + (2)(5t_f)\right) = \left(1 + 15t_w\right)$$

Alternatively,

$$b_f \leq 0.7d$$

For steels for which $\sigma_0 > 414$ MPa, the length of the members can be reduced by $\sqrt{414/\sigma_0}$.

9.3.2.4.4 Verification of Compliance with Design Criteria

Having obtained a first estimate of the section scantlings, it is necessary to check compliance with the following design criteria:

I. $Fp_d \leq p_{ck}$ (9.21a)

II. $\sigma_f + \sigma_{fbn} \leq \sigma_0$ for $1.8p_d$ or $2.0p_d$ (9.21b)

III. $\sigma_t > \sigma_0$ (9.21c)

If these criteria are not complied with, the preceding calculations need to be repeated.

Example 9.1

A submarine whose pressure hull has a diameter of 8 m and a compartment length of 12 m is to be constructed of high-tensile steel with a minimum 0.2% proof stress of 540 MPa. The safe operating depth is 300 m, and the maximum out-of-circularity of the pressure hull is 0.5%R. The T-bar ring frames are fabricated using built-up sections (no annealing).

1. Propose suitable scantlings for the ring frames given that $b_f < 0.7d_w$.
2. Propose values for the shell plating thickness, the frame spacing, and the web height.

Plate thickness is available in increments of 2 mm.

Solution

For a yield stress of 540 MPa, the lengths of the members will be reduced by $\sqrt{414/540} = 0.876$.
Hence $d_w \leq (0.876)20t_w = 17.5t_w$
The stiffener depth is then:

$$d = d_w + t_f = (17.5 + 1.5)h_w = 19t_w$$

The flange breadth, b_f, is then:

$$b_f \leq \left(1 + (2)(5)(0.876)(1.5)\right)t_w \leq 14.14t_w$$

The proposed value $b_f = 0.7$, $d = 13.3t_w$ is thus acceptable.

From the above results, the following relationships are obtained:

$$\bar{d} = 0.727d \quad I_f = 0.0099d^4$$

$$A_f = 0.104d^2 \quad I_z = 0.0023d^4$$

where I_z is the moment of inertia of the frame about its toe. To calculate the geometric properties of the section that includes the effective width of the shell plating, L_e, we employ the following relationships:

$$I_{sfe} = I_f + A_f(\bar{z} + t/2)/(1+\gamma_e) + L_e t^3/12 \tag{9.22a}$$

$$\bar{d} = d + (\gamma_e \bar{z} + t/2)(1+\gamma_e) \tag{9.22b}$$

$$\gamma_e = A_s / L_e t \tag{9.22c}$$

Instead of using (9.22a), we can use expression (9.18).

Preliminary selection of scantlings

From expression (9.15), we have:

$$t = 33.5 \cong 34 \text{ mm}$$

The frame spacing is given by (9.16). Substituting we obtain $L_f = 682$ mm

The second moment of area, I_{sfe}, is given by (9.17). Substituting we obtain:

$$I_{sfe} = 6.36 \times 10^8 \text{ mm}^4$$

The stiffener depth d is given by (9.19) and is found to be:

$$d = \sqrt[4]{6.36/0.05} = 336 \text{ mm}$$

Summarising we have:

$$d = 247 \text{ mm} \quad \bar{z} = 252 \text{ mm} \quad A_f = 12.558 \text{ mm}^2$$

$$L_e = 0.75 \text{ mm} \quad L_f = 511.5 \text{ mm} \quad I_{sfe} = (6.67)(10^8) \text{ mm}$$

$$I_z = (0.349)(10^8) \text{ mm}^4 \quad L = 682 - 18 = 664 \text{ mm}$$

$$I_f = (1.385)(10^8) \text{ mm}^4$$

Shell interframe strength

$$\alpha L = 2.314 \quad G = 0.758 \quad N = 1.007$$

$$p_y = 6.26 \text{ N/mm}^2 \quad p_m = 11.80 \text{ N/mm}^2 \quad p_{ck} = 4.60 \text{ N/mm}^2$$

The safety factor against interframe collapse $F = 4.60/3.015 = 1.53$ (>1.50) is thus found to be acceptable (just).

Frame bending

If $p = 1.8$, $p_d = 5.427$ N/mm², then $p_{yf} = 8.42$ N/mm², and the circumferential compressive stress in the ring frame is then equal to 348 N/mm², whereas the out-of-circularity $\delta_n = 20$ mm. The stresses that correspond to $n = 2, 3, 4$ are included below:

n	p_n	σ_{bf}	σ_{tf}
2	32.4	39	387
3	27.7	125	473
4	47.9	123	471

We note that the critical mode is $n = 3$ in which case the maximum stress is equal to 473 N/mm² that is much less than the yield stress. We could consider the use of cold-formed sections ($p = 2.0p_d$ and $\sigma_{tf} = 529$ N/mm²).

Frame tripping

$$\sigma_t = 572 \text{ N/mm}^2$$

Since this value is higher than the yield stress, it is acceptable (design criterion III). The average section thickness is then:

$$t_{av} = t + \frac{A_s}{L_s} = 52.41 \text{ mm}$$

One important reason for which the safety factor is chosen to be greater than unity is that the possibility of exceedance of the operating depth arises during operations. An additional source of uncertainty arises in the exact determination of the collapse strength of the section.

We note that, from the point of view of strength, exterior frames are preferred although there are other questions with regard to fabrication that need to be addressed. The strength of the cylindrical pressure hull is aided by the presence of transverse bulkheads whose presence results in a reduction in the unsupported length of the (long) pressure hull structure that may suffer overall buckling since imperfections that result from the fabrication procedure are ever-present.

9.4 USE OF NUMERICAL PROCEDURES IN PRESSURE HULL DESIGN

9.4.1 Submarine Design Formulas and Numerical Methods

The design of submarine pressure hulls based on the combined use of analytical formulas and laboratory test data has been described in the previous sections. That approach is based on designing for failure in modes that are well-understood and accurately predicted. For more complex failure modes (ring-stiffener failure and cylindrical shell overall collapse), larger safety factors are applied. This approach is fast and economical but has a number of shortcomings [20]:

1. It is overly conservative because failure is based on first yield, and the worst case is considered when allowing for OOC and residual stresses.
2. It cannot allow for variations in structural arrangements that include internal structure, as well as in loading and extent of damage.
3. It cannot lead to improved structural arrangements when needed, such as in the case of penetrations, because it is not a tool that can be used at the initial, synthesis design stage.
4. Reliability analyses cannot be carried out with confidence because extensive parametric studies have been carried out only for interframe collapse. There is insufficient data for other modes of failure.

TABLE 9.2
Comparison of Traditional Design Methods and NLFEA [20]

Area of Comparison	Traditional Methods	NLFEA
Modelling and computational costs	Cheap	Expensive
Requirement for expertise in analysis method	Low/medium	Medium/high
Variability of results for different analysts	Low	Analyst dependent
Modelling of structure and geometric imperfections	Conservative	Realistic
Failure criteria	First yield	Limit states

Nonlinear finite element procedures can address items 1 and 2 above. This is possible by making use of limit state design, allowing for structural complexity (OOC, overall modes, internal structure) as well as the effects of corrosion and in-service damage (dents). Analytical design methods are compared with NLFEA in Table 9.2.

The use of numerical procedures in submarine pressure hull design is thus seen to be desirable, but it is necessary beforehand to have accurate quantitative estimates of the differences between the numerical predictions of strength and the actual strength. The established way to obtain such data is to conduct purpose-designed tests that will thus act as benchmarks for numerical analysis procedures. To this end, the British and the Canadian navies have reported the results from such tests as well as comparisons with numerical results. Such comparisons make it possible to determine the value of a partial safety factor that allows for the uncertainty introduced when using numerical analysis to predict collapse strength. The partial safety factor that is thus determined can be introduced in a design procedure that includes numerical analysis software.

It has been found that NLFEA can provide accurate predictions of collapse strength of pressure hulls if realistic imperfections (OOC, residual stresses) are modelled with sufficient precision. The accuracy of numerical procedures is thus dependent on the quality of the modelling. Benchmark studies can thus provide guidance on the required standards of modelling as well. These include material response as well as the imperfections inherent in the structure (OOC, overall mode, residual stresses).

9.4.2 Determination of Partial Safety Factor for Uncertainty in Predictive Model

In order to determine the partial safety factor that accounts for uncertainty in a predictive model, a survey of published results was carried out [21]. The uncertainty inherent in these results, expressed in terms of bias and coefficient of variation (COV), is included in Table 9.3.

In Table 9.3, the sample size is the total number of comparisons presented in the corresponding source publications. The results from references [11, 12, 16] have been considered separately in the last row of the table because they allow for the experimental error more precisely, leading to a bias near to unity and a reduced COV. This indicates that if standardised testing and numerical procedures are used, the quality of the results can be improved.

The *bias* is the mean value of the *modelling uncertainty factor* X_m that is defined as the ratio of a single experimental value to the value predicted by a model. It is thus a measure of the systematic modelling error.

The scatter that results from the uncertainties that are related to the true geometric and material properties of the test specimens is expressed in quantitative terms by the *coefficient of variation*, COV, this being the ratio of the standard deviation to the bias.

In addition to the above, it is necessary to establish a degree of confidence, which acts a measure of the level of accuracy of a sample of results. This is expressed in terms of the probability (*confidence interval*) that the results are accurate, so that a statement of the form "the methodology

TABLE 9.3
Experimental-Numerical Comparisons for Pressure Hull Collapse Load

Source	Model Type	Sample Size	Bias	COV (%)
Agharaj et al. [1]	3D	4	0.908	2.2
Boote et al. [4]	3D	8	0.924	8.8
Bushnel [6]	Axisymmetric	71	0.959	5.8
Graham [11, 12]	3D	14	0.979	4.1
Jiang et al. [16]	3D	13	1.009	3.4
Morandi et al. [23]	3D	4	0.986	3.8
Ross and Johns [25]	Axisymmetric	3	0.973	18.8
All of the above		117	0.964	6.3
3D FE Models		43	0.972	6.0
Axisymmetric models		74	0.960	6.5
References [11, 12, 16]		27	0.993	4.0

is accurate to within 5% with a confidence level of 99%" can be issued. In this manner, different samples of results can be compared on the basis of bias and confidence level.

In [12], following a comparison of collapse predictions using NLFEA and large-scale test models, it was concluded that if certain modelling procedures are followed, a partial safety factor of 1.06 can be used. The use of a more detailed material model led to a collapse pressure that exceeded the test value by 8.5%, so that in this case, the partial safety factor is 1.085. This value is based on the maximum percentage discrepancy between numerical and experimental collapse pressures using a standardised approach to numerical modelling; that is, it is based on a lower-bound methodology, and the corresponding confidence level has been found to be 90% [21].

In [21], the corresponding partial safety factor for a 99.5% confidence level is found to be equal to 1.14.

9.4.3 Methods to Obtain a PSF from the Bias and the COV

One method of deriving a PSF from a given sample is to determine the type of distribution of X_m and to use that distribution to establish a lower bound for a future observation based on a specified level of confidence. A heuristic approach to determining the type of distribution of a sample is to plot a histogram of the data. The shape of the histogram is suggestive of the distribution. A histogram of the modelling uncertainty data included in Table 9.3 is included in Figure 9.12, and it appears to be approximated by the normal distribution.

This distribution can then be used to establish a lower bound for a future observation based on a specified level of confidence [21]. In this manner, the statistical distribution of the results is accounted for, in contrast to the lower bound traced out in Figure 9.11 that is a lower bound to all data points, including outliers.

In cases in which the data are limited, a degree of uncertainty arises with regard to the nature of the distribution. A more exact method that can be used in this case is the *probability plot*. This provides a means of determining whether sample data truly conforms to a proposed distribution on the basis of visual examination of the data, as described above.

Probability plots are distribution-specific, so that a hypothesis is made on the form of the distribution and the plot is made on specific axes for that distribution. The modelling uncertainty factor X_m is plotted against the observed cumulative frequency of the data, as shown in Figure 9.13.

The straight line through the data was plotted using the trendline option provided by Microsoft Excel. This confirms that the data conform to the normal probability distribution.

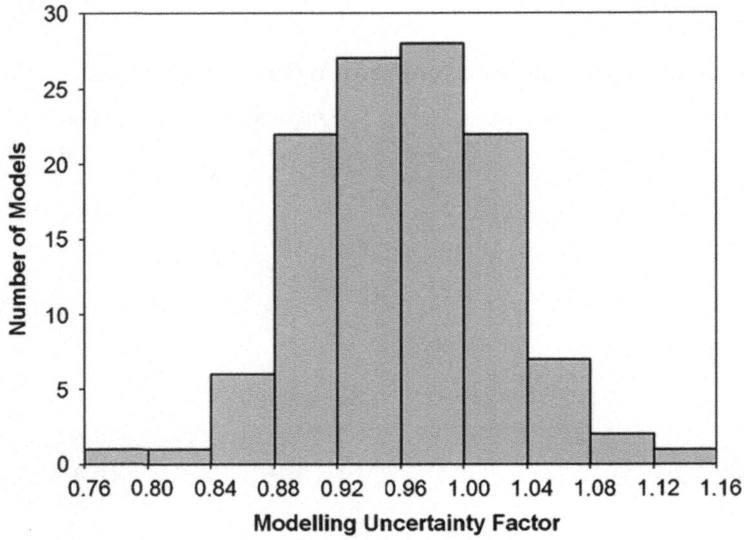

FIGURE 9.12 Histogram of modelling uncertainty factor X_m for data included in Table 9.3 [21].

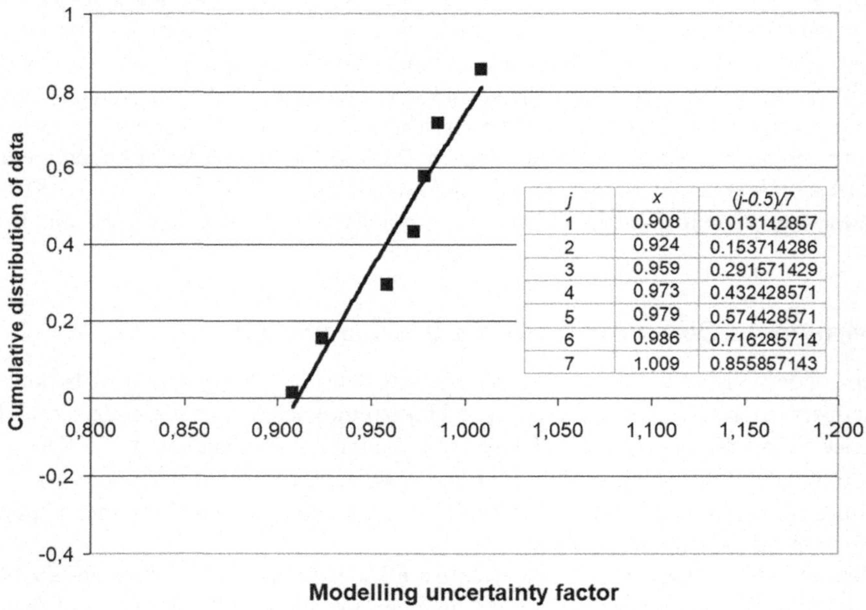

j	x	(j-0.5)/7
1	0.908	0.013142857
2	0.924	0.153714286
3	0.959	0.291571429
4	0.973	0.432428571
5	0.979	0.574428571
6	0.986	0.716285714
7	1.009	0.855857143

FIGURE 9.13 Normal probability plot for data included in Table 9.3.

A normal probability plot can also be constructed by plotting the standardised normal scores z_j against the modelling uncertainty factor, X_m, where:

$$\frac{j-0.5}{n} = P(Z \leq z_j) = \Phi(z_j) \tag{9.23}$$

For example, if $(j - 0.5)/n = 0.716$, $\Phi(z_j) = 0.716$ gives $z_j \approx +0.57$ from the normal probability table.

To increase the robustness of studies of individual structural components such as a ring-stiffener and the attached shell plating, a much more extensive model of the structure ought to be analysed as well. Examples of such models are a single compartment and the entire pressure hull including end closures, the internal structure, and transverse bulkheads. This practice is followed in the structural analysis of surface ships, in which case a number of cargo spaces are modelled or possibly the entire hull girder, in addition to a local detail. This enables an accurate determination of local boundary conditions to be carried out. It also provides a better understanding of interaction effects of different failure modes as well as that of structural components.

It was seen previously that in analytical approaches, the lower bound of the experimental results is used as the design curve (Figure 9.9). If, however, partial safety factors are used to quantify modelling uncertainty, the mean curve has to be used to determine maximum scatter, rather than the lower-bound curve.

Modelling uncertainty can also be calculated using reliability-based methods. This involves specifying the target reliability and performing reliability analyses for typical designs [23].

In order to harness the strengths of the various approaches that are at the disposal of the designer, the concept of a *hierarchical high-fidelity analysis* was introduced by Arbocz and Starnes [2]. This was used to predict the critical buckling load of thin-walled shells loaded in compression, in the aerospace field. In this, a number of levels of fidelity are identified.

In the case of submarine structures, a hierarchical analysis is initiated by calculating the classical elastic buckling stress of the structure assuming simple boundary conditions. At the next level, nonlinear geometric behaviour is introduced, as well as more accurate boundary conditions. Following this material nonlinearity can be introduced, so that the effects of material imperfections (residual stresses) and also failure mode interaction are modelled. The results at all levels considered are compared with experimental results and the assumptions made at each level assessed. At the lower levels, parametric studies may be carried out leading to an efficient design following which nonlinear numerical analysis can be used to determine strength and reliability.

9.5 USE OF STANDARDS AND CLASSIFICATION SOCIETY RULES IN PRESSURE HULL DESIGN

The two factors that have contributed to the wider dissemination of information on the design of naval submarines are firstly that design and construction are being carried out in more countries around the world and secondly that non-naval entities such as classification societies are nowadays involved in these activities (Figure 9.14). There exist many design codes that remain classified, whereas in other cases, they are openly available. The European Standard for externally loaded pressure vessels is based on the UK BS5500 code that is not classified because it is based on results that are accessible in the open literature. In parallel, a number of classification societies have rules for submarine pressure hull design.

In this section, we shall present the results of a statistical analysis of the design curves included in analytical formulations such as Figure 9.9, following which we shall discuss the results obtained from two comparative studies involving classification society rules for pressure hull design.

9.5.1 Accuracy of Design Curves for Interframe Collapse: The UK Standard

The statistical methods used in Section 9.5.2 to assess the quality of results obtained using numerical methods will be used to assess the design curves used in analytical formulations [21]. This study is applied to a modified mean empirical design curve applied to the data points included in Figure 9.15. The difference between this curve and the mean curve included in Figure 9.9 is that the latter is not a true mean, being more conservative in the elastic region. The true mean curve is used in this study for the data shown. The aim of this study is to enable accuracy assessment to be made for future predictions.

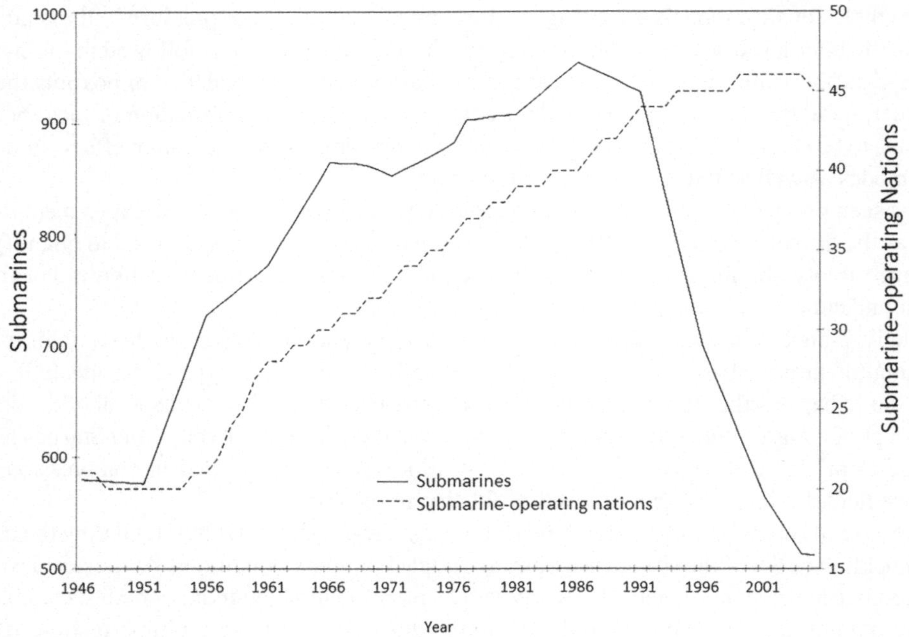

FIGURE 9.14 Submarines and submarine-operating nations [29].

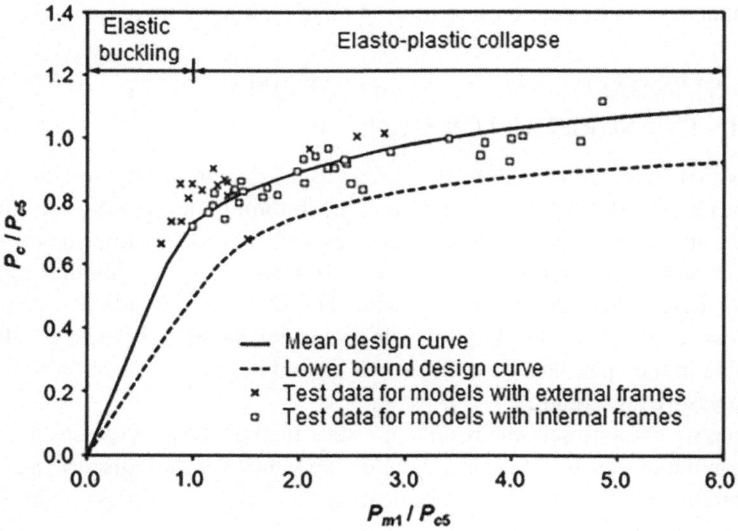

FIGURE 9.15 Experimental results and design curves for interframe collapse.

For every experimental point included in Figure 9.15, a modelling uncertainty factor is calculated for each of the two design curves (mean and lower bound). These are used to calculate the bias and COV for each curve. Results are summarised in Table 9.4.

The mean curve plotted in Figure 9.15 has a bias equal to 1.022, that is, it underpredicts the collapse pressure by 2% on average. The lower-bound curve on the other hand has a bias of 1.305 which indicates that it underpredicts the collapse strength by 31% on average. The scatter related to the lower-bound curve is also much greater, as seen from the relative values of the COV for the two curves (17.2% for the lower-bound curve and 8.52% for the true mean curve).

TABLE 9.4
Accuracy of Design Curves for Interframe Collapse[a] [21]

Method	n	Bias	COV (%)
Lower bound curve	50	1.305	17.2
Mean curve	50	1.022	8.52

[a] n = sample size. The data has been obtained from [22].

Both curves are seen to underpredict the collapse strength, with the lower-bound curve being much more conservative than the mean curve.

In the UK Standard published in 2001, [7], the operating depth is termed the Deep Diving Depth (DDD) and this represents the limit of safe operation. However, the ultimate strength of the pressure hull has to exceed the strength that corresponds to the DDD by a margin that is expressed as a partial safety factor γ_D. This is introduced to allow for errors in depth measurement as well as the possibility of overdiving.

Pressure hull strength is defined by the DDD divided by two partial safety factors, γ_c and γ_i. The first of these is introduced to account for uncertainties in hull fabrication such as misalignment, low-quality welds, and undersized sections, whereas the second allows for errors in the predictive model, both systematic and random. The collapse pressure, p_c, is associated with the collapse depth, DCD, so that

$$\gamma_D \text{DDD} \leq \frac{1}{\gamma_c \gamma_i} \text{DCD} = \frac{1}{\gamma_c \gamma_i} \frac{p_c}{\rho_w g} \tag{9.24}$$

where ρ_w is the density of seawater within the operational area and depth of the submarine.

The procedure that is followed in [7] for pressure hull design, includes analytical, empirical, and numerical methods. The analytical expressions relate to the results included in Chapters 4 and 5, whereas the empirical expressions relate to the design curves for interframe collapse based on test results. The Standard requires the use of a nonlinear finite difference program, [18], to predict the overall collapse pressure by analysing a single ring-stiffener and the attached cylindrical shell plating. To the predicted collapse pressure correction factors are applied to account for finite compartment length, interframe deformation of the pressure hull, and failure mode interaction with interframe collapse. A parametric study of overall collapse pressure for a range of sinusoidal OOC shapes is required, and the lowest value is used for design purposes. This approach allows the DDD to be predicted on ultimate collapse rather than yield initiation and is thus less conservative.

9.5.2 Comparison Study of Classification Society Rules

In this section, we shall present the combined results of two comparison studies reported in [28], [27]. These were carried out to assess the rules of several classification societies (ABS, GL, and DNV). Calculations were also carried out using an analytical approach, the rules of the API, and the finite element program MAESTRO. The results obtained from these methods were compared with a series of tests results made available by the U.S. Navy (NAVSEA) that were obtained for small-scale ring-stiffened cylinders and were thus treated as a benchmark.

The tests results selected are representative of the different types of failure experienced in ring-stiffened cylindrical shells having diameter-to-thickness ratios in the range $112 < D/t < 198$, which is typical of submarine pressure hulls. Four tests cases were selected, two of which had external stiffeners (Figure 9.16a) and two had internal stiffeners (Figure 9.16b).

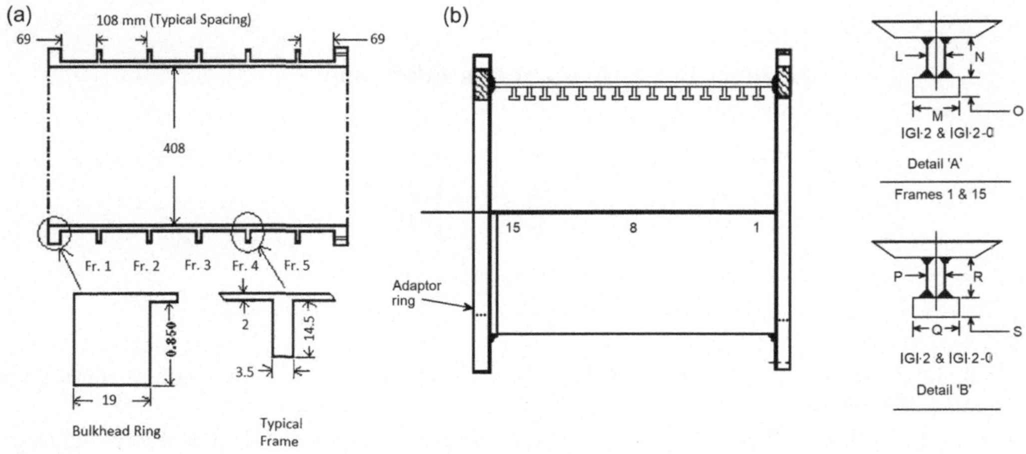

FIGURE 9.16 Test cylinders (dimensions in mm). (a) Cylinder 1d (externally stiffened). (b) Cylinder 1f (internally stiffened).

TABLE 9.5
Comparison of Design Rule Calculations with Test Results (Pressure in N/mm²)

	1.d		1.f		2.a		2.c	
Cylinder	Pressure	Mode	Pressure	Mode	Pressure	Mode	Pressure	Mode
Test	4.40	L	15.28	i/GI	7.39	AX	25.28	i/GI
ABS	4.20	L	14.16	AX	5.86	AX	29.24	AX
GL	4.21	L	20.35	AX	7.15	AX	31.71	AX
DNV	1.14	L	7.56	GI	3.38	GI	17.06	GI
API	3.10	L	12.76	GI	4.93	L	26.28	GI
Anal. Soln.	4.20	L	14.86	AX	6.08	AX	28.33	AX
MAESTRO	3.93	L	13.33	GI	5.53	L	28.93	GI

Note: API, DNV and MAESTRO only address L and GI for ring-stiffened cylinders.
L, Asymmetric buckling; AX, Axisymmetric yielding; GI, General Instability; e/I, Elastic/inelastic.

External hydrostatic pressure was applied including axial line load to simulate load on the end plate. Results are summarised in Table 9.5.

9.5.2.1 American Bureau of Shipping (ABS)

In the 1990 edition of the *ABS Rules for Building and Classing Underwater Vehicles, Systems and Hyperbaric Facilities*, use is made of analytical expressions to predict interframe, ring-stiffener, and overall buckling strength of stiffened cylindrical shells, as discussed in Chapters 4 and 5. A rapid calculation of strength can thus be carried out which, however, underpredicts collapse strength.

For the cases in which failure occurred by interframe yielding (cylinder 2.a) and interframe buckling (cylinder 1.d) in the tests, the calculated pressures were found to be accurately estimated by the ABS Rules. However, for the two cases in which failure occurred by overall buckling (general instability), the Rules predicted failure by axisymmetric yielding. The Rules account for elastic general instability alone, whereas in the test inelastic overall buckling was observed. This arose at a much lower pressure than that at which elastic overall buckling is predicted by the Rules. In the case of cylinder 1.f, the failure pressure due to elastic overall buckling was estimated accurately enough, but in the case of cylinder 2.c, the estimated failure pressure was more than 100% in excess of that measured during the test. The difference in this case can be attributed in part to the in-built OOC circularity of the test cylinder, which is not allowed for in the Rules.

9.5.2.2 Germanischer Lloyd

In the 1988 edition of the Germanischer Lloyd *Rules for Underwater Technology*, all three failure modes are addressed and flowcharts are provided to aid in programming for computer use. The Rules allow for an OOC up to a nominal value of 0.5%*R* by specifying a reduced allowable pressure.

In the case of cylinder 2.a that failed by interframe yielding, the failure pressure was over-estimated by a significant amount, namely 11.8%. This failure mode is that which is most dependent on the yield strength of the material, and therefore, a small difference between the assumed yield strength and the actual yield strength of the test specimen may have contributed to the discrepancy between the estimated and the actual failure pressure. An iterative operation is followed to determine the failure pressure in the Rules, similar to that used in the analytical methodology.

Cylinder 1.f failed by inelastic overall buckling, whereas the GL Rules predicted interframe yield failure at a much higher pressure. This is because inelastic overall buckling was not predicted by the Rules. A comparison of the elastic overall buckling pressure with that of the analytical procedure showed good agreement.

Cylinder 2.c failed by inelastic overall buckling. The GL Rules were used to estimate the elastic overall buckling pressure that, however, exceeded the prediction of the analytical method. This is attributed to the fact that perfect circularity was assumed in the Rules, whereas the test model had a two-wave sinusoidal imperfection.

9.5.2.3 DNV

As in the case of the API code, the DNV rules for the design of stiffened shells are intended for use primarily by the offshore industry. The document used in this comparison study was *Buckling Strength of Shells, Recommended Practice DNV-RP-C202*. In this, buckling stability is based on the limit state design (LSD) or load and resistance factor design format (LRFD). The approach followed is semi-empirical, and an assumed level of imperfections is implicit in the code, so that no allowance needs to be made for initial distortions and residual stresses by the user. The failure modes considered are interframe buckling and overall buckling. The reason for which interframe yielding is not allowed for is because the *R/t* ratios that are encountered in the offshore industry are significantly larger than those in submarine pressure hull design so that this failure mode does not arise.

The procedure described in the DNV document predicted the failure mode correctly in the cases of cylinders 1.d, 1.f, and 2.c, whereas in the case of cylinder 2.a that failed by interframe yielding, it predicted failure by overall buckling. This difference is because interframe yielding is not included as a failure mode for the reason mentioned above. The failure pressures were underpredicted to a significant extent. The latter differences are attributable to the semi-empirical nature of the equations in the Rules.

9.5.2.4 API

The comparison was based on the 2001 edition of the API Bulletin on *Stability Design of Cylindrical Shells* that was drafted primarily for use by the offshore industry. It includes sections on the design of ring-stiffened steel cylindrical shells where the plating is cold or hot-formed.

Strength is evaluated in the case of local shell buckling, overall buckling, ring-stiffener buckling as well as column buckling. It is not, however, evaluated for interframe yielding. The formulations are semi-empirical and are based on classical linear theory that is modified by reduction (knock-down) factors to account for the effects of imperfections, boundary conditions, nonlinear material properties, and residual stresses. The reduction factors are determined from measurements on shells of representative size and initial imperfections. This enables a rapid estimation of strength to be carried out which is nonetheless on the conservative side.

In the case of cylinders 1.d and 1.f, the API procedure predicted the failure mode correctly, but the failure pressure was underpredicted. In the case of cylinder 2.c, the failure pressure was overpredicted although the failure mode (inelastic overall buckling) was correctly predicted. Cylinder 2.a failed by interframe yielding, and the procedure did not predict this correctly because this failure mode is not catered for in the procedure. The differences between the predicted and the actual results are attributed to the fact that firstly the test data used to derive constants used are not necessarily similar to the cases considered in the present study and secondly that the knock-down factors applied do not correspond to the test cylinders that were machined.

9.5.2.5 Analytical Solution

The closed-form analytical solution based on the theory described in Chapters 3–5 correctly predicted the failure mode for two of the four test cylinders and predicted a failure pressure within 4% for those two cases. Both test cylinders that failed by inelastic general instability, were incorrectly predicted to fail by axisymmetric yield. This indicates that the two modes of failure are very close together and that the closed-form solution has difficulty in discerning the two failure modes.

9.5.2.6 MAESTRO

In the version of the MAESTRO™ program used, limit states are implemented in general to determine proximity to failure. In the case of ring-stiffened cylinders subjected to hydrostatic pressure alone, the limit states are replaced by formulas that are used to account for interframe buckling, overall buckling, and column buckling, as in the API code.

The results were therefore expected to be similar to those for the API code. Local shell failure and general instability were accurately predicted in all four of the test cylinders, whereas the failure pressure was calculated to within 15% of the test results. The MAESTRO™ solution was thus found to be more accurate than the design rules because it allows for the use of non-uniform stiffeners and spacing. Additionally, the cylinder can be modelled in whole, including the end plates and boundary conditions. These additional features appear to have contributed to the improved accuracy of the predicted failure pressures when compared to those of the classification society design rules.

NOTES

1 Resistance tests performed at DTMB in 1950 were based on a minimum resistance Series 58 hull form having $L/D = 7$ [10].
2 It is easiest to use the Model Basin formula (3.45), which is accurate for the intermediate range of values of $L/(Rt)^{1/2}$ that are of practical interest.
3 Expression (6.33b).
4 Pressures p_{c1} and p_{c3} correspond to other, non-critical regions of the pressure hull (Wilson).

REFERENCES

1. Aghajari, S., Abedi, K., Showkati, H. (2006). Buckling and post-buckling behavior of thin-walled cylindrical steel shells with varying thickness subjected to uniform external pressure. *Thin-Wall Struct.* Vol. 44, 904–909.
2. Arbocz, J., Starnes, J.H. (2003). "A hierarchical high-fidelity analysis procedure for buckling critical structures. (AIAA 2003-1849)." *Proceeding of 44th AIAA/ASME/ASCE/AHS Structures, Structural Dynamics and Materials Conference*, AIAA, Norfolk, VA.
3. Binns, S. (2008). "Meeting the challenge of designing high capability SSKs." UDT Conference Proceedings: Underwater Defense Technology Conference, San Diego, CA.
4. Boote, D., Mascia, D., Monti, M., Rizzuto, E., Tedeschi, R. (1997). Elastic instability of thin cylindrical shells: numerical and experimental investigation. *Ocean Eng.* Vol. 24, 133–160.
5. Busby, R.F. (1974). "Manned Submersibles." Office of the Oceanographer of the Navy USA.
6. Bushnell, D. (1976). Buckling of elastic–plastic shells of revolution with discrete elastic–plastic ring stiffeners. *Int. J. Solids Struct.* Vol. 12, 51–66.

7. Defence Procurement Agency (2001). *SSP 74 Design of Submarine Structures.* Sea Technology Group, UK.
8. European Committee for Standardisation. (2009). *European Standard EN 13445:3 Unfired fusion welded pressure vessels. Part 3: Design.*
9. Faulkner, D. (1983). *The Collapse Strength and Design of Submarines.* London: *RINA Symposium on Naval Submarines.*
10. Gertler, M. (1950). "Resistance experiments on a systematic series of streamlined bodies of revolution for application to the design of high-speed submarines." DTMB Report C-297, David Taylor Model Basin, Washington, DC.
11. Graham, D. (2007). Predicting the collapse of externally pressurised ring-stiffened cylinders using finite element analysis. *Mar. Struct.* Vol. 20, 202–217.
12. Graham, D. (2008). "A partial safety factor for pressure hull collapse prediction using finite element analysis." *Warship 2008: Naval Submarines 9.* Glasgow: RINA Conf.
13. Graham, D., Keron, I., Mitchell, G., Creswell, D. (1974). DRA structural research in submarines and submersibles. *Mar. Struct.* Vol. 7, 231–256.
14. Griffiths, C.H. (1984). Soviet double-hulled submarines. *Submarine Rev.* Vol. 1984, 45–51.
15. Heller Jr., S.R. (1962). A personal philosophy of structural design of submarine pressure hulls. *Naval Eng. J.* Vol. 1962, 223–236.
16. Jiang, L., MacKay, J.R., Wallace, J., Smith, M.J., Norwood, M., Bosman, T. (2008). "Finite element modeling of collapse experiments of ring stiffened cylinders with simulated corrosion damage." *Warship 2008: Naval Submarines 9.* Glasgow: RINA Conf.
17. Kendrick, S. (1964). "Structural design of submarine pressure vessels." *NCRE Report R483*, Naval Construction Research Establishment, Dunfermline.
18. Kendrick, S. (1979). The influence of shape imperfections and residual stress on the collapse of stiffened cylinders. *Proc. Conf. Signif. Dev. Des. Shape Inst. Mech. Eng.* Vol. 1979, 25–35.
19. Kendrick, S. (1986). "Design of Submarine Structures". *ARE Conference: Advances in Marine Structures*, Dunfermline.
20. Mackay, J.R. (2007). "Structural analysis and design of pressure hulls: the state of the art and future trends." *DRDC Atlantic* TM 2007–188.
21. MacKay, J.R., van Keulen, F., Smith, M.J. (2011). Quantifying the accuracy of numerical collapse predictions for the design of submarine pressure hulls. *Thin-Walled Struct.* Vol. 49, 145–156.
22. McCormick, J. (2009). "An Overview of Submarine Design". In: *Design of Submarine Structures*, ASRANet Online Course.
23. Morandi, A.C., Das, P.K., Faulkner, D. (1996). Finite element analysis and reliability based design of externally pressurized ring stiffened cylinders. *Trans RINA.* Vol. 138, 171–88.
24. Reis, A.J. (1987). Interactive buckling in thin-walled structures. Chap. 7. In: Rhodes, J., Walker, A.C. (Eds). *Developments in Thin-Walled Structures,* Vol. 3, London: Elsevier Applied Science Publishers.
25. Ross, C.T.F. Johns, T. (1998). Plastic axisymmetric collapse of thin-walled circular cylinders and cones under uniform external pressure. *Thin-Wall Struct.* Vol. 3, 35–54.
26. Sutton, H.I. (2018). "*Single versus Double Hull.*" Webpage: *Covert Shores.* Available at: https://www.hisutton.com/Single-versus-Double_Hull.html (Downloaded 2024-02-15).
27. Temme, M.W. (2003). "Comparison of analytical and numerical approaches for determining failure of ring-stiffened cylindrical shells." *M.Sc. Thesis,* Dept Mech. Eng. Mass Inst. Tech., Cambridge, MA.
28. *The Evolution of the Albacore.* Available at: https://www.ussalbacore.org/building-and-changing. (Downloaded 2024-02-15)
29. Tingle, C. (2009). Submarine Accidents. A 60-year statistical assessment. *Profession. Safety.* Vol. 2009, 31–39.

10 Corrosion and Fatigue in Submarine Structures

10.1 INTRODUCTION

In this chapter, we shall consider the two most important sources of deterioration of submarine structures: corrosion and fatigue. In common with all other structures that operate in the marine environment, submarine structures deteriorate as a result of exposure to the environment through the mechanisms of corrosion, fatigue, and corrosion–fatigue.

The first part of the chapter deals with corrosion, whereas the second with fatigue and corrosion–fatigue.

10.2 CORROSION

The term *corrosion* is used to denote the physical and chemical changes to bodies that result from interaction with the environment. According to ISO 8044 (1986),

> Corrosion is the interaction of a metal with its environment which may cause changes to its properties and which frequently may cause a degradation of the performance of the metal, its environment or the system that comprises them.

Depending on the manner in which corrosion occurs, it is classed as *electrochemical corrosion* and *chemical corrosion.*

- Electrochemical corrosion concerns cases of degradation due to chemical processes in the presence of an electric current.
- Chemical corrosion denotes metal degradation that results from the reaction with substances in the absence of a liquid electrolyte.

Electrochemical corrosion is the most commonly encountered type of corrosion in marine structures.

10.2.1 Electrochemical Corrosion

When two strips of different metals, for example, iron and copper, are immersed in a sodium chloride solution and the ends that are not immersed are joined with a wire, the solution converts the energy of the chemical reaction into electrical energy, thus acting as an *electrolyte*. In such a system, the metal strips are called *electrodes*, and their parts that are not immersed are called *poles* (Figure 10.1).

An electric current then flows from one metal, the *cathode,* to the other, the *anode*. During this chemical action, termed *electrolysis*, the metal that forms the anode corrodes.

An electric current then flows from one metal, the *cathode,* to the other, the *anode*. During this chemical action, termed *electrolysis*, the metal that forms the anode corrodes.

When a metal is immersed in an electrolyte such as seawater, the potential difference between the metal and the electrolyte causes the formation of a local *corrosion current* and corrosion begins. The ionisation reaction is:

$$Fe \rightarrow Fe^{2+} + 2e^{-}$$

DOI: 10.1201/9781003489153-10

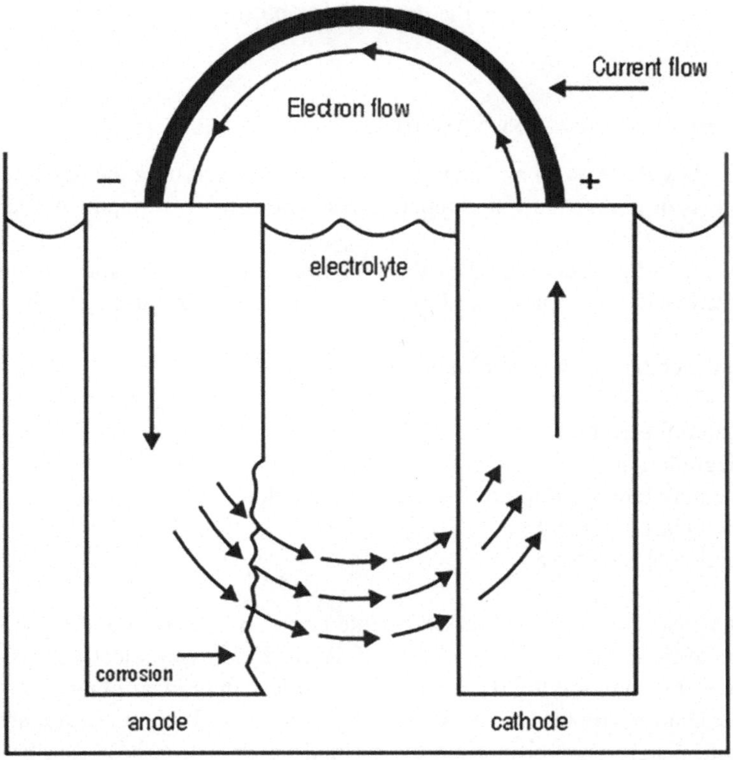

FIGURE 10.1 Electrochemical corrosion [2].

Electrons flow within the metal towards the cathode (electrolyte solution) and give rise to the corrosion current. At the cathode, the corrosion current has two possible ways of acting:

a. $2H^+ + 2e^- \rightarrow H_2\uparrow$
b. $2H_2O + O_2 + 4e^- \rightarrow 4OH^-$

The most important result of these two chemical reactions is the formation of hydroxyl ions (OH^-) that react with the ferrous ions resulting in iron oxides (rust). If the electrolyte has a low oxygen content, *ferrous oxide* Fe_3O_4 is formed (*black rust*). If the oxygen content of the electrolyte is substantial, *ferric oxide*, (*brown rust*), which also contains water molecules, is formed.

The types of corrosion that occur in the marine environment are:

1. Uniform corrosion
2. Pitting corrosion
3. Stress-corrosion cracking
4. Cavitation erosion
5. Microbial corrosion

In the case of submarine structures, the important type of corrosion that occurs is uniform corrosion, which leads to thickness reduction over the entire area that is exposed to the electrolyte. In many marine applications, uniform corrosion is not of concern because of the slow rate of penetration.

However, in the case of pressure hulls that are weight-optimised and safety of life is critical, due attention needs to be exercised.

10.2.2 Corrosion in Submarine Structures

A report published by the Australian Navy provides details concerning the corrosion of submarine structures, the costs of maintenance and repair, the sources of corrosion, as well as proposed remedies [3].

The types of corrosion encountered in the pressure hull, forgings, and equipment were joint electrochemical/crevice corrosion as well as pitting corrosion, examples of which are shown in Figure 10.2.

The corrosion encountered was attributed to:

1. Design/material selection
2. Normal degradation
3. Inadequate maintenance routines and ship's husbandry
4. Inadequate Quality Control
5. A combination of the above

Over 50% of the total corrosion repair and maintenance costs were found to be directly related to design and material selection factors (Table. 10.1). In these cases, electrochemical and crevice corrosion mechanisms featured highly and occurred in less than ideal physical configurations. If therefore suitable changes are introduced at the design stage, involving structural arrangements and material selection, a significant proportion of the total number of failures can be prevented. The benefits accrued concern repair and maintenance cost reduction as well as reduced downtime of the vessel.

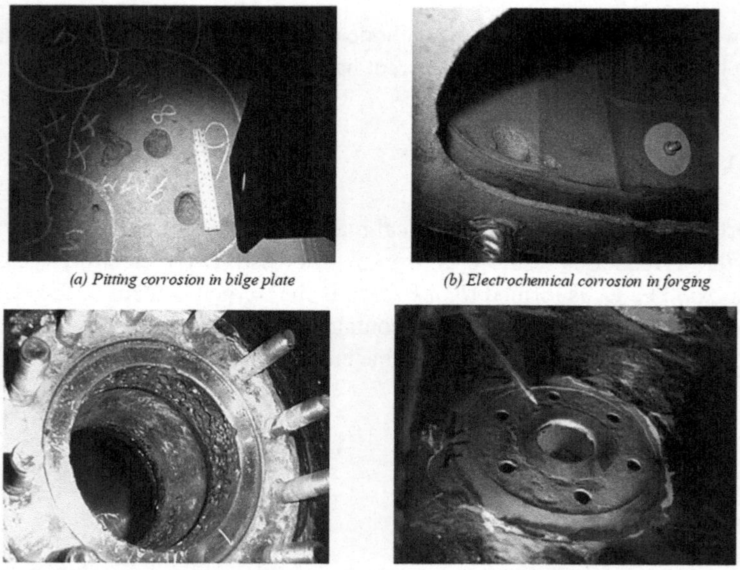

(a) Pitting corrosion in bilge plate *(b) Electrochemical corrosion in forging*

(c) Pitting corrosion in mechanical hull penetration *(d) Electrochemical corrosion in electrical hull penetration*

FIGURE 10.2 Corrosion in submarine structural and mechanical components. (a) Pitting corrosion in bilge plate. (b) Electrochemical corrosion in forging. (c) Pitting corrosion in mechanical hull penetration. (d) Electrochemical corrosion in electrical hull penetration.

TABLE 10.1
Overview of Major Corrosion Issues at Full Cycle Dockings[a]

Corrosion Issue(s)	Structural Integrity or Safety Impact?	Corrosion Mechanisms	Causes of Corrosion
General corrosion in tanks	Yes	• Electrochemical corrosion • Differential aeration	• Coating defects • Sharp edges • Mechanical damage • Previous coating repairs breaking down
Marconi PHPs (3 mm thickness reduction)	Yes	• Electrochemical corrosion • Crevice corrosion	• Detail design not suited for paint adhesion on the hull/stud interface
Bilge plating (10 mm thickness reduction)	Yes	• Electrochemical corrosion • Differential aeration • Coating damage • Suspected biological corrosion	• Design does not factor in damage to coatings during maintenance • Design does not factor in access for inspection and ease of coating repairs • Bilges not cleaned regularly
Flood areas (several mm)	Yes	• Electrochemical corrosion • Differential aeration • Coating detachment	• Complex geometries and sharp edges • Poor choice of protective schemes (SSE's)
Body-end valves (>10 mm)	Yes	• Electrochemical corrosion	• Detail design is not suitable for protection using coatings
Battery compartments	Yes	• Electrochemical corrosion • Pitting corrosion	• Detail design does not allow battery acid to drain freely • Maintenance routines need refinement • Mechanical damage of protective coatings
Hatches	Yes	• Electrochemical corrosion • Mechanical damage to sealing face	• Detail Design uses low corrosion resistant materials • Hatches not greased and protected with covers
Shaft seal	Yes	• Electrochemical corrosion	Under review

[a] A *Full Cycle Docking* takes place every 72 months (6 years).

Ranked in order of importance, from the point of view of repair and maintenance costs, the locations of corrosion were:

- *Electrical Hull Penetrations PHPs*: bolted joint-galvanic and crevice corrosion
- *Bilges*: severe pitting corrosion due to localised coating damage
- *Battery Compartments*: severe pitting corrosion due to localised coating damage
- *Body end Valves*: bolted joint-galvanic and crevice corrosion
- *Seawater Flood Areas*: Stern Tube, SSEs, Torpedo Tubes, Free Flood Spaces
- Hatches
- Tanks
- Shaft seal
- Mechanical Hull Penetration Forgings

Of the above, the ones related to design/material selection factors are listed below:

- Hull Valve Forgings
- Pressure Hull Penetrators (PHPs)
- Bilges
- Battery Compartments
- Body end Valves
- Seawater flood areas
- Stern Tube, SSEs, Torpedo Tubes, Free Flood Space
- Hatches
- Tanks

Inadequate maintenance was found to relate to the following:

- Bilges
- Hatches
- Tanks

A cost–benefit analysis helped prioritise problem areas. The top priorities were found to be:

- Electrical penetrations
- Bilges
- Battery compartments
- Body-end valves

The remedies proposed were:

1. Design changes
2. Material selection changes
3. Use of improved protective coating systems
4. Improved quality control by conducting survey inspections and performing prompt coating repairs where needed. Quality control involves the use of quantitative reporting systems in order to optimise survey inspection intervals, repair methods, priority surveys, and effective cost management
5. Processing of historical data concerning coating and corrosion repairs in order to gain an improved of the current and the future state of submarine structures.

10.2.3 Inspection and Repair of Submarine Structures for Corrosion

The challenge of assessing the strength of the damaged hull can be avoided by repairing the corrosion damage. The first task, however, is finding the damaged areas, a task that is not as simple as in the case of surface ships, because the entire outer hull surface is covered with anechoic tiles. These do not provide complete watertightness, and thus, corrosion may arise beneath them and is thus not visually detectable. It therefore becomes necessary to survey the outer hull surface. Removing the tiles is an expensive and time-consuming procedure, and in order to circumvent this problem, an NDT measurement procedure has to be used. Several methods have been tried, and one that appears to be of promise is the use of an eddy current probe [4]. Eddy current testing is used in several industries (nuclear, aerospace) to detect flaws.

Corroded pressure hull material, once detected, is removed by grinding; if the damage is significant, it may be corrected by weld build-up of the plate thickness (weld “cladding” or “buttering”) or by replacement of the shell plating.

Clad metals are defined as two or more distinct metals or alloys—with very different properties—metallurgically bonded together to achieve functional benefits unachievable with a single metal. Almost any metal can be cladded, including alloys. End-use requirements such as corrosion- and abrasion-resistance or strength dictate which metals or alloys to use. Cladding procedures involve welding, whereby residual stresses, deformations, and geometric imperfections are introduced in the hull. Furthermore, the material properties of the hull may be changed as a result of the heat input from welding. *Weld overlay* is currently used to repair corrosion pits on submarine hulls. In welding overlay, a cladding layer is melted onto the surface of a similar metal to form a stronger base material.

As these factors may be detrimental to strength and stability, and because of the high cost of replacing damaged shell plates, it may be preferable to operate the pressure hull with a reduced plating thickness rather than carry out repairs. For this reason, since corrosion thinning of the hull material reduces the collapse pressure and thereby the safe operational limits of the vessel, the effect of thinning needs to be thoroughly understood.

In order to understand the effects of weld overlay, an investigation was carried out on behalf of the Canadian Navy (DRDC) by the Canadian Institute for Neutron Scattering, the aim being to define the maximum size of the corroded area over which weld overlays could be used effectively. The data were then used by DRDC to validate two different finite element approaches that simulate the effect of welding on the properties of the hull plating.

Apart from the pressure hull itself, exostructures are also subject to the corrosive action of seawater, especially ballast tanks (Figure 10.3). While a submarine is underway, the environment in a ballast tank is very aggressive because seawater is continually circulated in and out of the tanks, creating fertile grounds for metal corrosion. The need therefore arises to survey and maintain the coated surfaces periodically. The recoating of a ballast tank is done while the vessel is in drydock and is often a long process. Usually, two-part epoxy coatings are used and their curing time extends to 7 days, diminishing the vessel's battle-readiness. In addition, two-part epoxy paints contain solvents and many volatile organic compounds (VOCs).

A replacement coating was provided by industry that could be rapidly dried yet provide the same anti-corrosion protection. To this end, an energy-cured (visible light) repair coating that retains the protective properties of two-part epoxy coatings was developed and provided protection for a number of U.S. Navy submarines for up to 7 years without the need for recoating [16].

(a) Submarine exterior

(b) Ballast tank interior

FIGURE 10.3 Submarine exterior and ballast tank interior after a period at sea.

The assessment of the effect of corrosion on pressure hull strength has been studied using:

1. Finite element and finite difference analysis
2. Laboratory tests

We shall consider each in turn.

10.2.4 The Use of Numerical Tools to Predict Pressure Hull Strength

The application of numerical procedures to the analysis and design of pressure hull structures has formed the basis of design procedures that have been used in recent decades. The complexity of pressure hull failure means that all features of structural behaviour need to be modelled in order to obtain results that are sufficiently accurate for design purposes. Sources of inaccuracy, however, arise with the software and are related to numerical inaccuracy and the solver.

The path followed involves the use of numerical tools that are compared and calibrated against carefully executed laboratory tests (Chapter 8). These have proved invaluable in understanding pressure hull failure as well as acting as benchmarks for numerical and analytical tools.

Software used in the past includes:

- *Finite Difference Method*: BOSOR4, BOSOR5. This program, developed by Lockheed has been used in the practical design of shell structures including ring-stiffened shells
- *Finite Element Solvers*: ADINA, ASAS, ANSYS, STAGS, ABAQUS, PATRAN as well as in-house software especially developed for pressure hull analysis. The solvers listed are general 3D solvers that are capable of modelling the nonlinear response of ring-stiffened cylinders.

Nonlinear finite element analysis has shown that hull thinning reduces the interframe collapse pressure by reducing the bending and membrane stiffness of the shell, thereby leading to the early onset of yielding and collapse in the corroded region. Overall collapse pressure is also reduced, although to a lesser extent than interframe collapse, through a reduction in the bending stiffness of the combined shell-stiffener section.

The effect of corrosion thinning on overall collapse has also been studied using an elasto-plastic finite difference method for predicting overall collapse pressures. This method accounts for the nonlinear effects of out-of-circularity (OOC) and residual stresses due to cold rolling, which are critical factors when determining overall collapse pressures. The method allows the circumferential extent of one-sided thinning as well as the associated eccentricity to be included explicitly in the finite difference model. Corrosion thinning was applied uniformly to both sides of the shell.

In general, overall collapse pressure decreases with increasing depths of thinning; however, the effect of corrosion thinning was found to be dependent on its circumferential extent and its orientation with respect to OOC. Overall collapse strength does not monotonically decrease with the circumferential extent of thinning. The load-path eccentricity resulting from one-sided thinning effectively increases or decreases the imperfection magnitude depending on whether the corrosion is in-phase or out-of-phase with the OOC. This leads to cases whereby, as the thinning is extended around the cylinder, the collapse pressure actually increases relative to cases with smaller extents of thinning. In a similar manner, shell thinning that is out-of-phase with OOC (i.e., collocated with an outward bulge) can increase the collapse pressure, while out-of-phase shell thickening can decrease the collapse pressure with respect to a similar intact model. Furthermore, shell thinning was found to have a greater effect on collapse strength for models having small magnitudes of OOC, due to its relatively greater influence on OOC.

Nonlinear finite element analysis has also been used to determine the collapse pressure over a range of hull thinning, using experimentally obtained stress–strain behaviour of the material [19].

10.2.5 Pressure Hull Corrosion Rates

In [19], measurements of hull thickness reductions due to corrosion were reported on a submarine hull aged 10 years having a nominal thickness of 25.4 mm (1 in), as shown in Figure 10.4. Maximum and minimum thicknesses of 24.81 mm and 22.99 mm respectively were measured at opposite ends of the submarine hull. The average thickness was found to be 23.85 mm with a standard deviation of 0.41 mm. This gives an average corrosion thinning over the operational lifetime of the submarine of 1.55 mm.

In parallel, corrosion rates were determined using the potentiodynamic method and short-term immersion testing used to verify the former. The potentiodynamic method gave results of 0.17 mm/year for HY-80 steel and 0.18 mm/year for HY-80 heat-treated steel. Corresponding corrosion rates for test specimens used in immersion testing were 0.192 mm/year and 0.190 mm/year were estimated for as-received and heat-treated samples, respectively.

Since the experimental corrosion rate, based on the potentio-dynamic tests, was verified by the immersion tests, and since it was conducted under actual seawater environmental conditions, it was taken as equivalent to the actual corrosion rate for the submarine hull. The value of 0.18 mm/year allows us to estimate the coating lifetime. The number of years of operation (exposed to the sea environment) is equal to the age of the submarine minus the coating lifetime, minus the time in the dry-dock for maintenance. Hence, the *actual corrosion rate* is given by[1]:

$$\text{Actual CR} = \frac{\text{Average corrosion thinning}}{\text{Operational years}} = \frac{\text{Average corrosion thinning}}{\text{Age} - \text{coating lifetime} - \text{drydock time}}$$

For a 12 year-old submarine and a 2-year dry-dock maintenance period, the lifetime of the coating, using the actual (experimental) corrosion rate, was estimated to be 1.4 years. The duration of exposure to the corrosive environment was then equal to 8.6 years.

Finite element software capable of nonlinear analysis (material and geometrical nonlinearities) was used to model the effect of corrosion thinning on operation depth. For the intact case (no corrosion), it was predicted that the pressure hull would collapse at a depth of 567 m. The effect of corrosion thinning was found to be a reduction in collapse pressure of approximately 27 kPa per annum. This translates to a reduction in operating depth of 2.7 m/year, assuming a seawater density of 1025 kg/m^3 (Figure 10.5). On the basis of this information, operators need to adjust the depth of operation, as well as the annual decrease in depth of operation due to corrosion thinning.

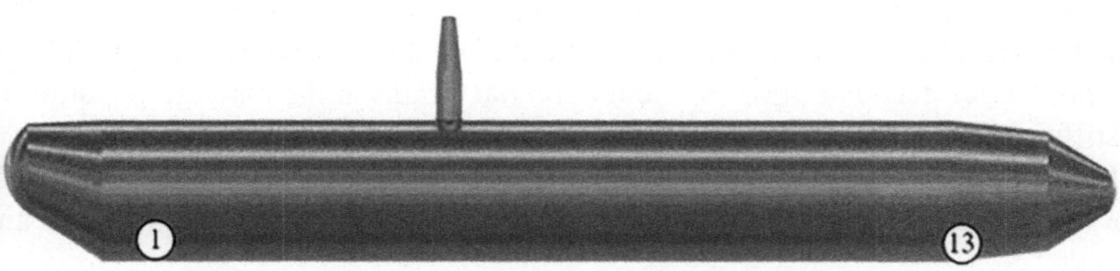

Position	1	3	5	7	9	11	13
Measurement [mm]	24.81	23.67	23.43	23.69	23.79	24.05	22.99
	24.72	23.68	23.40	23.94	24.00	-	-
Position	2	4	6	8	10	12	
Measurement [mm]	23.76	24.34	24.02	23.82	23.41	23.92	
	23.62	24.32	23.82	-	23.38	23.94	

FIGURE 10.4 Positions of hull thickness measurements.

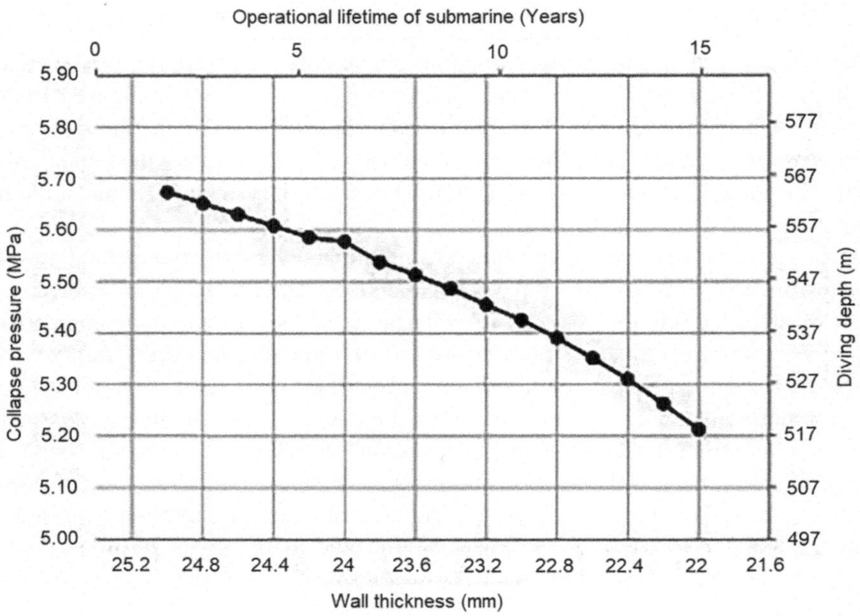

FIGURE 10.5 Effect of pressure hull corrosion thinning on collapse depth.

10.2.6 Tests and Strength Predictions for Intact and Corroded Ring-Stiffened Cylinders

In this section, a series of small-scale collapse tests on 20 ring-stiffened aluminium cylinders that were conducted by the Canadian Navy (DRDC) to study the effect of corrosion damage on hull strength and stability is described (Table 10.2) [10]. The test models were designed to fail by interframe collapse or overall collapse. In the former case, heavy ring-stiffeners were used, whereas in the latter case, light ring frames were used. Sample results for both types of failures are shown in Figure 10.6. Corrosion was simulated by use of a thinned patch that formed part of the shell.

Tables 10.3 and 10.4 include results for tests carried out on intact and corroded ring-stiffened cylinders respectively. The test data are compared with results obtained using (a) the empirical strength curve (Figure 9.9) and (b) the finite difference program

In Tables 10.3 and 10.4, the yield pressure, p_y, is that at which the von Mises stress exceeds the circumferential yield stress. The experimental number of lobes, n, is associated with the largest Fourier amplitude for decomposition of circumferential strains at the collapse pressure. When the shell and stiffener lobe numbers differ, this is shown. For example, 3/4 denotes dominant $n = 3$ and $n = 4$ in the shell and stiffener, respectively. The modelling uncertainty factor, X_m, is equal to the quotient of the experimental and predicted collapse pressures. Finally, the bias is the mean value of X_m for a group.

The artificial pressure hull thickness reductions applied were found to reduce the collapse strength of experimental models through high local stresses in the corroded region, leading to early onset of yielding and inelastic buckling. Bending associated with the eccentricity due to one-sided thinning was found to further increase the local stresses in the hull. Overall collapse pressures were more severely affected by corrosion damage than interframe collapse pressures. The percentage reduction in overall collapse pressure, compared with intact experimental models, was found to be closely related to the percentage depth of thinning (Figure 10.7). The accuracy of conventional collapse pressure predictions for the experimental models were significantly better for intact than for corroded cylinders.

TABLE 10.2
Test Program for Strength of Corroded Cylinders: Summary of Results

Phase	Aim	Results
1–3 [9]	Influence of corrosion depth and area on pressure hull failure modes	Hull thinning reduces collapse pressures by hastening the onset of yielding in the damaged shell
4	Influence of material properties, (yield strength) on overall collapse of damaged and intact models, as well as the effect of cyclic pressure loading past the yield limit	The relationship between yield stress and overall collapse pressure was found to be approximately linear in the range of yield stresses considered, so that a 10% increase in yield stress results in an 8%–9% increase in collapse pressure. That relationship was found to hold for both intact and corroded models
5–6	Influence of OOC	Overall collapse pressure decreases with increasing depths of thinning; however, the effect of corrosion thinning was found to be dependent on its circumferential extent and its orientation with respect to out-of-circularity

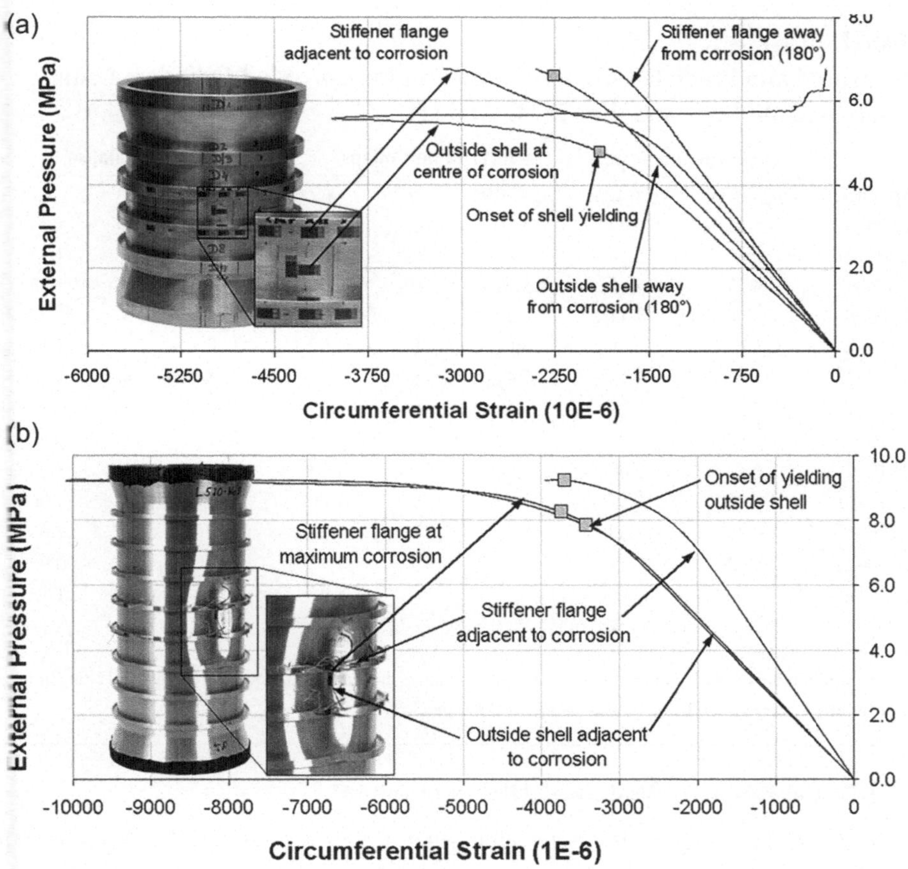

FIGURE 10.6 Pressure-strain curves for test specimens failing in two different modes [11]. (a) Interframe collapse mode. (b) Overall collapse mode.

TABLE 10.3
Measured and Predicted Collapse Pressures for Intact Cylinders Using Analytical Methods and FD Software [11]

	Experimental Results			Interframe Collapse			Overall Collapse		
Specimen	p_y (MPa)	p_c (MPa)	N	p_{ci} (MPa)	n	X_m	p_{co} (MPa)	n	X_m
L300–1	5.79	7.11	4	6.20	9	1.15	7.86	4	0.90
L300–2	n/a	7.87	n/a	6.20	9	1.27	7.86	4	1.00
L300–5	7.44	8.99	3	7.99	9	1.13	8.57	5	1.05
L300–6	7.53	9.14	3–4	7.99	9	1.14	8.57	5	1.07
L510–1	8.27	9.05	2–3	7.69	8	1.18	8.18	3	1.11
L510–4	8.12	9.79	3	8.56	8	1.14	9.11	3	1.08
L510–5	8.08	9.08	3	7.59	9	1.20	8.16	3	1.11
L510–6	7.41	8.48	3	7.59	9	1.06	8.45	3	1.00
Bias						1.16			1.04
COV						5.2%			6.6%

TABLE 10.4
Measured and Predicted Collapse Pressures for Corroded Cylinders Using Analytical Methods and the FD Method [11]

	Experimental Results		Interframe Collapse			Overall Collapse		
Specimen	p_y (MPa)	p_c (MPa)	p_{ci} (MPa)	n	X_m	p_{co}[a] (MPa)	n	X_m
L300–3	4.43	6.77	4.00	10	1.69	5.93	6	1.14
L300–4	n/a	6.94	4.00	10	1.74	5.93	6	1.17
L300–7	4.57	7.30	5.29	10	1.38	4.93	5	1.48
L300–8	4.56	7.11	5.29	10	1.34	4.93	5	1.44
L510–2	n/a	8.59	7.61	8	1.13	7.78	4	1.10
L510–3	7.85	9.25	7.61	8	1.22	7.78	4	1.19
L510–7	5.33	7.07	5.49	9	1.29	7.58	3	0.93
L510–8	4.97	7.21	5.82	9	1.24	7.81	3	0.92
L510–9	6.58	7.68	6.18	9	1.24	7.29	3	1.05
L510-10	6.29	7.81	6.53	9	1.20	7.38	3	1.06
L510–11	6.34	7.58	6.18	9	1.23	7.29	3	1.04
L510–12	5.93	7.29	6.53	9	1.12	7.38	3	0.99
Bias					1.32			1.13
COV					15.2%			15.8%

[a] The predicted overall collapse pressures p_{co} and n-values are based on Kendrick's elasto-plastic finite difference method, modified to allow for corrosion patches.

10.2.7 Comparison with Analytical Design Formulas

Table 10.5 provides a comparison of the submarine design formula (SDF) predictions for collapse pressure and mode with those observed during the experiments. The analytical collapse prediction for each cylinder was taken to be the lower of the mean empirical interframe collapse pressure, p_{ci} (mean), and the elasto-plastic overall collapse pressure, p_{co} (Tables 10.3 and 10.4). The mode shapes reported for interframe calculations are those associated with the elastic buckling pressure, p_m.

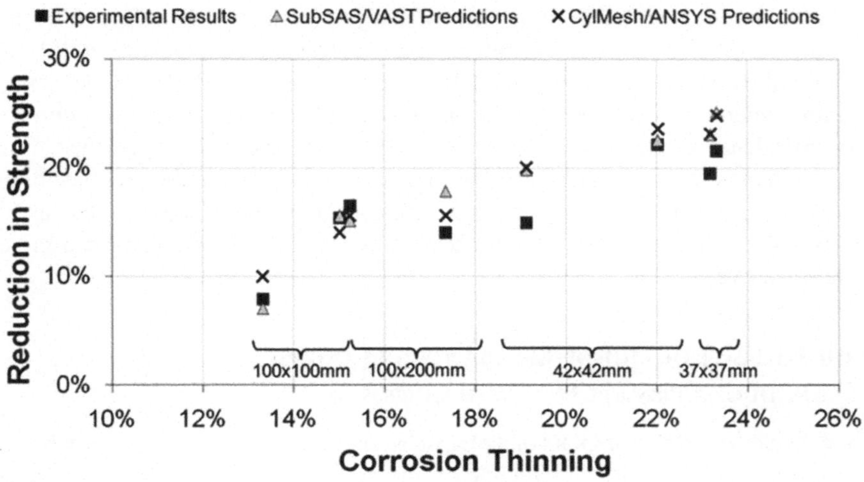

FIGURE 10.7 Reduction in overall collapse strength versus shell thinning for specimens with light internal stiffeners.[2]

TABLE 10.5
Comparison of SDF Collapse Predictions with Experimental Results

	SDF Prediction		Experimental Results	
Specimen	**Collapse Mode**	**Collapse Pressure (MPa)**	**Collapse Mode**	**Collapse Pressure (MPa)**
L300–2	Interframe ($n = 9$)	6.68	Interframe ($n = 7$)	7.87
L300–3	Interframe ($n = 10$)	4.28	Interframe ($n = 7$ or 8)	6.77
L300–4	Interframe ($n = 10$)	4.28	Interframe ($n = 7$ or 8)	6.94
L510–1	Interframe ($n = 8$)	8.24	Overall ($n = 3$)	9.05
	Overall ($n = 3$)	9.13		
L510–2	Overall ($n = 3$)	8.32	Overall ($n = 2$ or 3)	8.59

In general, the SDF shows poor correlation with the experimental collapse pressures, except in the case of the long T-section stiffened cylinders. They were able to predict whether interframe or overall collapse would occur in all cases but one (L510-No1). The overall elasto-plastic collapse pressure predicted for L510-No1 showed the best correlation with the experimental collapse pressure.

The interframe collapse pressures of the short cylinders were consistently underesti mated for the following reasons:

1. The analytical collapse pressure is for cylinders with up to 0.5% OOC and residual stresses due to fabrication.
2. The analytical predictions for the short corroded cylinders pessimistically assumed a uniform shell reduction, rather than the finite region of corrosion of the test specimens.

The overall collapse pressures predicted by the SDF were relatively accurate. The analytical predictions were within approximately ±3% of the experimental collapse pressures for the T-section stiffened cylinders. Overall collapse was under-predicted by 21% for the cylinder with penetrations, possibly because the cylinder was analysed as if no penetrations were present. The SDF successfully predicted an $n = 3$ collapse shape for the long T-section stiffened cylinders and the cylinder with penetrations.

In a parallel study using nonlinear finite element analysis (NLFEA), the effect of the presence of penetrations and castings in the pressure hull suffering corrosion was investigated [5]. In this, the pressure hull fitted with penetrations was modelled using a combination of shell and solid elements joined by beam elements. It was found that the presence of a casting increases the failure load above that of the corroded hull with no casting from 3% to 19%. The benefit of the presence of the casting increases as the thickness loss due to corrosion increases. Therefore, in the presence of corrosion, structural performance is not adversely affected. It is concluded that if possible the casting should be included in finite element models of submarine pressure hulls, especially when the effects of corrosion are considered.

10.2.8 The Influence of Out-of-Circularity on Collapse Strength of a Pressure Hull with Corrosion

The need to consider the effect of OOC on collapse strength is an important priority because finite element analysis has shown that a 0.5% OOC in the critical mode can bring about a significant reduction in collapse pressure (of the order of 25%) [10].

a. *Intact Specimens*

The various OOC criteria for individual test specimens, as summarised in Table 10.6, are plotted against the corresponding experimental collapse pressures in Figure 10.8.

It is seen from Figure 10.8 that, as expected, when the OOC increases, the collapse pressure decreases. This is seen to be true for all OOC criteria.

In Figure 10.9, the collapse pressure of intact specimens is normalised by dividing the actual collapse pressure by the boiler yield pressure for each specimen. In this manner, the effect of variation in yield stress of each specimen is factored out of the normalised collapse pressure p_c^* so that:

$$p_c^* = \frac{p_c}{p_0} \quad \text{where} \quad p_0 = \frac{\sigma_0 t}{R} \tag{10.1}$$

Figure 10.9 also includes curves of (a) the pressure to cause initial yield (dotted line) and (b) the elasto-plastic overall collapse pressure based on Kendrick's finite difference method, p_{co}.

b. *Corroded Specimens*

Corrosion damage interacts with OOC by effectively increasing or decreasing the imperfection amplitude depending on whether the corrosion is in-phase or out-of-phase with the OOC. Finite element analysis has indicated that out-of-phase corrosion damage has a much smaller effect on collapse pressure than in-phase corrosion.

TABLE 10.6
Ranges of OOC Measured on Specimens Based on Various Eccentricity Parameters Compared with Target (Nominal) Values

Eccentricity associated with nominal OOC[a]	e_{nom}	0.3%	0.5%
The greater of the inward and outward eccentricities	e_{max}	0.39–0.45	0.67–0.97
Average maximum eccentricity	e_{avg}	0.32–0.33	0.46–0.71
Fourier amplitude for $n = 3$	$e_{(n=3)}$	0.20–0.56	0.30–0.63

[a] The out-of-circularity is taken as the quotient of the radial eccentricity and the nominal mid-plane radius of the shell (121.5 mm).

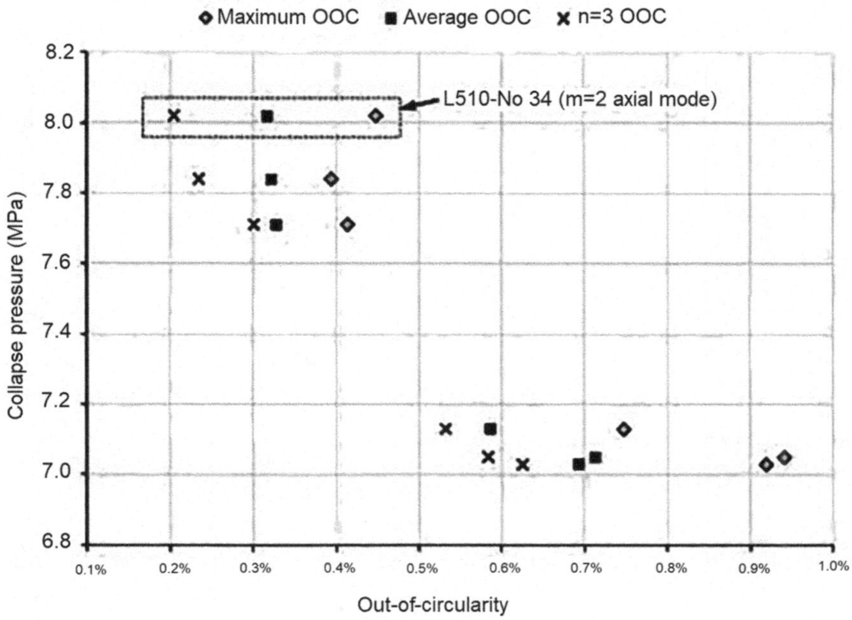

FIGURE 10.8 Effect of out-of-circularity on collapse pressure (intact specimens) [10].

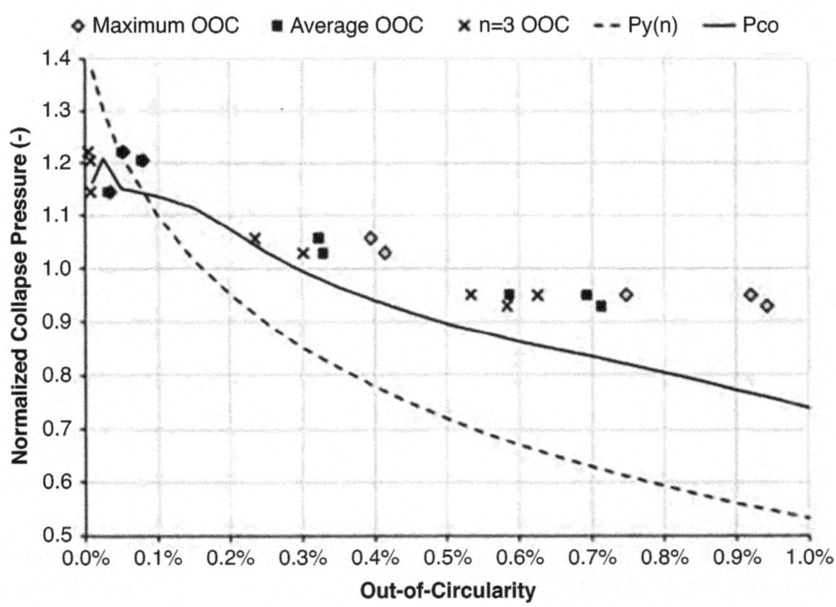

FIGURE 10.9 Effect of out-of-circularity on normalised collapse pressure (intact specimens) [10].

Figure 10.10 shows that, as in the case of intact specimens, the normalised collapse pressure reduces with increasing OOC. In this case, the measured OOC is replaced by the effective OOC. The strength of cylinders with out-of-phase corrosion is seen to be somewhat less than that of the intact specimens.

Test results are summarised in Figure 10.11. It is seen that on average, the test values obtained for the collapse pressure of the corroded specimens are between 10% and 15% below those of the intact specimens. Both intact and corroded specimens showed similar

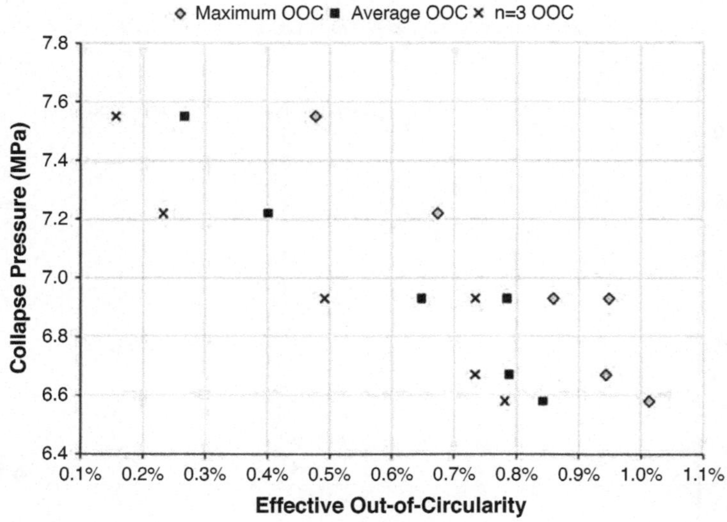

FIGURE 10.10 Effective out-of-circularity v. normalised collapse pressure (corroded specimens) [10].

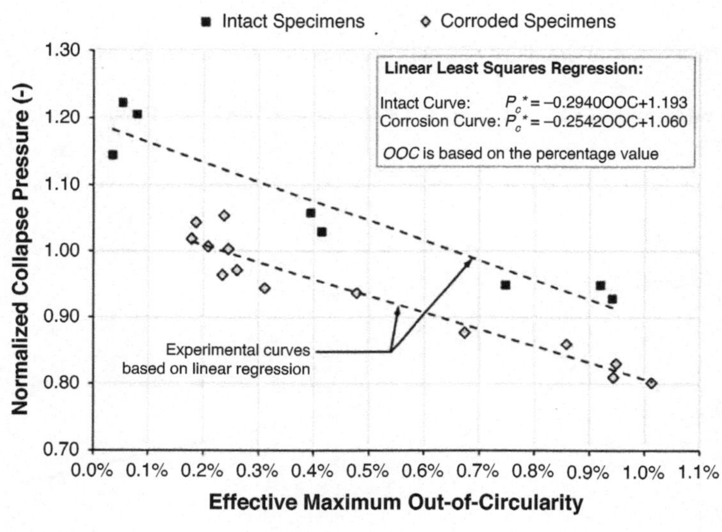

FIGURE 10.11 Effective maximum out-of-circularity v. normalised collapse pressure for all specimens [10].

sensitivity to OOC. The linear least squares regression indicates that intact specimens are slightly more sensitive to OOC. This arises because collapse of the corroded specimens was influenced by the corrosion thinning itself, not just its effect on OOC.

10.3 FATIGUE LOADING OF SUBMARINE STRUCTURES

Submarines and submersibles are vessels that are unique in several respects: environment, operating profile, hull form, materials, and fabrication methods are quite different to those of surface ships. In order to understand the effect that cyclic and variable loading acting over an extended period has on fatigue strength, it is necessary to consider each of these parameters in turn. Table 10.7 lists these and the factors that influence them most.

TABLE 10.7
Factors Affecting the Fatigue Life of Submarine and Submersible Structures

Parameter	Major Influence	Effect on Fatigue
Environment	–	Indirect
Operating profile	Mission requirements, officers	Indirect
Hull form (global level)	Environment, mission, strength requirements	Indirect
Fabrication methods	Design codes, yard practice	Direct
Materials	Strength and fatigue requirements	Direct
Structural design (local level)	Strength and fatigue requirements	Direct

10.3.1 Environment

Seawater temperature varies according to:

- Geographical location
- Season of the year
- Time of day
- Depth of water

Of importance to submarines that operate within the 250 m depth range (continental shelf) are geographical and seasonal variations. Geographical variations in way of the sea surface are shown in Figure 10.12, and it is seen that it varies from around the freezing point of water (–2°C to +2°C) to more than 30°C.

As regards depth, the temperature decreases gradually as shown in Figure 1.1 and it is only at depths that exceed 500 m that a steep drop to approximately 4°C is noted. In the case of deep-diving submersibles that operate at depths that exceed 1,000 m, the temperature is in the range 0°C–4°C.

The most important effect that the seawater environment has on the fatigue behaviour of metal structures is to cause *corrosion–fatigue*. This involves fatigue loading in a corrosive environment.

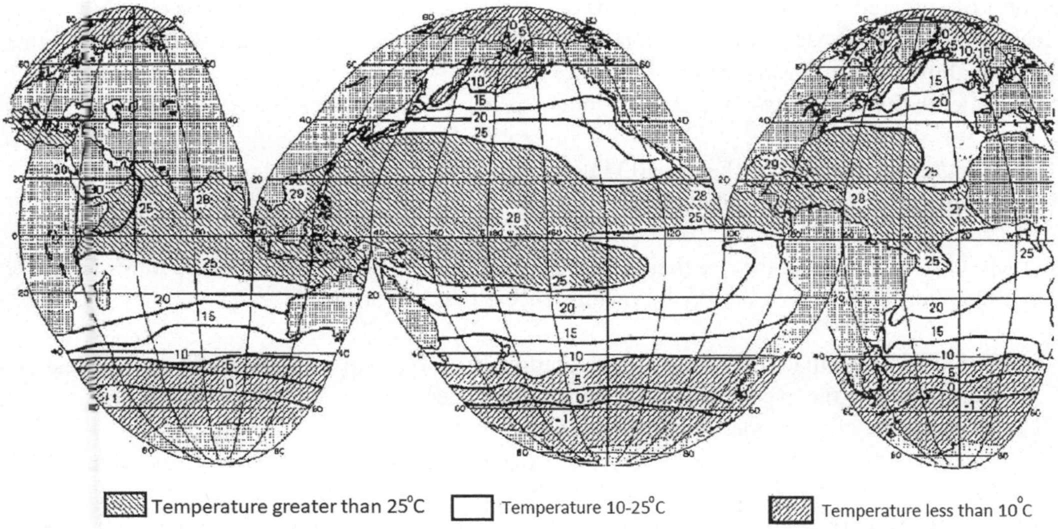

FIGURE 10.12 Geographical variation of sea surface temperature [2].

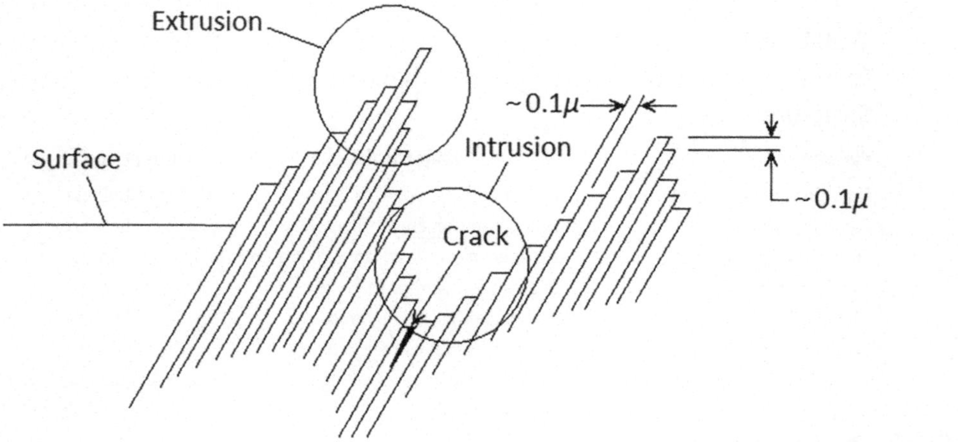

FIGURE 10.13 Persistent slip bands and crack nucleation.

To understand how the combined action of the two effects results in reduced fatigue life, it is necessary to consider the fatigue crack initiation mechanism. This involves micro-cracks arising in way of surface slip lines, as shown in Figure 10.13. In a new structure that is protected by a surface coating, no element of the surface area is exposed to the sea environment, and corrosion fatigue does not arise. If, however, some regions are exposed, local corrosion causes exposure of the surface, and the continuous loading in conjunction with thickness losses results in slip lines translating with respect to one another, so that the surface is no longer smooth. Micro-cracks can then form in way of intrusions and extrusions, thereby accelerating the fatigue crack initiation mechanism, resulting in reduced fatigue life of the structure.

Since corrosion results in the deterioration of the condition of the surface, the combined action of these processes is expected to lead to a reduced fatigue life than otherwise. Corrosion–fatigue leads to the formation of a fracture and is the result of cyclic loading in a corrosive environment. The environment may be solid, gaseous, or liquid, and this brings about a deterioration of the fatigue properties of the metal. This deterioration is due to a loss in ductility or, otherwise put, the *embrittlement* of the material. In the case of marine structures, it is seawater that acts as the embrittling medium. Metal embrittlement consists of the adsorption of atoms from the environment that disturb the ordered crystal lattice of the metal. In the case of submarines in which high-strength steels are used, the susceptibility to corrosion–fatigue increases only in part because the higher strength allows design for higher stresses.

In general, there exist two mechanisms for corrosion–fatigue to take place:

a. *Anodic slip dissolution* (Figure 10.14a)
b. *Hydrogen embrittlement* (Figure 10.14b)

Of these two mechanisms, the one that is most relevant to steels used in marine structures is anodic slip dissolution. This consists of the following process that repeats itself:

1. The oxide film that covers the metal ruptures at a crack tip as a result of tensile stress. This results in the exposure of bare metal to seawater.
2. The bare metal undergoes anodic dissolution.
3. A halide anion diffuses into the crystal lattice of the metal (in the case of seawater, it is Cl^-).
4. The metal corrodes and a thin oxide layer forms (*passivation*) which inhibits further dissolution and oxidation.

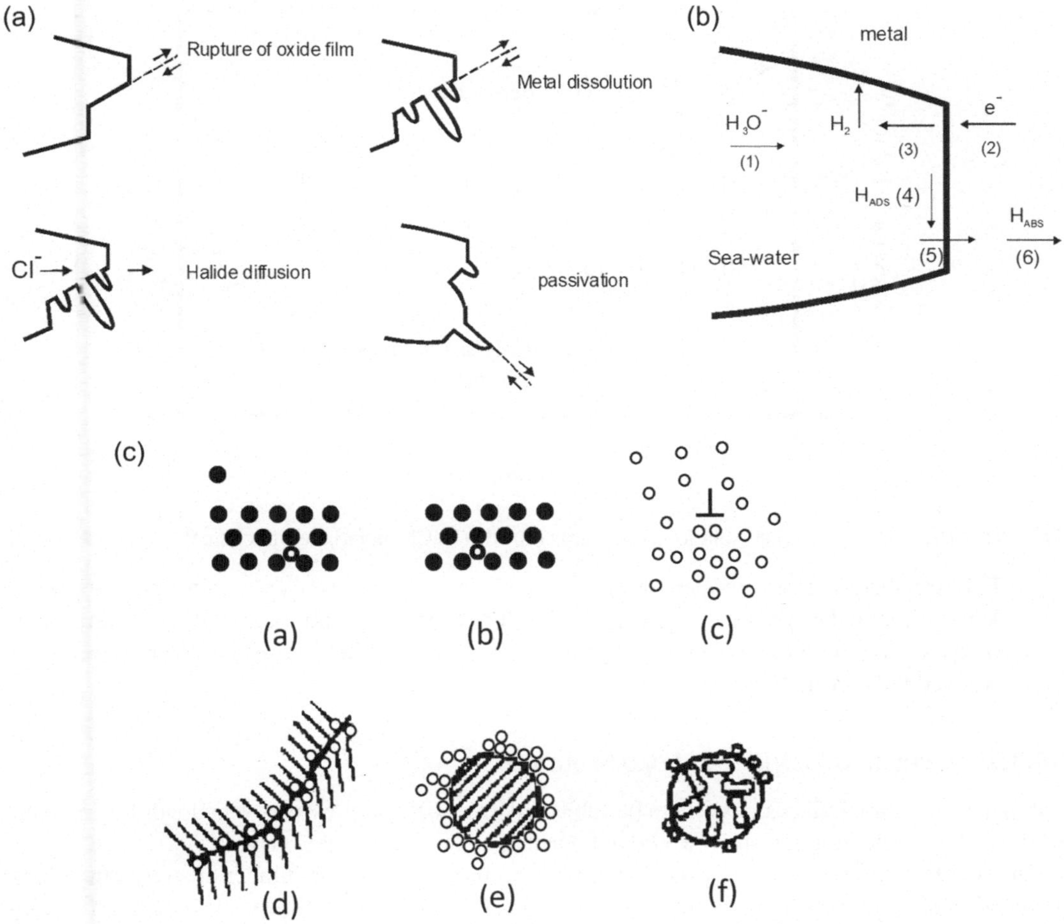

FIGURE 10.14 Corrosion fatigue mechanisms in seawater. (a) Anodic slip dissolution. (b) Hydrogen embrittlement. (c) Destinations of hydrogen in microstructure.[3]

Hydrogen embrittlement arises in the case of steels whose ultimate tensile stress exceeds 1000 MPa or else whose hardness is less than 30 Rockwell HRC. In this case, it is the adsorption of hydrogen atoms that causes the embrittlement of the metal, which results in a reduction in its fatigue life. Hydrogen embrittlement is enhanced by the following cathodic reaction:

$$H^+ + e^- \rightarrow H$$

This reaction takes place on the crack tip surface. The atomic hydrogen that is formed dissolves and migrates in the metal where its ions interact with the crystal lattice dislocations and cause a decrease in ductility. There are a number of possible mechanisms in which this may take place, of which the most investigated is *hydride-induced embrittlement.* Hydrogen can also be introduced into the metal during the steel-making process or else during fabrication.

In addition to the aforementioned, the presence of seawater has a deleterious effect on the fatigue life of structures for the following reasons:

1. Crack growth rates are higher in seawater than in air [14]. As a result, for a given stress range, the number of cycles to failure is lower in seawater than in air (Figure 10.15).

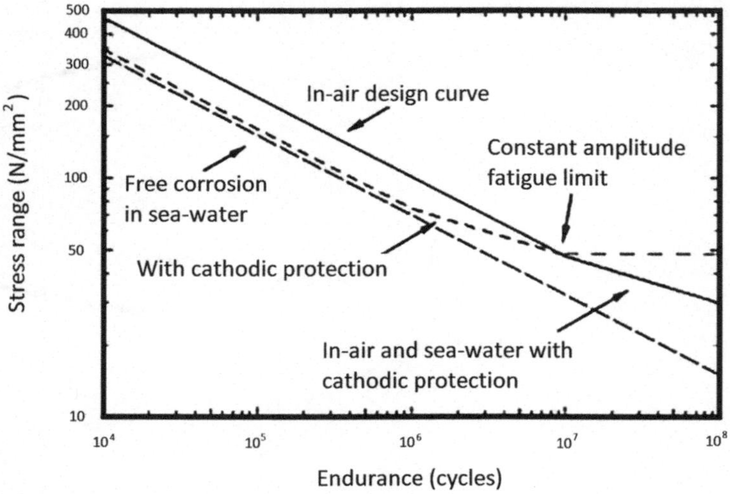

FIGURE 10.15 Effect of dilute solution of sodium chloride (NaCl) on fatigue life of steel.

2. For structures that are immersed in seawater, the range of stresses within which cyclic loading should be applied in order to avoid failure decreases continuously with the number of cycles *N*, in contrast to structures that operate in air. In the case of the latter, a limit is reached below which fatigue failure will not occur (Figure 10.15).

10.3.2 Operating Profile and Load Sequence

Submarine pressure hulls can expect to be subjected up to 10^4 compressive cyclic loads during their lifetime (low-cycle fatigue) which would not normally cause fatigue problems.

In order to estimate cumulative damage, these can be divided into a number of stress ranges and a corresponding number of load cycles. In doing this, the actual load sequence is ignored, although it is known that an overload can precipitate a *shakedown* of fatigue growth because of redistribution of residual stresses at crack tips. This effect has been reported in the case of surface ships [7].

10.3.3 Hull Form

Submarines structures consist of a number of different ring-stiffened axisymmetric shells that are welded to each other and are also welded to end closures at the extremities of the pressure hull. The junctions between two shell forms of different type, such as dome-cylinder and cone-cylinder[4], are potential sources of stress concentrations, especially within the vicinity of welds. The reason is that the response to hydrostatic pressure of the two pressure hull forms differs, so that at the junction, in order to ensure geometrical compatibility and continuity (Figure 10.16a), additional forces and moments act (Figure 10.16c).

The classic problem of this type is the cylinder–hemisphere junction whose solution is included in [6]. The results for the circumferential and longitudinal (meridional in the case of the hemisphere) stresses are reproduced here as Figure 10.16c. The *damping coefficients* κ and β for the sphere and cylinder and are given by:

$$\kappa = \sqrt[4]{3(1-\nu^2)}\sqrt{\frac{R}{t}} \tag{10.2a}$$

$$\beta = \frac{\sqrt[4]{3(1-\nu^2)}}{\sqrt{Rt}} \tag{10.2b}$$

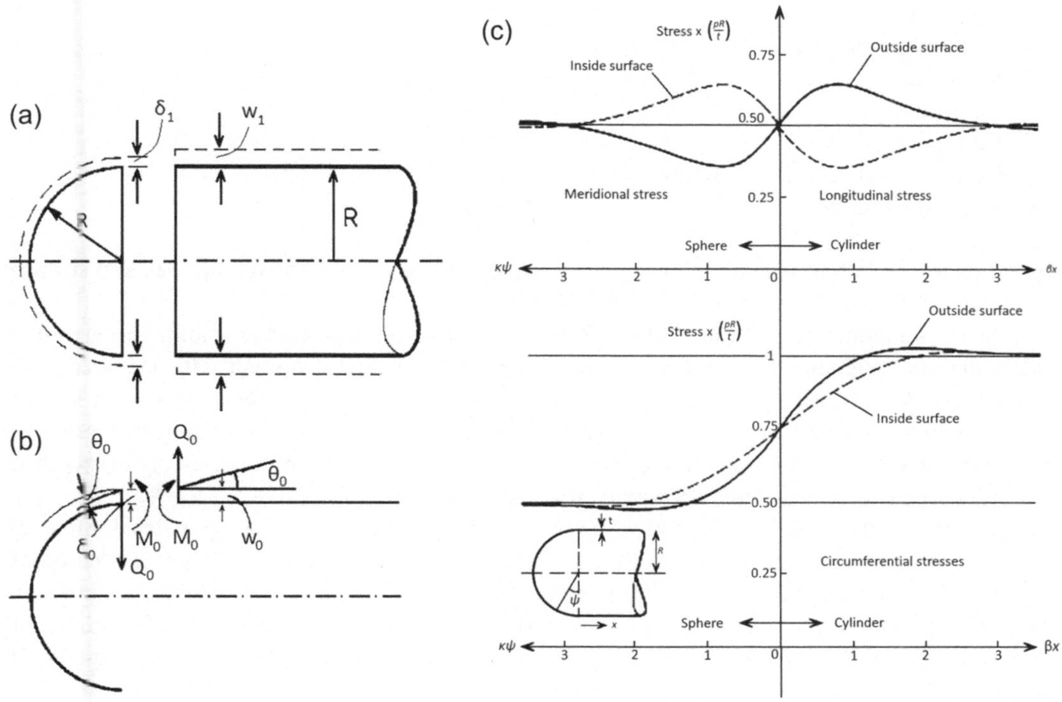

FIGURE 10.16 Cylinder–hemisphere intersection. Geometrical compatibility and force equilibrium [6]. (a) Geometry. (b) Loads at junction. (c) Resulting stress distributions in the axial direction.

From within, the pressure hull is supported by ring frames and bulkheads, whereas externally various items that make up the exo-structure (ballast tanks, fins etc) are welded to it. Thus, the most important potential fatigue failure sites will be the stress concentrations that arise in way of pressure hull penetrations and also in way of changes in geometry on the scale of the shell plating thickness.

Furthermore, in these regions of the structure, there are weld lines that are at tensile yield due to welding residual stresses.

It has been found that the most critical regions of the pressure hull from the point of view of fatigue strength are the intersections of axi-symmetric shell structures. Cracks have been detected in way of welds located close to such intersections. The largest stress concentrations arise in way of cone–cone and bulkhead–shell intersections. In the case of cone–cone junctions, compatibility forces and moments act, and there is no internal stiffening in way of the intersection. In the case of bulkhead–shell junctions, large shear forces act on the shell plating whilst on the bulkhead, the reaction force acts in the in-plane direction. In order to avoid very large stresses from developing in the cylindrical shell in way of the junction, the thickness is increased locally over a sufficient distance from the junction (Figures 9.4 and 9.6). Other means include (a) the use of curved transitions, (b) modifications to frame depth, and (c) modifications to frame spacing. The aim of these methods is to match the radial stiffness across the structural discontinuity.

10.3.4 Materials

In submarines and submersibles, the use of high-strength materials is essential in order to maintain the weight-to-displacement ratio at acceptable levels. Apart from high strength, the materials used have to be able to fulfil safety requirements under different conditions to those encountered on surface ships. The most important safety requirement for steel is the avoidance of *cleavage* fracture. The factors that influence cleavage fracture are:

- Temperature
- Material properties
- Stress level
- Thickness
- Fabrication practice
- Defect existence and distribution
- Applied strain rate

The following table that is sourced from [17] compares requirements for submarines and surface ships. The differences between the two types of structure are clear.

Of the factors that are listed in Table 10.8, we shall focus on material behaviour, although it is evident that this is related to a number of other factors. Thus, in order to specify the required level of toughness, it is necessary to specify stress levels and the maximum probable existing size of defect (crack) in the structure. Following this, the severity of loading is established using a selected fracture mechanics parameter. A test piece that has a defect (crack) is then subjected to the specified level of loading and is considered acceptable if no fracture ensues.

One of the major challenges that must be met when selecting or developing a structural steel for submarine pressure hulls is that as the yield strength increases, fracture toughness decreases (Figure 10.17a). This is the result of the increase in carbon content, which is added in order to increase the yield strength. The decrease in fracture toughness leads to a decrease in fatigue life of the structure. One other challenge that has to be met is that the fracture toughness of virtually all high-strength alloys decreases with increasing thickness[5] (Figures 10.17b and 10.18, Table 10.8).

The *thickness and size effect* is related to the transition from conditions of plane stress that persist in thin plating, to those of plane strain that characterise behaviour of thicker plating. There is no satisfactory explanation for the thickness dependence of toughness although some models have been proposed [1].

It is seen from Figure 10.18 that fracture toughness (as measured by the mode 1 stress intensity factor, K_{1C}) increases to a maximum for a particular thickness and then decreases to a value that is independent of thickness and corresponds to plane strain conditions.

TABLE 10.8
Factors that Affect Brittle Fracture Risk

Factor	Submarines	Ships (Warships)
Minimum service Temperature	Normally taken as −5°C	Normally 0°C (Air temp −30°C)
Material	High toughness quenched and tempered[a] *Q1N*(≡HY80) (550 MPa < σ_0 <650 MPa)	Mild steels and HTS
Fabrication	Welded ring-stiffened cylinder	Orthogonally stiffened grillages
Thickness	≈ 50 mm	10–15 mm (warships)
Stresses	Mainly compressive (pressure) σ_{max} = −400 MPa σ_{max} = +250 MPa	Tensile and compressive σ_{max} = +100 MPa
Defects	Weld defects in plate butts fatigue cracks at toes of ring frame stiffeners	Fatigue cracks due to poorly designed structural details
Strain rate effects	Ability to withstand underwater shock loads is required	Slamming, collision, explosive loading

[a] Carbon steel is heated and then rapidly cooled (quenched). This results in a deformed body-centred cubic (BCC) crystalline structure, termed *body-centred tetragonal* (BCT), with internal stresses. Quenched steel is hard but brittle. Tempering involves reheating quenched steel and then cooling. The elevated temperature restores ductility but reduces hardness [18, 20].

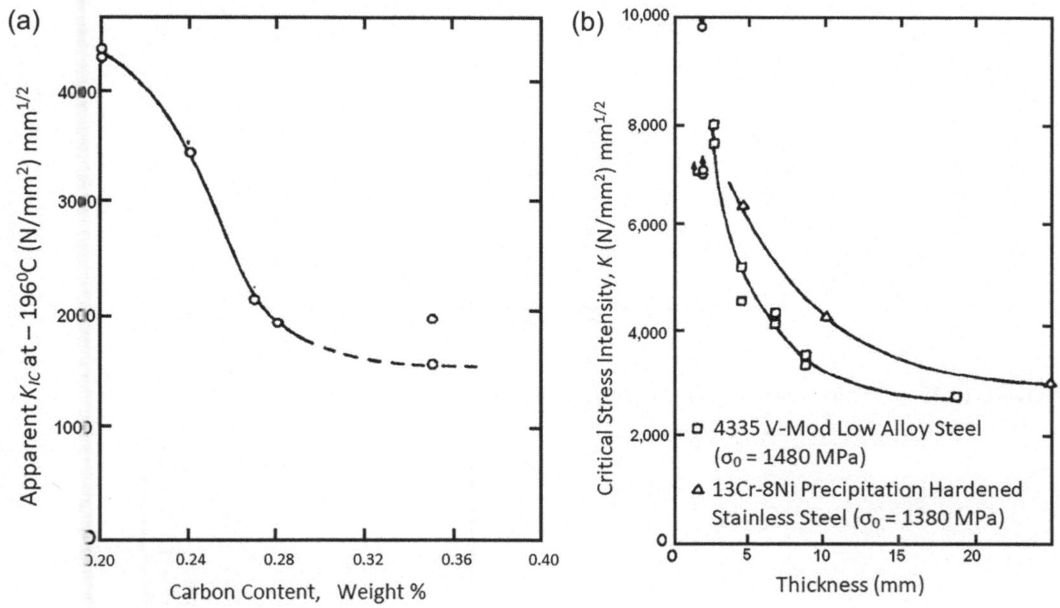

FIGURE 10.17 Effect of material and structural properties on fracture toughness of UHTS [12]. (a) Effect of carbon content. (b) Effect of thickness.

10.3.5 Fabrication Methods

As intimated earlier, residual stresses are an important contributory factor to fatigue failure. Following launching, the residual stress field that is measured is the result of the superposition of welding residual stresses upon cold-bending stresses. The residual stress fields in the shell plating between ring frames are shown in Figure 10.19.

The external load produces compressive stresses in the pressure hull that are superimposed on the residual stress field (Figure 10.18). If the yield stress in compression is not reached in way of the central zone of the plate, the net stress is simply the algebraic sum of the two values. If the yield stress is reached, fibres of the material become plastic, and the range of stresses extends from compressive yield to a finite stress.

In way of the heat-affected zone (HAZ), residual stresses are everywhere equal to the yield stress in tension. If an alternating stress field ranging from a tensile stress σ_1 to a compressive stress σ_2 is applied, the stress range is $\sigma_1 + \sigma_2$. Whilst loaded in tension, $(0{-}\sigma_1)$, the material in way of the HAZ yields and is thus subjected to a stress field ranging from 0 to σ_2. A redistribution of residual stresses will then take place.

Fatigue cracks originate in way of the toe of the fillet weld that joins the ring frame web to the shell plating as a result of the superposition of cyclic compressive stresses on the existing tensile residual stresses. This leads to a net tensile alternating load in way of the shell–ring frame intersection (Figure 10.20). They then propagate around the circumference of the pressure hull and develop into a long circumferential crack. At a later stage, they may also propagate through the thickness of the cylindrical shell leading to fracture failure of the pressure hull. Factors that affect the formation and propagation of cracks are the structural geometry of the pressure hull, the magnitude of the cyclic hydrostatic loading, and the size of latent defects in the welds.

Other factors that affect the fatigue life of submarine and submersible structures that are also related to the fabrication process are defects such as weld and member misalignment as well as the surface roughness of welds. Rules and codes for construction specify tolerances for misalignment and if these are not adhered to local bending occurs, giving rise to stress concentrations. Post-weld improvement methods go a long way toward ameliorating conditions.

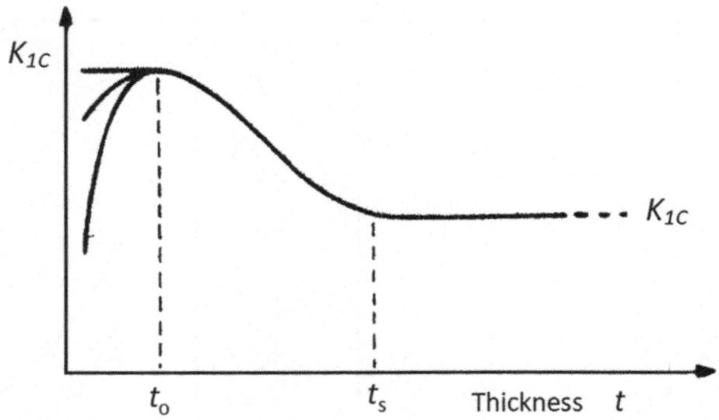

FIGURE 10.18 Fracture toughness (stress intensity factor) as a function of thickness [1].

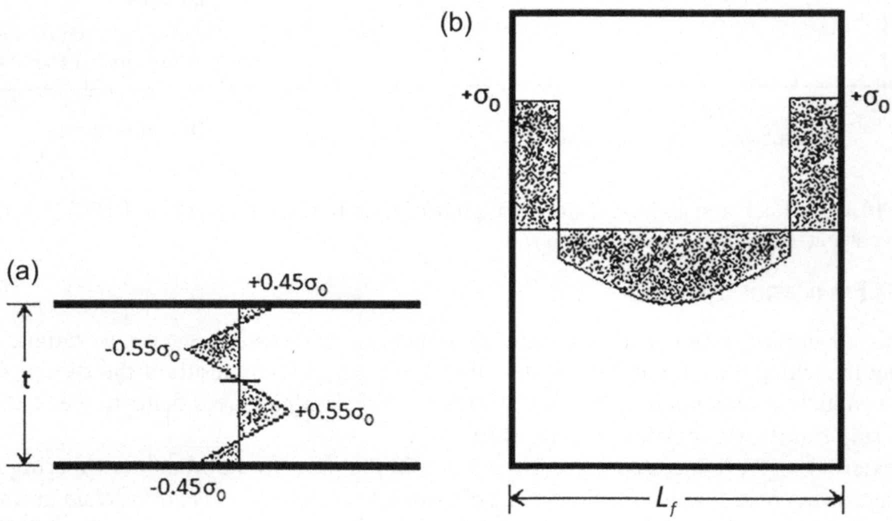

FIGURE 10.19 Residual stress fields in cylindrical shell plating. (a) Cold-bending stresses. (b) Welding residual stresses.

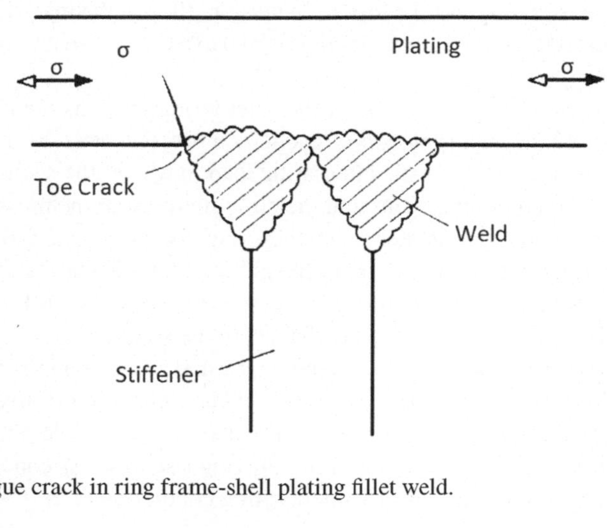

FIGURE 10.20 Fatigue crack in ring frame-shell plating fillet weld.

Other parts of the submarine structure that experience fatigue loading are the main ballast tanks, which are part of the exo-structure of the vessel. The main ballast tanks are subjected to internal pressure which produces a tensile field in the outer structure each time the tanks are blown.

10.3.6 Fatigue Life Prediction

In principle, the fatigue life prediction methods used for metal structures can be applied to submarine structures as well. We shall not elaborate on these but will describe a particular application of linear elastic fracture mechanics to the study of submarine hulls. It should also be mentioned that several navies have relied on test data in order to provide guidance on fatigue design of submarines. We shall address this question in Section 10.3.7.

The application of linear elastic fracture mechanics (LEFM) to a submarine structure is described in [13]. The method followed is based on the Paris model for crack growth according to which the rate of growth of the length of a crack, a, is dependent on the range of stress intensity factors (ΔK) and the properties of the material:

$$\frac{da}{dN} = C(\Delta K)^m \tag{10.3}$$

where N is the number of cycles, and $\Delta K = K_{max} - K_{min}$ (K = stress intensity factor). The constants C and m are used to describe the properties of the material. The stress intensity factor K is proportional to the magnitude of the applied stress. It also depends on crack geometry and mode of loading.

It was mentioned earlier that from the viewpoint of fatigue loading, a critical part of the submarine hull structure is the regions in way of intersections of conical sections that have different geometries, especially in the vicinity of welds. LEFM will be used to study one such case (Figure 10.21).

In this case, the alternating load is due to hydrostatic pressure. It is assumed to give rise to a crack of constant depth along the circumference of the pressure hull, and plane strain conditions are assumed. The fatigue life of the structure is determined according to the following sequence of calculations:

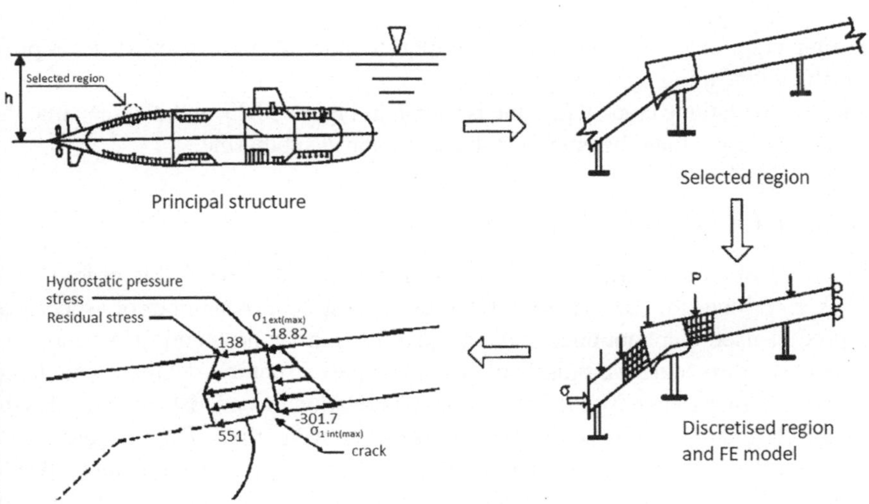

FIGURE 10.21 Fatigue model of cone-cone intersection [13].

1. Calculation of stress intensity factors K_t (=$f(a,t)$) for a range of operational depths and crack depths using[6]

$$K_t = K_n + K_f + K_r$$

where K_n relates to normal stresses, K_f relates to bending stresses and K_r to residual stresses
2. Calculation of ΔK, based on variation of hydrostatic pressure from zero to operating depth. For a given operating depth, the stress intensity factor is found as a function of crack depth, a.
3. The crack growth rate is given by the Paris law, equation (10.3). Therefore, the number of cycles N_k at a submerged depth k that are required for the crack depth to reach a final depth a_{fk} is found from:

$$N_k = \frac{1}{C}\int_{a_{ik}}^{a_{fk}} (\Delta K(a)_k)^{-m}\, da \tag{10.4}$$

where a_{ik} and a_{fk} are the initial and final values of the crack depth for the load cycle in question.
4. Definition of operation profile consisting of sequence of diving depths and corresponding number of dives for each depth.
5. Calculation of crack depth at the end of each dive.
6. The initial crack depth for each load cycle (dive) is taken as the crack depth found at the end of the previous cycle. The final crack depth can thus be found at the end of a load sequence.
7. The critical crack depth beyond which unstable fracture of the material will occur is found from $K_t(a_{cr}) = K_{IC}$.
8. For each depth range, the maximum permissible number of dives N_p so that fatigue failure does not occur is found.
9. The total permissible number of dives is then the sum of the number of dives performed for each depth range.

Following completion of these calculations, the total number of permissible dives is presented in graphical form as shown in Figure 10.22.

The growth of the fatigue crack with time is shown in Figure 10.23, and it is seen that the fatigue life is substantially larger than the operational life of a naval submarine.

10.3.7 Fatigue Testing of Submarine Structures

The understanding of fatigue failure in submarine structures that is needed in order to design new structures has for a long time been based on the use of test data. A number of aspects relating to the fatigue process itself were not understood or quantified in the past, and furthermore, it was not possible to analyse stress and strain distributions at complex structural details encountered in submarine structures. Other sources of uncertainty arise in relation to load histories and also the effects of load sequence (shakedown of residual stresses, as mentioned earlier). The environment, which as discussed earlier affects fatigue life in a multitude of ways, was not understood and indeed requires further study. Finally, if a submarine is designed to pressure vessel code requirements, the length of the fatigue life has been found to be insufficient [8].[7] For these reasons, it was considered necessary to generate and make use of test data. The UK Navy adopted an approach that is based on the use of scale models. It was found necessary to use fairly large-scale models in order to incorporate

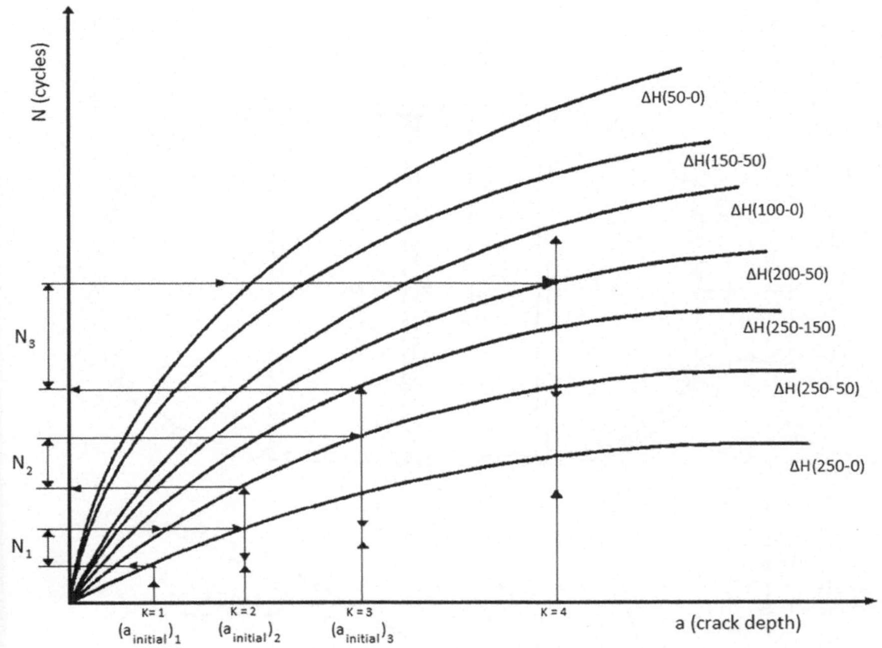

FIGURE 10.22 Calculation of permissible number of dives [13].

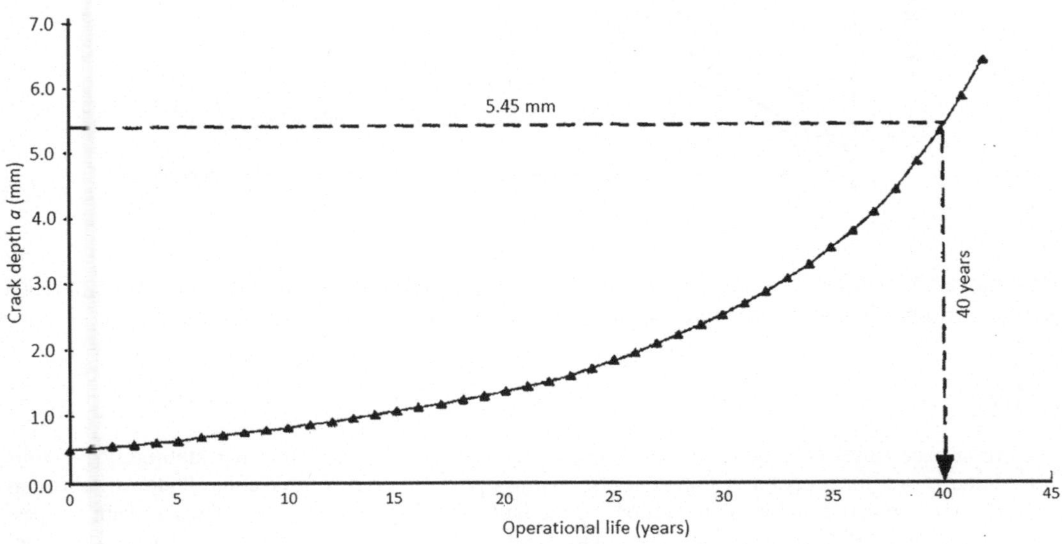

FIGURE 10.23 Growth of fatigue crack [13].

the effect of scale on residual and restraint stresses. The models are of a reduced diameter when compared to a full-scale structure but shell thickness and ring frame dimensions have been retained (Figure 10.24a).

Results for tests on two models are indicated in Figure 10.24b. It is seen that the crack size measured in both models is very close. The comparison with LEFM theory described earlier shows that the theoretical prediction is accurate up to 4,000–5,000 cycles; it then predicts a higher crack growth rate than that measured in either of the two models. From these results, it is concluded that predictions based on LEFM will provide a conservative estimate of crack size.

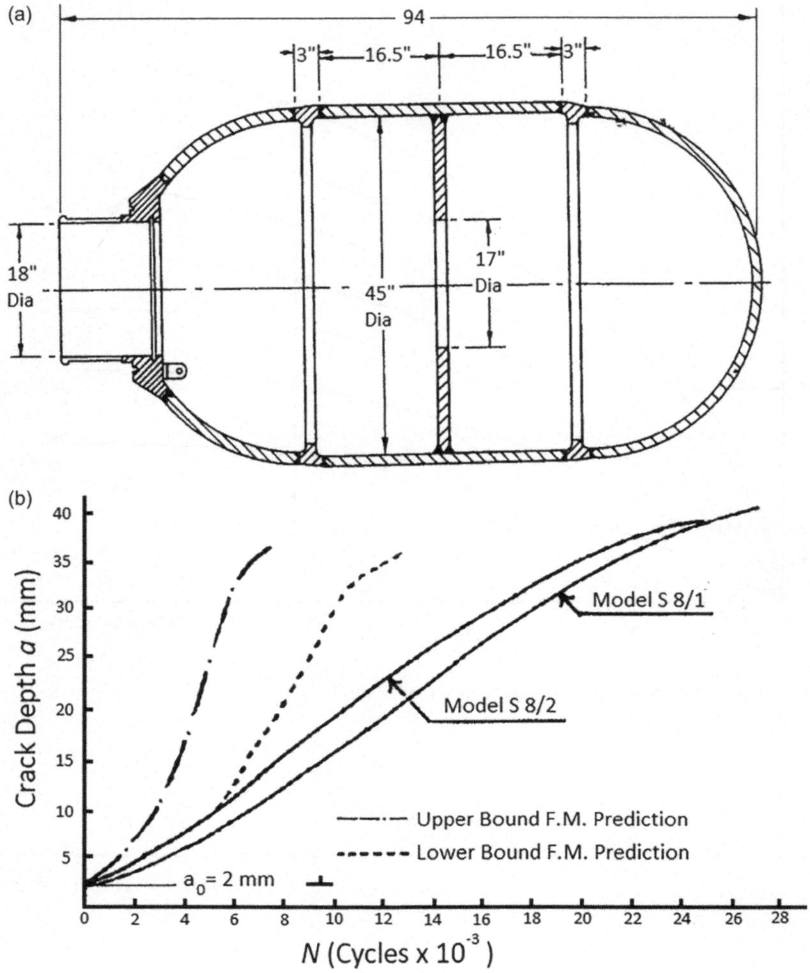

FIGURE 10.24 UK Navy fatigue testing of scale models [8]. (a) Fatigue test model. (b) Crack propagation and comparison with linear elastic fracture mechanics predictions.

NOTES

1 In surface ships, corrosion rates are significantly higher and vary during the lifetime of the ship. Initially and for a period of 6–8 years the coatings provide effective protection. Following this corrosion is observed and as more sections of the hull suffer the corrosion rate increases with the passage of time. After an operating life of approximately 15 years (third Special Survey) the hull requires detailed inspections and repairs. The reasons for this are several. Primarily, surface ships exposed to the solid-liquid-air interface that is rich in oxygen, thereby accelerating corrosion activity. Also, large internal regions of the structure such as ballast tanks and cargo spaces are exposed to corrosive activity. One other reason is that the maintenance standards of cargo ships in general are not at the same level as those of naval vessels.

2 The patch sizes shown at the bottom of the figure are specified by defining the circumferential and axial extents. SubSAS/VAST is the in-house finite element software package used by the Canadian Navy.

3 (a) Solid solution (b) solute-hydrogen pair (c) dislocation atmosphere (d) grain boundary accumulation (e) particle-matrix interface accumulation (f) void containing recombined hydrogen [18].

4 A detailed description of the method of edge coefficients applied to the cone-cylinder junction is included in Chapter 6.

5 Having said that, it has be added that full-scale model testing of submarine pressure hulls were carried out early on and it was found that even in a severely fatigue-cracked condition, HY-80 steel is very tough and does not crack in a rapid or catastrophic manner [15].

6 Addition (superposition) of stress intensity factors is permitted when they correspond to the same mode.

7 The application of the BS5500 1985 code to the design of the fillet weld used to attach the cylindrical shell to ring frames, assuming Q1N quality steel leads to a fatigue life that is less than 1,000 cycles, which is unacceptable.

REFERENCES

1. Broek D. (1986). *Elementary Engineering Fracture Mechanics*. Dordrecht: Martinus Nijhoff Publishers, 499 p.
2. Caridis P. (2009). *Inspection, Repair and Maintenance of Ship Structures*, 2nd Ed. Edinburgh: Witherby Seamanship International, 326 p.
3. Dankiw R. (2006). "Structural corrosion issues on collins class submarines." *Corrosion & Prevention* Austr. Corr. Assoc. Conf. Paper 063.
4. Eddiw-fi NDT Inc (2009). "Inspecting submarine hulls for corrosion." Available at: www.eddyfi.com (Downloaded 2024-02-15).
5. Gannon L. (2010). "Submarine pressure hull collapse considering corrosion and penetrations." Technical Memorandum DRDC Atlantic TM 2010–246.
6. Gill S.S (Ed). (1970). *The Stress Analysis of Pressure Vessels and Pressure Vessel Components*. Oxford: Pergamon Press, 405–511 pp.
7. Horn, A.M. et al (2009). *Fatigue and fracture*. In: Jang C.D.and Hong S.Y. (Ed). *Proceedings of the 17th International Ship and Offshore Structures Congress,* Volume 1 Committee III. 2 Report 475–585.
8. Kilpatrick, I.M. (1986). "The fatigue characteristics of submarine structures subjected to external pressure loading". In: Smith C.S. and Dow R.S. (Eds). *Advances in Marine Structures. Proceeding of ARE International Conference*. London: Elsevier Applied Science Publishers, 305–324 pp.
9. Mackay, J. (2007). "Experimental investigation of the strength of damaged pressure hulls – Phase 1." Tech. Mem. DRDC Atlantic TM 2006–304.
10. Mackay, J. (2011). "Experimental investigation of the strength of damaged pressure hulls – Phases 5 & 6. The influence of out-of-circularity on collapse." Tech. Mem. DRDC Atlantic TM 2010–239.
11. MacKay, J.R., Smith, M.J., van Keulen, F., Bosman, T.N., Pegg, N.G. (2011). Experimental investigation of the strength and stability of submarine pressure hulls with and without artificial corrosion damage. *Mar. Struct.* Vol. 23, No. 3, 339–359.
12. Parker, E.R., Zackay, V.F. (1973). Enhancement of fracture toughness in high strength steel by microstructural control. *Eng. Frac. Mech.* Vol. 5, 147–165.
13. Robles, L.B.R., Buelta, M.A., Goncalves, E., Souza, G.F.M. (2000). A method for the evaluation of the fatigue operational life of submarine pressure hulls. *Int. J. Fatigue*. Vol. 22, 41–52.
14. Shah Khan, M.Z., Burch, I.A. (1992). Effect of seawater on the fatigue life and crack growth behaviour of a new microalloyed steel for submarine hull application. *Int. J. Fat.* Vol. 14, No. 5, 313–318.
15. Sorkin, G., Pohler, C.H., Stavovy, A.B., Borriello, F.F. (1973). An overview of fatigue and fracture for design and certification of advanced high performance ships. *Eng. Frac. Mech.* Vol. 5, 307–352.
16. Spectra Photopolymers. "Repairing Corrosion in Submarine Ballast Tanks." Available at: https://www.sginc.com/articles/repairing-corrosion-in-submarine-ballast tanks/ (Downloaded 2024-02-15).
17. Sumpter, J.D.G. (1991). Fracture avoidance in submarines and ships. In: Smith C.S. and Dow R.S. (Eds). *Advances in Marine Structures – 2, Proceeding of ARE International Conference*. London: Elsevier Applied Science Publishers, 1–22 pp.
18. Thompson, A.W., Bernstein, I.M. (1980). The role of metallurgical variable in hydrogen assisted environmental fracture. In: Fontanna, M.G. and Staehle R.W. (Eds). *Advances in Corrosion Science and Technology*, Vol. 7, New York: Plenum, 53–173pp.
19. Viljoen, H.C., Nawaz, M., Cupido, L.H., Mitchell, G.P. (2022). Effect of corrosion thinning on depth of operation: Case study of an HY-80 steel submarine pressure hull. *Mar. Struct.* Vol. 81, 103103.
20. Wikipedia. *Carbon Steel*. Available at: https://en.wikipedia.org/wiki/Carbon_steel (Downloaded 2024-02-15).

11 Structural Analysis of Deep-Diving Submersibles

As can be seen from Appendix 1, the form of existing manned submersibles varies a great deal, as does that of unmanned submersibles, discussed in Chapter 1. A closer examination, however, reveals that the pressure chamber used in each case is based on either a sphere or a cylindrical shell with a hemispherical or other end closure. In addition to these, a number of configurations based on these primary forms have been considered and used in submersible pressure hulls. These include *multisegment* hulls that comprise a number of spherical or *barrelled* shells and *toroidal* shells that have been proposed for specific applications. Each of these pressure hull forms has advantages and disadvantages from the points of view of mathematical analysis, fabrication, and operational efficiency. In this chapter, we shall present the collapse behaviour of these structural types, based on tests and numerical studies described in recent journal articles. With regard to design issues, it has to be underlined that conflicting requirements arise and careful appraisal is required before selecting a suitable pressure hull form for a particular purpose. This question is addressed in Chapter 12.

11.1 SPHERES AND SPHERICAL CAPS

In the marine field, spherical shells have the following main applications:

- Pressure chambers of submersibles (Figure 11.1a)
- End closures of submarines (Figure 9.5)
- Cargo tanks on board gas carriers (Figure 11.1b)

They are also used to house equipment attached to submersibles. Spherical caps form part of a sphere and are used in way of access openings of submarine bulkheads and also as entire bulkheads (Figures 7.29 and 7.30).

For pressure hulls operating at great depths, the sphere is the form of choice because, as is shown in Section 12.2.1, it is the most efficient from the point of view of the weight-to-displacement ratio.

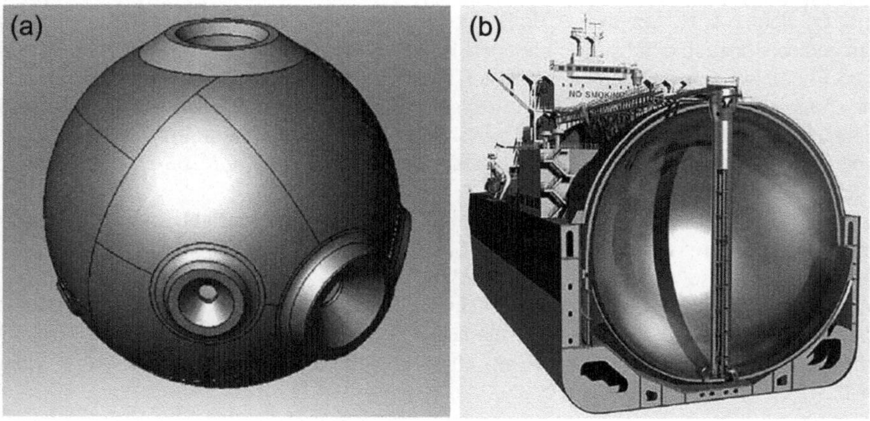

FIGURE 11.1 Marine applications of spherical shells. (a) Submersible pressure hull. (b) Membrane tank. (Moss) in LNG carrier.

DOI: 10.1201/9781003489153-11

The two primary failure modes of thin spherical shells that need to be considered are buckling and yielding. In Section 11.1.2, we shall consider the elastic buckling of a thin, perfect spherical shell subjected to external hydrostatic pressure. Following that, we shall describe tests carried out on buckling of spherical shells in order to investigate the validity of the theoretical result.

11.1.1 Yield Strength of Spherical Shells

Before discussing buckling, we shall briefly consider yielding. We consider a sphere of internal radius R and thickness t, subjected to external hydrostatic pressure p (Figure 11.2a). The external pressure gives rise to a radial force F_r and a circumferential force F_θ. For equilibrium, the external pressure has to be balanced by these forces.

For equilibrium. the external pressure on each hemispherical section is balanced by the internal forces F_θ. It is seen that the radial force is outward in Figure 11.2a and inward in Figure 11.2b. This violates the symmetry that we expect in the two free body diagrams. We conclude that no force acts in the radial direction, that is, $F_r = 0$ [22].

The total force normal to any circular section of a hemisphere due to external hydrostatic pressure p is equal to $p\pi R^2$. The resulting membrane stress σ_θ in the spherical shell is thus given by:

$$\sigma_\theta = \frac{F_\theta}{A} = \frac{p\pi R^2}{2\pi R t} = \frac{pR}{2t} \tag{11.1}$$

Expression (11.1) is termed the *boiler equation for spheres.* This equation was originally derived for cylindrical pressure vessels that are subjected to internal pressure. When external pressure is applied, buckling may take place, and this may precede yielding. The boiler yield stress is thus a theoretical upper bound to pressure vessel strength.

If $\sigma_\theta = \sigma_0$, the pressure to cause yielding, p_0, is given by:

$$p_0 = 2\sigma_0 \frac{t}{R} \tag{11.2}$$

We note that the pressure to cause yielding of a spherical shell is twice that of a cylindrical shell having the same radius-to-thickness ratio (3.49).

The application of external pressure results in a reduction of the radius. The constitutive relation is:

$$\varepsilon_\theta = \frac{1}{E}\left[\sigma_\theta - \nu\sigma_\theta\right] = \frac{1-\nu}{E}\sigma_\theta \tag{11.3}$$

where ε_θ, σ_θ are the membrane strain and stress respectively. Substituting from (11.2) in (11.3), we obtain the following relationship between ε_θ and the pressure p.

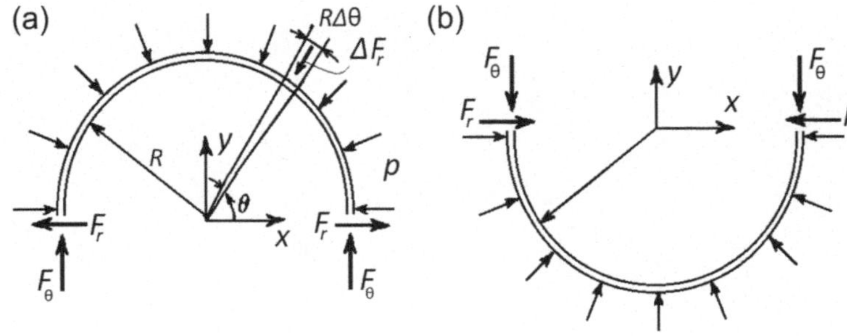

FIGURE 11.2 Forces acting on two hemispherical sections subjected to external hydrostatic pressure.

$$\varepsilon_c = \frac{pR}{2Et}[1-\nu] \tag{11.4}$$

Hence, the decrease in the length of the circumference is

$$(2\pi R)\frac{pR}{2Et}[1-\nu] \tag{11.5}$$

The new radius is:

$$2\pi R\left[1-\frac{pR(1-\nu)}{2Et}\right]/(2\pi) = R\left[1-\frac{pR(1-\nu)}{2Et}\right] \tag{11.6}$$

11.1.2 Classical Elastic Buckling Pressure of a Perfect Sphere

Because of their inherent strength, spheres are normally unstiffened. In the case of underwater vehicles, they are subjected to external pressure, whereas in the case of LNG tanks, the excess pressure is internal.

In this section, we shall outline the analysis that leads to the elastic buckling load of a spherical shell subjected to uniform external pressure, which is of importance in the first two applications. The contents of this section are an abridged version of the presentation included in [25]. The problem was originally treated by Zoelly in 1915 [27] and other researchers, assuming linear elastic *symmetric* behaviour. The solution of the buckling problem in which *asymmetric* buckling arises was given by van der Neut in 1932 [23]. The difference between symmetric and asymmetric behaviour lies in the fact that when buckling takes place, the shell initially buckles symmetrically but within a negligible time increment, the buckling mode changes to an asymmetrical form, as illustrated in Figure 11.6.

It is convenient to use a system of spherical coordinates whereby a point on the middle surface of the shell is defined by R, φ, and θ, as shown in Figure 11.3. R is the distance from a fixed point to the surface, which in the case of a spherical shell is its centre. It is therefore equal to the radial distance to the middle surface of the shell in the undeformed state, whereas φ is the angle of *elevation*, and θ is the angle of *azimuth* from a reference plane. The angle of elevation defines parallel circles, whereas the angle of azimuth defines meridians.

We consider an element OABC of a spherical shell, as shown in Figure 11.4. The element is defined by parallel circles at angles θ and $\varphi + d\varphi$ to the vertical axis and by meridians at an angle $d\theta$ apart. We define a local Cartesian coordinate system for the element with the origin at O and with

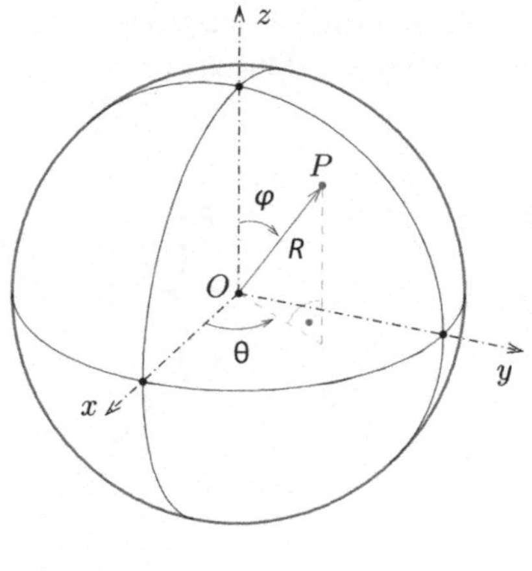

FIGURE 11.3 Spherical coordinate axis system.

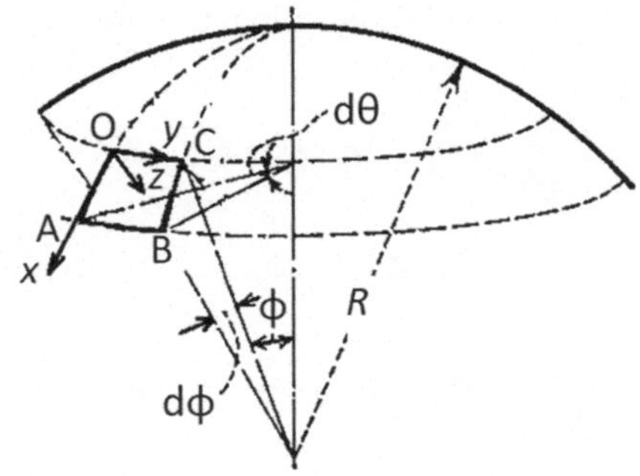

FIGURE 11.4 Element of spherical shell.

the Ox and Oy axes tangential to the meridian and the parallel circle, respectively (Figure 3.11). The Oz axis is taken to be perpendicular to the Oxy plane, positive inward. This local axis system is used to define translations of the element u, v, and w at O in the directions of the corresponding axes. The angle $d\theta$ can be obtained by rotating the meridional plane of side OA through angles $d\theta\sin\varphi$ and $d\theta\cos\varphi$ about the Ox and Oz axes, respectively (Figure 11.5).

Side AB is obtained by rotating side OC through the angle $-d\varphi$ with respect to the Oy axis (Figure 11.5). The in-plane normal forces N_θ, N_φ acting on the element OABC are shown in Figure 3.14, whereas the bending moments M_θ, M_φ and the transverse shear forces Q_θ, Q_φ are shown in Figure 3.17. No shear forces or twisting moments are included since the loading is rotationally symmetric, and the shell will fail in a symmetric manner. Equilibrium of the element is then satisfied by applying the following conditions:

$$\sum F_{Ox} = 0 \tag{11.7a}$$

$$\sum F_{Oz} = 0 \tag{11.7b}$$

$$\sum M_{Oy} = 0 \tag{11.7c}$$

Equilibrium in the direction of the Ox axis is satisfied by summing all force components as follows:

$$\left(N_\phi + \frac{\partial N_\phi}{\partial \phi} d\phi\right) R\sin(\phi + d\phi)d\theta - N_\phi R\sin\phi d\theta - N_\theta R d\phi \cos\phi d\theta - Q_\phi R\sin\phi d\theta d\phi = 0 \tag{11.8}$$

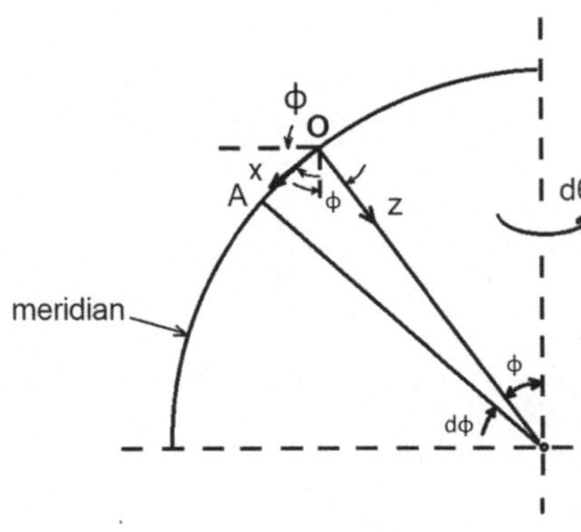

FIGURE 11.5 Generation of azimuthal angle, $d\theta$.

After applying the other two conditions and simplifying, the equations of equilibrium are obtained:

$$\left.\begin{aligned}
&\frac{\partial N_\phi}{\partial \phi}+(N_\phi - N_\theta)\cot\phi - Q_\phi = 0\\
&\frac{\partial Q_\phi}{\partial \phi}+Q_\phi\cot\phi + N_\phi + N_\theta + pR = 0\\
&\frac{\partial M_\phi}{\partial \phi}+(M_\phi - M_\theta)\cot\phi - Q_\phi R = 0
\end{aligned}\right\} \tag{11.9}$$

where p, the pressure load, is assumed to be perpendicular to the shell surface. As in the case of the cylindrical shell, we have to take into account the changes in angles between the sides of the element following deformation. Substituting the new angles in equations, we obtain

$$\frac{\partial N_\phi}{\partial \phi}+(N_\phi - N_\theta)\cot\phi - Q_\phi + N_\theta\left(\frac{u}{R}+\frac{1}{R}\frac{dw}{d\phi}\right) - Q_\phi\left(\frac{1}{R}\frac{d^2w}{d\phi^2}+\frac{w}{R}\right)=0 \tag{11.10a}$$

$$\frac{\partial Q_\phi}{\partial \theta}+Q_\phi\cot\phi + N_\phi + N_\theta + pR + N_\phi\left(\frac{1}{R}\frac{d^2w}{d\phi^2}+\frac{1}{R}\frac{du}{d\phi}\right)+N_\theta\left(\frac{u}{R}+\frac{1}{R}\frac{dw}{d\phi}\right)\cot\phi = 0 \tag{11.10b}$$

$$\frac{\partial M_\phi}{\partial \theta}+(M_\phi - M_\theta)\cot\varphi - Q_\phi R + M_\theta\left(\frac{u}{R}+\frac{1}{R}\frac{dw}{d\phi}\right)=0 \tag{11.10c}$$

If we eliminate Q_φ from these equations and substitute for M_φ, M_θ in terms of translations u, w using kinematic relations and also introduce the following constants,

$$\alpha_1 = \frac{D(1-\nu^2)}{R^2Et}=\frac{t^2}{12R^2} \qquad \alpha_2 = \frac{pR(1-\nu^2)}{2Et}$$

we obtain the equations of equilibrium in the following form:

$$\begin{aligned}
&(1+\alpha_1)\left[\frac{d^2u}{d\phi^2}+\cot\phi\frac{du}{d\phi}-(\nu+\cot^2\phi)u\right]-(1+\nu)\frac{du}{d\phi}\\
&+\alpha_1\left[\frac{d^3w}{d\theta^3}+\cot\phi\frac{d^2w}{d\phi^2}-(\nu+\cot^2\phi)\frac{dw}{d\phi}\right]-\alpha_2\left(u+\frac{dw}{d\phi}\right)=0
\end{aligned} \tag{11.11a}$$

$$\begin{aligned}
&(1+\nu)\left[\frac{du}{d\phi}+u\cot\phi-2w\right]+\alpha_1\left[-\frac{d^3u}{d\phi^3}-2\cot\phi\frac{d^2u}{d\phi^2}+(1+\nu+\cot^2\phi)\frac{du}{d\phi}\right.\\
&-\cot\phi(2-\nu+\cot^2\phi)u-\frac{d^4w}{d\phi^4}-2\cot\phi\frac{d^3w}{d\phi^3}+(1+\nu+\cot^2\phi)\frac{d^2w}{d\phi^2}\\
&\left.-\cot\phi(2-\nu+\cot^2\phi)\frac{dw}{d\theta}\phi\right]-\alpha_2\left(-u\cot\phi-\frac{du}{d\phi}+4w+\cot\phi\frac{dw}{d\phi}+\frac{d^2w}{d\phi^2}\right)=0
\end{aligned} \tag{11.11b}$$

After simplification by introducing the variable θ via the relation $u = d\theta/d\varphi$ equation (11.11a) becomes

$$\frac{d^2w}{d\phi^2}+\cot\phi\frac{d^2\theta}{d\phi^2}-(\nu+\cot^2\phi)\frac{d\theta}{d\phi}-(1+\nu)\frac{dw}{d\phi}$$
$$+\alpha_1\left[\frac{d^3w}{d\phi^3}+\cot\phi\frac{d^2w}{d\phi^2}-(\nu+\cot^2\phi)\frac{dw}{d\phi}\right]-\alpha_2\left(\frac{d\theta}{d\phi}+\frac{dw}{d\phi}\right)=0 \tag{11.12}$$

Recognising the repeated operation

$$\frac{d^2(...)}{d\phi^2}+\cot\phi\frac{d(...)}{d\phi}+2(...)$$

we use the symbol H for this and write (11.12) as follows:

$$\frac{d}{d\theta}[H(\theta)+\alpha_1H(w)-(1+\nu)(\theta+w)-\alpha_1(1+\nu)w-\alpha_2(\theta+w)]=0 \tag{11.13}$$

Performing the same substitution in (11.11b), simplifying and integrating both equations give the following:

$$H(\theta)+\alpha_1H(w)-(1+\nu)(\theta+w)-\alpha_2(\theta+w)=0 \tag{11.14a}$$

$$\alpha_1HH(\theta+w)-(1+\nu)H(\theta)-(3+\nu)\alpha_1H(w)+2(1+\nu)(\theta+w)$$
$$+\alpha_2[-H(\theta)+H(w)+2(\theta+w]=0 \tag{11.14b}$$

The system consisting of equation (11.14a and b) is solved using Legendre functions.[1] Substituting and after performing the necessary operations, two homogeneous equations are obtained. For buckling, the determinant of these equations has to be equal to zero. A characteristic equation is thereby obtained, and this leads to the following expression originally obtained by Zoelly [27] for the linear elastic symmetric buckling pressure of a perfect sphere:

$$p_{cr}=\frac{2E}{\sqrt{[3(1-\nu^2)]}}\left(\frac{t}{R}\right)^2 \tag{11.15}$$

11.1.3 Buckled Form and Collapse Pressure

Equation (11.15) concerns perfect spheres that buckle in a symmetric manner. The buckling pressure obtained from (11.15) is significantly higher than the buckling pressure of an equivalent cylindrical shell (same t/R), because the double curvature increases the shell stiffness.

The buckled form predicted by (11.15) is shown in Figure 11.6a and consists of a large number of uniformly distributed lobes over the entire surface. In the case of the test specimen shown in Figure 11.6a, a mandrel was placed internally, thus inhibiting the preferred deformation mode. Test results have revealed that:

1. The buckled form shown in Figure 11.6a is not observed in practice. At the very early stage, buckling does have the symmetrical mode but immediately changes to a single dimple, which is a strongly localised inward deflection at one pole and no deflection at the other pole, as shown in Figure 11.6b. In tests performed by Klöppel and Jungbluth, a high-speed sequence of pictures was obtained, and this showed that the dimple initiated near a clamped edge. The dimple grew at this position until the entire cap snapped through (Figure 11.6b).

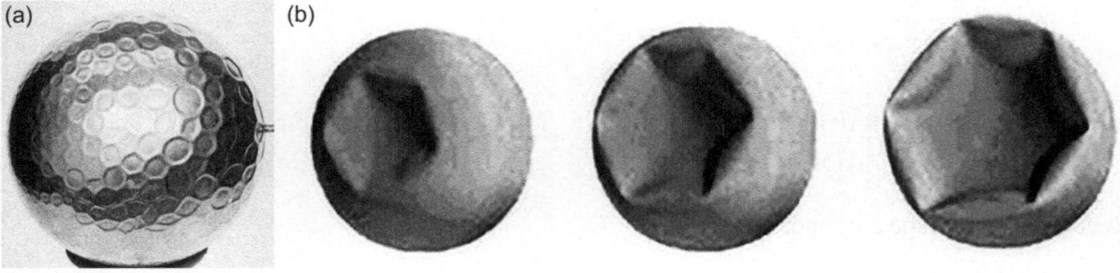

FIGURE 11.6 Elastic buckling of spherical shell due to uniform external pressure. (a) Symmetrical mode (theory). (b) Asymmetrical mode (single dimple).

2. Buckling occurs at significantly lower loads than that predicted by (11.15), as shown in qualitative terms in Figure 11.7a.

These differences from classical theory have been attributed to:

- Out-of-sphericity (initial geometrical imperfections)
- Thickness variations
- The effect of boundary conditions, including yielding along shell supports
- Variation in material properties

Of these, the most important contributing factor is initial geometrical imperfection. In 1939, von Kármán and Tsien analysed a spherical cap (Figure 11.8) assuming a buckled configuration such as that observed in tests [13]. The predicted response showed deformations in way of the dimple observed in tests ("snap-through" shown in Figure 11.7b). The analysis involved the use of an energy approach and a relationship that minimises the total energy at buckling was obtained.

Tests were subsequently carried out by research groups in several countries on spherical caps and selected data are included in Figures 11.21 and 11.22 (pp. 376–377).

A series of tests results at DTMB [18] on spherical caps that are included in Figure 11.21 (p.376) as black triangles showed that it is possible for the critical load to reach 70% of the classical value only in the case of test specimens with immeasurably small geometrical imperfections. For industrially produced structures, the critical load is much lower. Further details are included in Section 11.1.5.

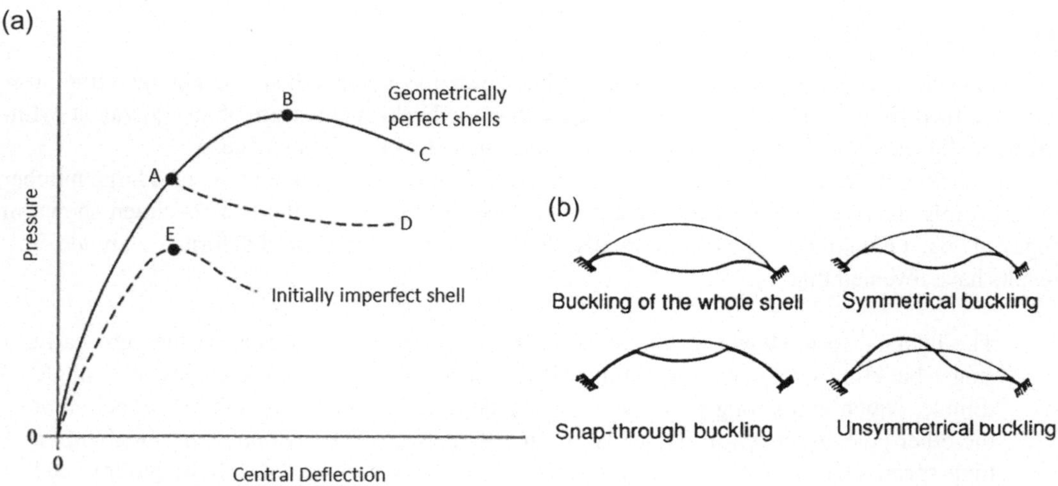

FIGURE 11.7 Hydrostatically loaded spherical caps. (a) Pressure–deflection form and (b) buckling modes.

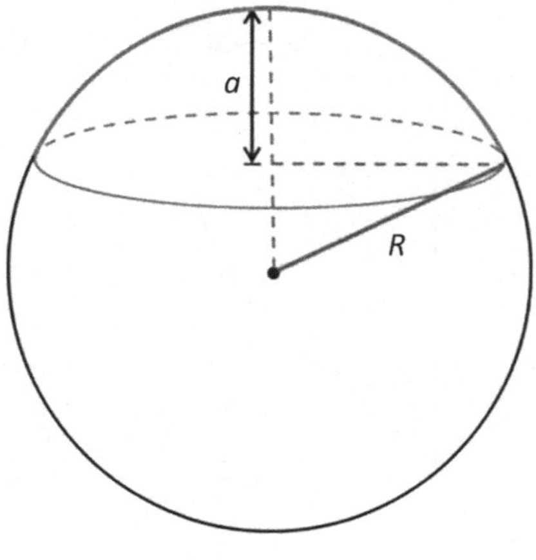

FIGURE 11.8 Geometry of spherical cap.

Figure 11.9 shows the effect of the shallowness parameter λ on the load–deflection curve of spherical caps loaded under external pressure [7] where

$$\lambda = 2\left[3(1-\nu^2)\right]^{1/4}\sqrt{\frac{a}{t}}$$

As the value of λ is increased, the response changes accordingly. For the flat plate ($\lambda = 0$), the stiffness increases as the pressure is increased, and membrane tension develops as it distorts under pressure. A slightly curved plate ($0 < \lambda < 3.5$) initially softens, then stiffens, but there is no local maximum load capacity. When $\lambda = 4$, nonlinear buckling with snap-through is observed. The stiffness reduces until it reaches zero at which point it snaps into an inverted position after which it stiffens with further increases in pressure. When $\lambda = 7$, bifurcation buckling is observed before axisymmetric snap-through takes place. Further increases in λ lead to similar behaviour but with an increasing degree of doubling-back, which characterises shells that are highly imperfection-sensitive.

In order to account for the reduction in collapse strength shown qualitatively in Figures 11.7a and 11.9, Koiter developed a general theory of elastic stability and performed a detailed analysis of the effect of initial geometrical imperfections [14,15]. These are important in the case of imperfection-sensitive structures such as cylindrical shells loaded in end compression and spheres under uniform pressure. The theory succeeded in explaining the divergence of the observed collapse pressure from that predicted by linear elastic buckling theory. An energy approach was followed, and the stability of equilibrium at the lower bifurcation point on the load–deflection diagram was determined using the principle of minimum potential energy. For a stable state, the second variation of the potential energy Π must be positive definite, that is, $\delta^2\Pi > 0$ (Figure 11.10a).

Koiter obtained relationships between the load and the deflection as well as imperfection sensitivity (Figure 11.11). It was found that asymmetric imperfections give rise to an asymmetric bifurcation load that is much lower than that predicted by classical theory (11.15). Figure 11.11a shows the classical buckling pressure p_{cr} of an inherently unstable structure such as a spherical shell. In a real structure, the observed collapse pressure p_c is significantly lower, and the relationship between p_{cr} and p_c is described by the imperfection sensitivity coefficient b.

This is plotted in Figure 11.11b for a sphere and a cylindrical shell subjected to end compression[2] as a function of the amplitude of the initial imperfection, w_0. It is seen that for highly imperfection-sensitive structures, ($b < -1.0$), the predicted collapse pressure p_c reduces to approximately 0.2–0.3p_c for $w_0 \approx t$.

The question of imperfection sensitivity of spherical shells and caps in particular is discussed at length in [7] in which several solutions of elastic buckling and postbuckling behaviour of spherical

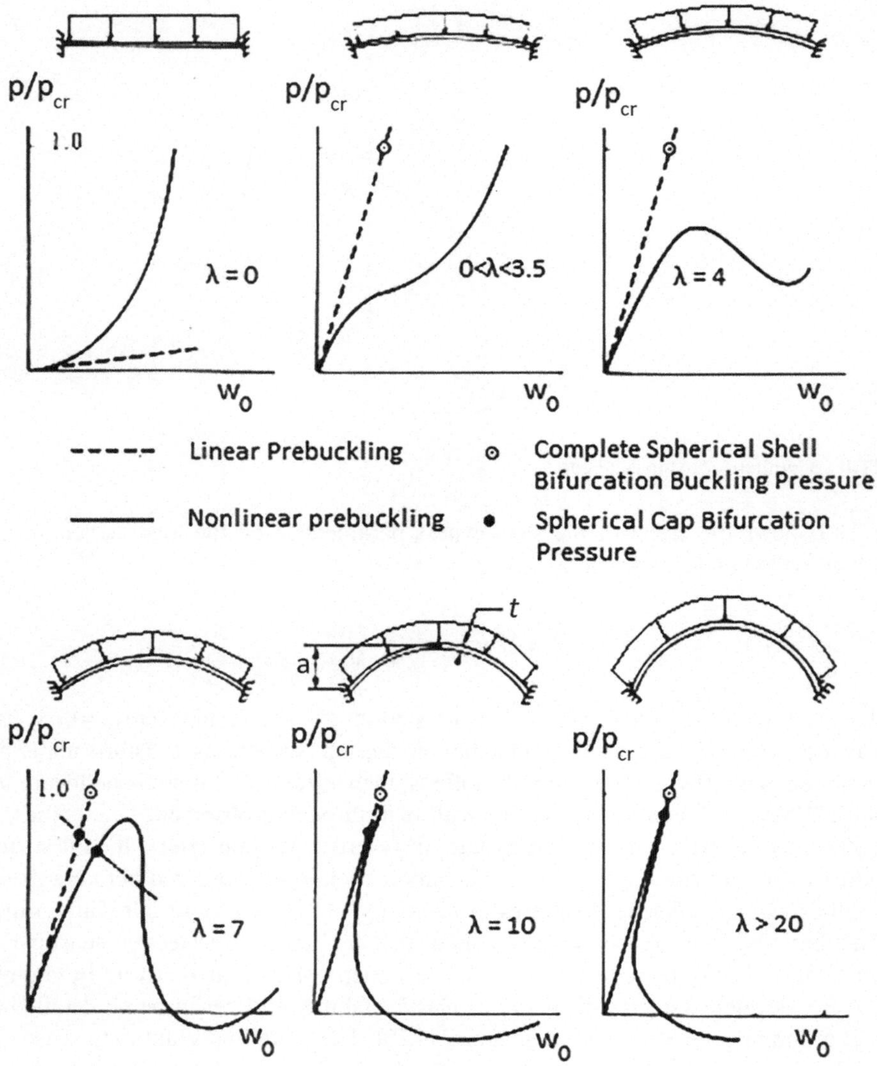

FIGURE 11.9 Effect of shallowness parameter λ on elastic buckling response [7].

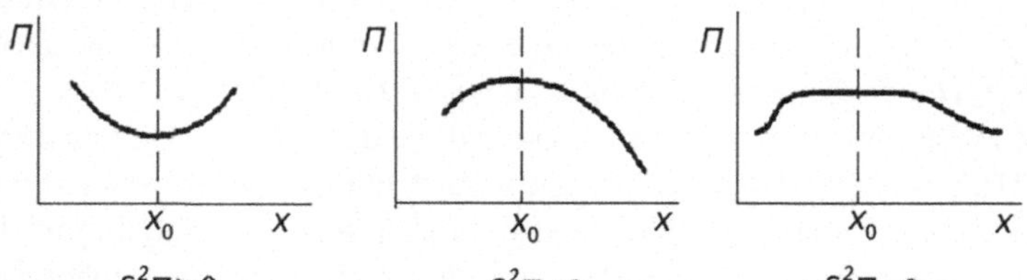

FIGURE 11.10 Equilibrium states of a body in a conservative force field. (a) Stable, (b) unstable, and (c) neutral.

shells with geometrical imperfections are described. It is noted that it is very difficult to apply Koiter's asymptotic theory to the case of spherical shells because a cluster of buckling modes exists at a critical pressure that is very slightly above the lowest critical mode [7].

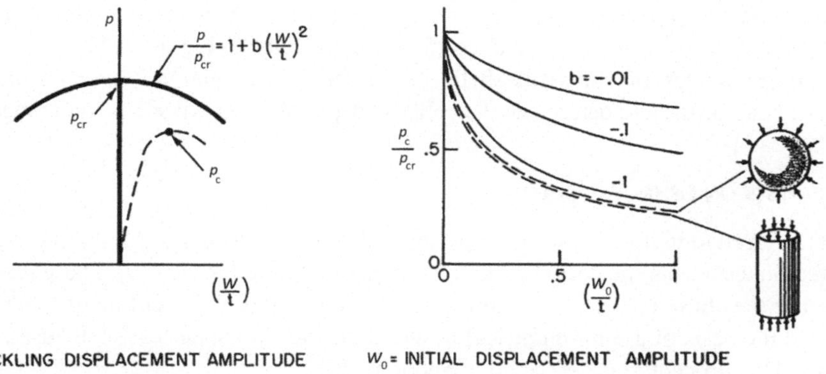

FIGURE 11.11 Relationship between classical and observed buckling pressure [6].

An expression was given by Mushtari and Surkin that predicts buckling loads lower than (11.15) by a factor of approximately one-third [9]:

$$p_{cr} = 0.36\frac{E}{\sqrt{1-\nu^2}}\left(\frac{t}{R}\right)^2 \tag{11.16}$$

In addition to this, there is now available a wide range of theoretical predictions for the elastic buckling load. The most well-known of these is that developed at DTMB, as follows [19]:

$$p_3 = 0.84E\left(\frac{t}{R_0}\right)^2 \tag{11.17}$$

where R_0 is the outer shell radius. This provides a reasonable estimate of strength provided that the initial departure of sphericity does not exceed 2.5% of the shell thickness (near-perfect shell).

11.1.4 Inelastic Buckling of Spherical Shells

The pressure to cause membrane yielding of a spherical shell is given by (11.2). We note that the failure pressure depends on material properties and the thickness-to-radius ratio, t/R. For HY-80 steel, we have $\rho = 7{,}800$ kg/m^3, $E = 207$ GPa, $\nu = 0.3$, and $\sigma_0 = 690$ N/mm^2.

Table 11.1 includes the failure pressures for elastic buckling and yielding for a spherical shell for t/R ratios in the range $0.005 < t/R < 0.05$. It is seen that in the case of very thin spherical shells, elastic buckling occurs for depths up to 1,000 m and is the critical failure mode. For $t/r \geq 0.02$, failure is

TABLE 11.1
Collapse Mode and Depth for a Spherical Shell

t/R	p_0 (MPa)	p_{cr} (MPa)	Collapse Depth (m)
0.005	6.9	1.95	195
0.010	13.8	7.80	780
0.020	27.6	31.25	2,760
0.040	55.2	125.0	5,520
0.050	69.0	195.0	6,900

inelastic and occurs at depths that exceed 1,000 m. It is therefore necessary to consider the inelastic collapse of spherical shells for use in deep-diving submersibles (Figure 11.12).

The failure modes of spherical shells are described in qualitative terms in Figure 11.13. Figure 11.14 can be used to determine the collapse depth of spherical shells for a range of materials.

11.1.5 Tests on Spherical Shells

The tests referred to in this section were performed at the David Taylor Model Basin laboratories in the United States during the period 1962–1965 and are reported in [16–19]. These results are important because they have provided an improved understanding of the buckling of spherical shells that has formed the basis of a semi-empirical design procedure for spherical shells used in underwater structures. This procedure is described in Section 12.3.1.

11.1.5.1 Buckling strength of hemispherical shells (elastic and inelastic)

The first group of tests that were reported in [16] consisted of 26 models of hemispherical shells each bounded by a ring-stiffened cylinder (Figure 11.15) machined from aluminium alloy bar stock.

Of these, 13 models were machined from 6061-T6 aluminium alloy with nominal yield strength of 43,000 psi, and 13 models were machined from 7075-T6 alloy with nominal yield strength of 80,000 psi. This series of tests covers thickness-to-radius ratios from about 1/10 to 1/100.

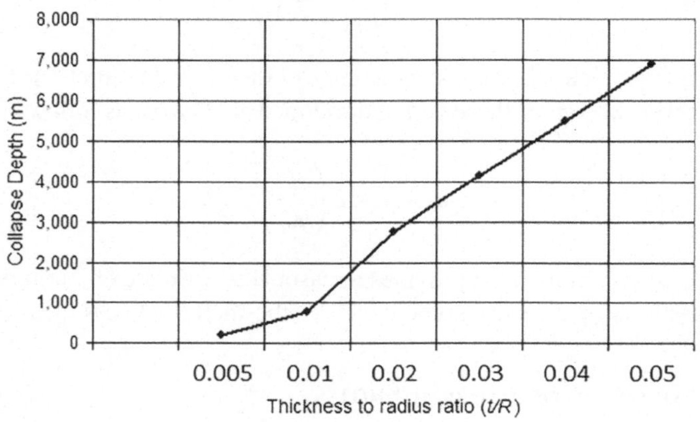

FIGURE 11.12 Collapse depth versus thickness to radius ratio (t/R).

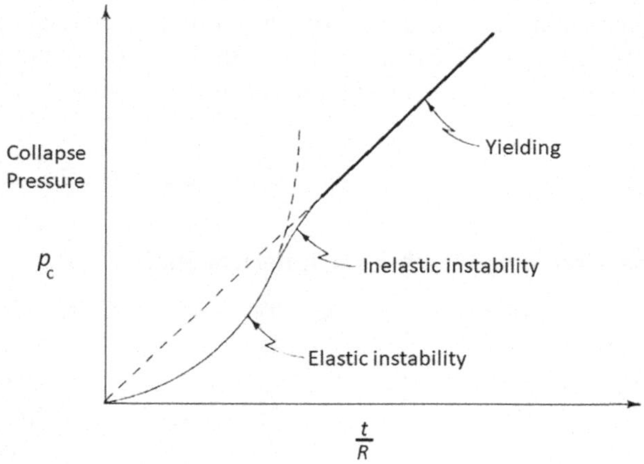

FIGURE 11.13 Failure modes of spherical shells used in deep-diving submersibles [8].

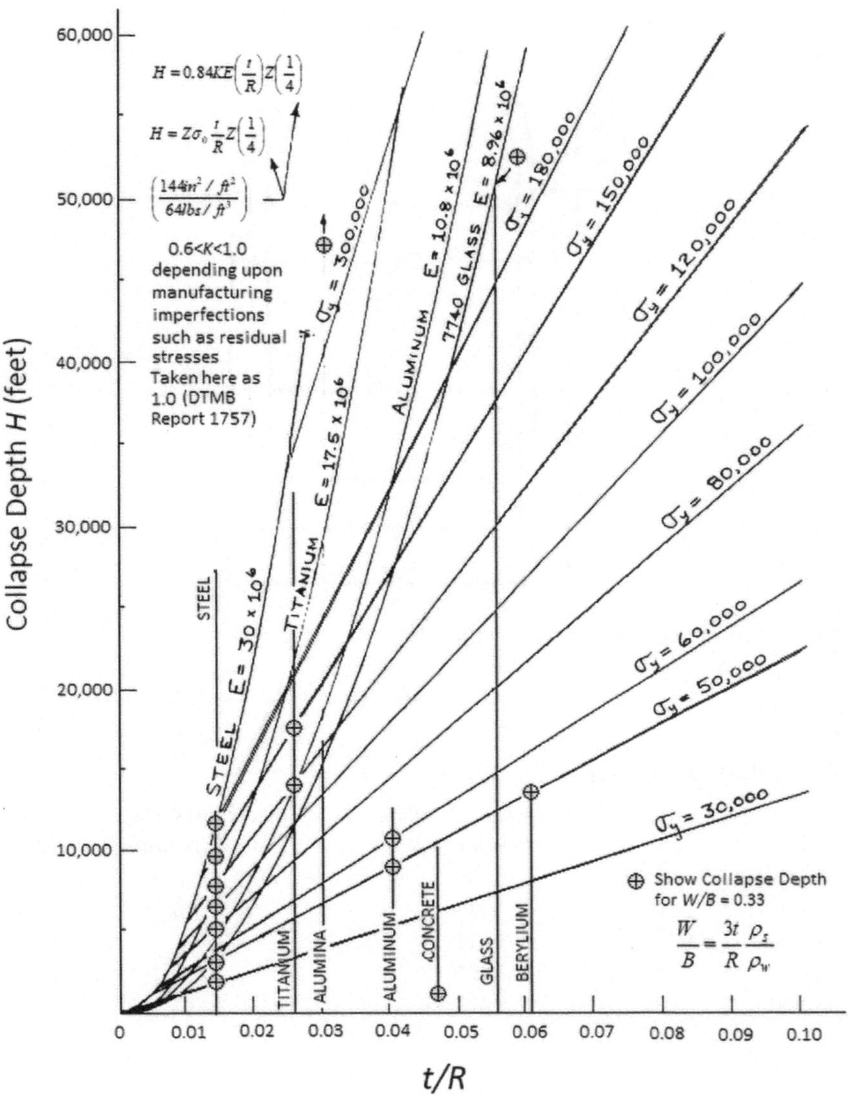

FIGURE 11.14 Collapse of spherical shells for a range of materials [8].

Table 11.2 includes test specimen particulars, the observed collapse pressure, and the elastic buckling pressure predicted using the Zoelly formula, (11.15), of the five models failing in the elastic range. These specimens were made of 7079-T6 aluminium having a yield stress of 43,000 psi (2,910 MPa).

It is seen that the ratios of the experimental collapse pressure to the theoretical collapse pressure according to the Zoelly formula lie in the range 0.53–0.69. These ratios are significantly larger than those obtained in earlier tests reported in the literature. The important difference between this series of models with earlier ones is that these models are machined, thereby reducing significantly the magnitude of initial geometrical imperfections, residual stresses, and adverse boundary conditions. In earlier tests, the specimens were formed from flat plates, a procedure that introduces initial imperfections (distortions and residual stresses) (Figure 11.16).

It is thus seen that the theoretical prediction of the Zoelly formula appears to be correct in the case of perfect spherical shells. In practice, it is not possible to manufacture spherical shells to the

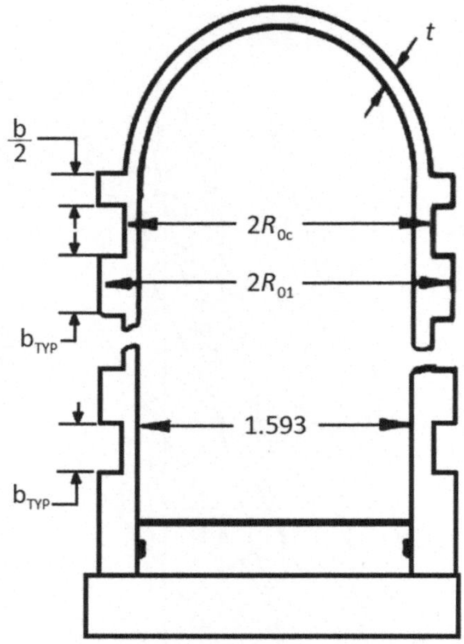

FIGURE 11.15 Group 1 model specimen arrangement.

TABLE 11.2
Specimen Particulars and Collapse Pressures

Model no	t (in)	R (in)	Experimental Collapse Pressure (psi)	Theoretical Collapse Pressure (psi)	Experimental Theoretical
M	0.0076	1.620	800	1,180	0.68
N	0.0077	1.630	830	1,210	0.69
O	0.0100	1.616	1,170	2,030	0.58
OO	0.0096	1.630	1,230	1,875	0.66
P	0.0106	1.616	1,215	2,285	0.53

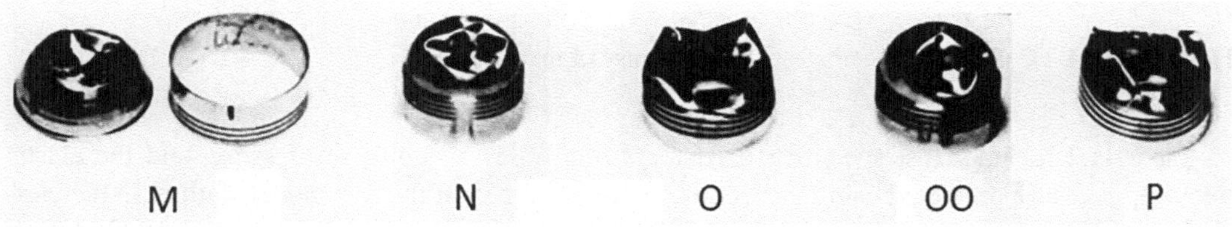

FIGURE 11.16 Hemispherical sections failing by elastic buckling [16].

required degree of perfection so that design formulas are needed to allow for the reduction and variation in strength resulting from initial distortions, residual stresses as well as the effects of boundary conditions.

In order to consolidate the findings of these tests, two more models (PS-1, PS-2) were tested [17]. In these, the specimens were fabricated with extreme care in order to reduce to the maximum possible extent the presence of initial distortions. The aim of these tests was to confirm that departures from the theoretical elastic buckling Zoelly formula are attributable to the presence of initial distortions. Results are included in Table 11.3.

TABLE 11.3
Collapse Pressures for Accurately Machined Spherical Shells under External Pressure (7075-T6 Aluminium, Yield Stress = 80,000 psi)

Model no	Experimental Collapse Pressure (p_{exp})	$\frac{p_{exp}}{p_1}$[a]	$\frac{p_{exp}}{p_2}$[b]
PS-1	347	0.73	1.06
PS-2	429	0.90	1.30

[a] p_1 = Zoelly formula (11.15).
[b] p_2 = DTMB empirical buckling pressure formula (11.17).

TABLE 11.4
Experimental and Predicted Collapse Pressures for Models Failing by Inelastic Buckling

Model no	Experimental Collapse Pressure p_{exp} (psi)	$\frac{p_{exp}}{p_{B/G}}$	$\frac{p_{exp}}{p_L}$	$\frac{p_{exp}}{p_E}$
A	890	0.89	0.89	0.95
B	935	0.89	0.89	0.96
C	1,810	0.84	0.84	0.89
D	1,790	0.83	0.83	0.88
E	3,350	0.87	0.87	0.94
F	3,375	0.86	0.86	0.93
G	4,950	0.82	0.82	0.96
H	5,150	0.86	0.86	0.99
I	5,900	0.77	0.87	0.89
II	6,300	0.83	0.83	0.95
J	6,450	0.80	0.80	0.94
K	8,200	0.83	0.83	0.96
L	8,200	0.80	0.80	0.94
Q	3,160	0.89	0.89	0.94
R	3,280	0.92	0.92	1.00
S	5,875	0.90	0.90	0.97
T	6,200	0.89	0.89	0.98
U	9,050	0.88	0.88	0.98
V	9,300	0.91	0.91	1.02
W	12,200	0.88	0.88	1.02
X	12,150	0.87	0.87	1.02

It is seen that in the case of model PS-2, a ratio of 0.90 was obtained, this being by far the highest ratio of experimental to theoretical collapse pressure recorded in the literature for spherical shells subjected to external pressure. This increase in buckling strength is attributed to the near-perfect sphericity of the test specimen.

The remaining 21 models of this first group of tests failed in the inelastic range. The collapse pressures and the ratios of experimental to predicted values are included in Table 11.4. The formulas used to predict inelastic collapse are given below:

Bijlaard and Gerard formula:

$$p_{B,G} = 1.154\sqrt{\frac{E_s E_t}{(1-\nu_p{}^2)}}\left(\frac{t}{R}\right)^2 \tag{11.17}$$

Lunchick formula:

$$p_L = 1.154\sqrt{\frac{E_s E_t}{(1-\nu_t)(1+\nu_s)}}\left(\frac{t}{R}\right)^2 \tag{11.18}$$

$$\text{where} \quad \nu_t = \frac{1}{2} - \left(\frac{1}{2} - \nu\right)\frac{E_t}{E}$$

$$\nu_s = \frac{1}{2} - \left(\frac{1}{2} - \nu\right)\frac{E_s}{E}$$

Empirical formula:

$$p_E = 0.8\sqrt{\frac{E_s E_t}{(1-\nu^2)}}\left(\frac{t}{R_0}\right)^2 \tag{11.19}$$

In these expressions, E_s = secant modulus, E_t = tangent modulus, ν = Poisson's ratio (elastic), ν_p = Poisson's ratio (plastic), and (ν_s, ν_t) are used to determine a composite Poisson's ratio for the plastic range in the Lunchick formula, R_0 = radius to outer surface of shell.

It is seen from Table 11.4 that in all cases, the available theories of inelastic instability of Bijlaard, Gerard, and Lunchick gave collapse pressures that are much higher than the corresponding experimental collapse pressures. The DTMB empirical formula (19.19) was based on the actual collapse strength of the models failing elastically (Table 11.3) and the use of a simple plasticity reduction factor. The effects of a variable Poisson's ratio were ignored (doing so leads to a more conservative result), and it is seen that this empirical formula gives more accurate values [16].

11.1.5.2 The effect of unsupported shell length on buckling strength of hemispherical shells

The aim of the third group of tests was to study the effect of unsupported shell length on both elastic and inelastic buckling strength. In total, 102 small machined spherical shells were tested, grouped as follows:

1. Four models representing complete spheres (Series S) (Figure 11.17).
2. Eight 300° spherical segments with clamped edges (Series DSS) (Figure 11.18).
3. Seventy three clamped spherical segment models with included angles in the range 5°–120° (Series SS) (Figure 11.19).
4. Seventeen externally stiffened machined hemispheres designed for an exploratory study of the effect of stiffening systems on the collapse strength of relatively unstable spherical shells (Series SSS). Nine models had circumferential stiffeners only, four had meridional stiffeners only, and four had both circumferential and meridional stiffeners (Figure 11.20).

11.1.5.2.1 Series S

Each model in Series S ruptured during collapse, with failure occurring in regions removed from the hatch. Table 11.5 presents the average stresses of the Series S model at collapse calculated using the following equation:

$$\sigma_{avg} = \frac{pR_0^2}{2Rt} \tag{11.20}$$

where R is the radius to the midsurface of the shell, and R_0 is the outer radius.

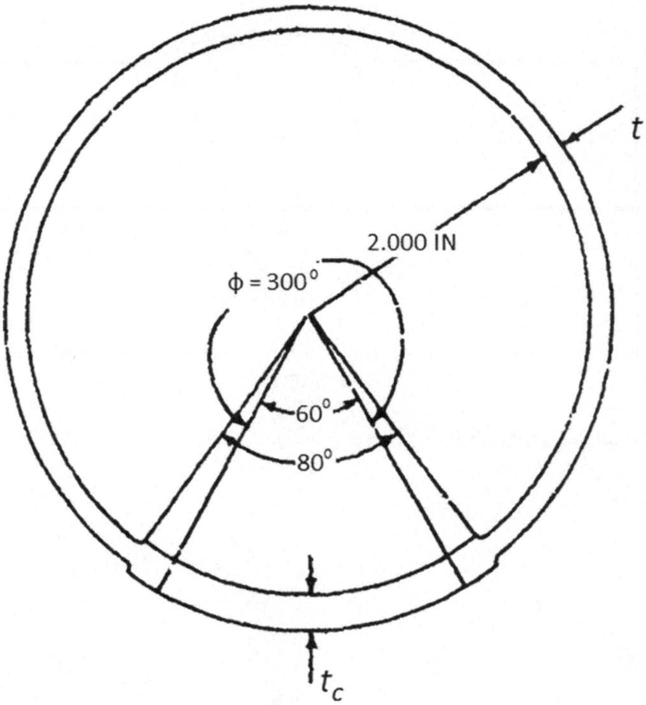

FIGURE 11.17 Complete sphere (series S).

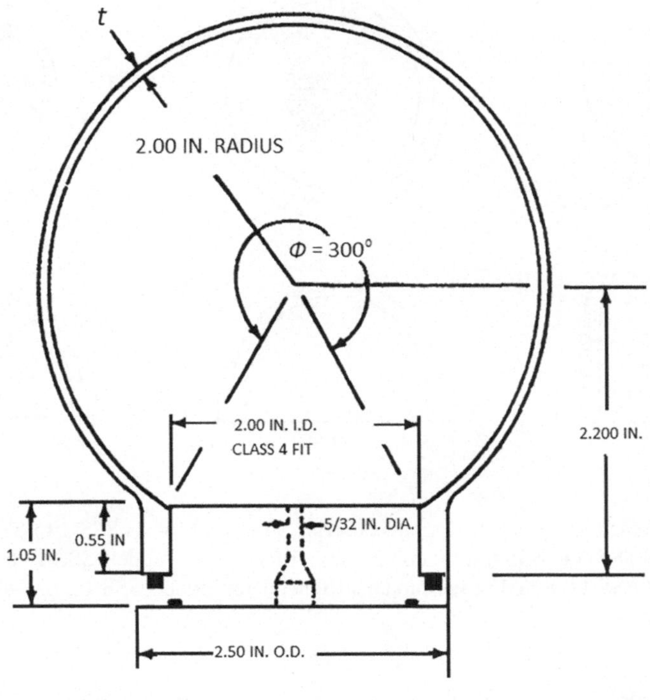

FIGURE 11.18 Spherical segment (series DSS).

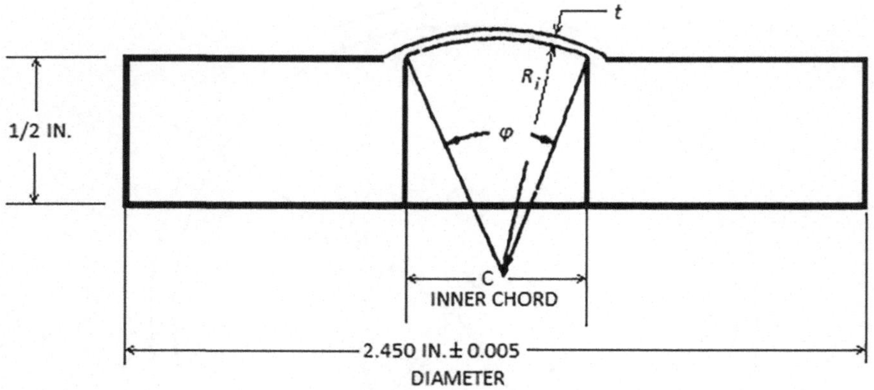

FIGURE 11.19 Spherical cap (series SS) (120° > φ > 5.3°).

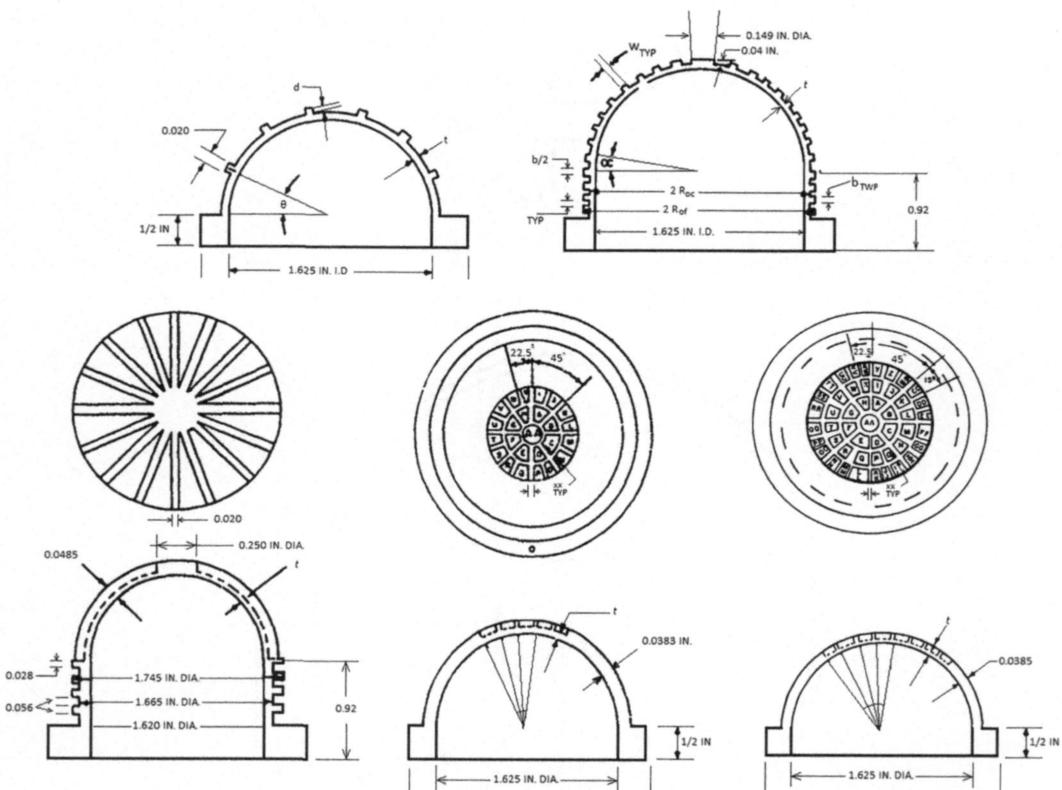

FIGURE 11.20 Stiffened spherical shell models (Series SSS). (a) Models SSS-1 to SSS-4 (circumferential stiffeners). (b) Models SSS-5 to SSS-9 (circumferential stiffeners). (c) Models SSS-10 to SSS-13 (meridional stiffeners). (d) Models SSS-14, SSS-15 (meridional stiffeners). (e) Models SSS-16, SSS-17 (circumferential & meridional stiffeners).

Each model collapsed at stress levels equal to or greater than the proportional limit of the material. Thus, each model collapsed by inelastic buckling.

Since the collapse of each model in Series S took place in the inelastic range, the empirical equation (11.19) is considered applicable in calculating their collapse strength. It may be repeated that (11.19) was proposed for use in calculating the inelastic as well as elastic buckling strength

TABLE 11.5
Geometric Parameters and Collapse Pressures for Series S Models (psi)[a] [18]

Model	φ (deg)	p_{exp}	σ_{avg}	p_3	p_E	$\frac{p_3}{p_E}$	$\frac{p_{exp}}{p_E}$
S-1	360	14,300	84,300	70,260	14,150	5.05	1.01
S-2	360	8,500	76,500	29,530	8,720	3.40	0.98
S-3	360	3,000	62,300	5,355	3,400	1.58	0.88
S-4	360	1,500	58,300	1,495	1,465	1.02	1.02

[a] p_3 is calculated using (11.17).

of machined deep spherical shells with ideal boundaries. The experimental collapse pressures of these models are compared in Table 11.5 with the pressures calculated using (11.19). Table 11.5 also presents the empirical elastic buckling pressure, p_E, calculated using (11.19) and indicates the relative stability of each model by presenting the ratio of the empirical elastic buckling pressure to the empirical inelastic buckling pressure.

11.1.5.2.2 Series DSS

Models DSS-1 through DSS-8 each ruptured during collapse. All models failed in the shell in areas away from the end ring except for Models DSS-7 and DSS-8 whose failure initiated in the shell at the junction with the heavy end ring. Details are included in Table 11.6.

11.1.5.2.3 Series SS

Models SS-1 through SS-42 and Models SS-57 through SS-73 each collapsed in a sudden, unmistakable manner. Some of the models collapsed in a non-symmetric mode, and others apparently collapsed in an asymmetric mode. Models SS-43 through SS-56 were not collapsed nor did rupture under external pressure.

11.1.5.2.4 Elastic Buckling of SS Models

The collapse pressures of the shallow spherical segment models that failed elastically (Models SS-25, SS-26, SS-33 through SS-36, and SS-57 through SS-73) are compared in Figure 11.21 with test results recorded in the literature and with nonlinear asymmetric theory of spherical shell buckling.

The ordinate is the ratio of the experimental collapse pressure p_{exp} to p_3, the empirical elastic buckling pressure for machined deep spherical shells with membrane boundaries, given by:

TABLE 11.6
Geometric Parameters and Collapse Pressure for Series DSS Models (psi)

Model	φ (deg)	θ	p_{exp}	σ_{avg}	p_3	p_E	$\frac{p_3}{p_E}$	$\frac{p_{exp}}{p_E}$
DSS-1	300	17.20	11,100	74,925	49,780	10,800	4.61	1.03
DSS-2	300	16.84	11,375	74,630	52,690	11,660	4.52	0.98
DSS-3	300	25.57	4,720	69,310	10,520	4,810	2.19	0.98
DSS-4	300	19.38	8,400	73,565	31,390	8,420	3.73	1.00
DSS-5	300	30.21	2,800	57,555	5,500	3,250	1.69	0.86
DSS-6	300	30.84	2,925	64,305	4,800	2,975	1.61	0.98
DSS-7	300	44.95	1,200	54,470	1,113	1,070	1.04	1.12
DSS-8	300	42.73	1,290	52,985	1,360	1,330	1.02	0.97

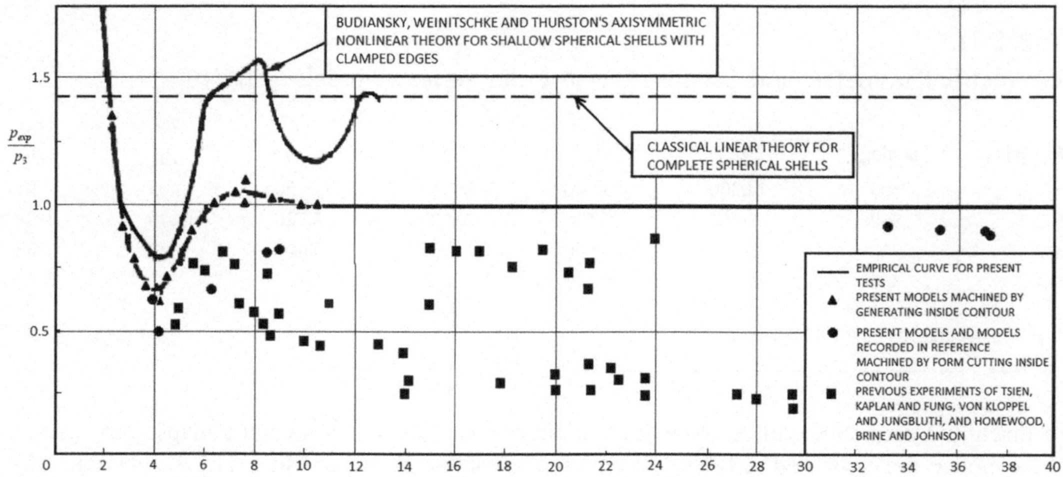

FIGURE 11.21 Elastic buckling pressure versus arc length parameter,[3] θ [18].

$$p_3 = 0.84E\left(\frac{t}{R_0}\right)^2 \quad (v = 0.3) \tag{11.21}$$

The abscissa is the nondimensional arc length parameter ϑ defined by:

$$\vartheta = \left[\frac{3}{4}(1-v^2)\right]^{\frac{1}{4}} \frac{L_a}{\sqrt{Rt}} = \frac{0.91L_a}{\sqrt{Rt}} \quad (v = 0.3) \tag{11.22}$$

where L_a is the unsupported arc length of a spherical shell

It is seen in Figure 11.21 that in the range $5.5 > \vartheta > 0$, the theoretical curve based on nonlinear theory for shallow spherical shells is of the same form as the curve drawn through the experimental results reported here. For $\vartheta > 5.5$, the discrepancies that arise between the theoretical curve and the experimental results have been attributed to the fact that the theoretical prediction is based on a symmetric buckling mode, whereas in practice, the mode can be asymmetric. One further observation is that small clamped segments ($\vartheta < 5.5$) fail at a lower load than longer segments ($\vartheta \gg 5.5$). It appears therefore that stiffening of spherical shells may not necessarily lead to an increase in strength. This question is addressed further in the next series of tests.

11.1.5.2.5 Inelastic Buckling of SS Models

The test results of those models that failed at stress levels above the proportional limit of the materials are presented in Figure 11.22 as a function of the arc length parameter ϑ. The results are quite consistent when plotted in contours of ratios of p_E to $p_{cr,}$ where p_{cr} is the classical elastic buckling pressure of a perfect sphere, given by (11.15). For those deep segments which had ratios $p_3/p_E = 1$ but had membrane stresses of collapse (σ_{avg}) of about 70% of the proportional limit, the observed collapse strength was about 20% less than would be expected for a near-perfect complete sphere or a deep machined sphere with ideal boundaries. As p_3/p_E increased (that is, as the shell became more stable), the effects of clamping the edges diminished for all except the very shallow, or short, shells.

Regardless of the value of the ratio p_E/p_{cr}, the buckling strength of all segments increased rapidly for values of ϑ below 2.2–2.5. Therefore, it is apparent that stiffeners must be placed at relatively close intervals in order to realise an increase in either elastic or inelastic buckling strength. Placing

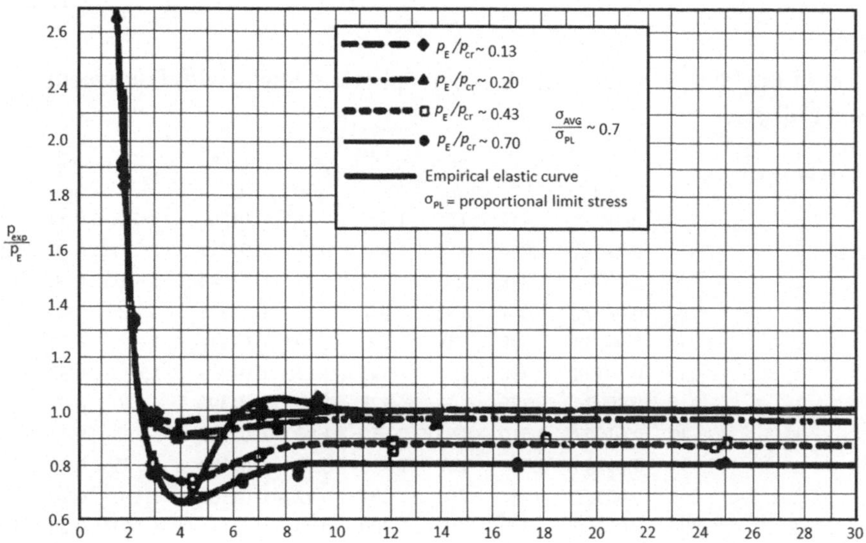

$$\vartheta = [\frac{3}{4}(1-\nu^2)]^{1/4}\frac{L_a}{\sqrt{Rt}} = \frac{0.91L_a}{\sqrt{Rt}}$$

FIGURE 11.22 Inelastic buckling pressure versus arc length parameter, ϑ [19].

stiffeners at spacing greater than the arc length corresponding to a ϑ approximately equal to 2.5 will not increase the local buckling strength of the shell and may possibly weaken it. This question is investigated further in the SSS series of tests.

Another conclusion obtained from these tests is related to initial departures from sphericity or local "flat spots". If a spherical shell contains a flat spot covering an arc length associated with a value of $\vartheta \geq 2.2$, the collapse strength of this shell appears to be relatively independent of the nominal radius of the shell. That is, collapse is a local phenomenon and collapse strength ought to be calculated on the basis of local geometry over a "critical length" rather than on nominal geometry, as practiced in the past.

If it is assumed that a critical arc length, L_c, is associated with $\vartheta = 2.2$, we can rearrange (11.22) to obtain an expression for L_c:

$$L_c = \frac{2.2\sqrt{R_l t}}{\left[3/4(1-\nu^2)\right]^{\frac{1}{4}}} = 2.42\sqrt{R_l t_a} \quad (\nu = 0.3) \tag{11.23}$$

where t_c is the average thickness over the critical length, and R_l is the local radius to the midsurface of the shell over a critical arc length for which $\theta = 2.2$.

A semi-empirical design methodology for deep spherical shells having initial geometrical imperfections based on this approach is discussed in Chapter 12.

11.1.5.2.6 Series SSS Models

The experimental collapse pressures for spherical shells stiffened externally in the circumferential, meridional, and both directions (Figure 11.20) are compared in Table 11.7 with:

a. The estimated collapse strength of the shells neglecting the stiffeners
b. The estimated collapse strength of equivalent unstiffened shells of the same weight

TABLE 11.7
Comparison of Buckling Pressures of Stiffened Spherical Shells with Equivalent Unstiffened Ones

Model	Stiffener Type[a]	p_{exp}	$\frac{t}{R_0}$	p_3	$\frac{p_{exp}}{p_3}$	p_E	$\frac{p_{exp}}{p_E}$	$\frac{t_e}{R_0}$,[b]	p_3'[c]	$\frac{p_{exp}}{p_3'}$	p_E'	$\frac{p_{exp}}{p_E'}$
SSS-1	1	1,100	0.01275	1,475	0.75	1,437	0.76	0.01502	2,045	0.54	1,850	0.60
SSS-2	1	875	0.01083	1,064	0.82	1,064	0.82	0.01311	1,555	0.56	1,505	0.58
SSS-3	1	700	0.00902	738	0.95	738	0.95	0.01244	1,405	0.50	1,390	0.50
SSS-4	1	920	0.01071	1,041	0.88	1,048	0.88	0.01526	2,115	0.44	1,905	0.48
SSS-5	1	950	0.00938	798	1.19	798	1.19	0.01418	1,825	0.52	1,705	0.56
SSS-6	1	930	0.00938	798	1.17	798	1.17	0.01409	1,800	0.52	1,690	0.55
SSS-7	1	915	0.00926	778	1.18	778	1.18	0.01397	1,770	0.52	1,665	0.55
SSS-8	1	2,740	0.01967	3,510	0.78	2,650	1.03	0.02423	5,325	0.51	3,440	0.80
SSS-9	1	2,390	0.01836	3,060	0.78	2,430	0.98	0.02294	4,775	0.50	3,200	0.75
SSS-10	2	265	0.00562	287	0.92	287	0.92	0.01107	1,110	0.24	1,110	0.24
SSS-11	2	290	0.00575	300	0.97	300	0.97	0.011198	1,140	0.26	1,140	0.26
SSS-12	2	435	0.00757	520	0.84	520	0.84	0.01881	3,210	0.14	2,510	0.17
SSS-13	2	825	0.01072	1,042	0.79	1,042	0.79	0.02057	3,840	0.22	2,810	0.29
SSS-14	3	1,325	0.00944	809	1.64	809	1.64	0.02003	3,640	0.36	1,875	0.71
SSS-15	3	1,345	0.00986	882	1.53	882	1.53	0.02076	3,910	0.34	1,815	0.74
SSS-16	3	1,125	0.00830	625	1.80	625	1.80	0.02187	4,340	0.26	1,660	0.68
SSS-17	3	870	0.00627	356	2.44	356	2.44	0.01993	3,600	0.24	1,395	0.62

[a] 1. Stiffeners parallel to hemispherical openings only
2. Meridional stiffeners only
3. Both stiffeners parallel to opening and meridional stiffeners

[b] t_e represents the shell thickness of an equivalent unstiffened sphere of the same weight.

[c] Prime designates use of equivalent shell thickness in calculations.

It was found that:

- Installing stiffeners at sufficiently close intervals increased the collapse strength of the spherical shells. However, this increase was in no case as great as would be expected for a machined unstiffened shell of the same weight (Table 11.7).
- Models having an unsupported arc length $\vartheta > 2.2$ collapsed at a pressure below that calculated for the unstiffened shell.
- Models SSS-1 through SSS-13 failed in a very sudden manner, and no drop in applied hydrostatic pressure prior to collapse was observed.
- The models with both circumferential and meridional stiffeners (Models SSS-14 through SSS-17) were considerably stronger than predicted for the unstiffened portion only

The difficulty in designing an efficient stiffening system stems from the stress field that arises when a shell is subjected to external pressure. A stiffener resists load in its axial direction only, whereas the external pressure produces a uniform biaxial stress field in an unstiffened spherical shell. Since a stiffener offers no resistance to load in the plane normal to the stiffener, it can be no more than half as efficient as the shell material, if buckling is neglected.

A stiffened spherical shell can buckle in various modes, depending on the stiffening system. Placing stiffeners in a single direction at relatively large intervals will have no strengthening effect on a spherical shell and may possibly have a slight weakening effect. The maximum pressure to be achieved from a spherical shell with stiffeners in one direction only is that pressure, which causes the stresses to reach the yield point of the material, given by:

$$p_u \equiv p_0 = \frac{2\sigma_0 Rt}{R_0^{\,2}} \tag{11.22}$$

For spherical shells uniformly stiffened in two directions, the maximum pressure to be achieved is that pressure, which causes the average stresses in the stiffener–shell combination to reach the yield point of the material. This pressure may be determined by substituting an equivalent thickness in (11.22):

$$p_u \equiv p_0 = \frac{2\sigma_0 Rt_e}{R_0^{\,2}} \tag{11.23}$$

11.1.5.2.7 General conclusions from the DTMB series of tests on spherical shells and caps [18]

Previous experiments on shallow spherical shells with clamped edges recorded in the literature showed lack of repeatability as well as significantly lower buckling strength. The models used in the earlier tests were fabricated using pressed and welded sheet metal so that initial distortions, residual stresses, and thickness variations were introduced. These contributed to significant strength reductions, compared to the theoretical elastic buckling pressure. Of these factors, the most important one is the presence of initial distortions. Figure 11.23 shows in a qualitative manner that increases in the magnitude of the initial distortions have a detrimental effect on buckling strength. In fact, the loss of strength is a function of *t*/*R*, so that as slenderness increases, so does the reduction in strength. In other words, thicker shells are less imperfection-sensitive. The results obtained from the tests reported in this section, which had been machined, follow a very definite pattern, as seen in Figure 11.21, with the buckling pressure approaching that given by the Zoelly formula.

In the case of initially stress-free complete spheres or of deep spheres[4] with ideal boundaries whose local variations in radii are less than about 2.5% of the shell thickness, (11.19) can be used to predict both the elastic and inelastic strength.

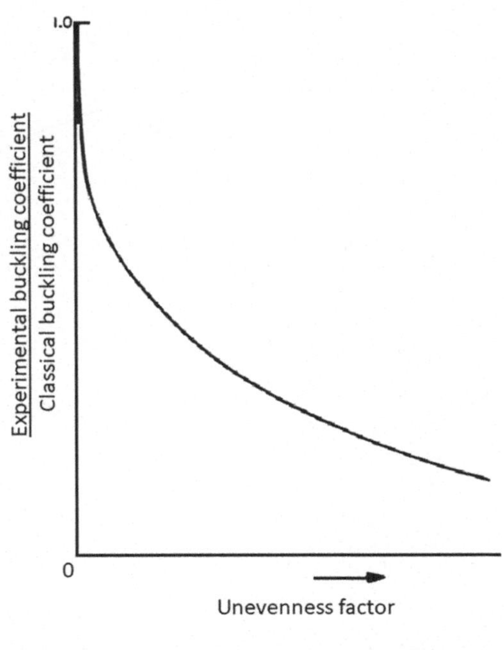

FIGURE 11.23 Effect of initial distortions on buckling strength of spherical shells [19].

A spherical segment with an arc length associated with a value of the geometric parameter θ of about 2.2 or greater is not strengthened by its boundary. Thus, stiffeners must be placed at very close intervals in order to increase the strength of spherical shells. Similarly, the strength of an initially imperfect sphere is mainly dependent on the local curvature over a "flat spot" of critical length rather than on the nominal curvature.

11.2 BARRELLED SHELLS

A barrelled shell has a positive Gaussian curvature, and its meridional curvature lies between that of a cylindrical shell (zero) and that of a spherical shell (finite) (Figure 11.24). As a result, it is expected to have a much higher buckling strength than a cylindrical shell but less than that of a spherical shell, when subjected to external pressure. The meridional shape of a barrel can take various forms ranging from generalised ellipses, circular segments, to approximations given by splines.

The response of barrelled shells has been studied in relation to axial compression and external radial pressure, which includes external hydrostatic pressure. Theoretical and experimental work has been carried out by a small number of research groups worldwide. In a paper published in 1987 [2], the optimal shape of a barrelled shell subjected to the combined action of axial compression and external pressure was sought. In [24], an experimental study related to gas carriers is reported. Other papers dealing with analysis and testing of barrelled shells subjected to external pressure are [1,3,5]. A series of papers have also been published by Jasion [10,11,12,21].

11.2.1 Geometry of Barrelled Shells

Figure 11.24 shows the geometry of a barrelled shell with a cylindrical shell and a spherical shell printed in dashed lines. Cylindrical and spherical shells are seen to be special cases of barrelled shells, for which the meridional radius of curvature $R_1 \rightarrow \infty$ and $R_1 = R_2$, respectively. The radius R_1 is identically equal to R. The geometric relation for the shell of revolution is [12]:

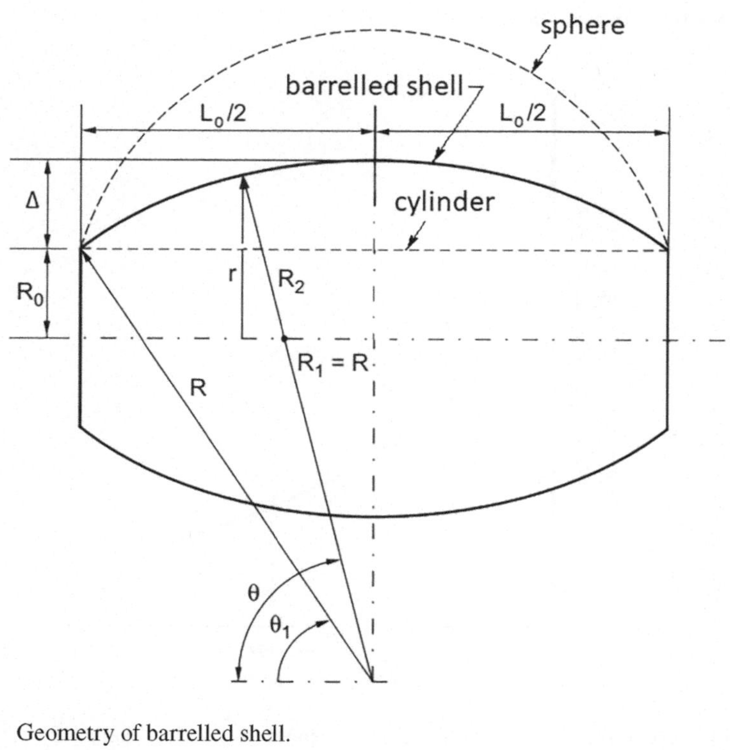

FIGURE 11.24 Geometry of barrelled shell.

$$\frac{\partial}{\partial\theta}(R_2 \sin\theta) = R\cos\theta \tag{11.24}$$

$$\text{where} \quad \theta_1 \le \theta \le \pi - \theta_1$$

11.2.2 Failure Modes of Barrelled Shells Subjected to External Pressure

In the case of relatively thick barrelled shells, yielding arises at one point, although this does not signify failure as the shell can sustain further loading. Failure arises either in the form of bifurcation buckling or axisymmetric collapse (Figure 11.25). The failure load is then the lower of the two.

A complete shell stability analysis comprises two parts: a linear analysis and a nonlinear analysis. The linear eigenvalue analysis is used to determine the critical elastic buckling load and the buckled form. A nonlinear analysis is used to determine the critical load and to plot the equilibrium path in the post-critical range. To determine the critical load, it is necessary to find a bifurcation point at which an initial stable equilibrium changes to an unstable equilibrium configuration. At this point, a change of buckled form takes place.

Nonlinear analyses are carried out in an incremental manner, as described in Chapter 8. Barrelled shells have been studied using both types of analysis, as reported below.

Laboratory tests have confirmed that barrelled shells can sustain much higher loading than mass equivalent cylinders. The effect of geometry (meridional radius) on bifurcation pressure and collapse pressure has been studied using finite element analysis as well as laboratory tests [3,11].

Błachut considered a specific geometry and studied the manner in which the magnitude of the meridional radius affects the nature of the failure mode and the critical pressure, using finite element analysis. For a barrelled shell for which $L_0/R_0 = 1$ and $R_0/t = 33.33$, results are plotted in Figure 11.26. The failure pressure (bifurcation, collapse) is plotted on the ordinate axis, whereas the amount of barrelling, Δ, is plotted along the abscissa in non-dimensional form (Δ/R_0). Figure 11.26 includes curves for four different boundary conditions (u = axial, v = circumferential w = radial displacement, β = mid-surface rotation). The point $\Delta/R_0 = 0$ corresponds to the cylindrical shell, and it is noted that the failure pressure rises with the amount of barrelling to 1.75 times that. For larger values of Δ/R_0, the difference in failure pressure between the two forms reduces. It is also noted that failure is characterised by bifurcation buckling for $0.1 > \Delta/R_0 > 0$, whereas for $\Delta/R_0 > 0.1$, elasto-plastic collapse arises. It is not clear, however, whether it is possible to achieve enhanced performance of a barrelled shell over a cylindrical shell in the case of a vessel made from several jointed barrels (multisegment structure). This requires further investigation in view of the large sensitivity of buckling load to radial constraints in way of the barrel edges [3].

The effect that the amount of barrelling has on the buckled form was studied by Jasion and Magnucki [11], and it was found that a strong dependence exists. As seen in Figure 11.27, the

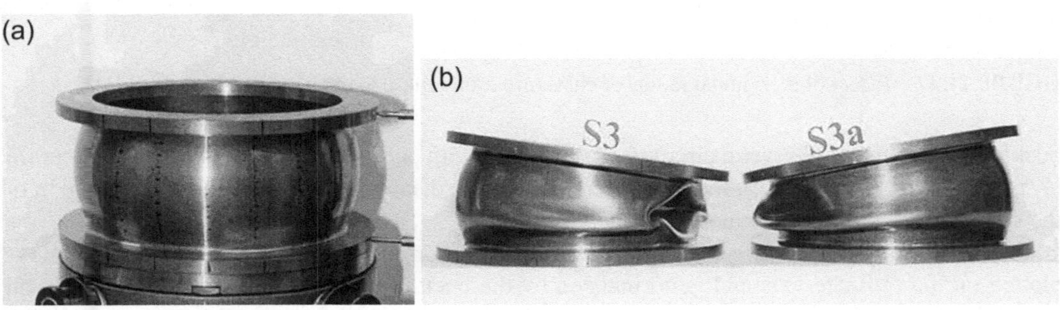

FIGURE 11.25 Test specimens subjected to external pressure prior to and following collapse [3]. (a) Intact barrel. (b) Collapsed barrels.

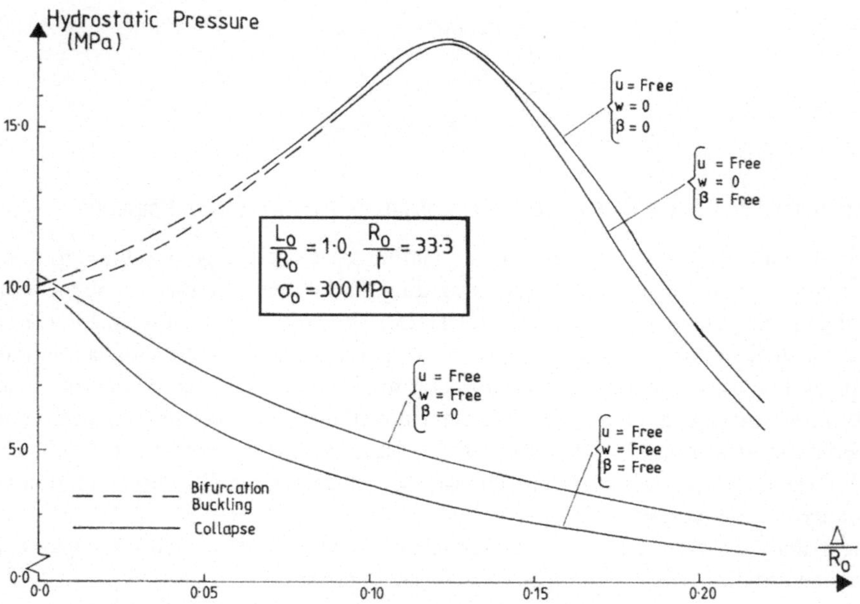

FIGURE 11.26 Effect of barrelling and boundary conditions on failure mode of a specific geometry [3].

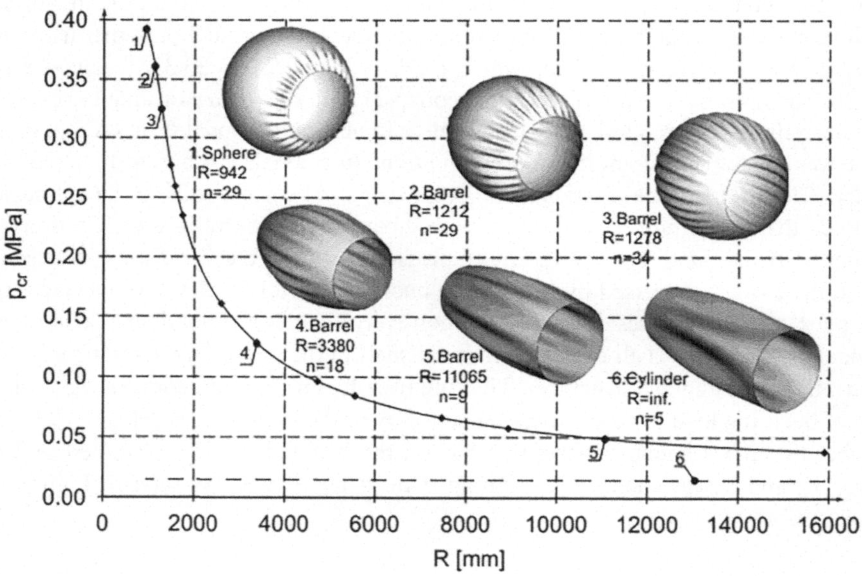

FIGURE 11.27 Effect of meridional radius of curvature R on failure load and on buckled form [11].

buckled form consists of several circumferential waves with one half-wave lengthwise. For $\theta_1 > 0.75$ (θ_1 is defined in Figure 11.24), the circumferential waves disappear in way of the central region of the shell and are present only in way of the edges.

Figure 11.28 shows the effect of shell slenderness R_0/t on the elastic buckling load, p_{cr}. This is plotted on the ordinate axis and is normalised by the buckling load of the equivalent cylindrical shell (p_{cr}^c). It is seen that the relationship between the geometry and the critical load is quasi-linear for all three slenderness ratios considered. When the meridional radius of curvature R is decreased, the critical load increases monotonically, as observed in Figure 11.27. It is seen that the margin of strength of the barrelled shell in excess of that of a cylindrical shell increases with shell slenderness (R_0/t). If the plot is enlarged (view A in Figure 11.28), a number of curves appear, each of which corresponds to a different buckled form with a distinct number of circumferential waves [10].

Results from nonlinear analysis are plotted in Figure 11.30. The two plots include load–deflection curves for two different types of initial distortions and extend into the post-collapse range. The magnitude of the initial distortions is defined as the maximum deviation from the nominal diameter and is expressed as the fraction of the shell thickness given by $(d_{max} - d_{nom})/2 = i$ (Figure 11.29). Collapse curves are included for $i = 0.01, 0.1, 0.2, 0.5$.

As can be seen from Figure 11.30, unloading takes place following collapse so that the response becomes unstable. At collapse, the deflection pattern changes from the single (symmetrical) mode shown as sketch A in Figure 11.30a to a single buckle (sketch B). In order to monitor response for a realistic, non-symmetrical pattern of initial distortions, an alternative form of initial distortions was introduced, and the results are plotted in Fig 11.30b. The alternative form was a linear combination of the first 10 eigenmodes with equal weights. In this case, the maximum deviation from the ideal form is $0.1t$. In both cases considered, the final deflected form is a local dent (Figure 11.30a and b).

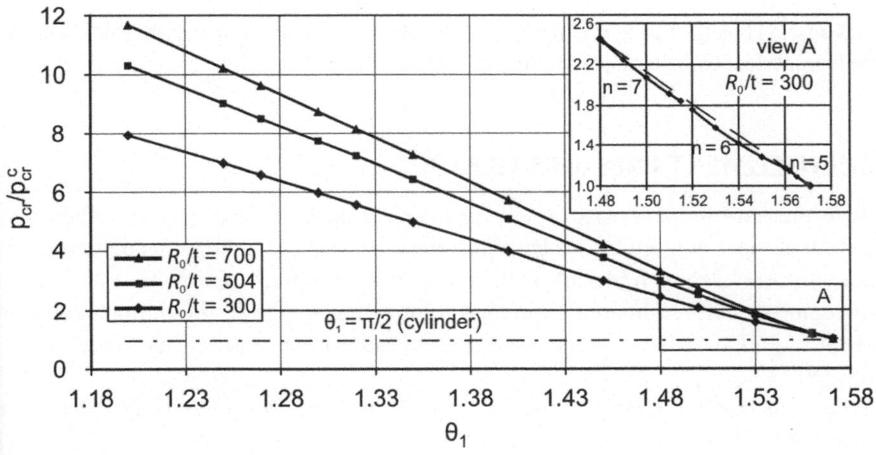

FIGURE 11.28 Effect of barrelled shell slenderness (R_0/t) on buckling strength [10].

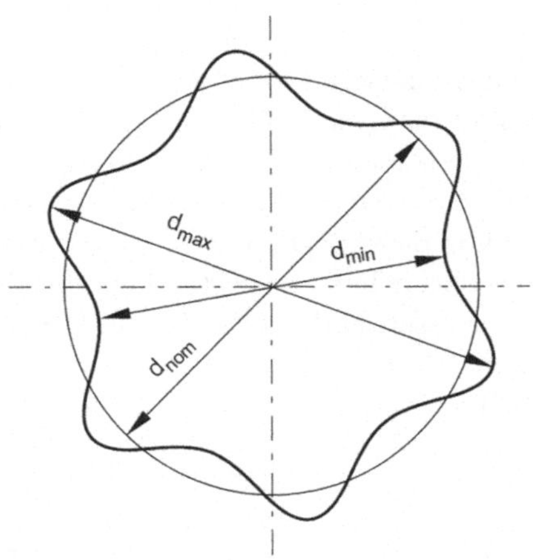

FIGURE 11.29 Definition of parameters for initial distortions of barrelled shells [10].

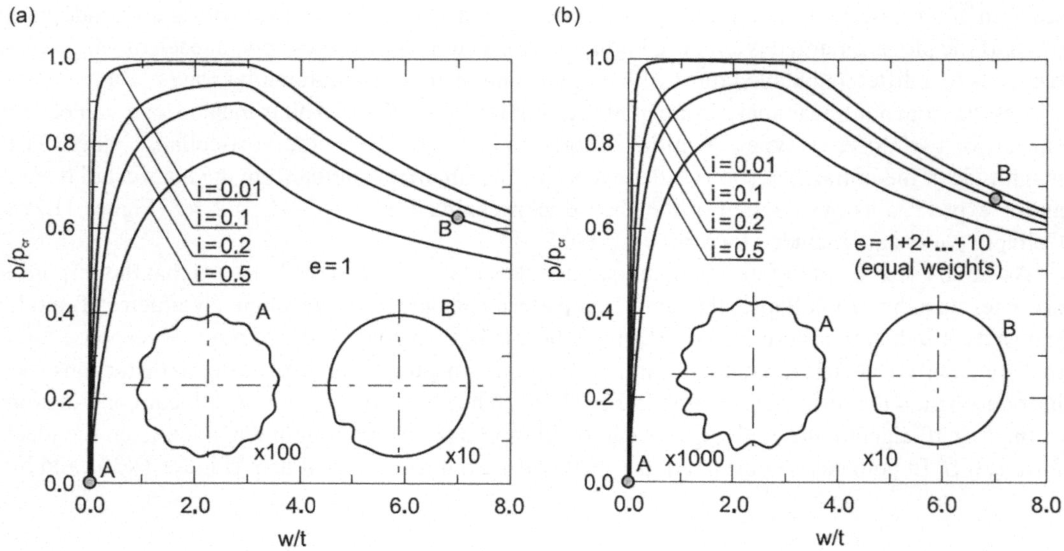

FIGURE 11.30 Load–deflection curves for barrelled shells with initial distortions ($\theta_1 = 1.32$). (a) Single symmetrical mode imperfection pattern. (b) Multimode imperfection pattern [10].

11.3 MULTISEGMENT PRESSURE HULLS

Multisegment pressure hulls consist of several intersecting hulls that may be either *spherical* or *barrelled*. Both of these configurations have positive Gaussian curvature, thus rendering them more efficient than an equivalent cylindrical shell from a structural point of view.

A multi-segment shell retains the advantages of a cylindrical hull, providing a convenient interior arrangement and also that of a spherical shell because it has a lower buoyancy factor than the cylindrical shell.

Concept designs involving series of intersecting spheres were considered early on for use in deep submergence rescue vehicles (DSRVs), leading to the design and construction of *Deep Quest* for the US Navy. In this section, we shall discuss the response of both multisphere and multi-barrelled pressure hulls.

11.3.1 Multisphere Pressure Hull

When two or more spherical shells intersect with one another, the buckling strength of each one is reduced in comparison to that of the intact shell, because of the loss of integrity. For practical purposes, a relatively large opening is required, and this reduction in strength has to be compensated for. To this end, a circular ring of suitable material and dimensions is fitted in way of each intersection, the aim being to restore the strength of the adjacent pressure hulls to that in the intact condition. The geometry of a two-sphere hull is shown in Figure 11.31.

The geometry of a multisphere pressure hull is defined by the following parameters (Figure 11.31):

- Thickness of spherical shell (t)
- Radius (internal) of spherical shell (R)
- Intersection angle (α)
- Overall length of the multisphere hull (L)
- Width of circumferential rib (L_r)
- Thickness of circumferential rib (t_r)

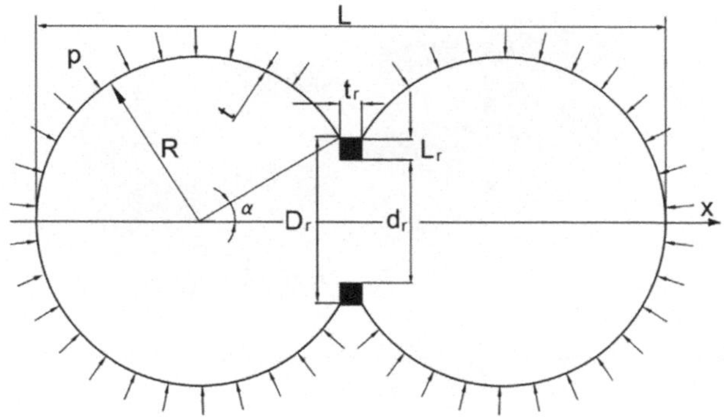

FIGURE 11.31 Geometry of a bispherical pressure hull.

- Outer diameter of circumferential rib (D_r)
- Inner diameter of circumferential rib (d_r)

11.3.2 Tests on Multisphere Pressure Hulls

11.3.2.1 Elastic Response of Titanium bi-Sphere Model

The earliest tests performed on intersecting spherical shells were carried out on titanium shells and were used to provide a benchmark for finite element analysis procedures that were subsequently used to design an advanced deep-diving submersible [20]. The aim of the test program was to evaluate the feasibility of a bi-sphere pressure hull and to assess the role of the intersection ring. Additionally, material behaviour was assessed with respect to creep and yield. Two different materials were used for the intersection ring (titanium and ceramic).

The model, shown in Figure 11.32a, was machined from a bar of titanium alloy 6AL-4V. Two series of tests were performed, with the pressure increasing from 0 to 9,000 psi (62.5 MPa). During the second test, creep measurements were made, but creep was found to not affect response. As seen from the pressure–strain plots (Figure 11.32b), the response is linear, so that yielding did not occur and failure was not observed. Strain gauge measurements were compared with predictions from finite element analysis, and it was confirmed that the predictions were satisfactory so that the software could be used in the design of a full-scale vessel. With regard to the geometry of the intersection ring, it was confirmed that it was not necessary to used a long tapered form on the adjacent shell and that the ring material can be limited to the intersection ring. The replacement of the titanium ring with a higher strength lighter ceramic ring was found to yield considerable weight savings. The weight-to-buoyancy ratio using titanium throughout was calculated to be equal to 0.726, but when ceramic was used for the intersection ring, this reduced to 0.70, offering an increase in payload of 9.5%.

11.3.2.2 Steel Bi-sphere Models Tested to Collapse

In a second series of tests, six bi-sphere models each fitted with intersecting rings of three different sizes were tested in order (a) to assess the effect of ring size on collapse pressure and (b) to act as a benchmark for a finite element procedure [26]. In these models, the intersection ring was welded to the adjacent spherical shells using TIG butt welding (Figure 11.33). The models were subjected to external pressure and were tested to collapse.

Numerical modelling was carried out using ABAQUS software and use was made of the modified Riks method. The finite element models consisted of approximately 7,100 elements for the bi-spherical shells and 4,538 elements for the single spherical shell, shown in Figure 11.34.

$$\text{DS1-1/DS1-2}: t_r = 0\left(\text{no ring}\right) \quad \text{DS2-1/DS2-2}: L_r = 6/t_r = 5 \quad \text{DS3-1/DS3-2}: L_r = 6/t_r = 10$$

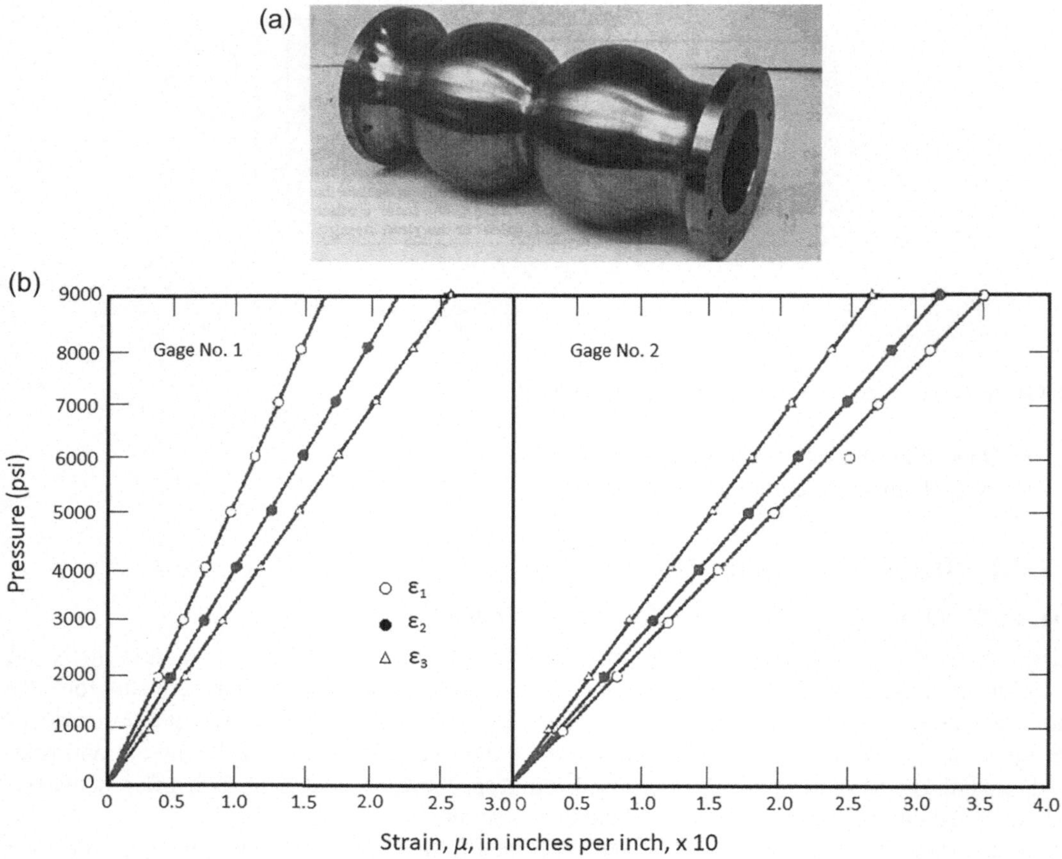

FIGURE 11.32 Titanium bi-spherical model [20].

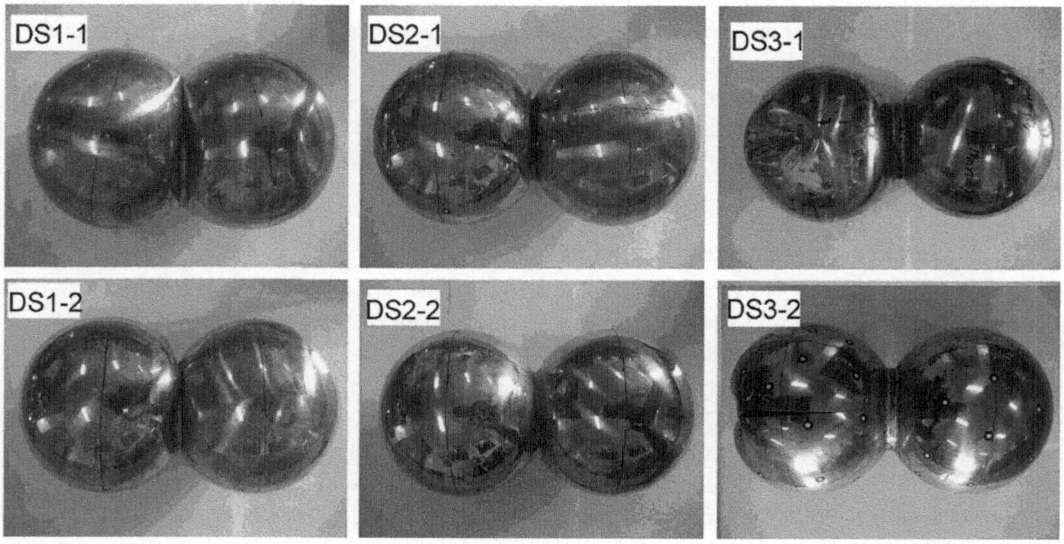

FIGURE 11.33 Bi-spherical pressure hull test specimens prior to testing [26].

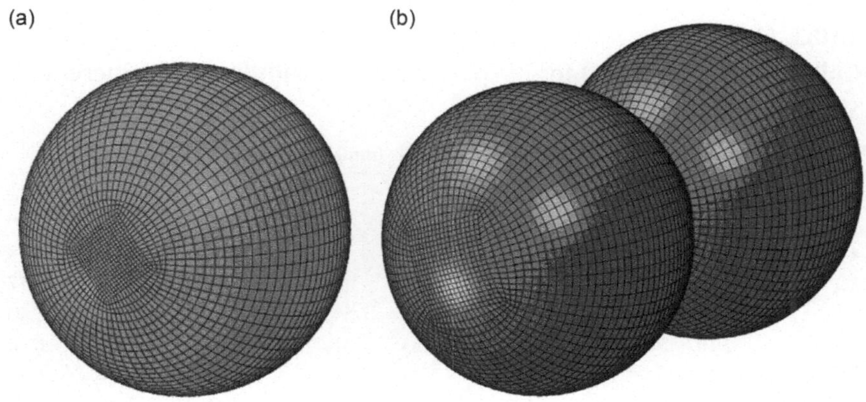

FIGURE 11.34 Finite element discretisation of (a) single and (b) bi-sphere model [26].

Two specimens with different geometries were manufactured to ensure repeatability of the experiment. After welding, the wall thickness of each individual spherical shell was measured using an ultrasonic probe, whereas the geometries of all shells were obtained using a three-dimensional optical scanner. The data obtained were used in the numerical analysis procedure.

All shells were tested to collapse under external hydrostatic pressure in a 20 MPa maximum pressure cylindrical pressure chamber. In all cases, failure occurred suddenly with rapid decreases in pressure accompanied by a small bang at the collapse pressure. In all cases, the collapse mode consisted of a local dent either near the rib-ring (DS1-1/DS1-2/DS2-1/DS2-2) or at the other end of the shell (DS3-1/DS3-2) as seen in Figure 11.35.

The length and thickness of the intersection ring were found to have a significant impact on the collapse load of the bi-spherical shell (Table 11.8). The collapse load of a bi-spherical shell fitted with a reinforcing ring was found to be approximately five times that of one without an intersection ring. The buckling behaviour of a bi-spherical shell fitted with an intersection ring appears to be similar to that of a single spherical shell, with collapse occurring in the form of a dent (Figure 11.35).

The collapse load of a bi-spherical shell with no intersection ring ($t_r = 0$) is 1.18 MPa.

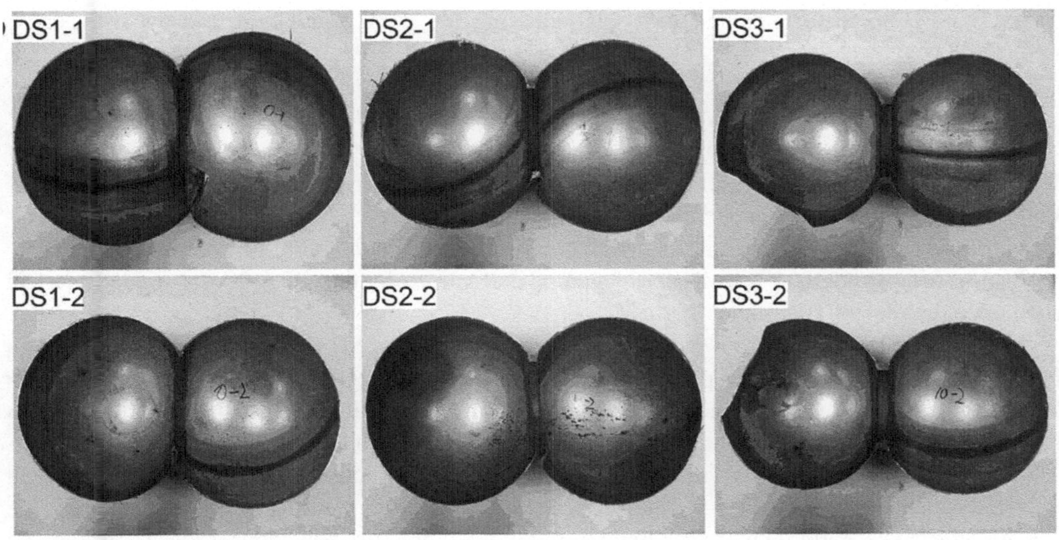

FIGURE 11.35 Bi-spherical pressure hull test specimens following testing [26].

TABLE 11.8
Collapse Loads (in MPa) for Bi-spherical Shells with Rings of Different Sizes [26]

	L_r (mm)							
t_r (mm)	1	2	3	4	5	6	7	8
2	1.44	1.69	1.90	2.10	2.30	2.50	2.70	2.90
4	1.57	2.05	2.45	2.98	3.22	3.61	4.00	4.38
6	1.75	2.40	2.98	3.55	4.11	4.67	5.22	5.77
8	1.93	2.74	3.48	4.21	4.94	5.66	6.34	6.40
10	2.10	3.04	3.94	4.83	5.71	6.40	6.38	6.40
12	2.23	3.26	4.31	5.35	6.34	6.41	6.41	6.42
14	2.31	3.44	4.62	5.80	6.40	6.40	6.40	6.40
16	2.34	3.55	4.86	6.15	6.40	6.40	6.40	6.40
18	2.40	3.58	5.10	6.36	6.40	6.40	6.40	6.40

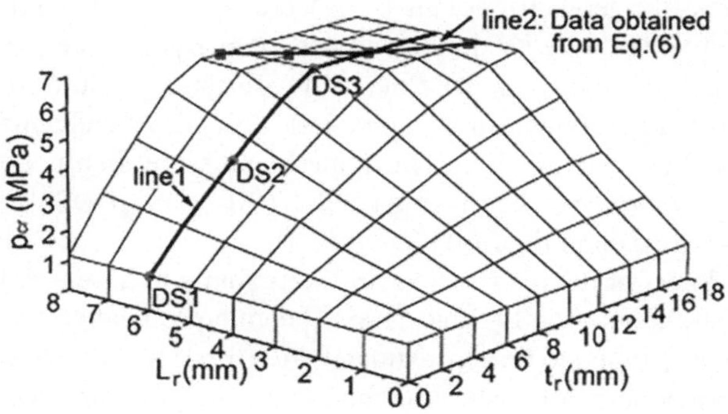

FIGURE 11.36 Effect of ring width (L_r) and thickness (t_r) on collapse load of a bi-spherical shell [26].

In Figure 11.36, the data plotted were obtained using numerical calculations and also using equation (12.38).[5]

$$\text{DS1}: t_r = 0\left(\text{no ring}\right) \quad \text{DS2}: L_r = 6/t_r = 5 \quad \text{DS3}: L_r = 6/t_r = 10$$

It is seen that the collapse pressure increases quasi-linearly with ring thickness and/or ring length. Following the initial increase in collapse load, a plateau is observed, and no further increase in collapse pressure is observed with subsequent increases in ring length and/or thickness.

The limiting strength of a bi-spherical shell is determined by the strength of a single spherical shell once the rib ring is sufficiently thick.

The load–deflection path for each of the three geometries DS1, DS2, and DS3 is included in Figure 11.37. In this, the applied pressure is normalised with respect to the initial applied load (p_0 = 1 MPa), whereas the maximum deflection (w_{max}) is normalised with respect to the shell thickness. It is seen that the load increases until the collapse load is reached, following which unloading occurs. The contribution of the intersection ring is apparent in the large increase in strength that is observed for DS-3 when compared to DS-1. The pressure to cause initial yielding, p_{fy}, is indicated by the triangle in each case. For loads exceeding the initial yield pressure behaviour is elasto-plastic, and it is observed that collapse occurs in the elasto-plastic range. Results are summarised in Table 11.9.

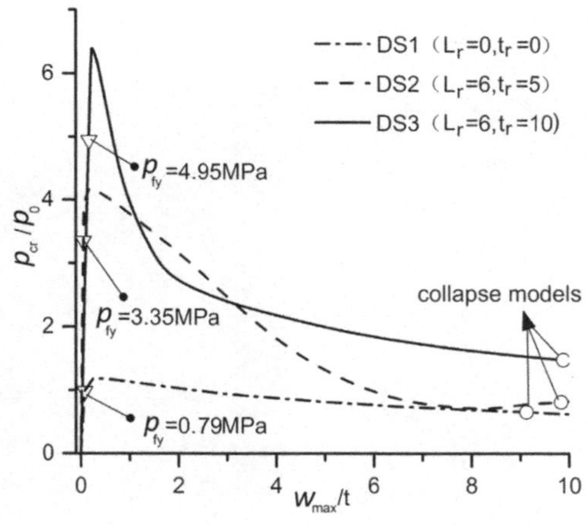

FIGURE 11.37 Load–deflection curves for three bispherical models with different ring dimensions (p_0 = initial applied load, p_{fy} = pressure to cause first yield) [26].

TABLE 11.9
Comparison of Predicted Bucking Pressures Using ABAQUS Models with Collapse Pressures Obtained Experimentally [26]

	p_{test} (MPa)	p_{ABAQUS} (MPa)	p_{test}/p_{ABAQUS}
DS1-1	1.74	1.16	0.67
DS1-2	1.67	1.05	0.63
DS2-1	3.55	3.32	0.93
DS2-2	3.73	3.45	0.92
DS3-1	4.39	4.35	0.99
DS3-2	5.06	4.69	0.93

11.3.3 Multi-Barrelled Pressure Hulls

In this section, we shall discuss numerical and experimental studies performed on multi-barrelled composed of two and four barrels. Each segment consisted of a barrel of the same geometry, shown in Figure 11.38. This geometry was selected as the result of an optimisation procedure, described in Section 12.7.2.

11.3.3.1 Two Barrel Assembly

The arrangement considered consists of two identical barrels connected by rings welded to each barrel and bolted to each other. The test specimens were produced from a thick, mild steel tube using a computer-controlled machining process. At the barrel ends, integral top and bottom flanges were machined (Figure 11.38a).

The boundary conditions at the ends of each barrel and in particular radial restraint have been found from previous work [3] to have a notable effect on barrel collapse load. Therefore, when joining barrelled shells using a ring-type connection, it is important to allow for this effect by providing sufficient radial restraint at the intersection.

At the ends of the assembly, for present purposes, the boundary conditions were:

- Radial displacements and rotations are fully restricted
- Ends are allowed to translate freely in the axial direction (insets of Figure 11.39).

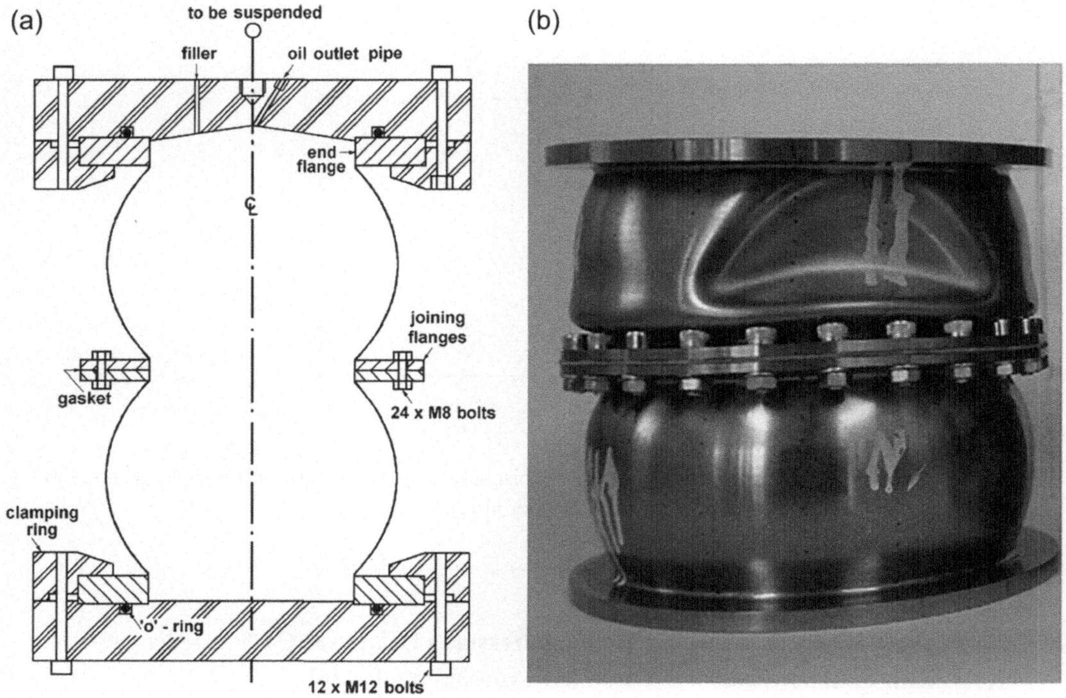

FIGURE 11.38 Two-barrel pressure hull test specimen. (a) Geometry and (b) Test specimen following collapse [4].

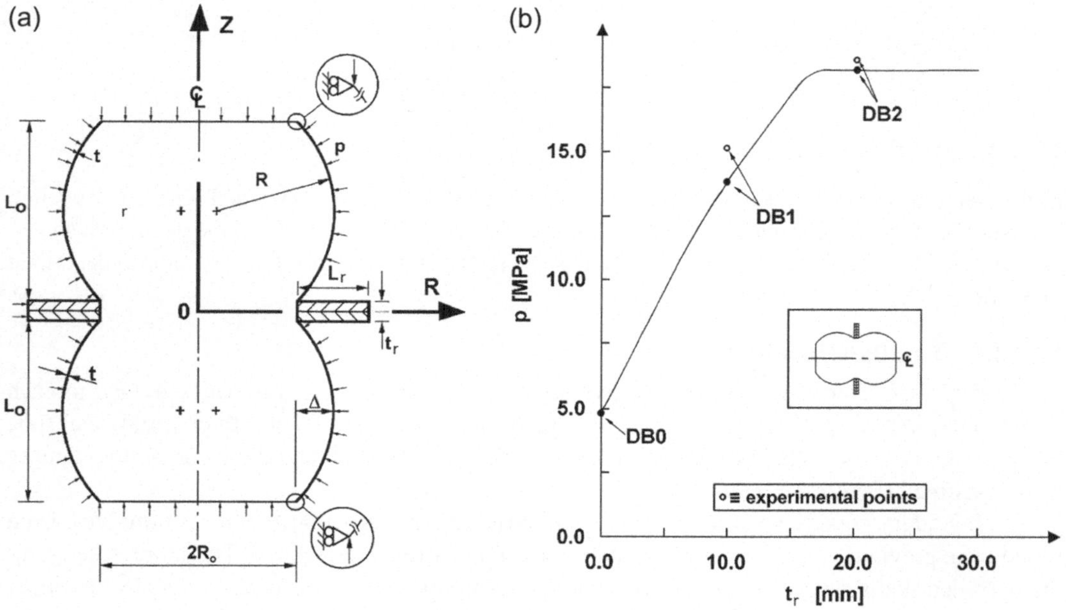

FIGURE 11.39 Bi-segment barrelled shell with ring consisting of two flanges. (a) Geometry and (b) Load versus ring thickness, t_r [4].

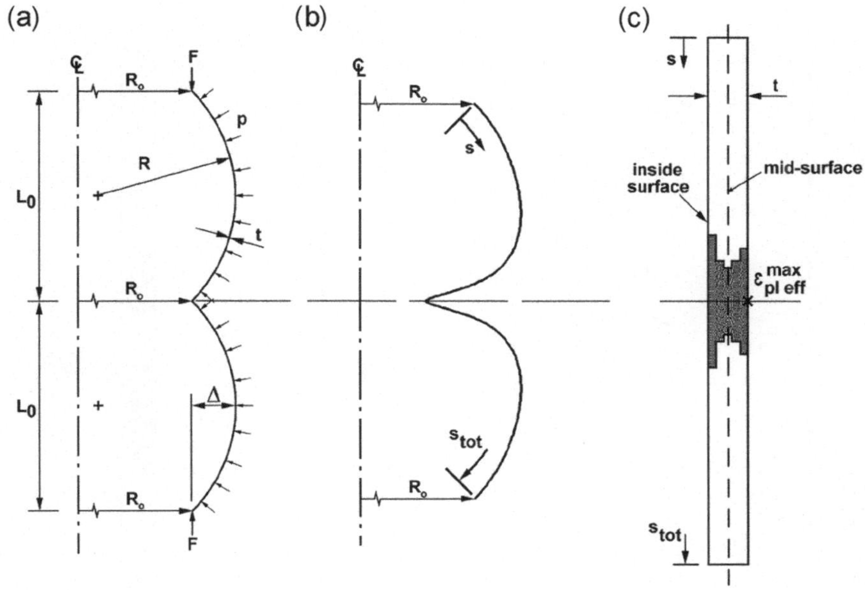

FIGURE 11.40 Two-barrel pressure hull without intersection ring (a) initial shape (b) at collapse pressure (c) spread of yielding at collapse pressure [4].

Figures 11.40 and 11.41 respectively illustrate the form of each barrel and the spread of plasticity at the collapse load, in the absence of a ring and in the presence of a ring. From these figures, it is seen that the maximum stresses arise in way of the junction of the two barrels. In the absence of a ring, collapse is characterised by yielding in way of the shells at this position. When the dimensions of a ring that has been introduced are increased, the extent of yielding in the ring reduces until a critical level in the value of the ring thickness t_F is reached beyond which collapse arises with yielding arising solely in the shell. A study was carried out to investigate the effect of ring length, L_r, on collapse pressure. The results of the effect of ring thickness t_r and ring length L_r are summarised in the interaction diagram shown in Figure 11.42. It is observed that following increases in both t_r and L_r, there is no further increase in the collapse load.

11.3.3.2 Multi-Barrel Assembly

The results of a numerical study on the effect of the number of barrelled segments in an assembly are included in Table 11.10, p.393. It is seen that the two computer procedures used predict that the collapse load is unaffected by the number of barrels that form part of the assembly. Also, the effect of the presence of a ring is independent of the number of barrels.

The analysis of the multi-segment models was carried out using BOSOR5 and the test measurements. In order to numerically model the test conditions as accurately as possible, the top end of the assembly was treated as being fully clamped, whereas the bottom end was treated as free.

Figure 11.43a shows a four-barrel assembly following collapse. The effect of shell thickness on the collapse pressure is shown in Figure 11.43b. It is observed that, as in the case of the two-barrel assembly, the collapse pressure does not increase beyond a certain point, with further increases in shell thickness.

A comparison of numerical predictions of collapse pressures for both two and four barrel assemblies with test measurements is included in Table 11.11. It is seen that in the case of the two-barrel assemblies, the finite element analysis underpredicts the collapse pressure, whereas in the case of the four-barrel assembly, it overpredicts it by approximately 10%. Possible reasons for these

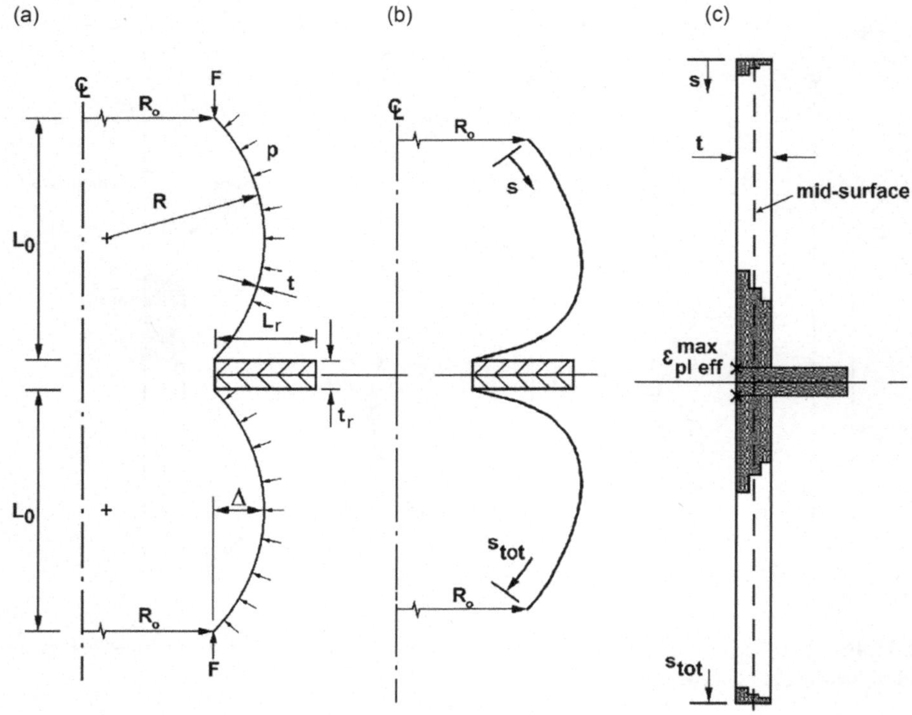

FIGURE 11.41 Two-barrel pressure hull with intersection ring (a) initial shape (b) at collapse pressure (c) spread of yielding at collapse pressure (effective ε_{pl}^{max} = 3.22%) [4].

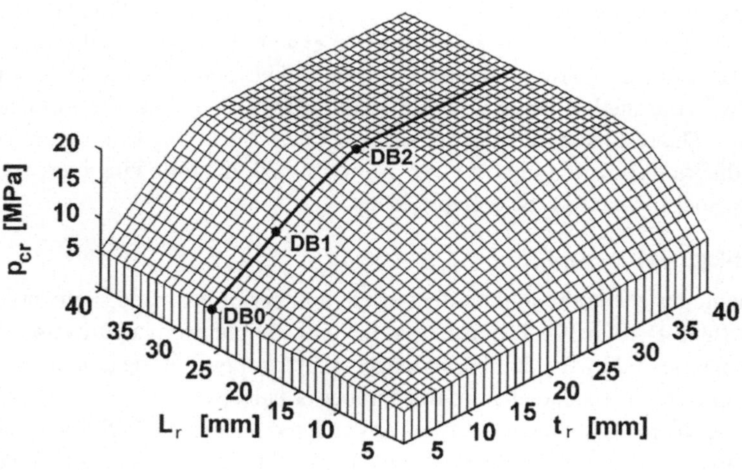

FIGURE 11.42 Effect of intersection ring dimensions (L_r, t_r) on collapse pressure [4].

discrepancies in the case of the two-barrel assemblies are the effect of initial distortions, whereas in the case of the four-barrel assembly, it is the lack of axial straightness, this being an issue that requires further investigation.

TABLE 11.10
Collapse Pressures for Multi-Barrel Hull [4]

Number of Segments		1	2	4	8	16
t (mm)		2.90	2.90	2.90	2.90	2.90
p_c (MPa)[a]	BOSOR5	17.80	4.72	4.56	4.56	4.56
	ABAQUS	17.80	4.61	4.61	4.61	4.61
p_c (MPa)[b]	BOSOR5	17.80	13.66	13.66	13.66	13.66
	ABAQUS	18.09	13.90	13.90	13.90	13.90

[a] No ring between segments.
[b] The dimensions of the connecting ring are: $t_r = 10$ mm, $L_r = 26$ mm.

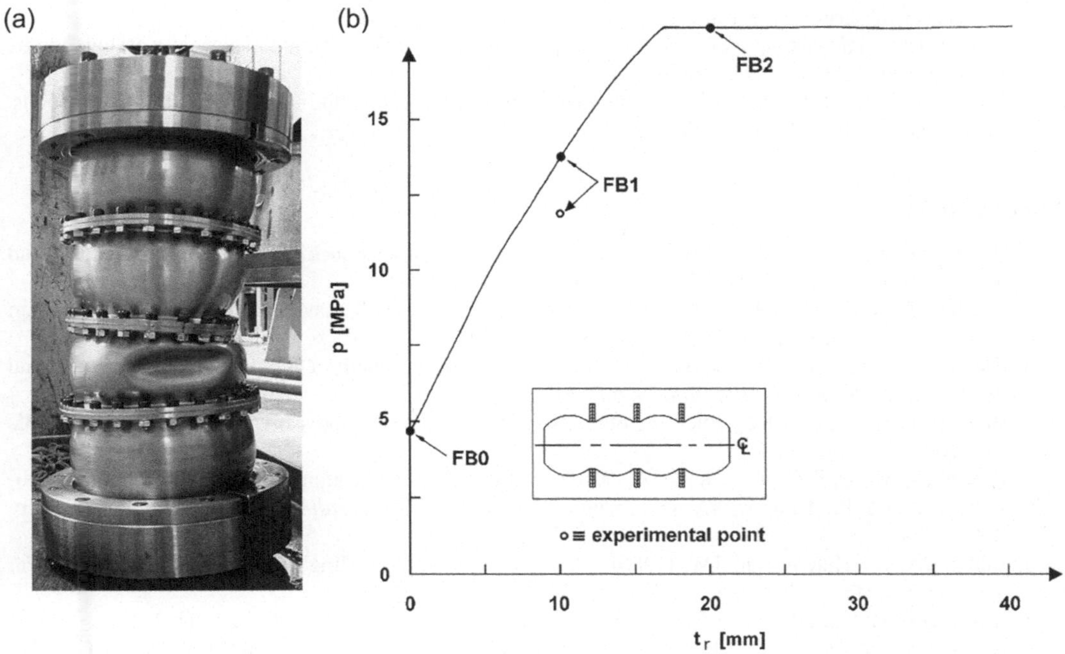

FIGURE 11.43 (a) Four-barrel assembly following collapse. (b) Effect of flange thickness on collapse pressure of four-barrel assembly [4].

TABLE 11.11
Comparison of Predicted Failure of Pressure Hulls Using BOSOR5 Models with Test Results [4]

	Lower Yield		Upper Yield		
	BOSOR5	BOSOR5 Test Result	BOSOR5	BOSOR5 Test Result	p_{exp} (MPa)
DB1	14.59	0.96	15.19	0.99	15.24
DB2	17.59	0.94	18.35	0.99	18.62
FB1	13.45	1.08	13.97	1.13	12.41/11.93[a]

[a] The two measurements given refer to a repetition of the test. The first value refers to the first test, whereas the second to the repeated test.

NOTES

1 Legendre functions are solutions of the Legendre ODE $(1-z^2)y''-2zy'+l(l+1)y=0$. The series solutions obtain the following forms:

$$y_1(z)=1-\frac{l(l+1)}{2!}z^2+\frac{l(l+1)(l-2)(l+3)}{4!}-+\cdots$$

$$y_2(z)=z-\frac{(l-1)(l+2)}{3!}+\frac{(l-1)(l+2)(l-3)(l+4)}{5!}-+\cdots$$

For integer $l=0, 1, 2, 3, \ldots$ one of series $y_1(z)$ or $y_2(z)$ terminates to yield a polynomial. The other solution is an infinite series $Q_l(z)$, convergent for $|z<1|$, known as a Legendre function of the second kind.

2 Much research has been carried out on cylindrical shells in end-compression. These are of interest in the aeronautical, aerospace and offshore industries, although the loading parameters, stiffening systems and geometries differ in each case.

3 The reports describing the test results by other authors are listed in [18].

4 In contrast to a *deep spherical shell*, in a *shallow spherical shell* the height is normally assumed to be less than one-eighth of its base radius.

5 This formula gives the ring thickness obtained on the basis of compatibility of shell and ring distortions. Its derivation is included in Section 12.5.2.

REFERENCES

1. Barski, M., Kruzėlecki, J. (2005). Optimal design of shells against buckling under overall bending and external pressure. *Thin Walled Struct.* Vol. 43, 1677–1698.
2. Błachut, J. (1987). Combined axial and pressure buckling of shells having optimal positive Gaussian curvature. *Comput. Struct.* Vol. 26, 513–519.
3. Błachut, J. (2002). Buckling of externally pressurised barrelled shells: a comparison of experiment and theory. *Int. J. Press. Vess. Pip.* Vol. 79, 507–517.
4. Błachut, J., Smith, P. (2008). Buckling of multi-segment underwater pressure hull. *Ocean Eng.* Vol. 35, 247–260.
5. Błachut, J., Wang, P. (1999). On the performance of barrel-shaped composite shells subjected to hydrostatic pressure. In: Toropov, V.V. (Ed). *Engineering Design Optimization.* Bradford: MCB University Press, 59–65 pp.
6. Budiansky, B., Hutchinson, J.W. (1966). "A survey of some buckling problems". AIAA/ASME *7th Structures & Materials Conference*, Florida, USA.
7. Bushnell D. (1981). Buckling of shells-pitfall for designers. *J. AIAA.* Vol. 19, No. 9, 1183–1226.
8. Evans, J.H., Adamchak, J.C. (1969). *Ocean Engineering Structures*, Vol. 1. Cambridge MA: MIT Press.
9. Galiev, Sh.U., Błachut, J., Skurlatov, E.D., Panova, O.P. (2004). Experimental and theoretical design methodology of hemispherical shells under extreme static loading. *Str. Mat.* Vol. 36, No. 5, 518–524.
10. Jasion, P. (2009). Stability analysis of shells of revolution under pressure conditions. *Thin-Walled Struct.* Vol. 47, 311–317.
11. Jasion, P., Magnucki, K. (2007). Elastic buckling of barrelled shell under external pressure. *Thin-Walled Strt.* Vol. 45, 393–399.
12. Jasion, P., Magnucki, K. (2012). Elastic buckling of horizontal barrelled shells filled with liquid - Numerical analysis. *Thin-Walled Struct.* Vol. 52, 117–125.
13. von Kármán, T., Tsien, H.S. (1939). The buckling of spherical shells by external pressure. *J. Aeronaut. Sci.* Vol. 7, 43–50.
14. Koiter, W.T. (1945). "Over der stabiliteit van het elastisch evenwicht". *Doctoral Thesis*, Delft University. [Eng. transl.: "On the stability of elastic equilibrium", Air Force Flight Dyn. Lab. Tech. Rep. AFFDL-TR-70-25, February 1970.]
15. Koiter, W.T. (1963). "Elastic stability and post-buckling behaviour". *Proceedings of Symposium on Non-Linear Problems.* Madison: Univ. Wisconsin Press, 257–275 pp.
16. Krenzke, M.A. (1962). "Tests of machined deep spherical shells under external hydrostatic pressure." DTMB Report 1601, Washington, DC.

17. Krenzke, M.A. (1963). "The elastic bucking strength of near-perfect deep spherical shells with ideal boundaries." DTMB Report 1713, Washington, DC.
18. Krenzke, M.A., Kiernan, T.J. (1963). "Tests of stiffened and unstiffened machined spherical shells under external hydrostatic pressure." DTMB Report 1741, Washington, DC.
19. Krenzke, M.A., Kiernan, T.J. (1965). "The effect of initial imperfections on the collapse strength of deep spherical shells." DTMB Report 1757, Washington, DC.
20. Leon, G.F. (1971). Intersecting titanium spheres for deep submersibles. *J. Eng. Mech. Div.* Vol. 97, 981–1006.
21. Magnucki, K., Jasion, P. (2013). Analytical description of pre-buckling and buckling states of barrelled shells under radial pressure. *Ocean Eng.* Vol. 58, 217–223.
22. Nash, W.A. (1979). *Theory and Problems of Strength of Materials*, 4th Ed. New York: McGraw-Hill, 430 p.
23. van der Neut, A. (1932). "De elastische stabiliteit van den dunwandingen bol" (The elastic stability of the thin-walled sphere). *Doctoral Thesis*, Delft University.
24. Odland, J. (1981). Theoretical and experimental buckling loads of imperfect spherical shell segments. *J. Ship. Res.* Vol. 25, 201–218.
25. Timoshenko, S.P., Gere, J.M. (1961). *Theory of Elastic Stability*, 2nd Ed. Tokyo: McGraw-Hill Kogakusha, 541 p.
26. Zhang, M., Tang, W.-X., Wang, F., Zhang, J., Cui, W.-C., Chen, Y. (2017). Buckling of bi-segment spherical shells under hydrostatic external pressure. *Thin-Walled Struct.* Vol. 120, 1–8.
27. Zoelly, R. (1915). "Über ein Knickungsproblem an der Kugelschale" (On a buckling problem concerning the spherical shell). *Doctoral Thesis*, ETH Zurich.

12 Structural Design of Deep-Diving Submersibles

12.1 DESIGN ISSUES CONCERNING DEEP-DIVING SUBMERSIBLES

Before discussing structural design, it is necessary to make brief reference to the overall design procedure followed in underwater vehicles in general, because of the close relationship between the two. The first question that needs to be answered with regard to design concerns mission requirements. These include the types of activities that are to be carried out and the region of the ocean where these are to be conducted. The operating depth and duration underwater need to be defined at this early stage. Once the operating depth has been established, the pressure at that depth will determine the dimensional and compositional characteristics of the vehicle and its components.

In the next stage, feasibility studies can be carried out in order to select the equipment necessary to perform the required activities and to provide an answer as to whether the vehicle needs to be manned or unmanned. Such studies can be aided by making reference to existing designs described in the literature as well as in-house studies. The various aspects of a potential design need to be specified. These include:

- The form of the pressure hull, within which personnel operate and critical instrumentation is installed
- Selection of material for the pressure hull and calculation of weight-to-displacement ratio
- Consider the possible use of buoyancy material/buoyancy chambers
- Propulsion system and ballast
- Means of energy control
- Communication
- Visual and other means of observation and equipment required to conduct specific activities (seabed sampling, etc.)

Comparisons of alternative solutions to the various aspects of a potential design are required, and a number of potential solutions short-listed. Computer-based techniques can be employed if possible, and the entire process can be treated as a multidisciplinary optimisation problem, as described in [27]. Alternatively, generic approaches can be used, such as genetic algorithms [10]. These can be applied to the entire design problem or limited to structural considerations alone. Such approaches can be used at the preliminary design stage, which provides accurate internal arrangemens as well as dimensions and scantlings that can be used to obtain weight estimates.

The final stage involves full-scale, or more frequently, model-scale testing. At this stage, any further modifications should be of a minor nature so that refined techniques such as the finite element method ought to be used. This will require as input accurate measurements of initial imperfections.

12.2 STRUCTURAL DESIGN OF SUBMERSIBLE PRESSURE HULLS

In common with all other types of structures, the structural design of deep-diving submersibles involves the need to satisfy a number of different design goals. The three basic areas that need to be considered are:

DOI: 10.1201/9781003489153-12

TABLE 12.1
Material Properties for Pressure Hull Optimisation [16]

Material	Specific Gravity	Modulus of elasticity (psi × 10^6)	Specific Modulus (10^6 in)	Compressive Strength (ksi)	Specific Strength (10^3 in)
HY-80	7.68	30	106	80	282
6Al-4V Ti	4.32	16.5	103	120	750
7075-T5 Al	2.65	10.5	107	73	745
6061-T5 Al	2.65	10.5	107	40	408
GR/EP #1 (2:1 Layup)	1.60	12.5	211	145	2,449

- Pressure hull form
- Pressure hull material(s)
- Issues related to the fabrication procedure

It has to be noted that these are inter-related so that decisions that are taken concerning one area have consequences with regard to the others, so that matters are not necessarily straightforward. These inter-relationships concern performance, cost, and weight.

There are, however, certain important differences in the structural response of underwater vehicles in general and that of other structural types. These concern the presence of (a) initial geometrical imperfections and (b) residual stresses due to forming. The first of these is important because of the high imperfection sensitivity of shells of revolution. Residual stresses due to forming are the result of the fabrication processes employed in underwater structures in general and result in reduction in collapse strength. One other difference that arises, and in particular with aerospace structures, which are also fabricated using shells of revolution, is that the slenderness ratio (R/t) of underwater pressure hulls is significantly smaller (<50) than that of aerospace structures (200–500). This leads to inelastic buckling, in contrast to aerospace structures that buckle elastically. However, since the pressure hulls of underwater vehicles can be classed as of intermediate slenderness, the assumptions of thin-walled structures theory are applicable in both cases.

The theoretical basis of submarine cylindrical pressure hull analysis and design, described in Chapters 3–6 and 9, is equally applicable to submersible structural design, in particular in the case of submersibles whose operating depth is limited so that shell buckling is a possibility, in which case the pressure hull is stiffened with ring frames. The major difference between submarines and submersibles, especially deep-diving ones, is that the use of high-strength materials needs to be considered in order to satisfy weight-to-displacement ratio requirements. It is therefore necessary to have available a computer-based design procedure that can be used to consider designs using materials having different properties (Table 12.1).

The optimum configuration will differ in each case, the aim of such a procedure being to determine this. Such a procedure was developed and used in the elastic design of deep-diving submersibles [16]. The steps in the procedure are given as follows.

STRUCTURAL DESIGN PROCEDURE FOR DEEP-DIVING SUBMERSIBLES

1. Define the pressure hull envelope shape, material, and shell thickness to meet a yield criterion using a specific yield safety factor γ_{yield}. This step involves a membrane shell theory analysis and internal space requirements.

2. Define the ring frame spacing to avoid interframe buckling using a specific safety factor γ_{buck}.
3. Determine the distribution of the hull envelope material as well as any required ancillary stiffening, to avoid overall buckling for a given safety factor γ_{overall}. This involves determination of required values for bending stiffness per unit meridian length (*EI*/*L*).
4. Define the requirements for intermediate stiffening members to resist local instabilities (or yielding) using a specific safety factor γ_{local}, while providing the required value of *EI*/*L*. This involves determining the shape and proportions of the stiffening arrangements used (frames, toroids, sandwich, etc).
5. Adjust membrane–stiffener proportions to account for local stress concentrations; payload dimensions (clear bore) requirements, or fabrication process limitations.
6. Calculate hull structure *W*/*D* for the selected membrane/stiffener geometry.

The selection of pressure hull form taking into account operating depth requrements and material properties is discussed in the following section. The shell thickness can be found by introducing the partial safety factor γ_{yield} ($1.0 < \gamma_{\text{yield}} \leq 1.5$) to the following expression for the circumferential membrane stress:

$$\sigma_\theta = \frac{1}{2} p \frac{R}{t} \tag{12.1}$$

so that

$$t = \frac{1}{2} \frac{pR}{\sigma_0} \gamma_{\text{yield}} \tag{12.2}$$

Expression (12.1) does not allow for the presence of ring frames and their effect on the membrane and bending stresses.

The ring frame spacing can be determined using the Taylor Model Basin formula for the elastic buckling pressure of an unstiffened cylinder (3.45). Rearranging this formula, the frame spacing, L_s, is obtained:

$$L_s = \frac{\sqrt{Dt_p}}{p\gamma_{\text{buck}}} \left[2.6E \left(\frac{t_p}{D} \right)^2 + 0.2h\gamma_{\text{buck}} \right] \text{ (in)} \tag{12.3}$$

where:

t_p = shell thickness (in)
h = depth of water (feet)
p = pressure (psi) (=0.445h)
D = pressure hull diameter (in)

Using the value of L_s, the ring frame scantlings are determined such that overall buckling is avoided. The Bryant equation can be used to this effect (5.29). Rearranging, the moment of inertia of the ring frame and attached shell plating, I_s, is obtained:

$$I_s = \frac{p_{cr}\gamma_{\text{buck}} L_s D^3}{8(n^2-1)E} - \frac{D^2 L_s t_p}{4(n^2-1)} \frac{\lambda^4}{\left(n^2 - 1 + \frac{\lambda^2}{2}\right)\left(n^2+\lambda^2\right)^2} \tag{12.4}$$

where $\lambda = \pi R/L$

A structural design procedure for deep-diving submersibles needs, however, to allow for inelastic failure, and appropriate techniques such as the use of analytical equations and finite element analysis need to be incorporated. The preliminary design procedure described in Section 9.3.2.5 for ring-stiffened cylindrical shells does include inelastic effects.

12.2.1 Pressure Hull Form: The Weight-to-Displacement Ratio as a Measure of Merit

The question of pressure hull form is of fundamental importance in submersible as well as pressure hull design. In general terms, the selection of pressure hull form depends on the following considerations:

- Structural efficiency
- Arrangement of internal spaces
- Collapse depth
- External dimensions
- Hydrodynamic performance

A qualitative comparison of pressure hull forms for underwater vehicles is included in Table 7.1 (p.234) from which the relative advantages of each type can be gauged, for both submarines and deep-diving submersibles. It is seen that the requirements of these two different types of underwater vehicles differ, leading to different hull forms.

For reasons discussed in Section 9.1 in relation to submarine pressure hulls, the weight-to-displacement ratio (W/Δ) is used as a measure of merit (hull efficiency) in the case of submersible pressure hulls as well. We seek to reduce the value of W/Δ in order to ensure sufficiency buoyancy of the vehicle, given that in addition to pressure hull weight, a range of equipment has to be included in the total weight calculation.

Expressions for W/Δ for different hull forms have been obtained on the basis of linear theory. The following expression has been derived for membrane shells (deep-diving submersibles) having any geometry [16]:

$$\frac{W}{\Delta} = C\left(\frac{\sigma_0}{\rho}\right)^{-1} 12h \tag{12.5}$$

where

C = shape efficiency factor

σ_0 = material uniaxial strength (psi)

ρ = material density (lb/in^3)

h = depth in feet

The ratio in brackets in (12.5) is the *material-specific strength*, so that in order to reduce W/Δ for a given value of this ratio, the lowest possible value of C is sought. Values of the shape efficiency factor for different forms are included in Figure 12.1, where C is plotted against a/b, this being the ratio of the major to the minor axis for each hull form.

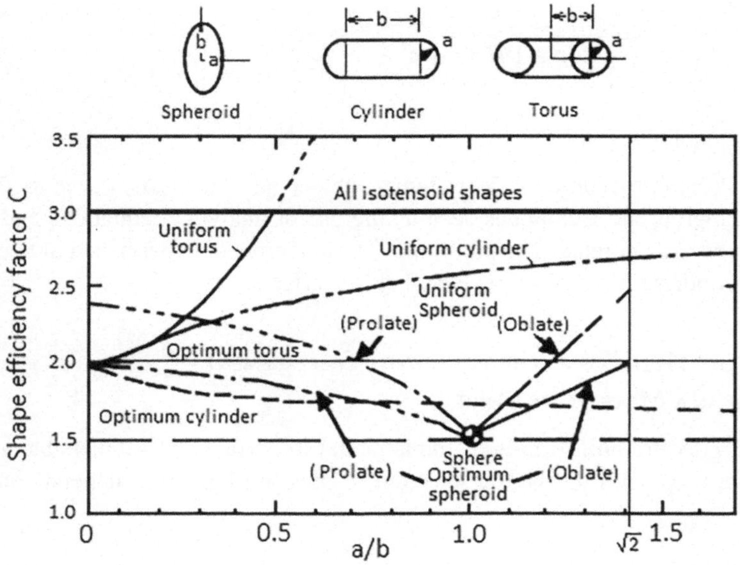

FIGURE 12.1 Shape efficiency factors C for different pressure hull forms.[1]

It is seen from Figure 12.1 that C assumes values in the range 1.5–3.5, with $C = 1.5$ for the sphere and $2.75 > C > 2$ for uniform cylinders of different values of a/b. It is seen from the figure that there exist other forms that may be used to combine hull efficiency with suitable internal arrangements. For spheres and cylindrical shells, the strength and buckling criteria are given in Table 12.2.

These expressions have been plotted for a variety of materials for these two structural forms in Figure 12.2a and b. It is seen that in the case of the sphere, initially and down to a critical depth, the design is buckling-dominated, beyond which the design becomes yield-critical. In the case of the cylindrical shell, the design is buckling-dominated for the entire depth range, for the high-yield materials considered.

In the case of a multisphere hull, the weight-to-displacement ratio is higher than that for a single sphere. The increase in W/Δ can be estimated by assuming that the induced strain in the spherical shells is equal to that in the reinforcing rings at the intersections (deformation compatibility). The area of the ring can then be calculated.[2] On this basis, the lower bound to the weight penalty for a

TABLE 12.2
Failure Criteria for Sphere and Uniform Cylinder [16] (h = depth in feet, γ_{buck} = safety factor against buckling[a])

Criterion	Sphere	Uniform Cylinder
$\left(\frac{W}{\Delta}\right)_{yield}$	$18\gamma_{yield}\left(\frac{\sigma_a}{\rho}\right)^{-1}h$	$24\gamma_{yield}\left(\frac{\sigma_a}{\rho}\right)^{-1}h$
$\left(\frac{W}{\Delta}\right)_{buckle}$	$3.856\gamma_{buck}\sqrt{6h\left(\frac{E}{\rho}\right)^{-1}\left(\frac{\rho}{\rho_w}\right)}$	$7.043\gamma_{buck}\sqrt[3]{h\left(\frac{E}{\rho}\right)^{-1}\left(\frac{\rho}{\rho_w}\right)^{2}}$

[a] For manned submersibles, the safety factor against buckling, γ_{buck}, assumes values less than 4.0.

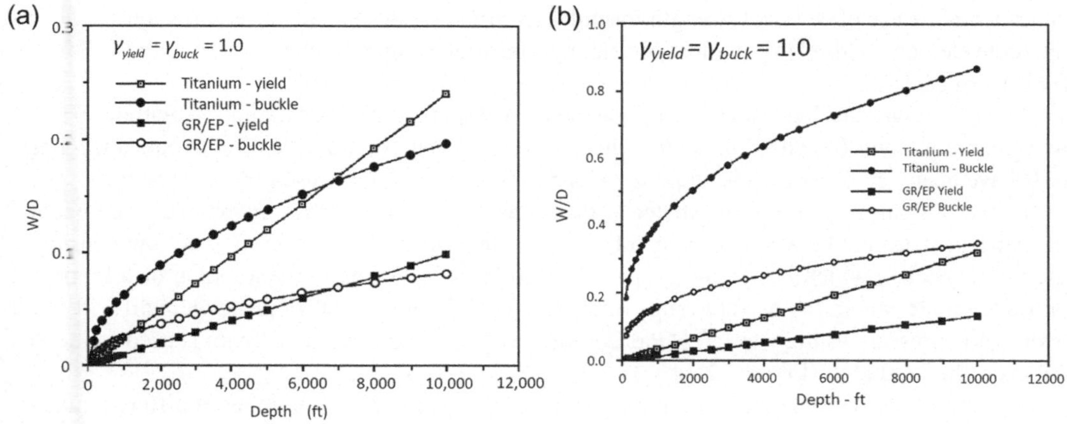

FIGURE 12.2 Weight-to-displacement ratios (W/Δ) for different materials ($\gamma = 1.0$ throughout) [16]. (a) Sphere. (b) Unstiffened cylindrical shell.

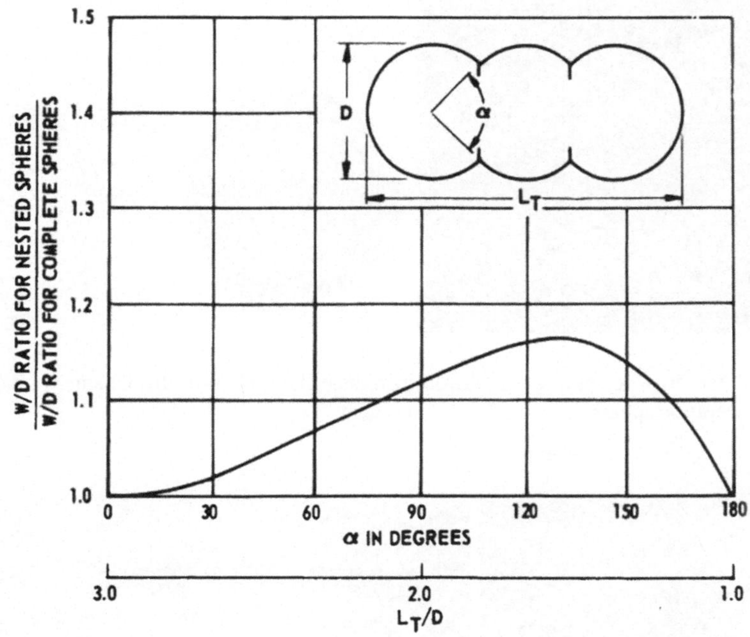

FIGURE 12.3 Lower bound on W/Δ ratio for a three-sphere arrangement [23].

three-sphere arrangement is shown in Figure 12.3. Estimates of W/Δ ratios for other arrangements of this type can be obtained using the same approach (Figure 12.3).

In addition to selecting a suitable combination of pressure hull form and material type, other ways that can be used to lower the W/Δ ratio are:

- The improved utilisation of the strength potential of materials
- The use of special-purpose *buoyancy materials*
- The addition of *structural floats*

Apart from the use of classical strength formulas in structural design, advanced procedures such as the finite element method in combination with optimisation techniques can be used to reduce weight for a given payload.

Buoyancy materials have been used since the early days of undersea exploration, with the *Trieste* being the most well-known example, for which gasoline was used. Since then, solid buoyancy materials have been developed, with *syntactic foam* being the most widely used one at present.

Syntactic foam is a composite material that consists of hollow glass microspheres embedded in a matrix that may be a polymer, metal, ceramic, or cementitious material. The low density of the microspheres, which lies in the range 250–400 kg/m^3, permits the fabrication of a family of buoyancy materials in the density range of about 550–700 kg/m^3 that, in principle, will withstand hydrostatic pressure as high as 150 MPa. Commercially available syntactic foam is used for depths not exceeding 6,000 m (Figure 12.4).

Structural floats assume the form of hollow spheres (Figure 12.5). A number of different materials have been used, but the most successful one is ceramic, which has a compressive strength that is five times that of steel and a density of one-third. These have been used for the *Nereus* AUV that

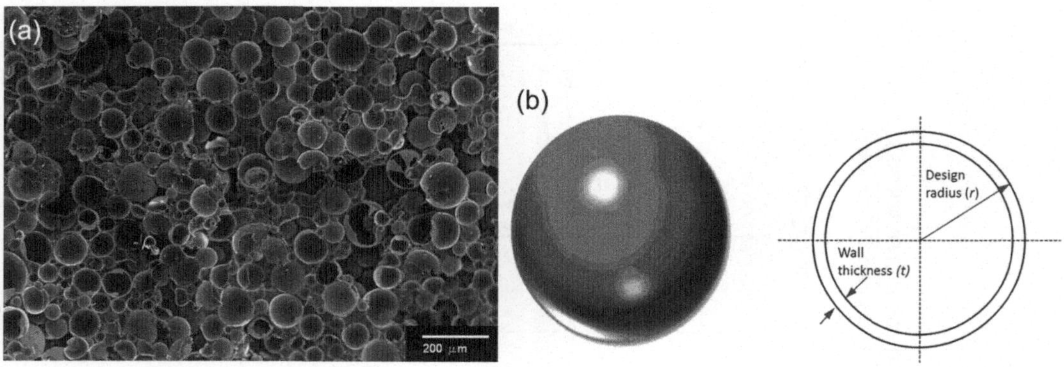

FIGURE 12.4 Alternative techniques employed to reduce the weight-to-displacement ratio. (a) Syntactic foam [34] and (b) ceramic sphere.

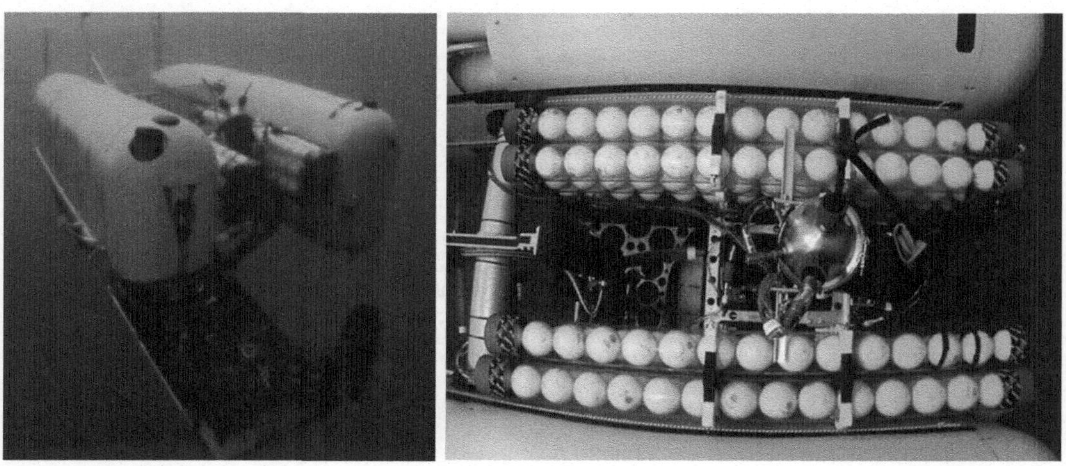

FIGURE 12.5 AUV *Nereus* (a) and ceramic hollow spheres (b).

reached the deepest part of the ocean in 2009. The spheres used each weigh 140 g and displace 404 g in sea water, for a weight-to-displacement ratio having a nominal value of 0.35. They are individually jacketed with 5 mm thick elastomeric polyvinyl chloride (PVC) that provides additional buoyancy as well as protection against impact loading. In total, 1,472 spheres are fitted, generating 417 kg of net buoyancy [7]. Ceramic spheres are also used in the offshore oil industry and are used as buoyancy modules for a variety of equipment. The size of the spheres used is determined by the desire to increase buoyancy, and this depends on being able to reduce the volume between the individual spheres (Figure 12.5).

12.2.2 Potential Hull Forms of Deep-Diving Rescue and Search Vehicles

We shall now turn to the results of an early study carried out at the David Taylor Model Basin in which a number of different pressure hull forms and their respective suitabilities were considered for potential use in deep-diving rescue and search vehicles [23]. This study is considered important because it implements a general framework that can be used a basis for any new design study even though certain tools used have been superseded. A series of charts giving values of the weight-to-displacement ratio for a range of materials used in combination with different pressure hull forms developed in this study are included in Appendix 7.

The mission requirements for the two classes of vehicles considered are listed in Table 12.3.

Comparisons of alternative combinations of pressure hull form and material used were carried out using the weight-to-displacement ratio (W/Δ) as a measure of merit.

The pressure hull forms considered were:

- Unstiffened spherical shell
- Ring-stiffened cylindrical shell
- Multisphere pressure hull
- Stiffened or unstiffened prolate spheroidal shell

The materials considered were:

- High-strength steels
- Titanium alloys
- Aluminum alloys
- Glass-reinforced plastics (GRP)
- Glass
- Glass-ceramics
- Alumina

TABLE 12.3
Mission Requirements for Rescue and Search Vehicles

Vehicle Type	Primary Mission Requirements	Secondary Requirements
Rescue vehicle	1. Rescue personnel from a sunken submarine up to collapse depth 2. Effect transfer of personnel from one submarine to another	1. Perform limited search and rescue operations 2. Accomplish oceanographic tasks at depths to 6,000 ft (≈1,800 m)
Search vehicle	1. Search the ocean floor up to 20,000 ft (≈6,100 m) for large or small objects 2. Recover small objects	1. Perform limited work and maintenance operations at depths to 20,000 ft (≈6,100 m)

TABLE 12.4
Material Properties Used in Comparative Study [23]

Material	Modulus of Elasticity (GPa)	Poisson's Ratio	Density (kg/m³)
High-strength steels	208	0.30	7,800
Titanium alloys	125	0.30	4,420
Aluminium alloys	75	0.30	2,770
GRP – solid fibres			
Cylindrical and spheroidal shells	44	0.15	2,050
Spherical shells	40	0.15	2,050
GRP – hollow fibres			
Cylindrical and spheroidal shells	21	0.15	1,505
Spherical shells	19	0.15	1,505
Glass	62	0.20	2,226
Glass-Ceramics	120	0.20	2,611
Alumina	312	0.20	3,730

TABLE 12.5
Selection of Material for Desired W/Δ Ratio [23]

		Material Properties		
Desired W/Δ Ratio	Material	Yield Strength (MPa)	Density (kg/m³)	Strength to Weight Ratio
0.5–0.6	HY-100 steel	690	7,800	0.088
0.4–0.5	HY-130 steel	930	7,800	0.120
<0.4	Titanium	765	4,420	0.170

The assumed material properties are listed in Table 12.4.

The diagrams included in Appendix 7 show that the W/Δ ratios are roughly the same for spherical, cylindrical, and spheroidal shells. Thus, the choice of pressure hull form lies with non-structural (weight) considerations. The one structural issue that arises is the design of large penetrations, which is simpler for spherical shells than for cylindrical and spheroidal shells, for reasons of symmetry.

With regard to the selection of pressure hull material, the preferred choice is much more clearly defined (Table 12.5). For $W/\Delta < 0.4$, titanium is the preferred material on the basis of the strength-to-weight ratio.

This study led to the design and construction of two multisphere vehicles, described briefly in Section 12.5.1 Weight to displacement ratio data relating to a variety of geometric forms for both types of vehicles are included in Tables 12.6 and 12.7.

12.3 DESIGN OF SPHERICAL SHELLS FOR DEEP-DIVING SUBMERSIBLES

In this section, we shall describe the procedures that are implemented in the design of spherical shells for deep-diving submersibles. In Section 12.3.1, we shall discuss a design procedure that was developed in the United States on the basis of the tests carried out at DTMB in the 1960s, summarily described in Section 11.1.5. Following that, we shall consider more recent approaches to the design of spherical shells and in particular the ways in which imperfections have been analysed and introduced in design codes. As an example of a design code we shall make reference to the

TABLE 12.6
Potential Hull Forms for Rescue Vehicle [23]

		Weight/Displacement		
Material	**Shape**	**Near-Perfect**	**$\Delta_8{}^a = 1/8$ in. Stress Relieved**	**$\Delta_8 = 1/8$ in. as Fabricated**
HY-100 steel	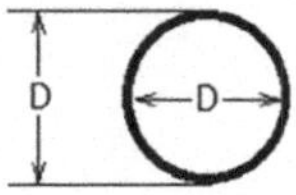	0.47	0.54	0.58
	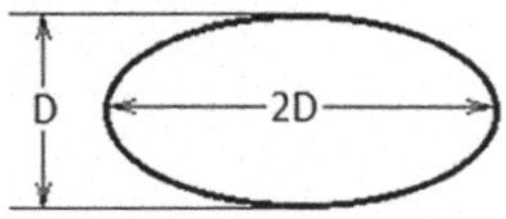	0.49	–	–
	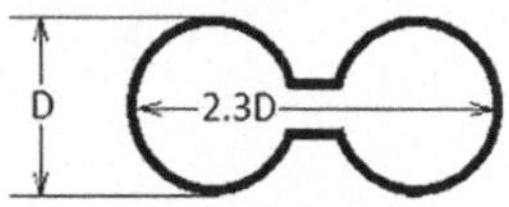	0.49	0.56	0.60
	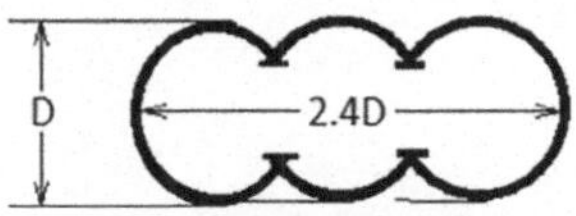	0.51	0.58	0.62
	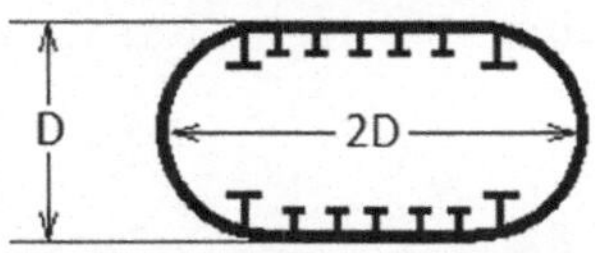	0.52	–	0.60
	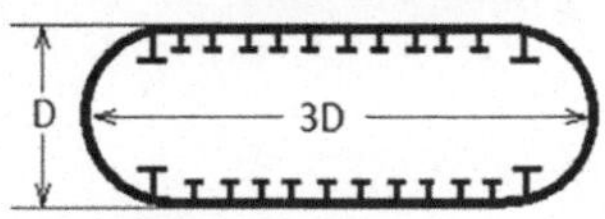	0.52	–	0.59
HY-140 steel	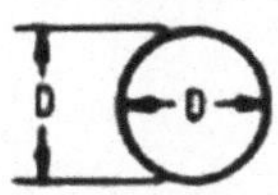	0.39	0.40	0.51
		0.40	–	–
	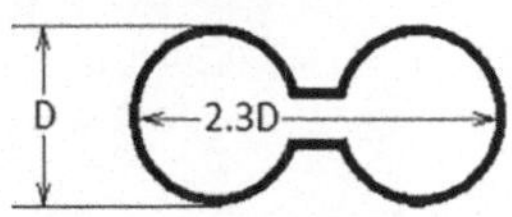	0.41	0.48	0.53
	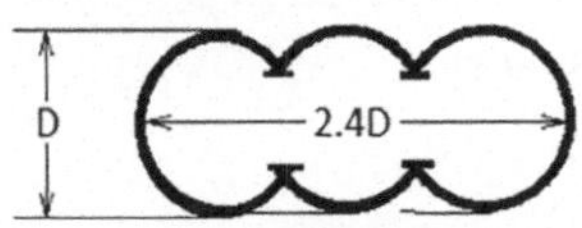	0.42	0.49	0.54
	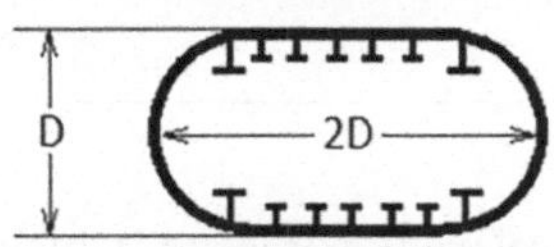	0.43	–	0.49
	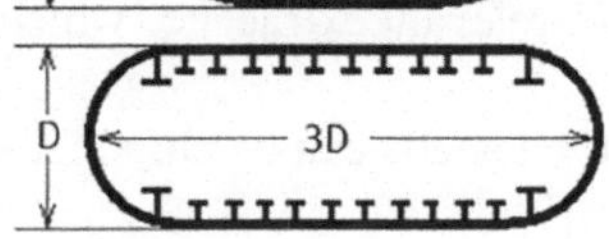	0.42	–	0.47

[a] Δ_8 = assumed maximum initial departure from sphericity for an 8 ft diameter shell.

TABLE 12.7
Potential Hull Forms for Search Vehicle [23]

		Weight/Displacement		
Material	**Shape**	**Near-Perfect**	**$\Delta_8 = 1/8$ in. Stress Relieved**	**$\Delta_8 = 1/8$ in. as Fabricated**
HY-150 steel	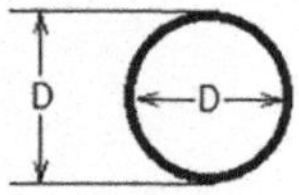	0.95	1.02	1.06
HY-180 steel	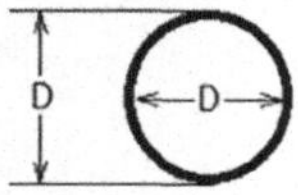	0.84	0.92	–
HY-110 Titanium	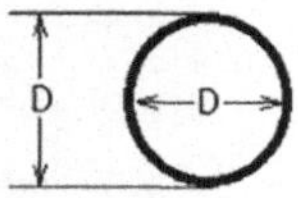	0.72	0.75	0.77
	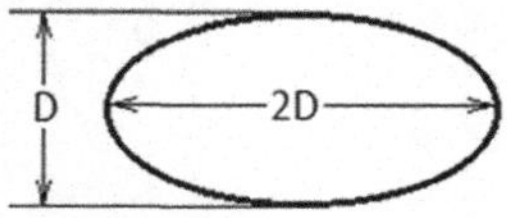	0.77	–	–
HY-110 Titanium	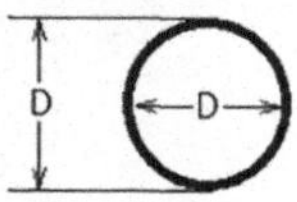	0.58	0.62	–
	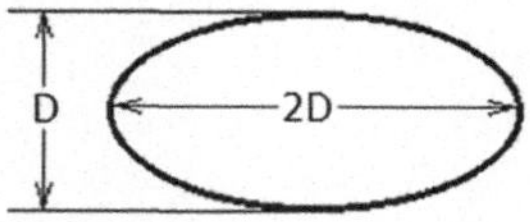	0.61	–	–
GRP Solid glass fibres	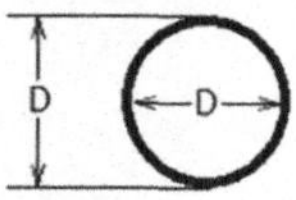	0.52	–	–
GRP Hollow glass fibres	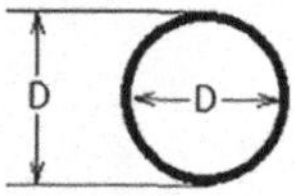	0.37	–	–
Glass	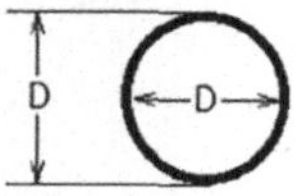	0.27–0.31	0.29–0.33	–
	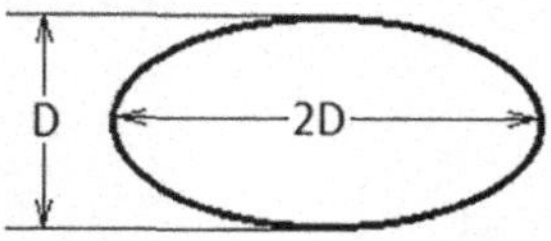	0.31–0.33	–	–

European code EN 1993-1-6 that was published in 2007, relevant extracts of which are included in Appendix 8.

For purposes of design, it should be noted that spherical shells can be classified as:

- Complete
- Deep
- Shallow

where “complete” refers to an entire sphere, “deep” refers to a spherical shell with part of its surface area missing, as in the case of a multisphere pressure hull and “shallow” when the ratio $a/R < 1/4$, where a is the rise of the apex of the cap above its base, and R is its radius (Figure 10.8) (Section 11.1.3)

12.3.1 A Design Procedure for Imperfect Spherical Shells

The design procedure described in this section is based on the collapse criteria for imperfect spherical shells that were developed by DTMB and are described in Section 11.1. The design procedure was presented by Heller in [19] and is recommended for use at a preliminary design stage. It can be used in conjunction with more advanced procedures such as the finite element method and optimisation software, by providing a starting point for these. Because it is iterative, some simple programming facilitates its use.

Before describing the details of the procedure, we need to derive the relationship between the out-of-circularity (OOC) (local defect) and the radius of the sphere. The geometry of a local defect in a spherical shell is shown in Figure 12.6. The initial radius is R and the radius of the local defect to the outer surface is R_l. The magnitude of the OOC is Δ, and the difference in the initial and the distorted chord height is X.

From Figure 12.6 we have

$$\Delta + X = H \tag{12.6a}$$

$$H = R(1 - \cos\theta) \tag{12.6b}$$

$$X = R_l(1 - \cos\alpha) \tag{12.6c}$$

Substituting (12.6b) and (12.6c) in (12.6a) and following algebraic manipulation, we obtain:

$$\frac{\Delta}{R} = 1 - \cos\theta - \left(\frac{R_l}{R}\right)(1 - \cos\alpha) \tag{12.7}$$

Alternatively,

$$\frac{\Delta}{t} = 1 - \cos\theta - \left(\frac{R_l}{R}\right)\frac{(1 - \cos\alpha)}{(t/R)} \tag{12.8}$$

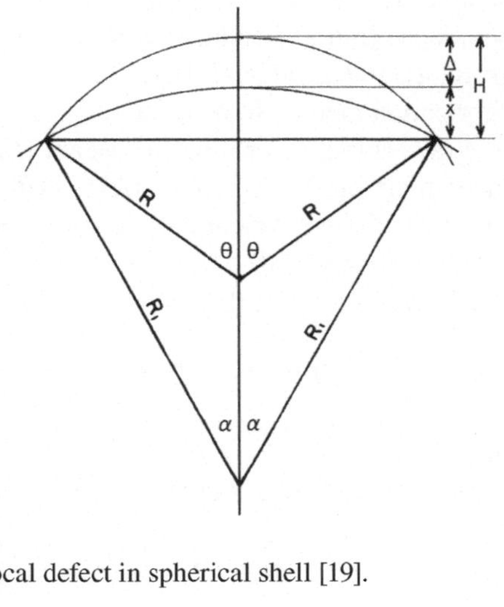

FIGURE 12.6 Geometry of local defect in spherical shell [19].

From Figure 12.6, we also have:

$$R\sin\theta = R_l \sin\alpha$$

or,

$$\theta = \arcsin\left[\left(\frac{R_l}{R}\right)\sin\alpha\right] \tag{12.9}$$

If the arc length of the local defect in Figure 12.6 is equal to the critical length, the expression for α is:

$$\alpha = \frac{L_{cr}}{2R_l} \tag{12.10}$$

In [22], the following expression for the critical length, L_{cr}, is given:

$$L_{cr} = \frac{2.2}{1-\nu^2}\sqrt{R_l t} \tag{12.11}$$

Substituting (12.11) in (12.10) gives:

$$\alpha = \frac{1.1}{1-\nu^2}\sqrt{\frac{t}{R_l}} \tag{12.12}$$

Substituting (12.10) and (12.12) in (12.8) gives:

$$\frac{\Delta}{t} = 1-\cos\left\{\arcsin\left[(R_l/R)\sin\left\{\frac{1.1}{1-\nu^2}\sqrt{\frac{t/R}{R_l/R}}\right\}\right]\right\} - \left(\frac{R_l}{R}\right)\left(\frac{1}{t/R}\right)\left[1-\cos\left\{\frac{1.1}{1-\nu^2}\sqrt{\frac{t/R}{R_l/R}}\right\}\right] \tag{12.13}$$

This is the relationship between the OOC Δ and the local radius R_l, plotted in Figure 12.7 for $\nu = 0.3$.

The design procedure comprises the following steps [19]:

1. Select initial estimate of inner radius, R_i, operating pressure P, maximum OOC Δ, and the safety factor against collapse (ratio of collapse pressure to operating pressure). If the maximum OOC is not given, assume a value of $0.01R_i$.
2. Select the material and fabrication method to be used
3. From the material stress–strain curve, determine the secant (E_s) and tangent moduli (E_t) and plot the square root of their product against stress. Definitions of the moduli are included in Figure 12.9, whereas the equivalent modulus for steel having a yield stress of 80 ksi (555 MPa) is included in Figure 12.10.
4. Determine a value for a trial thickness from the following expression:

$$t = 1.5PR_i(\sigma_0 - 0.75P) \tag{12.14}$$

   where σ_0 is the minimum yield stress.
5. Calculate the resulting nominal radius, R, from the expression

$$R = R_i + 0.5t \tag{12.15}$$

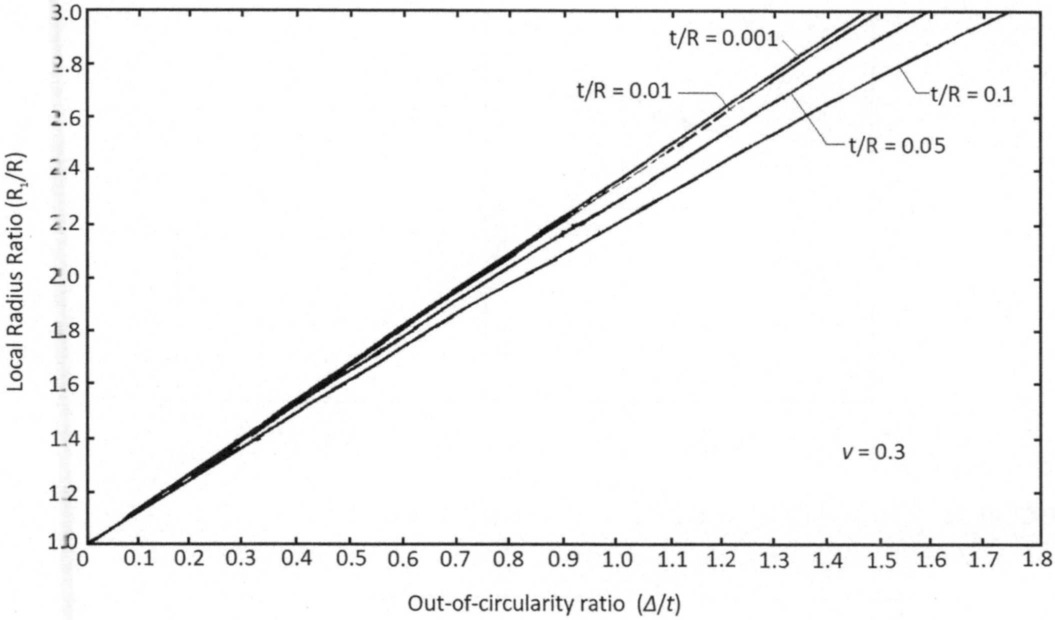

FIGURE 12.7 Out-of-circularity ratio versus local radius ratio for imperfect spherical shells [19].

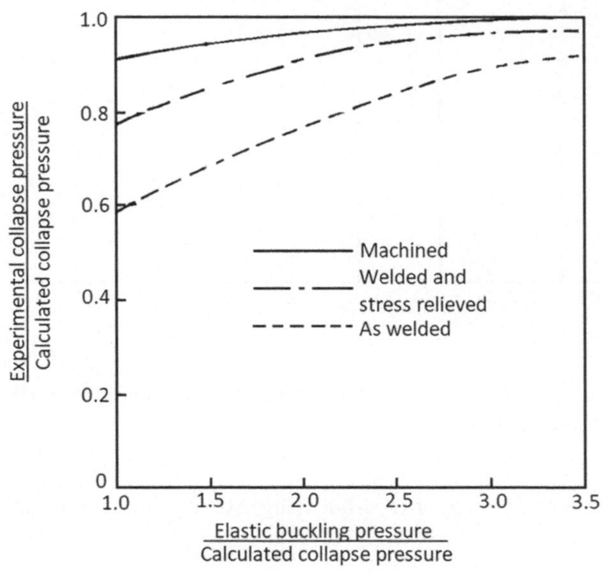

FIGURE 12.8 Lower limits of experimental data for various fabrication methods [19].

6. From Figure 12.7, determine the ratio of local radius to mid-thickness to nominal radius, R_l/R
7. Calculate R_l from the previous step. Then calculate the local outer radius R_{lo} from the expression

$$R_{lo} = R_1 + 0.5t \tag{12.16}$$

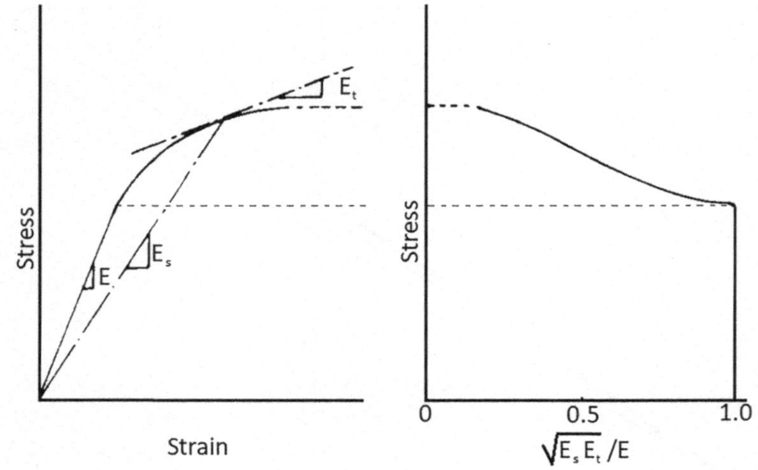

FIGURE 12.9 Stress–strain curve and stress–modulus relationship [19].

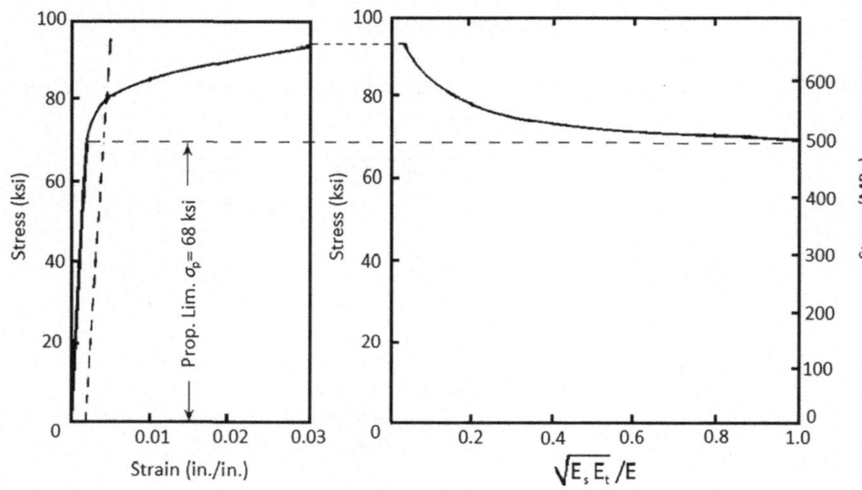

FIGURE 12.10 Material response curves for steel with yield stress = 80 ksi (555 MPa) [19]. (a) Stress–strain curve. (b) Stress–modulus relationship.

8. Determine a number of values of (pre-buckling) stress using the values of R_l and R_{lo} and the following expression:

$$\sigma = p\frac{(R_{lo})^2}{2tR_1} \tag{12.17}$$

9. For each of the stresses determined in the previous step, determine from Figure 12.10b the equivalent modulus. Using these values, determine the inelastic collapse pressure from:

$$p_i = 0.84\sqrt{E_s E_t}\left(\frac{t}{R_{lo}}\right)^2 \tag{12.18}$$

This expression gives the inelastic collapse pressure (when the pre-buckling stresses are between the proportional limit and the yield stress).

10. From the trial inelastic collapse pressures determined in the previous step, find the solution $P' = P_1'$ graphically. This is the calculated collapse pressure for the trial thickness determined in step 4.
11. Compare the calculated collapse from step 10 with the required collapse pressure. If it is less than the required collapse pressure, a new trial thickness must be assumed and steps 5–10 are repeated until the design requirement is fulfilled.
12. Calculate the elastic buckling pressure, p_E ($\equiv p_3$) using (10.17), where $R_0 = R_{lo}$:

$$p_E = 0.84E\left(\frac{t}{R_{lo}}\right)^2 \tag{12.19}$$

13. Using the value of the elastic buckling pressure, p_E, from step 12 calculate the ratio $p_E = p_1'$. Enter Figure 12.8 and read the value of the ratio experimental collapse pressure to the calculated collapse pressure, $p_{exp} = p_1'$.
14. Using the value of $p_{exp} = p_1'$, determine p_{exp}.
15. If $p_{exp} < p$ (from step 1), a new trial thickness must be assumed and steps 5–14 are repeated until a satisfactory value of the collapse pressure is obtained.

Example 12.1

Determine the shell thickness, excluding the corrosion allowance, of a spherical personnel capsule having an inner radius of 1,070 mm made of welded steel having a minimum specified yield stress of 555 MPa. The required operating depth is 275 m, and the maximum permitted out-of-circularity is 3.2 mm.

Solution

1. R_i = 1,070 mm, Operating pressure P = 2.75 MPa, Δ = 3.2 mm. The safety factor against collapse is taken equal to 1.5 so that the required collapse pressure is 4.125 MPa.
2. Data for the fabrication method are used (as welded)
3. The stress–strain curve is shown in Figure 12.10a, and the stress-equivalent modulus curve is shown in Figure 12.10b.
4. Trial thickness $t = \dfrac{(1.5)(2.75)(1{,}070)}{555-(0.75)(2.75)} = 8 \text{ mm}$
5. $R = 1{,}070 + (0.5)(8) = 1{,}074 \text{ mm}$
6. $\dfrac{\Delta}{t} = \dfrac{3.2}{8} = 0.4 \quad \dfrac{t}{R} = \dfrac{8}{1{,}074} = 0.0075$ From Figure 12.7 $\dfrac{R_l}{R} \cong 1.5$
7. $R_l = (1.5)(1{,}074) = 1{,}611 \text{ mm} \quad R_{lo} = 1{,}611 + (0.5)(8) = 1{,}614 \text{ mm}$
8. $\dfrac{R_{lo}^2}{2tR_1} = \dfrac{(1{,}614)^2}{(2)(8)(1{,}611)} = 101.06 \quad \sigma = 101.06P' \text{ (MPa)}$

P' (MPa)	4.0	4.5	5.0	5.5
σ (MPa)	404.24	454.77	505.03	555.83

9. From Figure 12.10b, the values of the equivalent moduli are read. These are converted to $\sqrt{E_sE_t}$ assuming E = 207 GPa.

σ	404.24	454.77	505.03	555.83
$\dfrac{\sqrt{E_sE_t}}{E}$	1.0	1.0	0.5	0.2
$\sqrt{E_sE_t}$	207,000	207,000	103,500	41,400

$$0.84\left(\frac{t}{R_{lo}}\right)^2 = (0.84)(8/1{,}614)^2 = (20.63)(10)^{-6}$$

and $P_i' = (20.63)(10)^{-6}\sqrt{E_s E_t}$

σ (MPa)	404.24	454.77	505.03	555.83
P_i' (MPa)	4.27	4.27	2.135	0.85

10. From Figure 12.11, it is seen that the inelastic trial collapse pressure $P_i' = 4.27$ MPa, which is in excess of the required collapse pressure of 4.125 MPa. Therefore, we can continue.
11. The elastic buckling pressure p_E is calculated from:

$$p_E = 0.84(207{,}000)\left(\frac{8}{1{,}614}\right)^2 = 4.27 \text{ MPa}$$

12. Hence, $\frac{P_E}{P_i'} = \frac{4.27}{4.27} = 1.0$
13. We use Figure 12.8 to determine P_{exp}/P_i'. For the "as welded" condition, the value obtained is 0.587.
14. Since the experimental collapse pressure $P_{exp} = 2.50$ MPa is less than the required collapse pressure, 4.125 MPa, we need to assume a shell thickness that is higher than 8 mm. To this end, Steps 5–13 are repeated assuming $t = 10$ mm.

5′. $R = 1{,}070 + (0.5)(10) = 1{,}075$ mm

6′. $\frac{\Delta}{t} = \frac{3.2}{10} = 0.32$ $\quad \frac{t}{R} = \frac{10}{1{,}075} = 0.0073$ $\quad$ From Figure 12.7 $\frac{R_l}{R} \cong 1.4$

7′. $R_l = (1.4)(1{,}075) = 1{,}505$ mm $\quad R_{lo} = 1{,}505 + (0.5)(10) = 1{,}510$ mm

8′. $\frac{R_{lo}^2}{2tR_l} = \frac{(1{,}510)^2}{(2)(10)(1{,}505)} = 75.75 \quad \sigma = 75.75P'$ (MPa)

P' (MPa)	4.0	4.5	5.0
σ (MPa)	303	340	379

9′. From Figure 12.10b, the values of the equivalent moduli are read. These are converted to $\sqrt{E_s E_t}$ assuming $E = 207$ GPa.

σ	303	340	379	
$\frac{\sqrt{E_s E_t}}{E}$	1.0	1.0	1.0	
$\sqrt{E_s E_t}$	207,000	207,000	207,000	207,000

10′. $0.84\left(\frac{t}{R_{lo}}\right)^2 = (0.84)(10/1510)^2 = (36.84)(10)^{-6}$

and $P_i' = (36.84)(10)^{-6}\sqrt{E_s E_t} = 7.62$ MPa

11′. This exceeds the required collapse pressure of 4.125 MPa.

12′. The elastic buckling pressure is

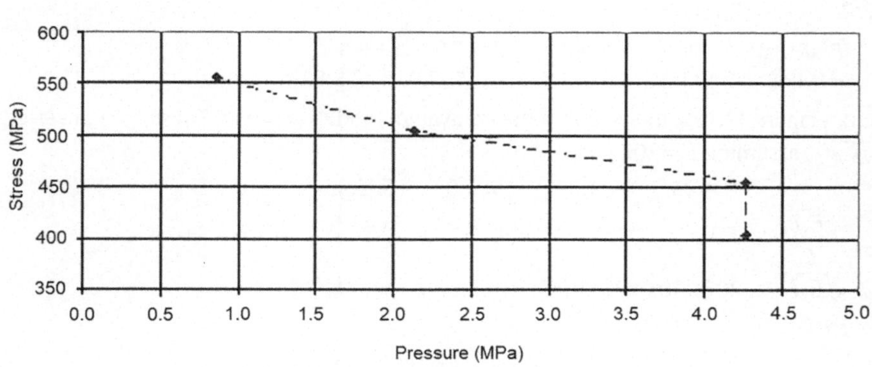

FIGURE 12.11 Graphical derivation of inelastic collapse pressure P_i'.

$$P_E = (36.84)(10)^{-6}(207{,}000) = 7.62 \text{ MPa}$$

Hence, $\frac{P_E}{P_1'} = 1.0$

13′. From Figure 12.8 for the as-welded condition, $\frac{P_{exp}}{P_1'} = 0.587$

Hence, $P_{exp} = (0.587)(7.62) = 4.47 \text{ MPa}$

14. Since this exceeds the required collapse pressure, a thickness of 10 mm is sufficient.

12.3.2 Alternative Definitions of Geometric Imperfections in Spherical Shells

In addition to the geometry of the initial imperfection described in the preceding section, a number of other methods have been implemented to this end [20]. Geometrical imperfections result from accidental damage and as such do not have a predefined form or location. It is, however, necessary to select what is considered to be a representative defect that can be used as a basis for design, without it being unduly pessimistic. To this end, position, dent amplitude, and shape have been variously used to define a typical defect. It is expected that such a defect will result in a reduction of the buckling pressure (p_{imp}) of a perfect spherical shell (p_{perf}) by 20%–50%. This reduction in load-carrying capacity is referred to as the *knockdown factor,* KDF, where KDF = p_{imp}/p_{perf}.

There are, broadly speaking, two different approaches currently used to define initial geometrical imperfections in spherical shells:

1. *Measured Geometric Imperfections* (MGIs): This approach is based on the use of optical instrumentation that is used with the full-scale model. This is expensive and time-consuming and as a result is not used in design work, but rather in laboratory testing, as described for example in [11].
2. The second approach consists of the use of several finite element analysis (FEA)-based techniques. These include:
   a. *Eigenmode Imperfection*: This involves assuming defects that have the form of the lowest eigenmodes of the structure in question. For spherical shells, these are shown in Figure 12.14.
   b. *Single-Load Indentation* (SLI) *Imperfection*: This can be defined in a number of ways. In the *single-perturbation load analysis* (SPLA), a lateral concentrated force is applied to the outer surface of the shell to simulate the dent/dimple shape of the structural imperfection. In the SLI, a displacement load is used instead of the lateral force to create the dent imperfection. Multiple lateral loads along the shell circumference have also been used to obtain the worst type of imperfection level. This technique is referred to as the *worst multiple-perturbation load approach* (WMPLA).
   c The *Reduced Stiffness Method* (RSM): This involves reducing or eliminating the components of the membrane stiffness of the shell, leaving only the bending stiffness. The membrane stiffness can also be reduced locally rather than globally (*locally reduced stiffness method*, LRSM).

12.3.3 Benchmark Study of PD5500 Design Code

Practical design of spherical shells subjected to external pressure involves the use of design codes, and the question arises as to how a code-based design compares with the strength of the actual structure. To provide an answer to this question, a study involving the use of a particular design code (PD5500) was carried out [20]. In this study, results from the use of PD5500 were compared with test results as well as numerical analysis. The test results used concerned 16 test specimens

TABLE 12.8
Geometry and Material Properties of Benchmark Models [20]

Type	ID	*R* (mm)	*t* (mm)	*R*/*t*	*E* (GPa)	σ_0 (MPa)	ν	References
Shallow shells	D1	1,816.5	1	1,816.5	207	303.5	0.28	Błachut
	D2	1,605.7	1.02	1,574.2				
	D3	878.6	1.03	853.01				
	D4	1,166.7	1.76	662.9				
	D5	759.3	1.76	431.42				
	D6	563.4	1.76	320.11				
Hemispherical shells	UNHS1	60	0.77	77.922	150.8	313.63	0.29	Zhang
	UNHS2	60	0.76	78.947				
	UNHS3	60	0.76	78.947				
	UNHS4	60	0.76	78.947				
	UNHS5	60	0.77	77.922				
Hemispherical shells	1#	58.84	0.432	136.2	193	205	0.28	Zhu
	2#	58.77	0.422	139.27				
	3#	58.99	0.423	139.46				
	4#	58.77	0.406	144.75				
	5#	58.75	0.415	141.57				

(six shallow shells and two sets of hemispherical shells) recorded in [6,38,40]. Details are included in Table 12.8.

In accordance with EN1993-1-6, relevant extracts of which are included in Appendix 8, the numerical model was assumed to be a perfect shell and geometric and material nonlinear analysis (GMNA) was carried out. The S4R element was used, and this is based on classical Kirchhoff shell theory and bending stresses with reference to Koiter–Sanders shell theory. This is a four-node shell element having six degrees of freedom. With regard to material nonlinearity, the elastic-rigid plastic model was assumed in conjunction with the *J2* flow plasticity theory. With regard to boundary conditions, a fully constrained/clamped boundary was assumed. Nonlinearity was allowed for using the arc-length method, described in Section 8.4.

Selected results of the benchmark study are included in Figure 12.12.

The following equation included in PD5500 was used to calculate the collapse pressure, p_{PD5500}:

$$\left(\frac{1}{p_{\mathrm{PD5500}}}\right)^2 = \left(\frac{1}{0.3p_{cr}}\right)^2 + \left(\frac{1}{p_{\mathrm{yield}}}\right)^2 \tag{12.20}$$

In this equation, p_{cr} and p_{yield} are the classical expressions for elastic buckling and yielding of a perfect spherical shell.

It is seen from the results tabulated in Figure 12.12 that the buckling stress calculated by PD 5500 underestimates the observed collapse pressures in all cases (for models D1–D6 by 46% on average, for models UNHS1 – UNHS5 by 33% on average, and for models 1#–5# by 10% on average), as reported in [6,37]. These results provide a measure of the conservativeness of the code. Indeed, in a comparison study of a number of design codes of spherical shells for use in deep-diving submersibles with existing designs, it was concluded that many of the existing designs do not comply with the code requirements nevertheless they have performed satisfactorily to date [26].

The results obtained using the nonlinear FEA code show good agreement with the test data, with the average percentage difference not exceeding 7%.

The load–deflection behaviour of spherical shells subjected to external pressure is shown in Figure 12.13. In these diagrams, the vertical deflection of the dome apex is plotted against external

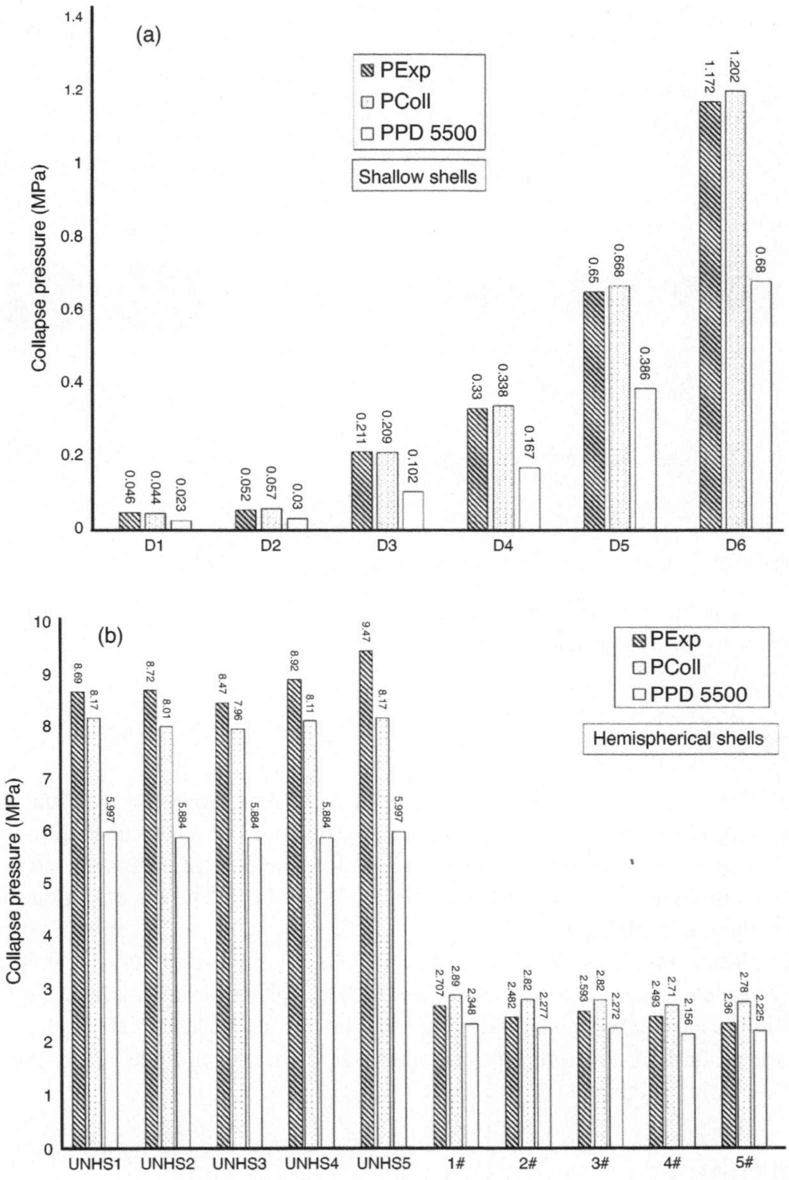

FIGURE 12.12 Comparison of predicted collapse pressure using the PD5500 code with test data and finite element analysis for (a) shallow shells and (b) hemispherical shells [20].

pressure. These results were obtained using FEA software. The pressures at which (a) yielding is initiated p_{Yield}, (b) collapse takes place, p_c and (c) post-collapse pressure, $p_{\text{post-collapse}}$, are indicated in each case. Figure 12.13a describes the behaviour of model D6, whereas Figure 12.13b that of model UBHS1.

It is observed that a significant pressure drop following collapse occurs in both cases, followed by a stable equilibrium path. A comparison of the two figures reveals that the hemispherical shell (UNHS1) having $R/t = 77.9$ exhibits significantly greater strength, with much lower post-collapse deflections than specimen D6 for which $R/t = 320.11$. The yield stresses are 313.63 and 303.5 MPa, respectively, so that this has a minimal effect on response.

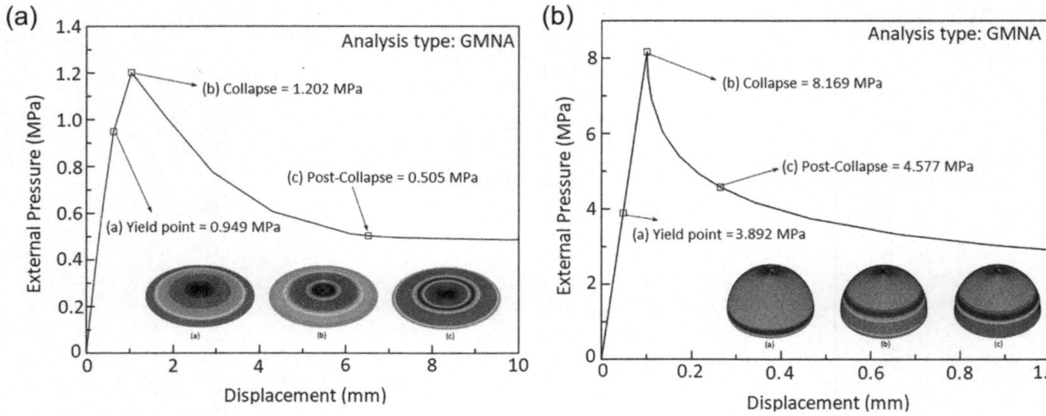

FIGURE 12.13 Load–deflection curve for (a) specimen D6 and (b) specimen UNHS1 [20].

12.3.4 Effect of Imperfection Type on Spherical Shell Response

In this section, we shall report on an investigation on the effect on buckling strength that different initial imperfections have [20]. The imperfection types considered were:

1. SLI
2. Eigenmode imperfections

Following EN1993-1-6, the Geometric and Material Nonlinear Imperfection Analysis (GMNIA) approach was used. The imperfections were introduced to a perfect spherical shell for amplitudes in the range $2.5 > w_0/t > 0.0$. Both hemispherical and shallow shells were analysed, and the geometries used are included in Table 12.8 (Model Nos. UNHS1, D6 and #1). The imperfection shapes applied to UNKS1 are illustrated in Figure 12.14.

Results are plotted in Figure 12.15a and b in which the KDF is plotted against imperfection amplitude (w_0/t) for each type of imperfection, for the hemispherical shell and the shallow shell. It is observed that in the case of the eigenmode shape imperfection, the loss of strength is much greater than in the case of the SLI, for both types of spherical shells. Eigenmode shape imperfections do not, however, occur in practice.

12.3.5 Structural Design of Acrylic Spherical Pressure Hulls

Acrylic has, together with other non-metallic materials, been considered for use in submersible pressure hulls from early on. Experience has shown that pressure hulls made of acrylic are as reliable as hulls made of conventional materials provided its particular mechanical properties are taken into account in design, fabrication, and also during operation at sea. These properties are discussed in Section 2.4. Acrylic can withstand much higher compressive than tensile stresses at failure rendering it a particularly suitable material for use in submersible pressure hulls, since these are subjected to external hydrostatic pressure.

The design of a penetration-free spherical hull is a relatively simple matter for both steel and acrylic hulls. To this end, the equations developed for thin shells of revolution can be applied. The assumptions upon which such an analysis can be based are:

- Axisymmetric structure
- Axisymmetric loading conditions

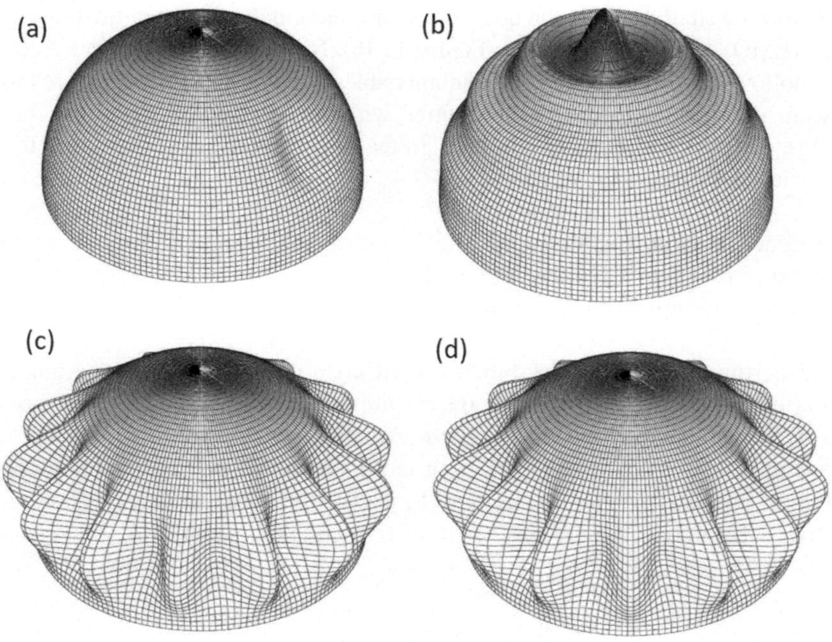

FIGURE 12.14 Imperfection shapes for UNHS1 model for (a) SLI midsection, (b) eigenmode $n = 1$, (c) eigenmode $n = 2$, and (d) eigenmode $n = 3$ [20].

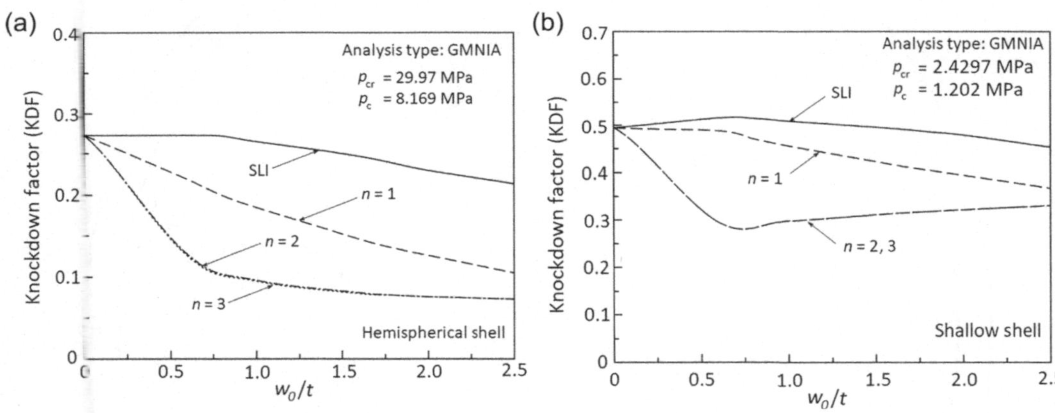

FIGURE 12.15 Knockdown factor versus imperfection amplitude for (a) hemispherical shell (UNHS1) and (b) shallow shell (D6) [20].

- Linear elastic response
- Infinitesimal strain
- Applicability of small deformation theory

The design of an acrylic pressure hull fitted with steel penetrations is, however, much more complicated because of the large difference in the moduli of the two materials; this results in relative displacements at the interface. For this reason, it is important to consider in detail the region of the penetrations in the hull where the continuity of a perfect sphere is interrupted. The most effective way to address the presence of penetrations and the differences in mechanical properties of the materials used is to model the entire structure using the finite element method.

In this section, we shall describe the questions considered in the design of the pioneering acrylic submersible NEMO, as presented in [32] (Figure 12.16). NEMO proved to be a successful design and has been followed by a number of other submersibles fitted with acrylic pressure hulls (Section 2.4.2). As seen in Figure 12.16, NEMO was fitted with openings at the two poles. In addition to these openings, there are a number of other design features that raise questions as to the suitability of axisymmetric shell analysis. These are:

- The presence of bonding adhesive joining the pentagonal members of the pressure hull
- The presence of bolt holes in the bottom plate
- The presence of scallops on the bottom plate flange

Since these departures from the axisymmetric shell arise periodically with the angle of revolution, the net effect is a zero displacement in the circumferential direction. It is therefore considered appropriate to treat the pressure hull as a shell of revolution.

The most significant structural consideration is the correct modelling of the interfaces of the steel penetrations and the acrylic pressure hull. In way of these interfaces, both normal and shear (friction) forces act, and because the coefficient of friction is unknown, two finite element models

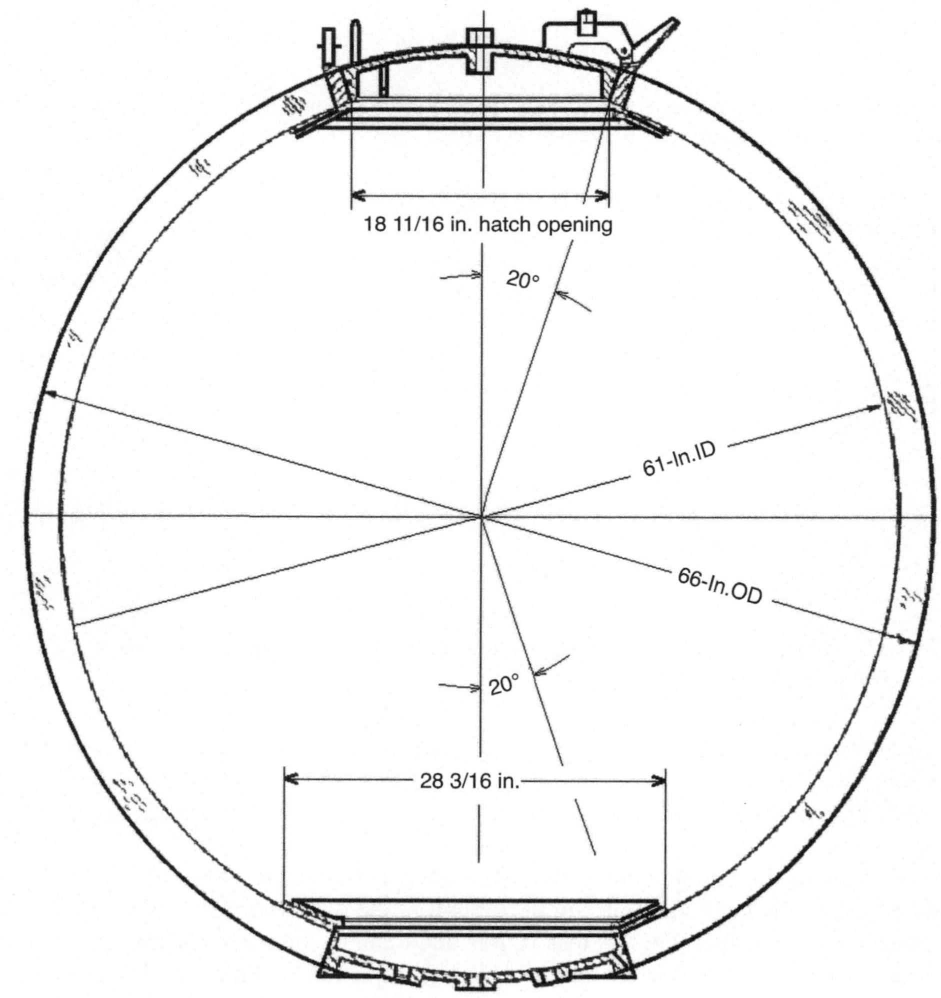

FIGURE 12.16 Acrylic spherical deep-diving submersible (NEMO) [32].

are considered. In the first, the boundary is assumed to be fixed (Figure 12.17a) so that no relative movement arises across the boundary, whereas in the second, a completely frictionless boundary is assumed (Figure 12.17b). These two different boundary conditions are treated as upper and lower bounds to the actual condition that arises across the interfaces.

The mechanical properties of the structural materials that comprise the spherical hull are listed in Table 12.9.

With regard to material failure, the von Mises yield criterion is used for metal alloys, but it has been found that it can also be used to monitor failure of structures made of acrylic. However, the time and temperature dependence of acrylic requires consideration, as discussed in Section 2.4.

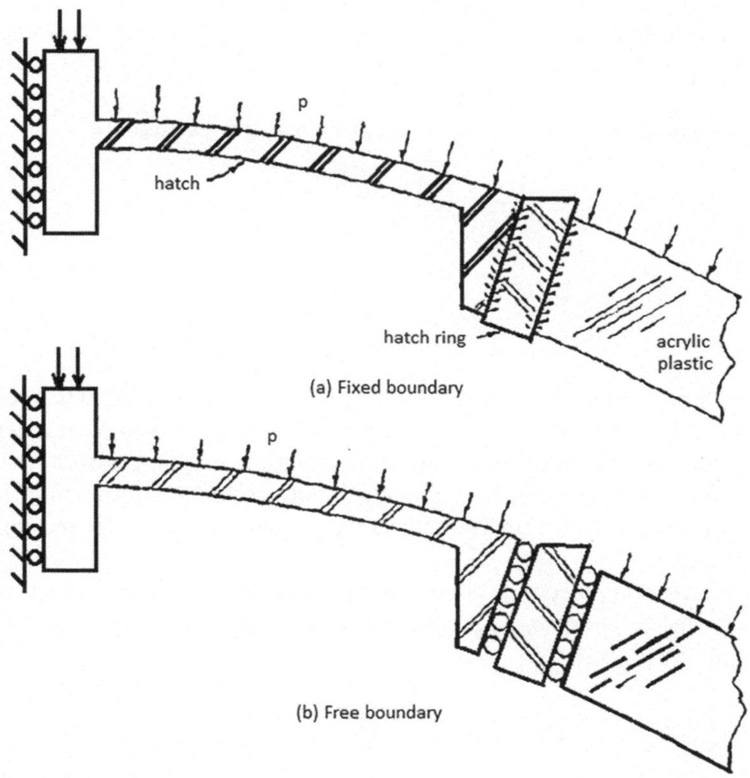

FIGURE 12.17 Finite element modelling of acrylic shell with steel plate boundary [32].

TABLE 12.9
Mechanical Properties of Material Comprising the Pressure Hull [32]

Component	Material	Modulus of Elasticity, E	Poisson's Ratio ν	Yield Stress σ_0
Top hatch and bottom plate	Steel AISI 4130 annealed	29,500,000 psi (205,000 MPa)	0.29	55,000 psi (382 MPa)
Pressure hull	Acrylic Plexiglas G	444,000 psi (3,085 MPa)	0.4	103 MPa (−) 67 MPa (+)

This information concerning the stress state of acrylic plastic is allowed for in the present design study.

It is therefore necessary, when analysing an acrylic pressure hull to allow for both the loading and path history. The following assumptions have been made:

1. Residual stresses. Under normal operations, there will be enough time between dives so that residual stresses will be negligible due to shakeout effects
2. With regard to load variation with time, it is assumed that the load increases from zero to the maximum operating depth instantaneously and is maintained for a specified duration at this level, following which it reduced to zero (step path). This simulates what may be termed a "crash" dive, for which the viscoelastic solution is equivalent to the elastic solution at $t = 0^+$. At $t = 0^+$, the induced stresses and stress concentrations are at maximum values because with the passage of time, these redistribute and relax. This, therefore, is a conservative approach to the load profile on the pressure hull.

For a thick-walled sphere,[3] the expressions for the principal stresses are:

$$\sigma_1 = \sigma_2 = \frac{-R_o(R_i^3 + 2r^3)}{2r^3(R_o^3 - R_i^3)} \tag{12.21a}$$

$$\sigma_3 = \frac{R_o^3(r^3 - R_i^3)}{r^3(R_o^3 - R_i^3)} \tag{12.21b}$$

where R_i, R_o are the inner and outer radii, and r is the radial coordinate. These equations were originally developed by Lamé and are used in conjunction with the von Mises yield criterion to obtain Mises stresses and to check for failure throughout the acrylic pressure hull, making due allowance for viscoelasticity effects. They provide a more accurate description of the stress field than the expressions used in thin-walled structures theory and are suitable for the regions in way of the acrylic–steel interface.

The following relationship between the operating depth h expressed as a function of time, $h(t)$, the time dependent yield criterion $\sigma_0(t)$, and the effective induced (Mises) stress per unit hydrostatic pressure, σ, is used:

$$h(T) = 2.25\frac{\sigma_0(T)}{\sigma} \tag{12.22}$$

This relationship enables planning the depth and duration of individual dives of the submersible given the particular time-dependent material properties and the characteristics of the individual design.

In addition to material failure, it is necessary to consider stability if $t/R < 0.1$. In the present case, $t/R = 0.076$, so that stability has to be investigated. It is, however, difficult to apply stability equations to such a problem because of the complex interrelationship of the material properties (E, ν) with time, temperature, and stress level. To overcome these difficulties, stability can be checked using finite element software.

For the design considered, operating depth curves that allow for load duration and temperature are included in Figure 12.18. The load duration used in the abscissa is directly influenced by the mission profile and can be determined using Figure 12.19, which accounts for different dive profiles. If a planned series of dives has a turn-around time less than the dive time shown in Figure 12.19a, the effect of creep has to be allowed for, and in this case, the dive times have to be added to

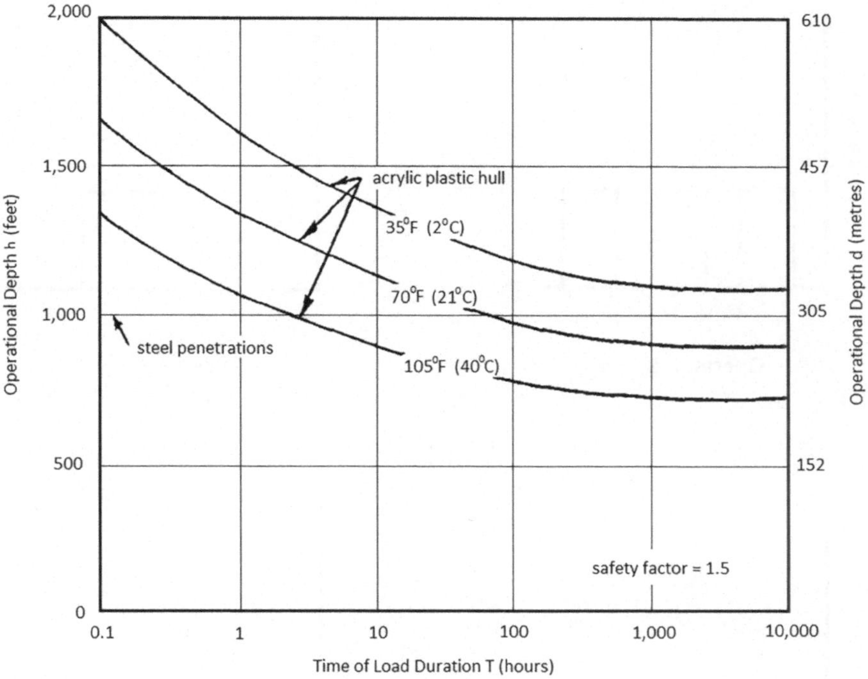

FIGURE 12.18 Operating depth curves for the submersible equipped with a particular acrylic pressure hull [32].

determine load duration. Otherwise put, sufficient time ought to be allowed between dives in order to eliminate the effects of creep.

12.4 DESIGN OF BARRELLED SHELLS

The geometry of a barrelled shell is illustrated in Figure 11.24. Before proceeding to consider what may be the optimum meridional radius of curvature of a barrelled shell for prescribed constraints, it is necessary to have available a procedure that may be used to define barrelled shells having specific properties such as constant mass or constant volume. The presentation given in this section follows closely [21].

12.4.1 Definition of Geometry

If the meridional radius of curvature R_1 of a cylindrical shell is gradually reduced, barrelled shells are created. At the other extreme, a spherical shell is formed for which the meridional and circumferential radii of curvature are equal ($R_1 = R_2$).

In Figure 11.24, θ_1 is an angle coordinate and is designated as the parameter to control shell geometry. By decreasing the value of θ_1, barrelled shells are created. The relationship between the length of a barrelled shell L, the angle coordinate θ_1, and the meridional radius of curvature R is:

$$L = 2R\cos\theta_1 \tag{12.23}$$

In the treatment that follows, the shell thickness t and the radius r_0 are maintained constant.

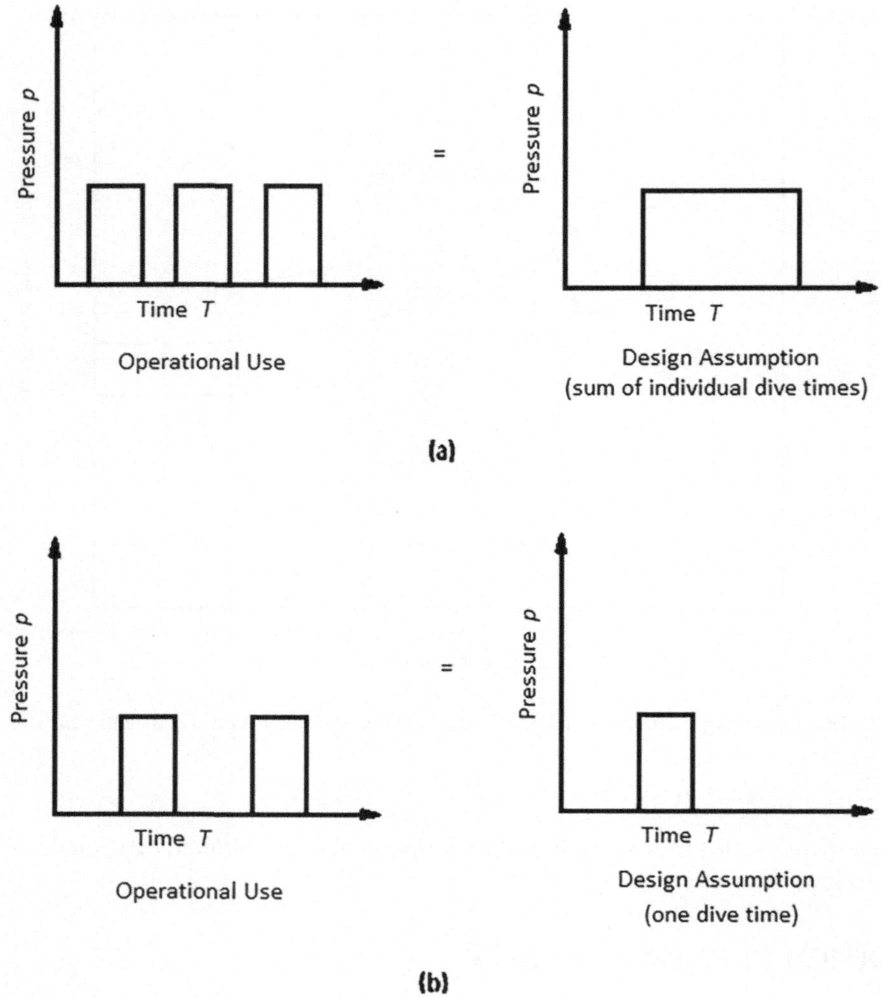

FIGURE 12.19 Determination of load duration time for the particular submersible [32]. (a) Turn-around time < dive time. (b) Turn-around time > dive time.

12.4.2 Barrels of Constant Mass

The mass of the barrelled shell shown in Figure 11.24 is given by:

$$m_s = 4\pi R^2 t\rho_s\left[\left(\frac{\pi}{2}-\theta_1\right)\left(\frac{r_0}{R}-\sin\theta_1\right)+\cos\theta_1\right] \tag{12.24}$$

The following variables are introduced:

$$x_1 = \frac{R_0}{t} \quad x_2 = \frac{R_0}{R} \quad c_m = \frac{m_s}{4\pi R_0{}^3\rho_s}$$

Substituting these variables into (12.24) gives, after rearranging,

$$(c_m x_1)x_2{}^2 + \left(\theta_1 - \frac{\pi}{2}\right)x_2 + \left(\frac{\pi}{2}-\theta_1\right)\sin\theta_1 - \cos\theta_1 = 0 \tag{12.25}$$

Solving this quadratic equation for x_2 leads to the following expression:

$$x_2 = \frac{1}{2c_m x_1}\left(\frac{\pi}{2} - \theta_1 + \sqrt{\left(\theta_1 - \frac{\pi}{2}\right)^2 - 4c_m x_1\left[\left(\frac{\pi}{2} - \theta_1\right)\sin\theta_1 - \cos\theta_1\right]}\right) \tag{12.26}$$

This expression gives x_2 as a function of c_m, θ_1, and x_1. If c_m is kept constant, a single curve relating x_2 to θ_2 is obtained. By varying c_m, a family of curves corresponding to shells of constant mass is obtained (Figure 12.20).

In the case of the spherical shell, the value of x_2 is given by the following expression (dashed line in Figure 12.20):

$$x_2 = \sin\theta_1 \tag{12.27}$$

Expressions (12.26) and (12.27) can then be used to obtain an expression for the angle θ_1 for the spherical shell that has the prescribed mass value, c_m:

$$\theta_1 = a\cos\left(\frac{\sqrt{4c_m^{\,2}x_1^{\,2} + 1} - 1}{2c_m x_1}\right) \tag{12.28}$$

This expression is plotted in Figure 12.20 and is represented by the dashed curve. Thus, for any value of c_m, the corresponding sphere is represented by the intersection of the dashed line and the corresponding c_m curve.

It thus becomes evident that if θ_1 is specified, the remaining parameters of the shell can be obtained (x_2, L and R). The length L is obtained from (12.23), whereas the radius R is obtained from

$$R = r_0 x_2 \tag{12.29}$$

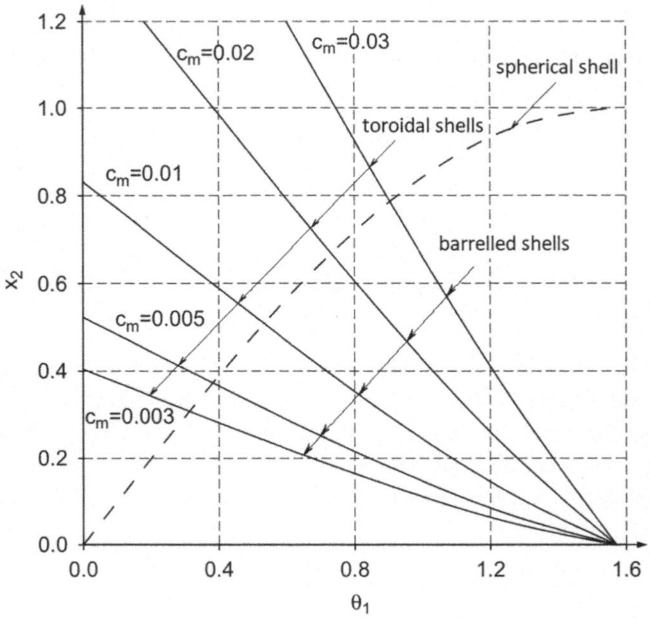

FIGURE 12.20 A family of constant mass curves for different values of c_m [21].

12.4.3 Barrels of Constant Volume

The next step is to create a family of shells of revolution of constant volume. The volume of the barrelled shell shown in Figure 11.24 is given by:

$$V_s = 2\pi R^3\left[\frac{1}{3}(2+\sin^2\theta_1)\cos\theta_1 + \left(\frac{r_0}{R}-\sin\theta_1\right)\left(\frac{r_0}{R}\cos\theta_1+\frac{\pi}{2}-\theta_1\right)\right] \tag{12.30}$$

The following variables are introduced:

$$x_2 = \frac{R_0}{R} \qquad c_v = \frac{V_s}{2\pi R_0^{\,3}}$$

Introducing these variables into (12.30) gives, after rearranging:

$$\begin{gathered} x_2^{\,3} - \left(\frac{\cos\theta_1}{c_v}\right)x_2^{\,2} + \frac{1}{c_v}\left(\frac{1}{2}\sin 2\theta_1 + \theta_1 - \frac{\pi}{2}\right)x_2 \\ + \frac{1}{c_v}\left[\sin\theta_1\left(\frac{\pi}{2}-\theta_1\right) - \frac{1}{3}(2+\sin^2\theta_1)\cos\theta_1\right] = 0 \end{gathered} \tag{12.31}$$

This is a cubic equation in terms of x_2, having the following form:

$$x_2^{\,3} + ax_2^{\,2} + bx_2 + c = 0$$

where

$$a = -\frac{\cos\theta_1}{c_v} \qquad b = \frac{1}{c_v}\left(\frac{1}{2}\sin 2\theta_1 + \theta_1 - \frac{\pi}{2}\right)$$

$$c = \frac{1}{c_v}\left(\sin\left(\frac{\pi}{2}-\theta_1\right) - \frac{1}{3}(2+\sin^2\theta_1)\cos\theta_1\right)$$

The cubic is solved using Cardano's method, and the following expression is obtained for x_2:

$$x_2 = \frac{p}{3u} - u - \frac{a}{3} \tag{12.32}$$

where

$$u = \sqrt[3]{\frac{q}{2} \pm \sqrt{\frac{q^2}{4}+\frac{p^3}{27}}} \qquad p = b - \frac{a^3}{3} \qquad q = c + \frac{2a^3 - 9ac}{27}$$

The only parameter that influences the value of x_2 is c_v. Changing it, a family of curves of constant volume may be created. Figure 12.21 presents the relation between values of the variables θ_1 and x_2. Intersection points of the curves of constant volume and the dashed line in Figure 12.21 represent spherical shells. Expression (12.32) may be now used to calculate x_2. Hence and assuming the value of θ_1, it is possible to calculate the values of L and R using expressions (12.23) and (12.29).

12.5 DESIGN OF MULTISPHERE PRESSURE HULLS

The idea of using multisegment pressure hulls for deep-diving submersibles has been considered from the earliest investigations carried out on the suitability of different pressure hull forms.

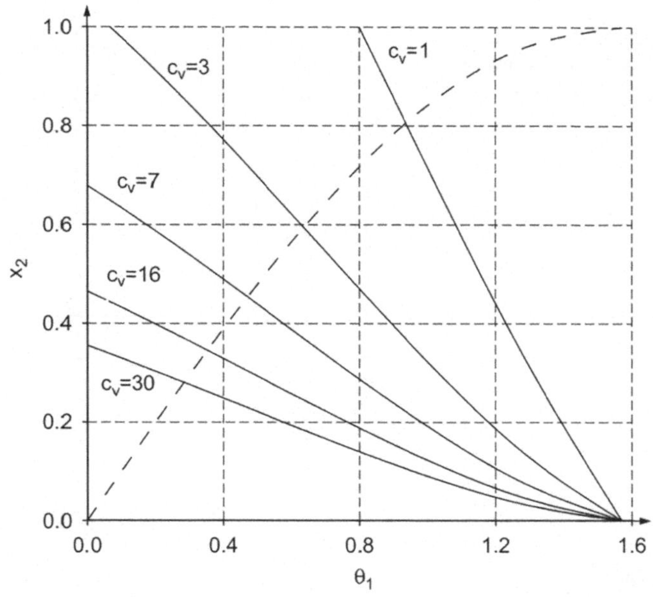

FIGURE 12.21 A family of constant volume curves for different values of c_v [21].

Different combinations of spheres attached to each other were considered, in conjunction with different metal alloys (HY-100, HY140 steels as well as titanium) [23]. Details are given in Section 12.2.2.

12.5.1 A Brief Literature Review

The first multisphere pressure hulls designed and built were the bi-sphere *Deep Quest* (Figure 12.22) and the tri-sphere *DSRV* (Deep Submergence Rescue Vehicle) (Figure 12.23) [14]. In these, steel alloys were used for both the spherical shells and the intersecting rings. Their principal particulars are included in Table 12.10.

In a later project, tests were carried out on a bi-sphere pressure of 1.2 m diameter fabricated using titanium, described in Section 11.3.2 [24]. The tests focussed on the use of alternative materials for the intersection ring, and it was concluded that using Al_2O_3 instead of titanium the payload could be increased by 10%.

The combined use of composite materials (graphite epoxy) with titanium intersection rings was studied for a 1.2 m diameter bi-sphere pressure hull, and it was found that this led to a weight saving of the pressure hull of 46%, compared to the weight of a steel pressure hull having the same diameter [18].

The validation of FEA for the design of multi-sphere pressure hulls is described in [39]. Tests on bi-sphere shells were carried out, described in Section 11.3.2, and predictions using FEA were compared (Table 12.11).

It is seen from Table 12.11 that in the case of test specimens DS2 and DS3, good agreement was obtained, with the predicted collapse pressure lying in the range 0.92–0.99 of the test results. In the case of specimen DS1, a larger discrepancy was observed (0.63, 0.67), and this was attributed to the effect of the weld joint, which had not been considered in the numerical analysis.

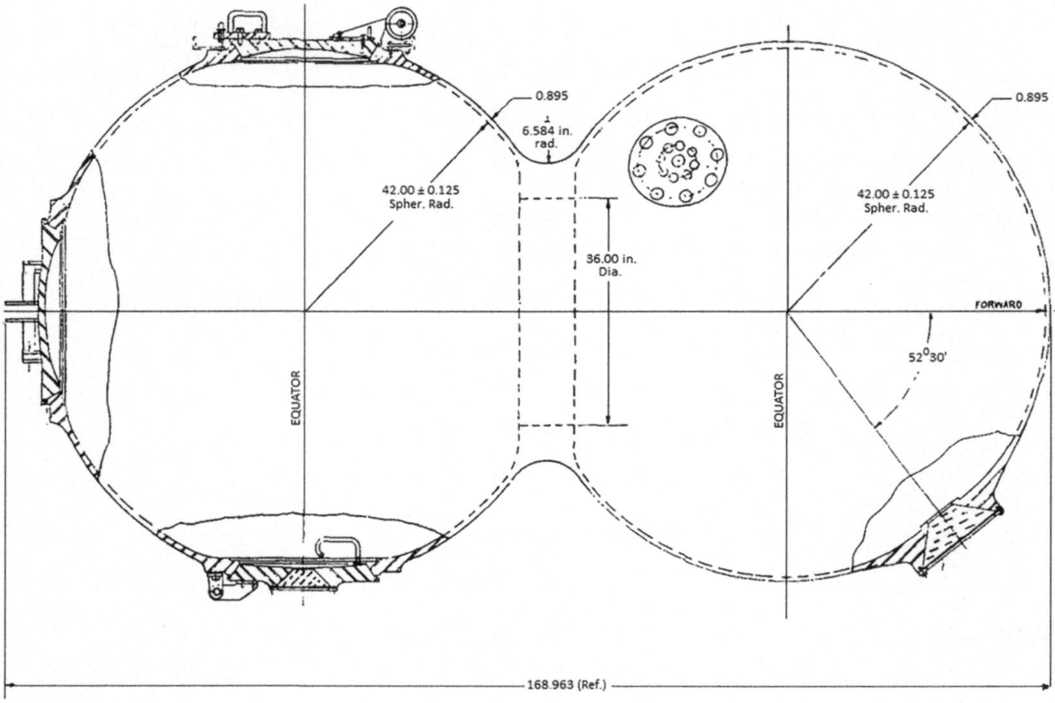

FIGURE 12.22 Profile of *Deep Quest* [14].

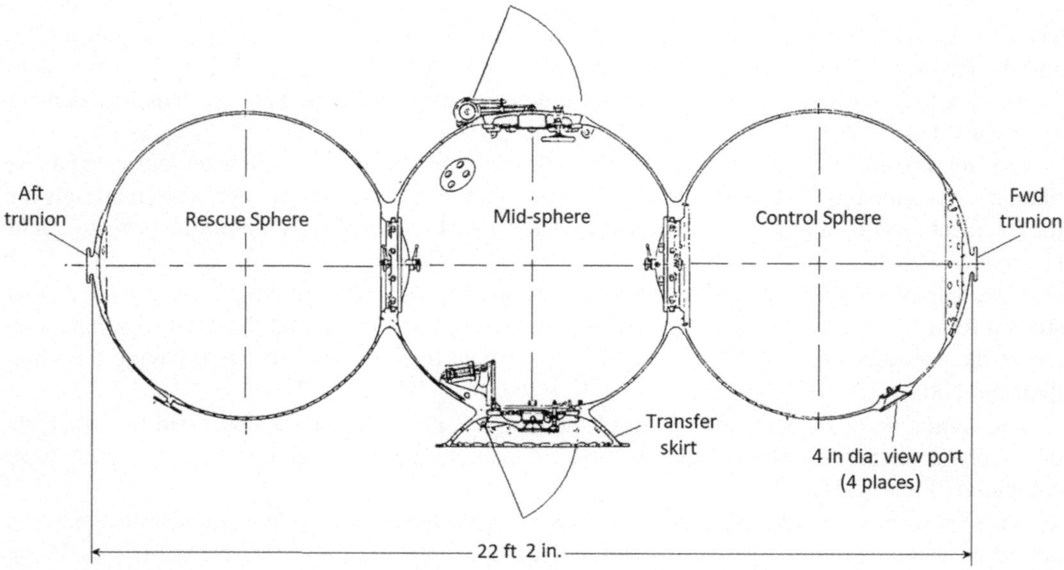

FIGURE 12.23 Profile of deep submergence rescue vehicle [14].

TABLE 12.10
Principal Particulars of *Deep Quest* and *DSRV*

	Deep Quest	DSRV
Sphere diameter	84 in. (2,134 mm) (inner)	90 in (2,286 mm) (outer)
Sphere thickness	0.895 in. (22.7 mm)	0.738 in.
Length of hull	14 ft 1 in. (4.29 m)	22 ft 2.36 in. (6.765 m)
Weight of hull	15,000 lb (6,804 kg) (in air)	20,525 lb (9,310 kg) (in air)
Operating depth	8,310 ft (2,533 m)	5,000 ft (1,524 m)
Factor of safety	2.0	>1.5
Test depth	8,500 ft (2,591 m)	–
Material	18-Ni maraging steel, 200	HY-140 steel
Yield strength	220–235 ksi (1,528–1,632 MPa)	140 ksi (972 MPa)

TABLE 12.11
Comparison of Predicted Bucking Pressures with Test Collapse Pressures [39]

Test Specimen No	p_{test} (MPa)	p_{ABAQUS} (MPa)	p_{ABAQUS}/p_{test}
DS1-1	1.74	1.16	0.67
DS1–2	1.67	1.05	0.63
DS2-1	3.55	3.32	0.93
DS2-2	3.73	3.45	0.92
DS3-1	4.39	4.35	0.99
DS3-2	5.06	4.69	0.93

12.5.2 Design of Intersection Rings in Multisphere Pressure Hulls

It was mentioned in Section 12.3.5 dealing with acrylic pressure hulls that an important challenge in the design of deep-diving submersibles is allowing for the properties of different materials and in particular the effect this has on stress distributions in way of interfaces of different materials. Such a problem arises also in the case of multi-sphere pressure hulls in which spheres are attached to each other using intersection rings for which the material used may be different to that of the spherical shells (Figure 12.24). The role of the rings is all-important since they do not only ensure watertightness but are also used to supplement the loss of strength of the adjacent spherical shells that results from the removal of part of the shell.

In this section, we shall outline a procedure described in [38] that can be implemented to ensure geometrical compatibility of the ring with the adjoining spheres by selecting appropriate ring dimensions. The ring is assumed to have a rectangular section of width L_r and thickness t_r. The outer and inner diameters of the ring are D_r and d_r, respectively.

In order to ensure that following application of external hydrostatic pressure no bending arises in way of each intersection, the radial compression stiffness of the ring has to be equal to that of the adjacent spherical shells, so that:

$$\delta_s = \delta_r \tag{12.33}$$

where δ_s δ_r are the radial displacements of the shell and ring, respectively.

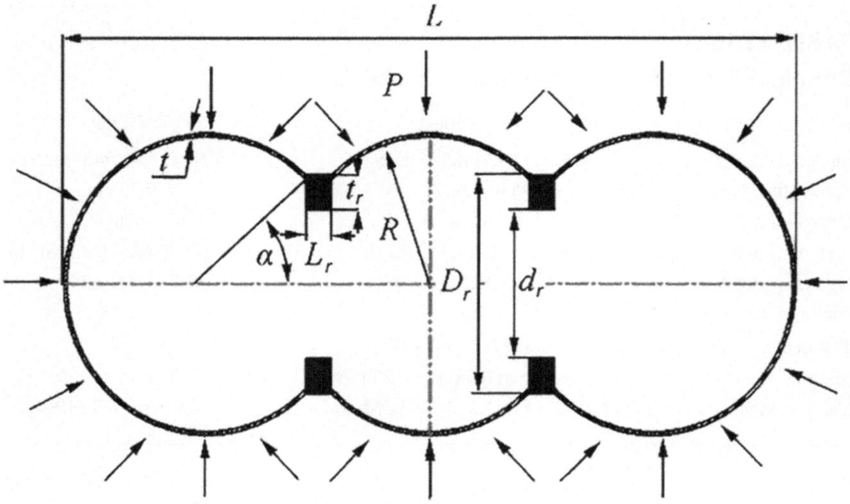

FIGURE 12.24 Three-sphere pressure hull [38].

The radial displacement of the spherical shell shown in Figure 12.24 is given by:

$$\delta_s = \frac{p_0 R^2 \sin\alpha}{2Et}(1-\nu) \tag{12.34}$$

where p_0 = uniform external pressure

The radial displacement of the ring, whose outer diameter is equal to D_r, is given by:

$$\delta_r = \frac{p_r D_r}{2E}\left(\frac{D_r^2 + d_r^2}{D_r^2 - d_r^2} - \nu\right) \tag{12.35}$$

where p_r is the uniform pressure acting on the ring, given by:

$$p_r = \frac{pR\cos\theta}{2t_r} + \frac{pR\cos\theta}{2t_r} + p = p\left(\frac{R\cos\theta}{t} + 1\right) \tag{12.36}$$

The first and second terms in the above expression are the pressure components acting on the ring in the radial direction by the adjacent shells. The following relationship between D_r, t_r, and L_r is thereby obtained [38]:

$$L_r = \frac{R\cos\alpha}{\dfrac{R^2\sin\alpha(1-\nu)}{tD_r}\dfrac{1}{\dfrac{D_r^2 + 2t_r^2 - 2D_r t_r}{2D_r t_r - 2t_r^2} - \nu} - 1} \tag{12.37}$$

where $D_r = \sqrt{4(R+t)^2} - 4R^2\cos^2\alpha$

From (12.33) to (12.36), the following expression is obtained for the ring thickness, t_r [39]:

$$t_r = \left[\frac{D_r^3}{2R(1-\nu)\tan\theta}\right]\left[\frac{1}{L_r(1-L_r)}\right] - \left[\frac{D_r(1+\nu)}{R(1-\nu)\tan\theta}\right] \tag{12.38}$$

12.6 BRIEF COMMENTS ON THE DESIGN OF CYLINDRICAL SHELLS USING COMPOSITES

Deep-diving submersibles are subjected to large hydrostatic pressures and as a result have thick shells (high t/R ratio). As a result, they can be designed without the use of stiffening, in contrast to shallow-water vessels that, having slenderer sections, are prone to buckling and require some sort of stiffening. A lighter structure would result were a stiffening system to be fitted to deep-diving submersibles, but the weight penalty of the unstiffened cylindrical shell is offset by the simplicity of fabrication and the absence of local bending and shear stress effects [31]. Deep-diving pressure hulls have been built of thermoset and thermoplastic cylinders having a shell thickness 50–100 mm and have operated satisfactorily.

In order to improve the weight-to-displacement ratio, a number of stiffening arrangements can be considered (Figures 2.16b, c and 12.25). The sketches show arrangements that are fabricated in different ways. Figure 2.16b and c show monolithic structures in which the ring stiffeners are integrated with the cylindrical shell and the case in which a sandwich cylinder is produced. From the production point of view, it is preferable to fabricate identical individual units that are assembled at a subsequent stage. Figure 12.25a and b show geometries comprising short cylindrical and truncated spherical segments, so that in the latter case, a multi-sphere pressure hull results.

The use of composite materials in general opens up a large range of design possibilities because, in contrast to design using metal alloys, the material itself can be designed in conjunction with the structure. In principle, this leads to a much more efficient design but having said that, it has to be borne in mind that the design work and expertise required for a complex structural arrangement such as that of an underwater vehicle require much effort and time. It is much more practical to select materials that have a proven history of performance in specific applications before proceeding to the design of new materials for use in a new arrangement.

The formulation of the equations of equilibrium of a laminate consisting of a number of laminas is described in many texts dealing with the nature, mechanics, strength, design, and failure of composites. The earliest strength formulation that was developed for composites is termed *classical laminate theory* (CLT), and this is analogous to the theory of thin-walled flat isotropic plates. CLT is a simplification of 3-D anisotropic elasticity in which the properties for a laminate are determined by assembling individual laminas each of which are each assumed to be orthotropic. Later on, it was recognised that transverse shear effects need to be included because composites have much lower shear stiffness than metal alloys. To this end, *first-order shear deformation theory*[4] (FSDT) was proposed, and this is now the established means of modelling the structural performance of composite materials.

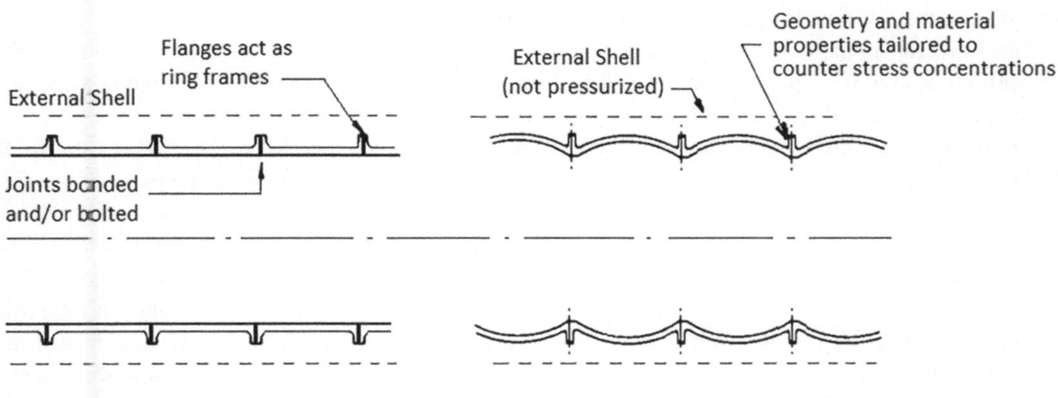

FIGURE 12.25 Segmented composite stiffened pressure hull configurations [31]. (a) Short cylindrical segments. (b) Truncated spherical segments.

12.7 OPTIMUM STRUCTURAL DESIGN OF UNDERWATER VEHICLES AND THEIR COMPONENTS

Optimisation methods have been used in the design of underwater vehicles with early efforts focussing on ring-stiffened cylindrical shells used in submarines. In these publications, classical optimisation methods were used in minimum weight design.

12.7.1 Ring-Stiffened Cylinders and Multisphere Pressure Hulls

In 1961, Gerard carried out minimum weight analyses of unstiffened and ring-stiffened cylinders loaded under external pressure [15]. The constraints used were buckling stability and compressive yield strength. The results were compared to establish the relative efficiencies of different types of materials, and conclusions were drawn of relevance to deep-diving submersibles.

In 1972, Simitses [30] presented a methodology aimed at the minimum weight design of a stiffened cylinder of specified material, radius, and length that is subjected to a given hydrostatic pressure. The solution involves two stages, in the first of which design charts are obtained, based on a simplified formulation of the objective function. In the second stage, these design charts are used to evaluate the design variables. The aim of this approach was to enable the designer to introduce needed changes and avoid failure mode interaction.

In 1973, Pappas described the use of the Direct Search-Feasible Direction (DSFD) algorithm[5] to develop a optimisation procedure aiming at minimising the weight of cylindrical shells fitted with T-bar ring frames, used in submersibles [28]. It was found that the ratio of the structural weight to the weight of displaced water (W/Δ) is independent of the diameter of the shell and that the optimal shell dimensions are a function of the diameter (proportional to it). Interframe buckling was found to be the critical failure mode in optimal designs, whereas if the OOC is significant, the critical failure mode is frame yielding. The value of using high-strength steels was also noted for two reasons: firstly, the adverse effect of eccentricity is reduced, and secondly, the elastic range of shell plating when high pressure is applied is extended.

In 1975, a paper published with aerospace structures in mind [9] considered the optimum structural design of cylindrical shells reinforced with T-bar ring frames. Three optimality criteria were considered: (a) minimum weight, (b) maximum separation of the lowest two natural frequencies of vibration, and (c) maximum separation of the lowest two natural frequencies of vibration with axial component. The structure was modelled as an orthotropic panel, and the optimisation procedure used was sequential unconstrained minimisation (SUMT). Solution constraints were imposed in order to prevent shell buckling, shell yielding, ring-stiffener bucking and yielding. It was found that the second and third optimality criteria were satisfied with a moderate increase in weight above the minimum weight solution (criterion a). Maximum frequency separations are obtained when the second and third lowest frequencies approach each other.

In a more recent paper (2010), gradient-based methods and genetic algorithms were used to reach an optimal design of a ring-stiffened cylindrical shell subjected to external pressure under strength, buckling as well as fabrication technology constraints [33].

With regard to multisphere pressure hulls, two recent studies have been reported. In the first of these (2004), the optimum design of multi-segment spherical shells under deep-water conditions was treated using a procedure that combines the *extended interior penalty function method* (EIPF) with the *Davidon–Fletcher–Powell* (DFP) method [25]. The design variables selected were shell thickness, width and inner radius of intersection ring, and the angle of intersection of the spherical shells. A sensitivity analysis revealed that shell thickness is most important with regard to interframe buckling, whereas the scantlings of the intersection ring are critical with regard to ring strength. In the second study [17], two typical failure modes were considered, and three new multiple intersecting forms presented.[6]

12.7.2 Barrelled Shells

A glance at Figure 11.24 immediately raises the question as to what ought to be the optimum meridional radius of curvature of a barrelled shell. Clearly, the larger the radius, the higher will be the stiffness and strength, but from the designer's perspective, it is necessary to compare geometries using an acceptable baseline. For barrelled shells, mass and volume have been used as a basis for comparison. In the case of mass, as the meridional radius is increased, the shell thickness has to be decreased in order to maintain a constant mass. An optimum meridional radius from the point of view of static strength can thereby be defined, assuming a constant thickness for a shell subjected to uniform external pressure. If, in addition, the thickness is allowed to vary to account for the non-uniform distribution of stresses, the optimisation becomes much more complicated. Furthermore, optimisation may be required against more than one load variable, in the case multiple loads act. Such is the case of shells forming part of underwater vehicles since they are subjected to both radial pressure and axial compression.

The procedure that we shall apply is termed *shape optimisation,* and this is used to determine the optimal geometry of a structure such that the highest possible structural efficiency is achieved, subject to static and/or dynamic constraints. In the case of shells, it is of particular importance, as shells, having a finite curvature in more than one direction, can be self-equilibrating structures, as, for example, in the case of underwater vehicles. Shape optimisation of shells has been investigated extensively in a number of fields such as civil engineering, architecture, mechanical and aerospace engineering, and a large number of techniques have been developed and applied.

In this section, we shall describe a methodology developed by Błachut [3] for barrelled shells that are subjected to both radial pressure and axial compression, bearing in mind underwater vehicles. In effect, *one-dimensional parametrical optimisation* is carried out in which the meridional curvature is selected as the control variable.

We consider a cylindrical shell having radius R_0, length L_0 and a constant thickness t. The material behaviour is assumed to be elastic-rigid plastic, and the loading consists of axial compression F_0 and external pressure p_0, acting concurrently (Figure 12.26).

In this treatment, we assume a constant volume basis, that is, the optimum shape is to have a volume equal to that of the cylindrical shell that acts as the starting point:

$$V_{bar} = V_{cyl} \tag{12.39}$$

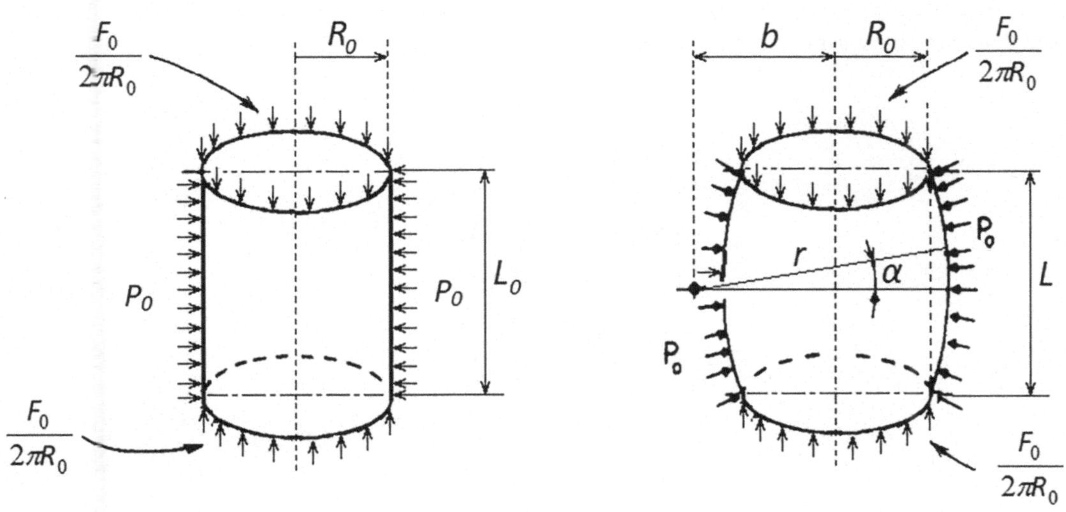

FIGURE 12.26 Geometry and loading of (a) cylindrical and (b) barrelled shell [3].

The action of a load system that consists of two loads can be represented in a two-dimensional polar coordinate system (R, β) where

$$R = \sqrt{\overline{F}^2 + \overline{p}^2} \quad \text{and} \quad \beta \in \left[0, \frac{\pi}{2}\right] \tag{12.40}$$

where $\overline{F}$ is the normalised critical load in axial compression (bifurcation or collapse load, whichever is the lowest) given by:

$$\overline{F} = \frac{F}{F_{CYL}} \tag{12.41}$$

and $\overline{p}$ is the normalised critical pressure (bifurcation or collapse pressure, whichever is the lowest) given by:

$$\overline{p} = \frac{p}{p_{\text{cyl}}} \tag{12.42}$$

where F_{cyl}, p_{cyl} are the critical collapse (bifurcation) loads for a cylinder under axial compression alone and radial pressure alone, respectively.

We proceed on a constant volume basis ($V = V_{cyl}$) and seek the meridional radius of curvature r of the barrelled shell (Figure 12.26b) that possesses the same volume, thickness, boundary conditions, truncation radius R_0 and maximises R for a given angle β, i.e

$$R^*(\beta) = \max_{c \in [0,1]} R(\beta) \quad \text{for} \quad \beta \in \left[0, \frac{\pi}{2}\right] \tag{12.43}$$

where $c = R_0/(R_0 + b)$. It is seen from Figure 12.27 that

- If $\beta = 0$, the axial compression $\overline{F} = 0$ and a solution is obtained for maximum pressure
- If $\beta = \pi/2$, the pressure $\overline{p} = 0$ and a solution is obtained for maximum axial compression
- For $\pi/2 > \beta > 0$, interaction curves are obtained for combined loading conditions, one of which corresponds to the optimum solution

The purpose of the exercise is to obtain an interaction curve for an optimum geometry, that is, the one that is the most efficient from the point of view of structural stability.[7]

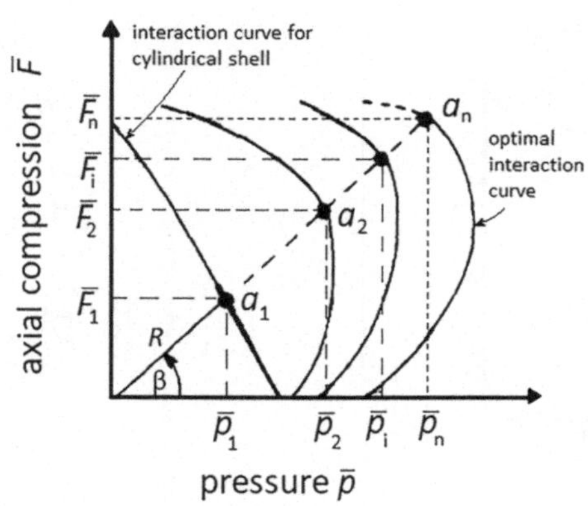

FIGURE 12.27 Procedure to obtain interaction curve [3].

We start with a cylindrical shell that has geometrical parameters (R_0/t, L_0/R_0) and determine the locus of point a_1, from relationship (1) for R, for a given angle β. At point $a_1 c = 1$. We then consider gradually decreasing values of c, that is, increasing degrees of barrelling, and calculate the corresponding buckling loads ($\overline{F}$, $\overline{p}$) for each geometry. Point a_n represents the optimal solution for a given angle β so that we have defined a single point on the optimum interaction curve for the particular geometry. By varying angle β, we can find the whole optimal interaction curve that defines stability when a combination of axial compression and external hydrostatic pressure acts.

Since we are implementing a constant volume approach, the geometry of the barrelled shell that corresponds to the particular value of the control variable c_i is obtained by applying the equality $V_{bar} = V_{cyl}$:

$$2\left[r^2 \sin\alpha - \frac{R_0(1-c_i)}{c_i} r\alpha\right] = R_0 L_0 \tag{12.44}$$

The appropriate stability problem is then solved using numerical analysis software.

It is seen from Figure 12.27 that domains enclosed by the optimal interaction curves are significantly larger than those for the cylindrical shells of the same volume. In other words, a given volume can be subjected to combinations of much larger loads than the the correspondding cylindrical shell. It has been found that this applies to both elastic and elasto-plastic buckling [3].

12.7.3 Optimum Design of Torispheres

Externally pressurised torispheres can lose their stability through axisymmetric snap-through, this being the limit point load, or else by asymmetric bifurcation with a number of wrinkles arising in way of the knuckle/crown junction. Both instability modes are considered in design, with the lower of the two being treated as the failure load.

Torispherical shells are widely used as end closures of both externally and internally pressurised shells. In this section, we shall discuss the results presented in studies dealing with the optimal design of externally pressurised unstiffened torispherical shells that may be used as end closures of underwater vehicles. The issues considered are:

- Optimisation of shell form with respect to buckling strength
- Sensitivity to initial distortions
- Optimum shape of a torisphere of constant thickness and given weight in order to maximise the failure pressure
- Selection of optimal form for minimum weight design

In the design of end closures, conflicting requirements arise: from the fabrication standpoint they should be as shallow as possible, whereas from the stress analysis standpoint, a higher contour for the end closure is required in order to achieve a smooth transition in the membrane stresses in way of the junction with the cylinder. This also minimises the bending effects at the junction. The former is satisfied by flat plates, whereas the second by hemispherical or ellipsoidal end closures. Structures whose geometry lies between the flat plate and the hemispherical end closure include conical, ellipsoidal, and torispherical end closures.

12.7.3.1 Geometry of a Torisphere Using the Four-Centred Ellipse Method

The geometry of a torisphere can be defined using graphics design techniques that are also used to define the shapes of ellipses, hemispheres, and ellipsoids (Figure 12.28a).

The use of the *four-centred ellipse* method to define the approximate geometry of an ellipse is described in [13] and has been applied to the case of a torisphere in [35].

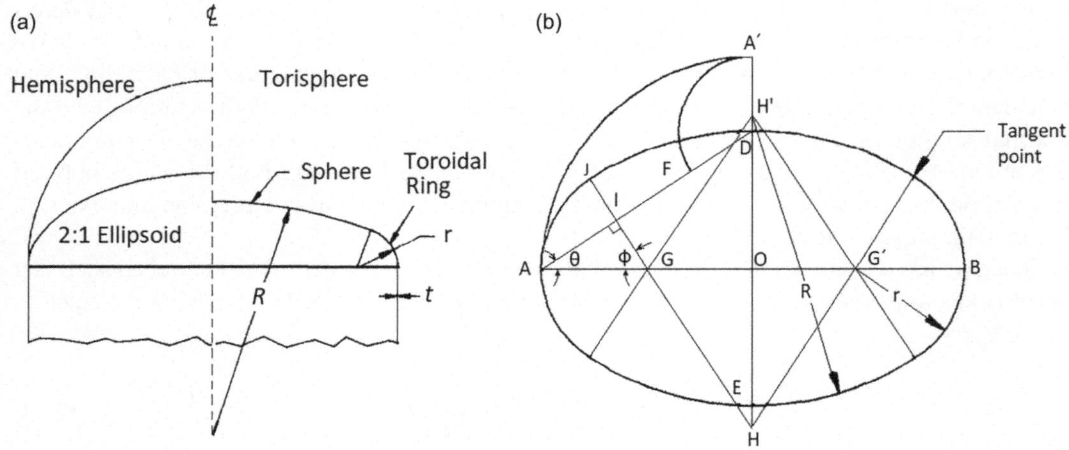

FIGURE 12.28 (a) End closures used in pressure vessels [35]. (b) Construction of a four-centred approximate ellipse [13].

In Figure 12.28b, a quadrant with centre at O and radius equal to $\overline{OA}$ is drawn such that point A′ is defined. Points A and D are then joined with diagonal $\overline{AD}$. With point D as centre, a circle segment having radius $\overline{DA'}$ is drawn that meets $\overline{AD}$ at point F so that $\overline{DF} = \overline{AO} - \overline{DO}$. Line segment $\overline{AF}$ is bisected such that $\overline{AI} = \overline{IF}$ and a perpendicular to $\overline{AD}$ is drawn at I crossing $\overline{AO}$ at G and intersecting $\overline{DE}$. If necessary, $\overline{IE}$ is produced and point H defined.

Line segment $\overline{OG'}$ is made equal to $\overline{OG}$ and line segment $\overline{OH'}$ equal to $\overline{OH}$. Then G, G′, H, and H′ will be centres of four tangent circle arcs forming a curve approximating the shape of an ellipse. The following geometric relations hold:

$$\overline{OA} = \text{major axis} = a$$

$$\overline{OD} = \text{minor axis} = b$$

$$\overline{DF} = \overline{AD} - \overline{OD} = a - b$$

$$\overline{OA} = \overline{OD} + \overline{DA'}$$

$$\overline{DA'} = \overline{DF} = a - b$$

$$\overline{AI} = \overline{FI}$$

$$\overline{OG'} = \overline{OG}$$

$$\overline{OH'} = \overline{OH}$$

The length of line segment $\overline{AI}$ is given by:

$$\overline{AI} = \overline{FI} = \frac{\overline{AD} - \overline{DF}}{2} = \frac{1}{2}\left[\frac{a}{\cos\theta} - (a - b)\right] = \frac{a}{2}\left[\frac{1 + \sin\theta - \cos\theta}{\cos\theta}\right]$$

Let r be the radius of the toroidal ring and R the radius of the spherical portion (Figure 12.28a). Both r and R can be derived from the above relationships as follows:

$$\begin{aligned} r &= \overline{GA} = \overline{GJ} \\ &= \frac{AI}{\cos\theta} \\ &= \frac{a}{2}\left[\frac{1+\sin\theta-\cos\theta}{\cos^2\theta}\right] \end{aligned} \tag{12.45}$$

$$\begin{aligned} R &= \overline{HG} + \overline{GJ} \\ &= \frac{\overline{GO}}{\sin\theta} + r = \frac{a-r}{\sin\theta} + r = \frac{r\sin\theta + a - r}{\sin\theta} \\ &= \frac{a}{2\sin\theta}\left[1 + \frac{1-\sin\theta}{\cos\theta}\right] \end{aligned} \tag{12.46}$$

From (12.45) and (12.46), r and R are seen to be functions of a, θ. We let K be the ratio of the minor axis to the major axis:

$$K = \tan\theta = \frac{b}{a} \tag{12.47}$$

Hence,

$$\varphi = \frac{\pi}{2} - \theta \tag{12.48}$$

and

$$r = f(a,\theta) = f(a,K) = f(a,b) \tag{12.49}$$

$$R = f(a,\theta) = f(a,K) = f(a,b) \tag{12.50}$$

For a hemispherical dome $\theta = 45°$. Then $K = 1$ and $r = R = a = b$

For a 2:1 ellipsoid, $b = a/2$. Then K=1/2

If the ratio K is specified, θ can be calculated from (12.47), the angle φ can be calculated from (12.48), and r and R can be calculated from (12.45) and (12.46), respectively. The torispherical end closure can then be sketched using r, R, θ, and φ. This method of defining the geometry of a torisphere is used in Section 12.7.3.5.

12.7.3.2 Optimisation of Shell Form for Given Buckling Strength

In [4], the optimum shape of a torisphere was sought in order to increase the failure pressure. The shape was varied by redistributing the mass in a step-wise manner in the meridional direction, using an arbitrary number of segments. Three different approaches to mass location were implemented, as shown in Figure 12.29.

The results obtained for each of the mass distribution approaches using the BOSOR5 program are compared to the shell of uniform thickness (t_0) in Table 12.12.

In the next stage, optimisation was carried out for geometries for which $0.5 \le \overline{\lambda} \le 2.0$ for case C. Results are plotted in Figure 12.30 from which it is seen that the buckling pressure increases from 15% to 50% over that of the shell with uniform thickness. Optimisation was conducted using a discrete version of *dynamic programming* [4].

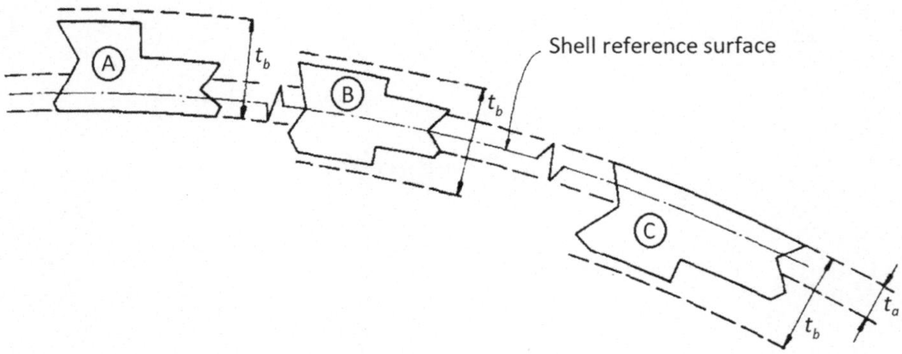

FIGURE 12.29 Mass distribution methods [4].

TABLE 12.12
Effect of Mass Distribution on Failure Pressure Increase

	Mass Distribution		
	A	*B*	*C*
Segment	t/t_0	t/t_0	t/t_0
1	0.91	0.96	0.96
2	1.10	0.96	0.96
3	1.30	1.10	1.10
4	1.10	1.30	1.30
5	0.70	0.70	0.70
Pressure increase (%)	22	29	34

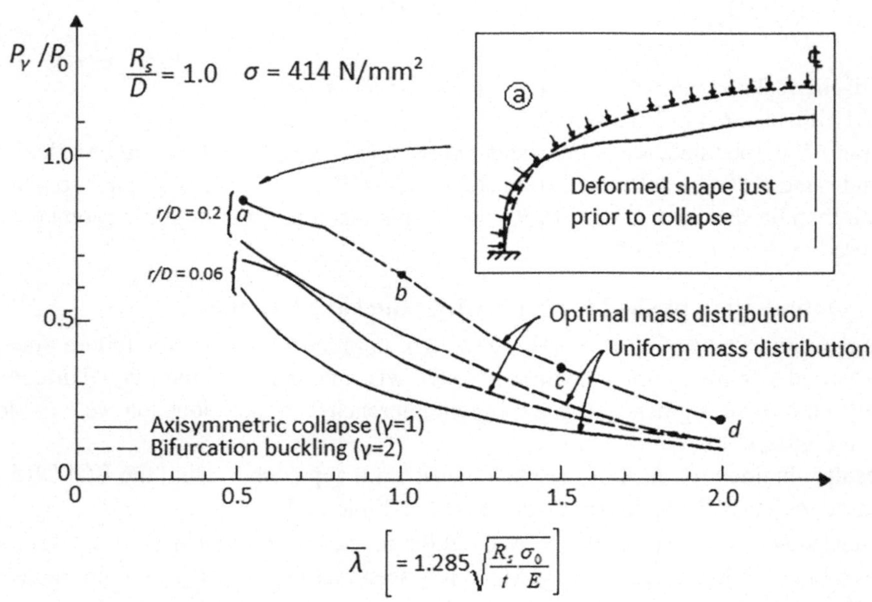

FIGURE 12.30 Pressure versus slenderness $\bar{\lambda}$ for optimised shell and for shell with uniform thickness.[8]

It is seen from Figure 12.30 that for the optimised shells, bifurcation becomes the controlling failure mode over a wider range of slenderness than for shells of uniform thickness.

12.7.3.3 Sensitivity to Initial Distortions

The results of a study on the effect of axisymmetric imperfections on the strength of optimally shaped torispheres are summarised in Figure 12.31 [4]. The imperfection considered was axisymmetric flattening in way of the apex (Figure 12.31 insert).

The curves plotted in Figure 12.31 constitute lower bound envelopes to the strength for both uniform and optimal torispheres. It is observed that the imperfection produces a larger strength reduction in the case of optimally shaped torispheres than for those having a uniform thickness.

The effect of torisphere geometry was also considered, and it was found that shallow torispheres ($r/D = 0.06$) are more sensitive to geometric imperfections than a corresponding hemisphere.

When the magnitude of the imperfection reaches a critical value for a particular geometry, the effects of optimisation erode. For $r/D = 0.06$ and for $r/D = 0.2$, the critical imperfection magnitudes were found to be $\delta_0 / t \geq 0.16$ and $\delta_0 / t \geq 0.25$, respectively. Torispheres that have a constant thickness can support a larger pressure once the amplitude of geometrical distortion exceeds the above values.

12.7.3.4 Optimum Shape of a Torisphere in Order to Maximise Failure Pressure

In [5], a method to determine the geometry of a torisphere having a constant thickness and weight such that the failure pressure is maximised is described. The failure pressure of the optimised torisphere was compared to that of a hemisphere having the same diameter and weight. The optimisation problem was then formulated as follows: Determine the geometry of a torisphere, described by the variables r/D, R_s/D, D/t such that the failure pressure is maximised. These variables are defined in Figure 12.32.

Failure may arise as a result of bifurcation or axisymmetric snap-through with the failure pressure and the corresponding failure mode being determined using the BOSOR5 code. The following side constraints were selected:

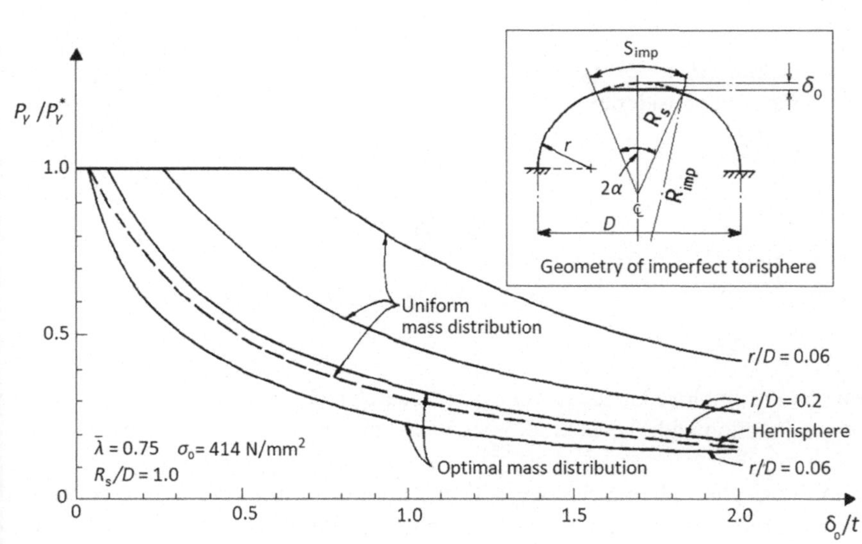

FIGURE 12.31 Torisphere imperfection sensitivity to axisymmetric distortion at the apex [4].

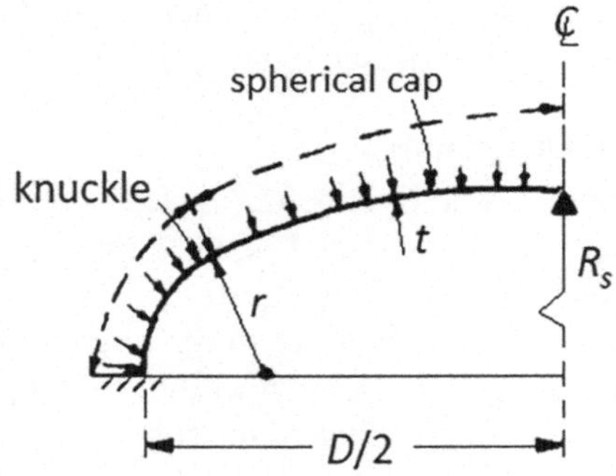

FIGURE 12.32 Definition of design variables.

$$0.6 \leq \frac{W_T}{W_H} \leq 1.0 \tag{12.51a}$$

$$100 \leq \frac{D}{t_0} \leq 1{,}000 \tag{12.51b}$$

where W_T is the weight of the torisphere given by:

$$W_T = 2\pi\gamma t\left[r\left(\frac{D}{2}-r\right)\left(\frac{\pi}{2}-\phi\right)+(r^2\text{-}R_s^{\,2})\cos\phi + R_s^{\,2}\right] \tag{12.52}$$

where

$$\phi = \arcsin\left(\frac{(D/2)-r}{R_s-r}\right)$$

and W_H is the weight of the hemisphere given by:

$$W_H = \frac{1}{2}\pi\gamma D^2 t_0 \tag{12.53}$$

where γ is the specific gravity of the material used. The following parameter ranges were selected on the basis of practical considerations:

$$D/t > 0 \tag{12.54a}$$

$$0.05 \leq r/D \leq 0.45 \tag{12.54b}$$

$$0.55 \leq R_s/D \leq 1.25 \tag{12.54c}$$

The solution method employed made use of a direct search technique, known as the *complex method of Box* [8] in conjunction with the BOSOR5 code used to calculate the failure mode and the critical pressure.

The optimal torisphere, being the one with the highest failure pressure, was determined and results are compared with the equivalent hemisphere in Figure 12.33. It is seen that the failure load for the geometry considered can be increased by 53%, by making optimum use of the material.

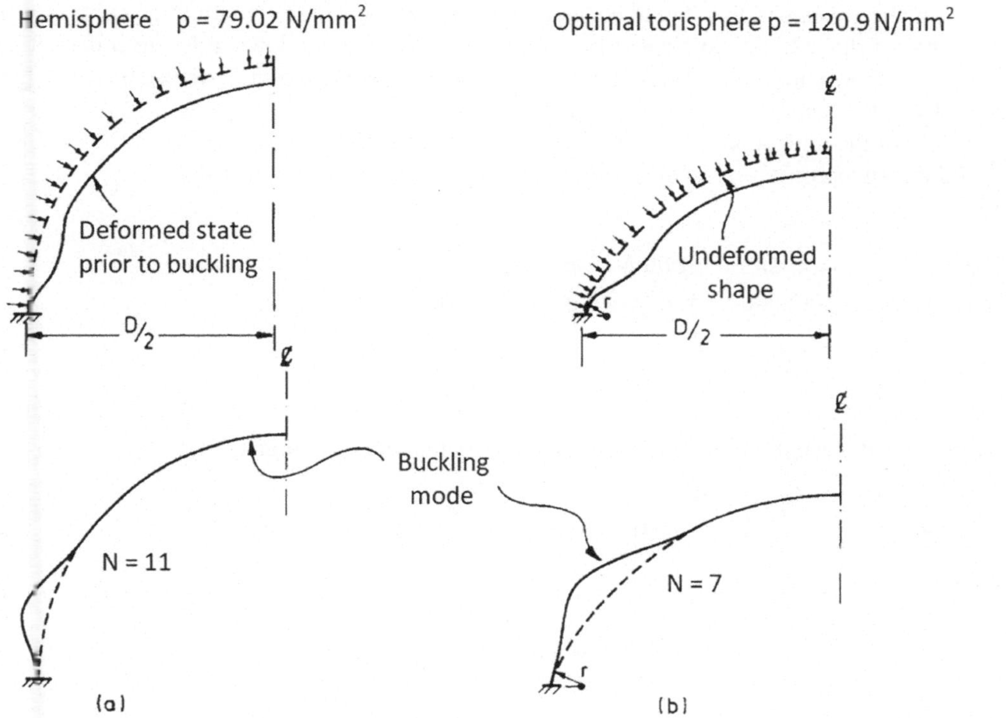

FIGURE 12.33 Hemispherical and torispherical end-closures having the same weight and diameter [5].

12.7.3.5 Optimal Form for Minimum Weight Design

In [35], the minimum weight elastic design of torispherical end closures of an externally pressurised vessel was pursued using the *discrete backtrack programming* method in order to obtain optimal forms that fulfil buckling strength requirements. The main idea of the backtrack programming method is that a partial search is carried out for each variable, and if all possibilities are calculated, a backtrack cycle is made and a new partial search will follow.[9]

In order to simplify the optimisation process, the four-centred ellipse method described in Section 12.7.3.1 was used to define the geometry of the torisphere. This enables the use of only two variables, the ratio of the lengths of the major to the minor axes, K, and the thickness t, so that optimal geometries are defined in terms of these two variables alone.

The steps followed in the procedure are:

1. Specification of design variables (K, t) on the basis of the four-centred ellipse method
2. Definition of geometry for a given value of K. From (12.47) we have:

$$\theta = \tan^{-1}(b / a) = \tan^{-1} K$$

also, $\varphi = \pi/2 - \theta$

The values of r and R are given by (12.45) and (12.46), respectively.

3. Specify constraints (material properties and manufacturing limitations on geometric parameters)
4. Selection of side constraints (range of values for K, t, and increments in discrete values)

The knuckle radius should be large in order to minimise the circumferential stresses in way of the cylinder–torisphere junction (that is, $K > 0.21$, following design codes). It has been shown that an unexpected drop in the buckling strength occurs when $0.95 < K < 1.00$. Thus, the range selected is $0.21 < K < 0.95$.

For the thickness, the selected range is $50 \le D/t \le 150$.

5. Perform optimisation leading to the minimum value of the ratio W_T/W_H

12.7.3.5.1 Outline of the Optimisation Procedure

The objective function to be minimised is $F(\bar{x})$ where:

$$F(\bar{x}) = \frac{W_T}{W_H}$$

where W_T is the weight of the torisphere having constant thickness t given by:

$$W_T = 2\pi\gamma t\left[r\omega\left(\frac{\pi}{2} - \Phi\right) + r^2\cos\Phi + (1-\cos\Phi)R^2\right] \tag{12.55}$$

where

$$\omega = \frac{D}{2} - r \quad \Phi = \sin^{-1}\left(\frac{\omega}{R-r}\right)$$

and γ is specific gravity. W_H is the weight of a hemisphere having the same diameter, given by:

$$W_H = \frac{1}{2}\pi\gamma D^2 t \tag{12.56}$$

The *design variables vector* is

$$\bar{x}_{opt} = (K, t) \tag{12.57}$$

12.7.3.5.2 Constraints

For the buckling pressure, a range of acceptable values is proposed on the basis of rule requirements (ASME, BS5500)

12.7.3.5.3 Upper and Lower Bounds of the Design Buckling Pressure

$$g_1 = (P_b - P_U)/P_U \le 0 \quad \text{where } P_U = 19.4P_w\left(P_w = \text{max. working pressure}\right) \tag{12.58a}$$

$$g_2 = (P_L - P_b)/P_b \le 0 \quad \text{where } P_L = 8.33P_w \tag{12.58b}$$

12.7.3.5.4 Upper and Lower Bounds of the Design Variables

$$g_3 = (x_i - x^{U_i})/(x^{U_i} - x^{L_i}) \le 0 \tag{12.59a}$$

$$g_4 = (x^{L_i} - x_i)/(x^{U_i} - x^{L_i}) \le 0 \tag{12.59b}$$

where x_i are design variables ($i = 1,2$), x^{U_i} is the upper bound value of design variable x_i and x^{L_i} is the lower bound value of design variable x_i.

TABLE 12.13
Results of Optimisation Process for Two Submarine Torispherical End Closures

	Model 1		Model 2	
Parameter	**Reference Dome**	**Optimum Dome**	**Reference Dome**	**Optimum Dome**
r (m)	0.743	0.292	0.872	0.523
R (m)	2.436	4.931	3.985	5.856
t (mm)	32	38	45	48
θ	31.383°	16.699°	28.013°	19.799°
ϕ	58.617°	73.301°	61.987°	72.201°
Critical buckling pressure (MN/m^2)	56.65	26.2	43.25	27.13
Weight (N)	4,061	3,885	11,062	10,433

The above procedure was implemented in the design of two submarine end closures having the form of torispheres. Details of the reference geometries as well as the corresponding optimised geometries and the weight saving are included in Table 12.13.

NOTES

1 For a prolate spheroid $a/b < 1$ whereas for an oblate spheroid $a/b > 1$. An *isotensoid shape* refers to a structural form made of a composite material, in which a constant (tensile) stress is induced along the length of all filaments when subjected to external loading.
2 A method to determine the required cross-sectional dimensions of the intersection ring is included in Section 12.5.2.
3 A sphere is considered to be *thick-walled* if the thickness exceeds 0.1 times the outer radius ($t/R > 0.1$).
4 The basis of first-order shear deformation theory is described in [10] with regard to isotropic materials (metals) and in [2] for composites
5 Details concerning the optimisation methods used throughout are given in textbooks on optimum structural design, for example [1,12,29]. In [28] a modified version of the Hooke and Jeeves algorithm was used.
6 This study was published in the *Journal of Ship Mechanics* in the Chinese language, making detailed information unfortunately inaccessible.
7 *Simultaneous mode design* (SMD) is in effect being implemented, that is, collapse will occur when both load types reach critical values. It is well-known, however, that in SMD nonlinear interaction arises, resulting in a lowering of the actual collapse load. This is not taken into account in the approach described in this section, but can be allowed for by selecting the critical mode and introducing safety factors having different values to ensure that collapse occurs in the preferred mode.
8 *Notation*

D shell diameter
p_γ failure pressure of shell with imperfections (limit load or bifurcation buckling)
p_γ^* failure pressure for perfect and uniform hemisphere or torisphere
p_0 membrane yield pressure of a spherical shell

$$= 2\left(\frac{t}{R_s}\right)\sigma_0$$

r radius of knuckle of a torispherical shell
R_s radius of crown portion of a torispherical shell
t_0 thickness of uniform torispherical shell
$\bar{\lambda}$ torispherical shell slenderness

$$= 1.285\sqrt{\frac{R_s}{t}\frac{\sigma_0}{E}}$$

9 For further information on the implementation of the backtrack programming method in the optimal design of a torispherical shell [35] may be consulted. A detailed theoretical background of the method is provided in [12,36].

REFERENCES

1. Arora, J.S. (2011). *Introduction to Optimum Design*, 3rd Ed. Oxford: Elsevier Academic Press, 880 p.
2. Barbero, E.J. (2017). *Introduction to Composite Material Design*, 3rd Ed. Boca Raton, FL: CRC Press, 570 p.
3. Błachut, J. (1987). Combined axial and pressure buckling of shells having optimal positive Gaussian curvature. *Comp Struct.* Vol. 26, No. 3, 513–519.
4. Błachut, J. (1988). Optimally shaped torispheres with respect to buckling and their sensitivity to axisymmetric imperfections. *Comput. &. Struct.* Vol. 29, 975–981.
5. Błachut, J. (1988). Search for optimal torispherical end closures under buckling constraint. *Int. J. Mech. Sci.* Vol. 31, 623–633.
6. Błachut, J. (2005). Buckling of shallow spherical caps subjected to external pressure. *J. Appl. Mech. Trans. ASME.* Vol. 72, 803–806. https://doi.org/10.1115/1.1993667.
7. Bowen, A.D., Yoerger, D.R., Tatlor, C. et al (2008). The nereus hybrid underwater robotic vehicle for global ocean science operations to 11,000 m depth. *OCEANS 2008*, Quebec City, QC, Canada.
8. Box, M.J. (1965). A new method of constrained optimization and a comparison with other method. *Comp. J.* Vol. 8, No. 1, 42–52.
9. Bronowicki, A., Nelson, R.B., Schmidt, L. (1975). Optimization of ring stiffened cylindrical shells. *AIAA J.* Vol. 13, No. 10, 1319.
10. Caridis, P.A. (2025). *Global Strength of Ships. Analysis and Design Using Mathematical Methods.* Cambridge: Cambridge University Press, 848 p.
11. Cho, S.-R., Muttaqie, T., Do, Q.-T., Park, S.-H., Kim, S.-M., So, H.-Y., Sohn, J.-M. (2019). Experimental study on ultimate strength of steel-welded ring-stiffened conical shell under external hydrostatic pressure. *Mar. Struct.* Vol. 67, 102634.
12. Farkas, J. (1984). *Optimum Design of Metal Structures*. New York: John Wiley, 219 p.
13. French, T.E., Vierck, C.J. (1978). *Engineering Drawing and Graphic Technology*, 12th Ed. New York: McGraw-Hill Inc, 968 p.
14. Garland, C. (1968). "Design and fabrication of deep-diving submersible pressure hulls." *SNAME Annual Meeting*, New York.
15. Gerard, G. (1961). Minimum weight design of ring stiffened cylinders under external pressure. *J. Ship Res.* Vol. 5, 44–49.
16. Gorman, J.J., Louie, L.L. (1991). Submersible pressure hull design parametrics. *SNAME Trans.* Vol. 99, 119–146.
17. Gou, P., Cui, W.-C. (2009). Structural optimization of multiple intersecting spherical pressure hulls based on Kriging model. *J. Ship Mech.* Vol. 13, No. 1, 100–106 (in Chinese).
18. Hall, J.C., Leon, G.F., Kelly, J.J. (1991). "Deep submergence design of intersecting composite spheres". *Proceedings of the 8th International Conference on Manufacturing*, Honolulu, HI, 2F1–2F12.
19. Heller Jr., S.R. (1974). Structural design of spherical shells subjected to uniform external pressure. *Naval Eng. J.* Vol. 86, 45–50.
20. Ismail, M.S., Mahmud, J., Jailani, A. (2023). Buckling of an imperfect spherical shell subjected to external pressure. *Ocean Eng.* Vol. 275, 114118.
21. Jasion, P., Magnucki, K. (2007). Elastic buckling of barrelled shell under external pressure. *Thin-Walled Struct.* Vol. 45, 393–399.
22. Kiernan, T.J., Nishida, K. (1966). "The buckling strength of fabricated HY-80 steel spherical shells." DTMB Report 1721, Washington, DC.
23. Krenzke, M., Hom, K., Proffitt, J. (1965). "Potential hull structures for rescue and search vehicles of the deep submergence systems project." DTMB Report 1985, Washington, DC.
24. Leon, G.F. (1971). Intersecting titanium spheres for deep submersibles. *J. Eng. Mech. Div.* Vol. 97, 981–1006.
25. Liang, C.C., Shiah, S.W., Jen, C.Y., Chen, H.W. (2004). Optimum design of multiple intersecting spheres deep-submerged pressure hull. *Ocean Eng.* Vol. 31, No. 2, 177–199.
26. Pan, B.-B., Cui, W.-C. (2011). A comparison of different rules for the spherical pressure hull of deep manned submersibles. *J. Ship Mech.* Vol. 15, No. 3, 276–285.

27. Pan, B.-B., Cui, W.-C. (2020). *Multidisciplinary Design Optimization and Its Application in Deep Manned Submersible Design*. Singapore: Springer Nature Singapore & Zhejiang Science and Technology Publishing House Co Ltd, 289 p.
28. Pappas, M., Allentuch, A. (1973). Automated optimal design of frame reinforced submersible, circular, cylindrical shells. *J. Ship Res.* Vol. 17, 208–216.
29. Rao, S.S. (2009). *Engineering Optimization. Theory and Practice*, 4th Ed. Hoboken, NJ: John Wiley & Sons, 813 p.
30. Simitses, G., Aswani, M. (1977). Minimum-weight design of stiffened cylinders under hydrostatic pressure. *J. Ship Res.* Vol. 21, 217–224.
31. Smith, C.S. (1991). Design of submersible pressure hulls in composite materials. *Mar. Struct.* Vol. 4, 141–182.
32. Snoey, M.R., Katona, M.G. (1971). "Structural analysis of a full-scale spherical acrylic plastic pressure hull." Tech. Rept. R-716 Naval Civil Eng. Lab., Port Hueneme, CA.
33. Wan, Z.-Q. (2010). Optimum design of cylindrical shells under external hydrostatic pressure. *J. Ship Mech. (in Chinese).* Vol. 18, 239–252.
34. Wikipedia. *Syntactic F*oam. Available at: https://en.wikipedia.org/wiki/Syntactic_foam (Downloaded 2024-02-15).
35. Yang, M.F., Liang, C.C., Chen, C.H. (1992). A rational shape design of externally pressurized torispherical dome ends under buckling constraints. *Comput. & Struct.* Vol. 43, 839–851.
36. Yuan, K.Y., Liang, C.C., Ma, Y.C. (1990). The estimation of the accuracy and efficiency of the backtrack programming method for discrete-variable structural optimization problems. *Comput. Struct.* Vol. 36, 211–222.
37. Zhang, J., Huang, C., Wagner, H.N.R., Cui, W.-C., Tang, W. (2020). Study on dented hemispheres under external hydrostatic pressure. *Mar. Struct.* Vol. 74, 102819. https://doi.org/10.1016/j.marstruc.2020.102819.
38. Zhang, J., Li, S.-L., Cui, W.-C, Xiang, K., Wang, F., Tang, W.X. (2020). Buckling of multiple intersecting spherical shells. *J. Mar. Sci. App.* Vol. 19, 634–641.
39. Zhang, M., Tang, W.-X., Wang, F., Zhang, J., Cui, W.-C., Chen, Y. (2017). Buckling of bi-segment spherical shells under hydrostatic external pressure. *Thin-Walled Struct.* Vol. 120, 1–8.
40. Zhu, Y., Zhang, Y., Zhao, X., Zhang, J., Xu, X. (2019). Elastic–plastic buckling of externally pressurised hemispherical heads. *Ships Offshore Struct.* Vol. 14, 829–838. https://doi.org/10.1080/17445302.2018.1564541.

Appendix 1
Manned Submersibles

The following table includes data concerning manned submersibles the majority of which are in operation at this point in time. The data included here have been extracted from [1].

Name	Figure	Dimensions	Pressure Hull Type & Material	Operating Depth (m)
C-Explorer 3	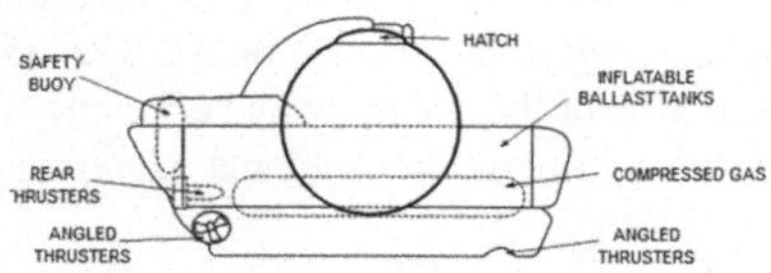	$4.23 \times 3.21 \times 2.33\,m^3$	Sphere Acrylic	300
Deep Flight Dragon	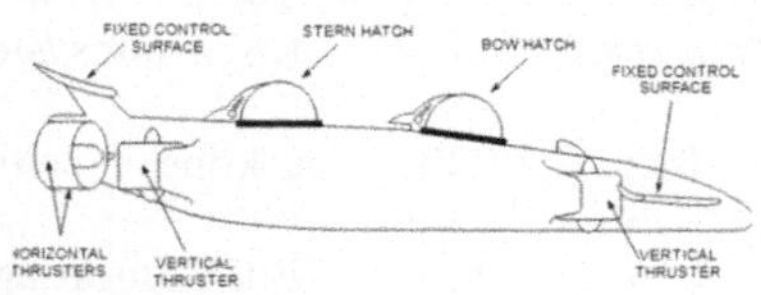	$5.0 \times 1.19 \times 1.1\,m^3$		100
Deep Flight 1	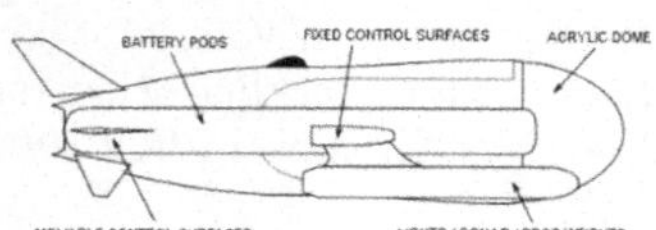	$3.96 \times 2.44 \times 0.91\,m^3$	Cylinder with Glass FRP hemispherical caps	1,000
Deep Flight Super Falcon	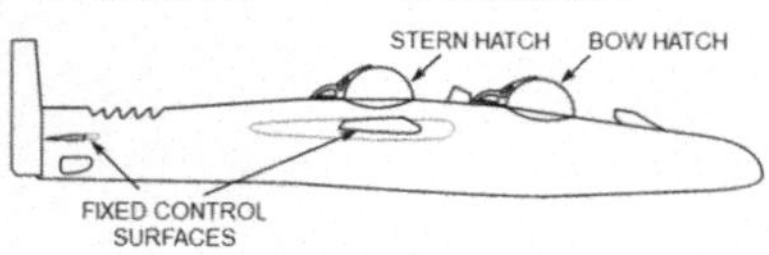	$6.2 \times 2.6 \times 1.2\,m^3$		305
Deepsea Challenger	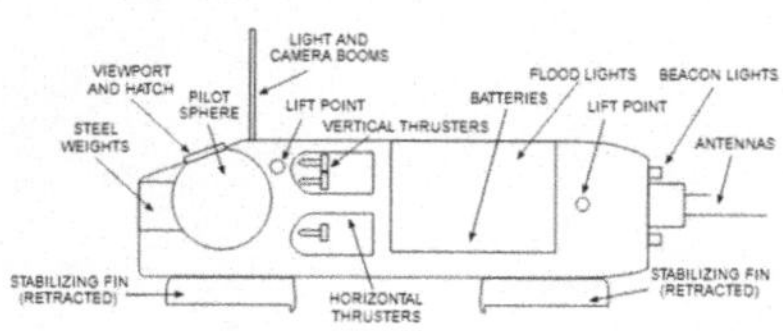			10,000
Euronaut	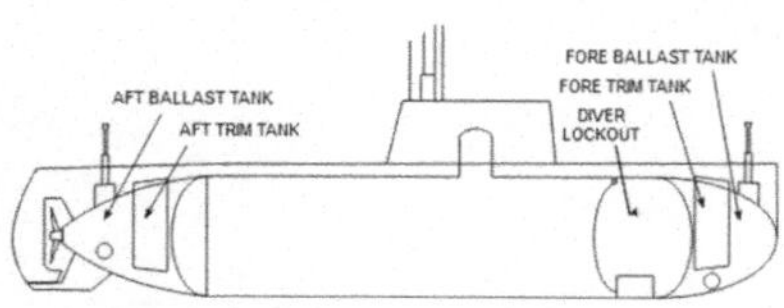	$16.01 \times 2.5 \times 4.17\,m^3$	Steel	250
Eurosub	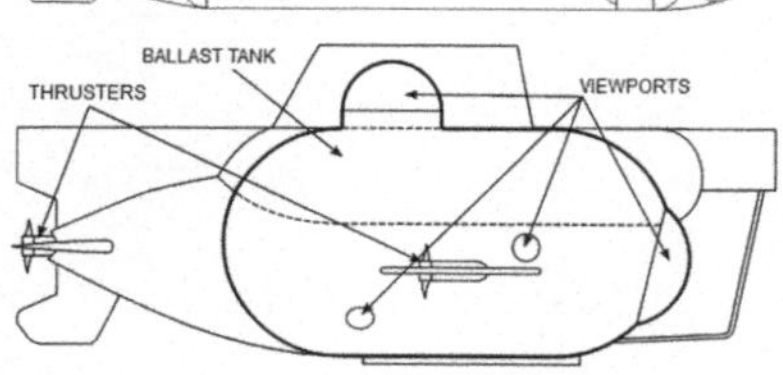	$4.5 \times 1.9 \times 1.8\,m^3$	Cylinder with hemispherical caps Steel S 355 J2 G3	250

(*Continued*)

Name	Figure	Dimensions	Pressure Hull Type & Material	Operating Depth (m)
Hiper Sub2	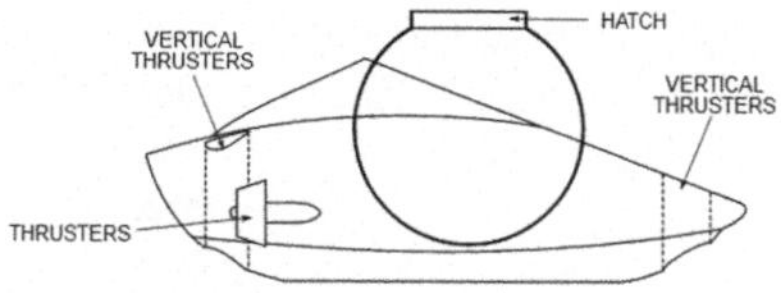	3.16×2.73×1.48 m^3	Sphere Acrylic	100
Jago	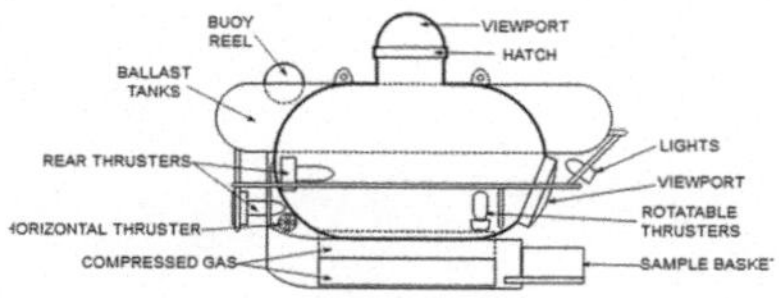	3×2×2.5 m^3	Cylinder with hemispherical caps Steel	400
Jiaolong	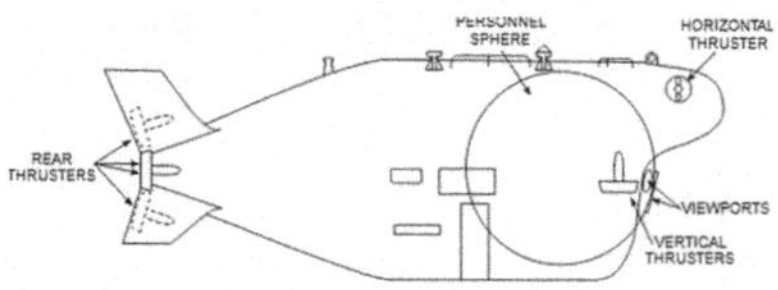	8.3×3×3.2 m^3	Sphere	7,000
Marlin S-101	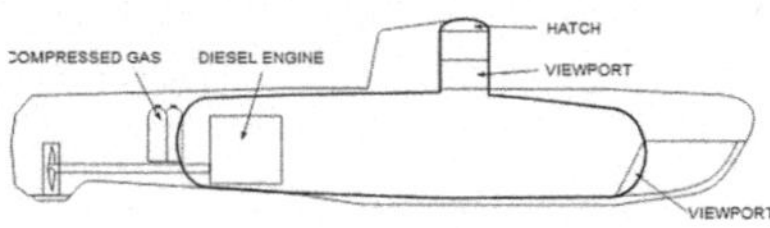	8.55×1.2×2.1 m^3	Cylinder with hemispherical caps 43A steel	200
Mir 1	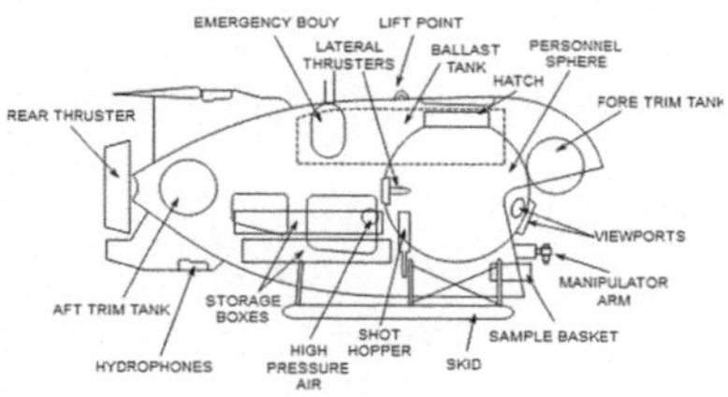	7.8×3.8×3.65 m^3	Sphere Maraging steel	6,000
Ocean Pearl	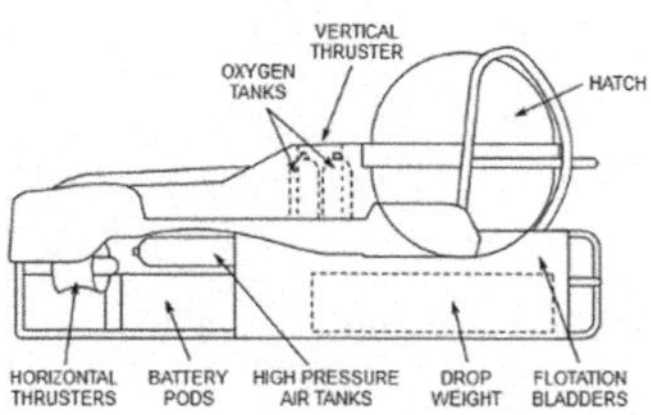	4.53×2.44×2.25 m^3	Sphere Acrylic	150
R300	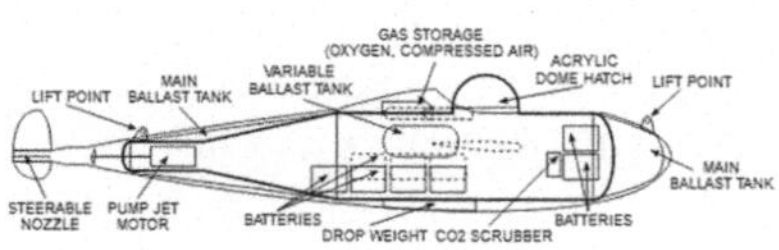	5.8×2×1.2 m^3	Cylindrical One cap and conical section A516 Grade 70 steel	91
Saga	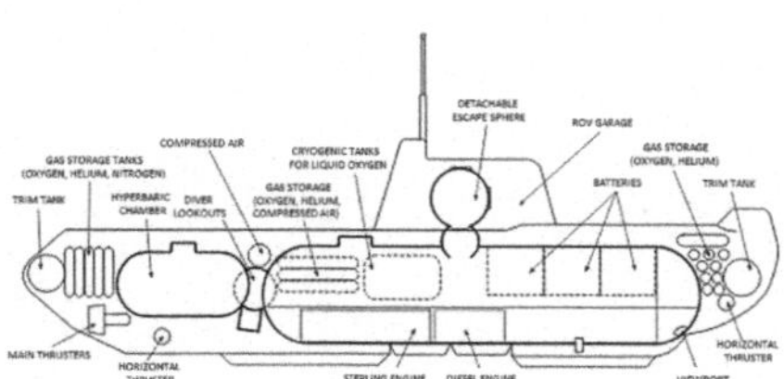	28.06×7.4×8.5 m^3	Cylinder with hemispherical caps + Escape sphere	600

(Continued)

Name	Figure	Dimensions	Pressure Hull Type & Material	Operating Depth (m)
Sgt. Pepper's	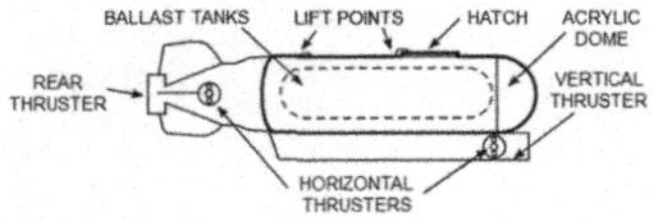	$2.98 \times 0.9 \times 0.9\,m^3$	Cylinder with hemispherical caps Steel	150
Shenhai Yonshi	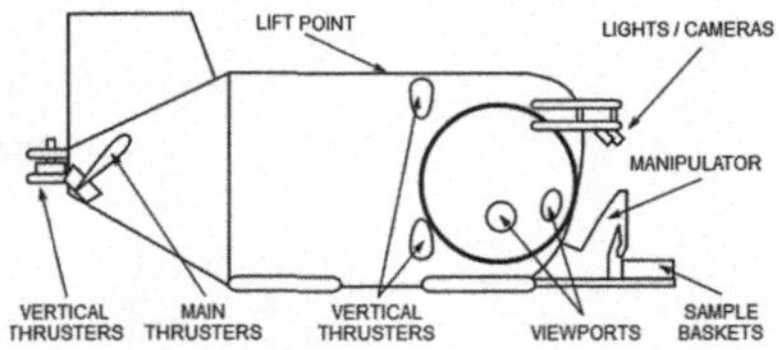	n/a	Sphere	4,500
Shinkai 6500	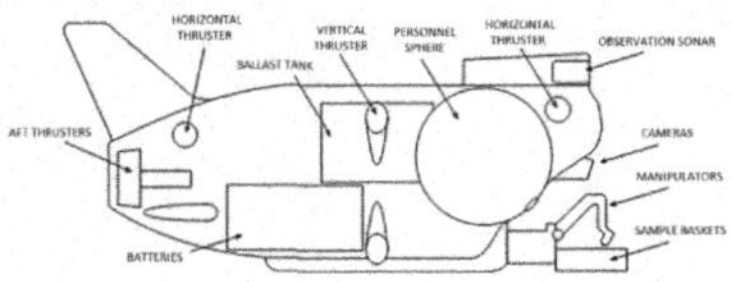	$9.5 \times 2.7 \times 3.2\,m^3$	Sphere Titanium	6,500
Triton 36,000/2 (Limiting Factor)	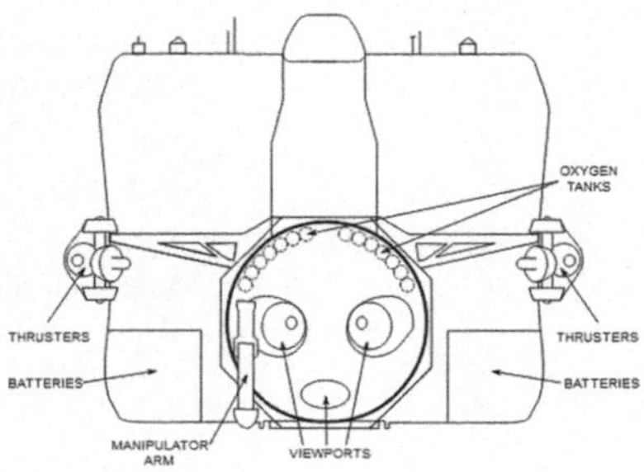	$4.6 \times 1.9 \times 3.7\,m^3$	Sphere Titanium	11,000
UC1 Freya	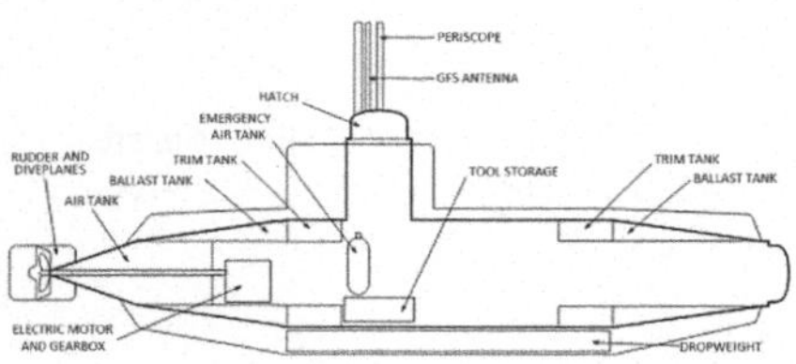	$7 \times 1 \times 2.9\,m^3$	Ordinary Steel	15
UC2 Kraka	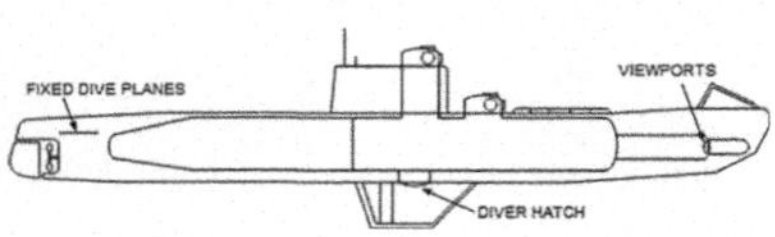	$12.6 \times 1 \times 2.9\,m^3$	n/a	40
UC3 Nautilus	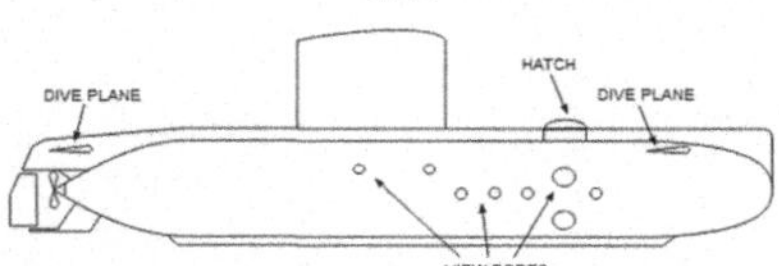	$17.76 \times 2\,m^3$	n/a	100
Vas 525/60 LE Mk 2	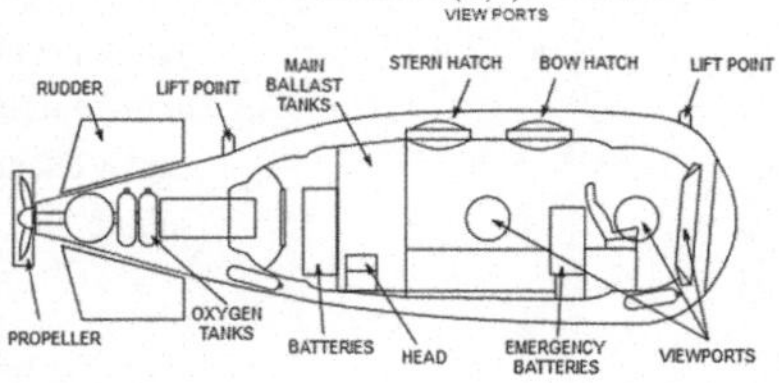	$8.4 \times 2.4 \times 2.6\,m^3$	n/a	160

REFERENCE

1. https://mannedsubmersibles.org/

Appendix 2

Coefficients for Elastic Buckling of Ring-Stiffened Cylindrical Shells with Initial Distortions

A2.1 EXACT THEORY

The total potential of a ring-stiffened cylindrical shell with initial distortions subjected to hydrostatic pressure is given by equation (4.75), repeated below:

$$\Pi = \left\{\frac{\pi E t L_b}{4R(1-\nu^2)}\right\}\left(A_{11}A^2 + A_{22}B^2 + A_{33}C_1^{\,2} + 2A_{12}AB + 2A_{13}AC_1 + 2A_{14}AC_0\right.$$

$$\left.+2A_{23}BC_1 + 2A_{24}BC_0 + 2A_{34}C_1C_0\right)$$

The coefficients included in the above expression are:

$$A_{11} = \lambda^2 + \left(\frac{n^2}{2}\right)(1-\nu)(1+k) - n^2\varphi_1 + n^2\left(\frac{w_p}{R}\right)\left\{k + (1-\nu^2)\left(\frac{A_f}{tL_f}\right)\left(1+\frac{e}{R}\right)^2\right.$$

$$\left.+(1-\nu^2)\left(\frac{I_f}{R^2 tL_f}\right) - (1-\nu^2)\left(\frac{2A_f}{tL_b}\right)R_2\left(1+\frac{e}{R}\right)\right\}$$

$$A_{22} = n^2 + \left(\frac{\lambda^2}{2}\right)(1-\nu)(1+3k) - \frac{\lambda^2\varphi_2}{2} - 2n^2\left(\frac{w_p}{R}\right) + n^2(1-\nu^2)\left(\frac{2R_1A_f}{tL_b}\right)\left(1-\frac{2w_p}{R}\right)$$

$$A_{33} = 1 + k\left\{\lambda^4 + (n^2-1)^2 + 2\lambda^2 n^2\right\} + (1-\nu^2)\left(\frac{2R_1A_f}{tL_b}\right)\left\{1-(n^2-1)\left(\frac{e}{R}\right)\right\}^2$$

$$+(1-\nu^2)\left(\frac{2R_1I_f}{R^2 tL_b}\right)\left(n^2-1\right)^2 + \varphi_1\left(1-n^2\right) - \left(\frac{\lambda^2}{2}\right)\varphi_2$$

$$+n^2\left(\frac{w_p}{R}\right)\left\{k + \left(1-\nu^2\right)\left(\frac{A_f}{tL_f}\right)\left(1+\frac{e}{R}\right)^2 + \left(1-\nu^2\right)\left(\frac{I_f}{R^2 tL_f}\right)\right.$$

$$\left.-\left(1-\nu^2\right)\left(\frac{2R_1A_f}{tL_b}\left(1+\frac{e}{R}\right)\right)\right\}$$

$$A_{12} = -\left(\frac{n\lambda}{2}\right)(1+\nu) - \left(\frac{n\lambda}{2}\right)\varphi_1 + n\lambda\nu\left(\frac{w_p}{R}\right)$$

$$A_{13} = \lambda\nu + k\left\{\lambda^3 - \left(\frac{n^2\lambda}{2}\right)(1-\nu)\right\} + \lambda\varphi_1$$

$$A_{14} = \lambda\varphi_1 + \lambda\nu\left(\frac{w_p}{R}\right)$$

$$A_{23} = -n - \left(\frac{\lambda^2 nk}{2}\right)(3-\nu) - n(1-\nu^2)\left(\frac{2R_1A_f}{tL_b}\right)\left\{1-(n^2-1)\left(\frac{e}{R}\right)\right\}$$

$$-n\left(\frac{w_p}{R}\right)\left\{k - 1 - (1-\nu^2)\left(\frac{2R_1A_f}{tL_b}\right)\left[1-(n^2-1)\left(\frac{e}{R}\right)\right]\right.$$

$$+(1-\nu^2)\left(\frac{A_f}{tL_f}\right)\left(1+\frac{e}{R}\right)^2 - (1-\nu^2)\left(\frac{2R_1A_f}{tL_b}\right)\left(1+\frac{e}{R}\right)$$

$$\left.+(1-\nu^2)\left(\frac{I_f}{R^2tL_f}\right)\right\}$$

$$A_{24} = -n\varphi_1 - n\left(\frac{w_p}{R}\right)\left\{1+(1-\nu^2)\left(\frac{2R_1A_f}{tL_b}\right)\right\}$$

$$A_{34} = -\varphi_1(n^2-2) - \left(\frac{\lambda^2}{2}\right)\varphi_2 + \left(\frac{w_p}{R}\right)\left\{1+(1-\nu^2)\left(\frac{2R_1A_f}{tL_b}\right)\right\} + (n^2-1)\left(\frac{w_p}{R}\right)\{k$$

$$+(1-\nu^2)\left(\frac{A_f}{tL_f}\right)\left(1+\frac{e}{R}\right)^2 - (1-\nu^2)\left(\frac{2R_1A_f}{tL_b}\right)\left(1+\frac{2e}{R}\right) + (1-\nu^2)\left(\frac{I_f}{R^2tL_f}\right)\right\}$$

$$\lambda = \frac{m\pi R}{L_b}$$

$$R_1 = \sum_{r=1}^{N} \sin^2\left(\frac{m\pi rL_f}{L_b}\right)$$

$$= \frac{N+1}{2} \quad \text{when} \quad \frac{m}{N+1} \neq 1$$

$$= 0 \quad \text{when} \quad \frac{m}{N+1} = 1$$

$$R_2 = \sum_{r=1}^{N} \cos^2\left(\frac{m\pi rL_f}{L_b}\right) = N - R_1$$

$$k = \frac{t^2}{12R^2}$$

e = eccentricity of centroid of frame section (see sketch below)

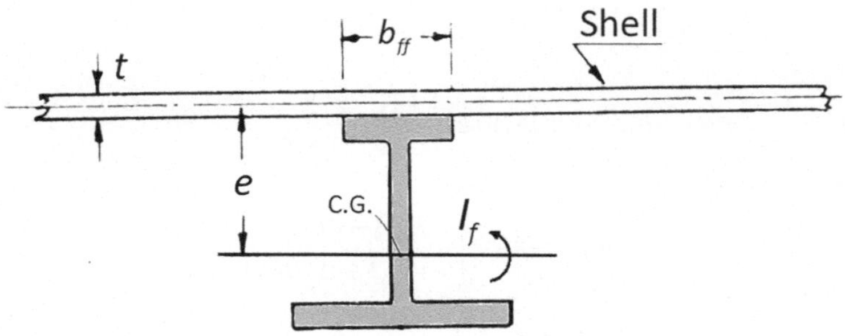

A2.2 APPROXIMATE THEORY

In the case of uniform external pressure ($P = p$) such that $\phi_1 = \phi_2$, an approximate theory can be used. The approximation involves ignoring the terms that include k, I_f, and e in the expressions for ϕ_1 and w_2/R and to write

$$R_1 = R_2 = \frac{N+1}{2}$$

The validity of this approximation has been checked using a numerical example [1]. The expressions for coefficients A_{11}, A_{12}, etc., included above are then replaced by the following:

$$A_{rs} = a_{rs} - \varphi_1 b_{rs} \quad \text{where} \quad r = 1,2,3 \quad s = 1,2,3,4$$

$$a_{11} = \lambda^2 + (n^2/2)(1-\nu)(1+k)$$

$$a_{22} = n^2\left\{1 + (1-\nu^2)(A_f/tL_f)\right\} + (\lambda^2/2)(1-\nu)(1+3k)$$

$$a_{33} = 1 + (1-\nu^2)(A_f/tL_f) - 2(n^2-1)(1-\nu^2)(A_f/tL_f)(e/R) + k(\lambda^4 + 2\lambda^2 n^2)$$
$$+(n^2-1)^2\left\{k + (1-\nu^2)(I_f/R^2 tL_f) + (1-\nu^2)(A_f/tL_f)(e/R)^2\right\}$$

$$a_{12} = -(n\lambda/2)(1+\nu)$$

$$a_{13} = \lambda\nu + k\left\{\lambda^3 - (n^2\lambda/2)(1-\nu)\right\}$$

$$a_{23} = -n\left\{1 + (1-\nu^2)(A_f(tL_f)\right\} - (\lambda^2 nk/2)(3-\nu) + n(n^2-1)(1-\nu^2)(A_f/tL_f)(e/R)$$

$$a_{14} = a_{24} = a_{34} = 0$$

$$b_{11} = n^2$$

$$b_{22} = (\lambda^2/2) + 2n^2(1-\nu/2)\frac{1 + (1-\nu^2)(A_f/tL_f)}{(1-\nu^2)(1 + A_f/tL_f)}$$

$$b_{33} = n^2 - 1 + \lambda^2 / 2$$

$$b_{12} = (n\lambda/2) - \frac{(n\lambda v)(1 - v/2)}{(1 - v^2)(1 + A_f / tL_f)}$$

$$b_{13} = -\lambda$$

$$b_{23} = -n(1 - v/2)\frac{1 + (1 - v^2)(A_f / tL_f)}{(1 - v^2)(1 + A_f / tL_f)}$$

$$b_{14} = -\lambda - \frac{\lambda v(1 - v/2)}{(1 - v^2)(1 + A_f / tL_f)}$$

$$b_{24} = n^2 + n(1 - v/2)\frac{1 + (1 - v^2)(A_f / tL_f)}{(1 - v^2)(1 + A_f / tL_f)}$$

$$b_{34} = n^2 - 2 + \lambda^2/2 - (1 - v/2)\frac{1 + (1 - v^2)(A_f / tL_f)}{(1 - v^2)(1 + A_f / tL_f)}$$

REFERENCE

1 Kendrick, S. (1953). "The deformation under external pressure of nearly circular cylindrical shells with evenly spaced equal strength nearly circular ring frames". NCRE Report R. 259.

Appendix 3
Bessel and Kelvin Functions

In this appendix, we shall provide some definitions of the special functions used in the Taylor–Wenk solution of the equations of equilibrium of conic sections, described in Section 7.2.3. More details on these special functions are available in [1] and also in the references at the end of [1].

A3.1 BESSEL FUNCTIONS

A Bessel function is a solution of the differential equation

$$x^2\frac{d^2y}{dx^2}+x\frac{dy}{dx}+(x^2-n^2)y=0 \tag{A3.1}$$

This is called *Bessel's equation.* Because the Bessel equation is a second-order linear differential equation, there must be two linearly independent solutions. Depending upon the nature of the physical problem in question, the solution can take the form of combinations of *Bessel functions.* These are called Bessel functions of the *second* or *third* kind.

Typically, the general solution is given as:

$$y=AJ_\nu(x)+BY_\nu(x) \tag{A3.2}$$

where the special functions $J_\nu(x)$ and $Y_\nu(x)$ are:

1. Bessel functions of the first kind, $J_\nu(x)$, which are finite at $x=0$ for all real values of ν
2. Bessel functions of the second kind, $Y_\nu(x)$, which are singular at $x=0$

The Bessel functions of the first kind are given by:

$$\begin{aligned}J_\nu(x)&=\sum_{k=0}^{\infty}\frac{(-1)^k(x/2)^{\nu+2k}}{k!\Gamma(\nu+k+1)}\\&=\frac{1}{\Gamma(1+\nu)}\left(\frac{x}{2}\right)^\nu\left\{1-\frac{(x/2)^2}{1(1+\nu)}\left(1-\frac{(x/2)^2}{2(2+\nu)}\left(1-\frac{(x/2)^2}{3(3+\nu)}(1-\ldots\right)\right)\right\}\end{aligned} \tag{A3.3}$$

By letting $x=ix$ (where $i=\sqrt{-1}$) in the Bessel equation, we can obtain the *modified Bessel equation of order* ν, as follows

$$x^2\frac{d^2y}{dx^2}+x\frac{dy}{dx}-(x^2+\nu^2)y=0 \tag{A3.4}$$

The solution to the modified Bessel equation yields *modified Bessel functions* of the first and second kind as follows:

$$y=CI_\nu(x)+DK_\nu(x)\quad x>0 \tag{A3.5}$$

where $I_\nu(x)$ is referred to as a modified Bessel function of the first kind, and $K_\nu(x)$ is referred to as a modified Bessel function of the second kind

A3.2 KELVIN FUNCTIONS

A more general form of Bessel's modified equation can be written as

$$x^2 \frac{d^2y}{dx^2} + x\frac{dy}{dx} - (\beta^2 x^2 + \nu^2)y = 0 \tag{A3.6}$$

where β is an arbitrary constant, and the solution is now

$$y = C\, I_\nu(\beta x) \ + D\, K_\nu(\beta x) \tag{A3.7}$$

If we let $\beta^2 = i$ where $i = \sqrt{-1}$, and we note

$$I_\nu(x) \ = i^{-\nu} J_\nu(ix) \ = J_\nu(i^{3/2}x) \tag{A3.8}$$

the solution is written as

$$y = C\, J_\nu(i^{3/2}x) \ + DK_\nu(i^{1/2}x) \tag{A3.9}$$

The *Kelvin functions* are obtained from the real and imaginary portions of this solution as follows:

$$ber_\nu = \text{Re}\ J_\nu(i^{3/2}x)$$

$$bei_\nu = \text{Im}\ J_\nu(i^{3/2}x)$$

$$J_\nu(i^{3/2}x) \ = ber_\nu x + i\ bei\ x \tag{A3.10}$$

$$\text{ker}_\nu = \text{Re}\ i^{-\nu} K_\nu(i^{1/2}x)$$

$$kei_\nu = \text{Im}\ i^{-\nu} K_\nu(i^{1/2}x)$$

$$i^{-\nu} K_\nu(i^{1/2}x) \ = ker_\nu x + i\ kei\ x$$

REFERENCE

1. Bessel Functions of the First and Second Kind. Available at: http://www.mhtl.uwaterloo.ca/courses/me755/web_chap4.pdf. (Downloaded 2024-02-15).

Appendix 4
Design Equations for Ultimate Strength of Ring-Stiffened Cylindrical Shells and Cones

A4.1 RING-STIFFENED CYLINDRICAL SHELLS

The semi-empirical design equation presented in this section has been developed such that the interaction of the failure modes is allowed for, and this is achieved by introducing them in a single strength formulation [1]. The equation was originally developed for ring-stiffened cylindrical shells and was later adapted for use with ring-stiffened cones. We shall first present the application to ring-stiffened cylinders following which we shall discuss the use of the equation in ring-stiffened cones [2].

A4.1.1 The Basis of the Design Equation

In the case of ring-stiffened cylindrical shells subjected to external hydrostatic pressure, the failure modes that need to be considered are shell yielding, interframe buckling, overall buckling, and ring-stiffener tripping (Chapter 6). The need to allow for failure mode interaction arises because one failure mode may lead on to another one, resulting in collapse of the section. Figure A4.1 illustrates the case of ring-stiffener tripping leading to overall buckling of a ring-stiffened cylindrical section.

The Merchant–Rankine formula that is used in structural design allows for the interaction of yielding with buckling and has the following form:

$$\left(\frac{p_c}{\rho p_m}\right)^2 + \left(\frac{p_c}{p_y}\right)^2 = 1 \tag{A4.1}$$

Where p_c = collapse pressure
p_m = local buckling pressure
p_y = yield pressure
ρ = knockdown factor

In the case of a ring-stiffened cylinder, the corresponding equation is as follows:

$$\left[\frac{p_c}{\rho_L p_m} + \frac{p_c}{\rho_T p_t} + \frac{p_c}{\rho_{OA} p_n}\right]^2 + \left[\frac{p_c}{p_y}\right]^2 = 1 \tag{A4.2}$$

where p_m, p_t, p_n, and p_y are the local buckling, stiffener tripping, overall buckling, and yield pressure, respectively. The expression for local interframe buckling, p_m, is that derived by von Mises (3.41). The expression for the pressure to result in ring frame tripping, p_t, is:

$$p_t = p_{fy} \frac{\sigma_t}{\sigma_y} \frac{R_S}{R_f} \tag{A4.3}$$

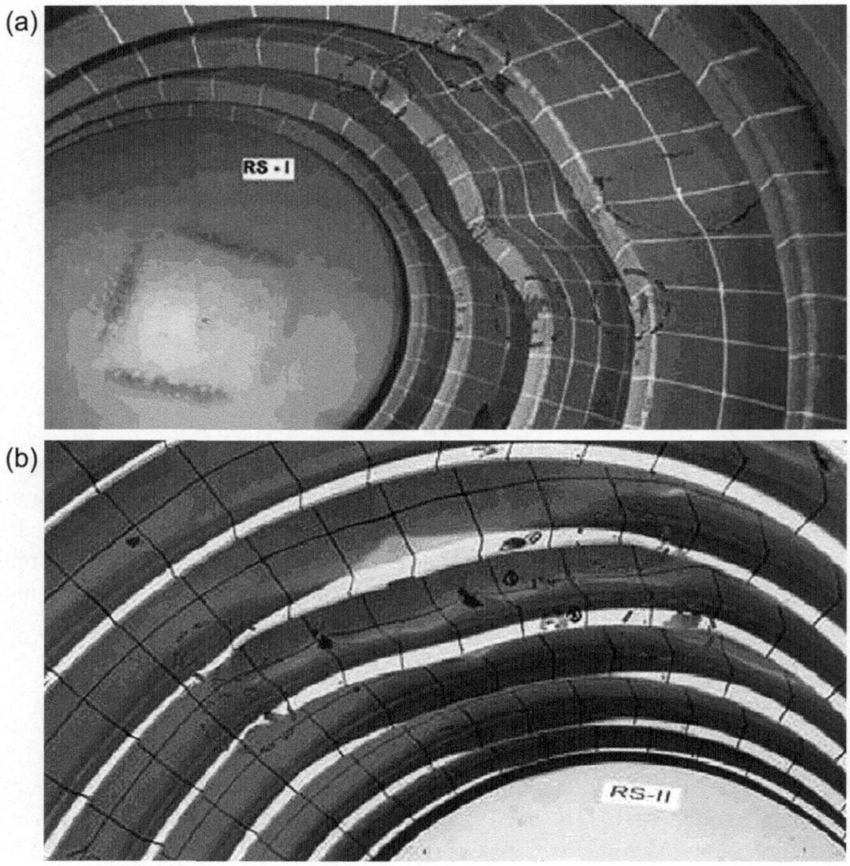

FIGURE A4.1 Consecutive failure modes (stiffener tripping leading to overall collapse) [1].

where p_{fy} is the pressure to initiate yielding in the ring-stiffener flange, derived in Section 6.2.2 and given below, whereas σ_t is the tripping stress of the ring frame and the attached shell plating.

$$p_{fy} = \frac{\sigma_0 tR}{R^2(1-\nu/2)}\left(1+\frac{A}{t_w t + \dfrac{2Nt}{\alpha}}\right) \tag{A4.4}$$

The formula for the stiffener tripping stress, σ_t, given below is based on an energy approach implemented by Faulkner [3] in which account is taken of pure torsional rigidity (GJ), longitudinal warping, and the rotational restraint k_{0n} that arises along the line of attachment of the stiffener to the shell plating.

$$\sigma_t = \frac{GJ + ET_p\left(\dfrac{n}{R}\right)^2 + k_{0n}\left(\dfrac{R}{n}\right)^2}{I_0 + \dfrac{k_{0n}}{\xi P_m}\left(\dfrac{R}{n}\right)^2} \tag{A4.5}$$

It is seen that σ_t is a function of n, the number of circumferential half-sine waves. It is thus necessary to select the critical value of n such that the minimum value of σ_t is obtained [4].

The overall elastic buckling pressure (p_n) is obtained from the Bryant formula (5.19), whereas the pressure to cause yielding in the shell plating at midbay in the circumferential direction, p_y, is given by (6.7).

A4.1.2 Derivation of Knockdown Factors ρ_T, ρ_{OA}, ρ_L

The presence of initial imperfections (initial distortions, thickness variations, and welding residual stresses) as well as variations in the modulus of elasticity and the yield stress result in reductions of the collapse pressures that correspond to each of the possible failure modes (p_m, p_{OA}, p_t). To account for this, *knockdown factors* for local buckling (ρ_L), overall buckling (ρ_{OA}), and tripping (ρ_T) are introduced in the design equation. In this section, we will provide a brief description of the approach followed in obtaining expressions for these. More details are included in [1].

The procedure followed to determine expressions for the knockdown factors is based on a regression analysis of test data when plotted against selected non-dimensional geometrical and material parameters. Best fit analyses were carried out in order to select the parameters that correlate best (minimum bias and COV) with each of the three knockdown factors.

As shown in Figure A4.2, an iterative approach is followed. To initiate the procedure, the values of two knockdown factors (ρ_L, ρ_{OA}) are assumed, and the third (ρ_T) is calculated by substituting in the design equation. Using this derived value, the value of ρ_{OA} is obtained and finally that of ρ_L.

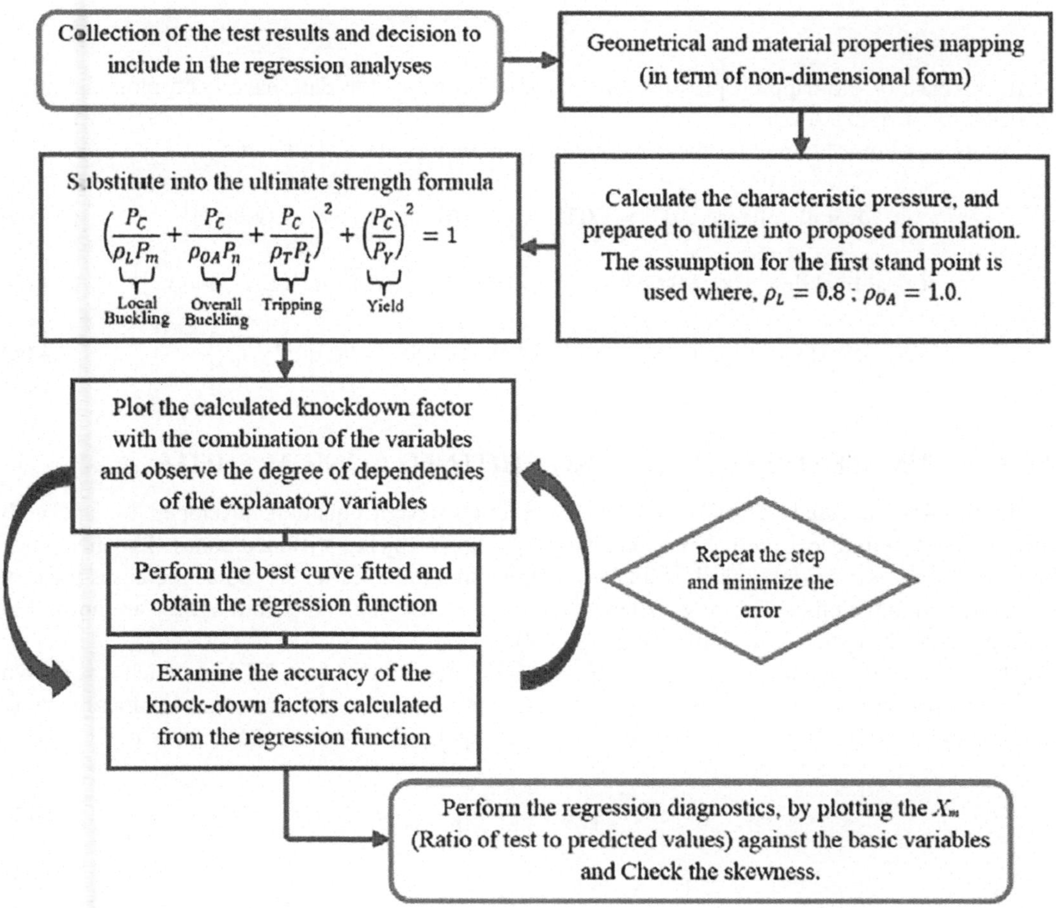

FIGURE A4.2 Derivation of knockdown factors [1].

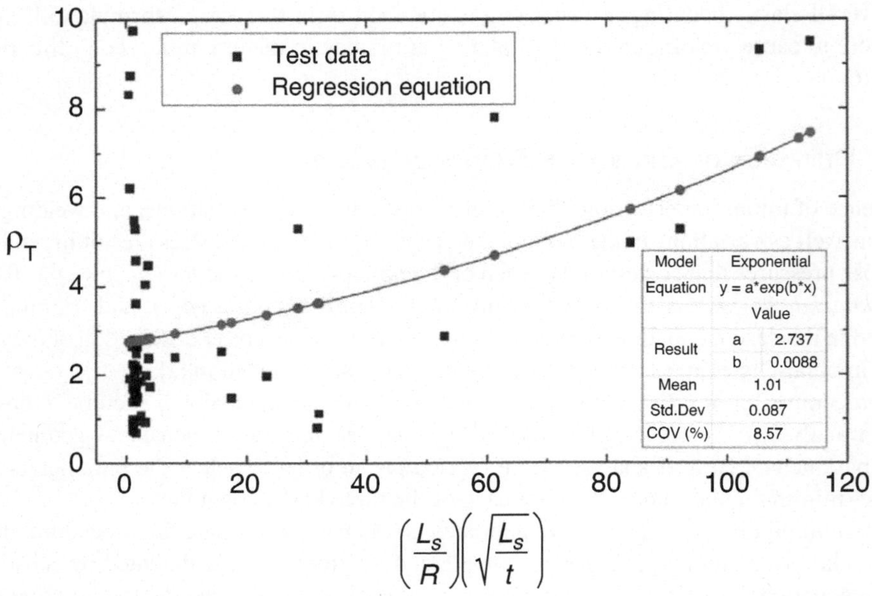

FIGURE A4.3 Best-fit curve for the stiffener tripping pressure knockdown factor (ρ_T) [1].

In the case of the tripping pressure knockdown factor, ρ_T, the data have been plotted against $(L_S/R)(L_S/t)^{1/2}$ (Figure A4.3).

The expressions obtained for the three knockdown factors are given below:

$$\text{Overall collapse} \quad \rho_{OA} = 1.055\ \exp\left\{0.167\left(h_w / t_w\right)\left(E_e/1{,}000\sigma_y\right)\right\} \tag{A4.6a}$$

$$\text{Local buckling} \quad \rho_L = 0.674\exp\left\{0.0006\left(h_w/t_w\right)\left(\sqrt{R_e L_S}/t_e\right)\left(E_e/1{,}000\sigma_y\right)\right\} \tag{A4.6b}$$

$$\text{Stiffener tripping} \quad \rho_T = 2.7374\ \exp\left\{0.088(L_S/R_e)\left(\sqrt{L_S/t_e}\right)\right\} \tag{A4.6c}$$

A4.2 ULTIMATE STRENGTH OF RING-STIFFENED CONICAL SHELLS

In this section, we shall describe how the interaction design equation developed for use with ring-stiffened cylindrical shells has been adapted for use with ring-stiffened cones. The difference between the two structural types lies solely in the fact that in the cone, the radius increases linearly from one end to the other. The ring-stiffeners are in both cases welded such that they are normal to the longitudinal axis of the structure (Figure A4.4).

The concept of the equivalent cylinder has been discussed in Section 7.2.5 in which it is shown that if certain approximations are made, the equation of equilibrium of a cone has identical form to that of a cylinder (5.28). Seide [5] has introduced the taper ratio function, $f(\Psi)$, given by

$$\Psi = 1 - \left(\frac{R_1}{R_2}\right) \tag{A4.7}$$

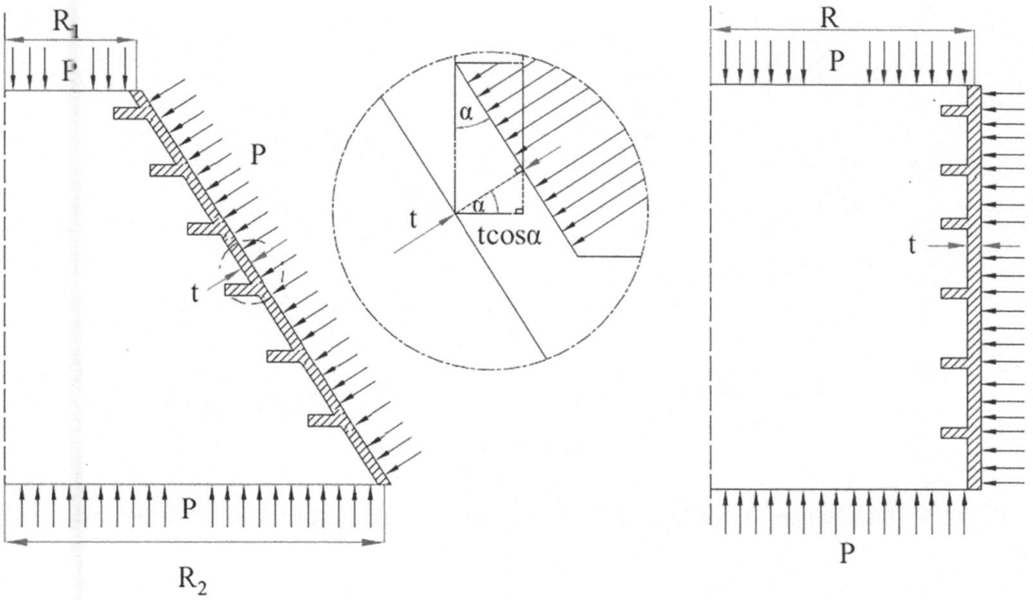

FIGURE A4.4 Geometries of ring-stiffened cylindrical shell and cone [2].

to provide a means of comparing the buckling pressure of a conical shell with that of an equivalent cylinder. This is introduced into the following relationship:

$$P_{eq} = \overline{P_m} f(\Psi) \quad \text{where} \quad \overline{P_m} = \frac{0.92}{\left(\frac{L_e}{R_e}\right)\left(\frac{R_e}{t}\right)^{2.5}} \tag{A4.8}$$

where P_{eq} is the buckling pressure of the equivalent cylinder. The radius of curvature R_e and length L_e of the equivalent cylinder are given by:

$$R_e = \frac{1}{2}(R_1 + R_2)\cos\alpha \quad L_e = \sqrt{L_C^{\,2} + (R_2 - R_1)^2} \tag{A4.9}$$

where α is the semivertex angle (Figure 7.1). The radius of curvature of the equivalent cylinder is thus seen to be equal to the average value of the radius of curvature of the cone, whereas L_e is equal to the distance along a meridional from one end of the cone to the other.

The taper ratio function is plotted against the buckling pressure in non-dimensional form in Figure A4.5. It is equal to zero for a cylindrical section and to unity for a cone having an apex. A verification of the conversion parameters from the conical shell can be carried out by performing linear-elastic buckling analyses for an equivalent cylindrical case. Results are plotted in Figure A4.6, and it is seen that the results obtained for the equivalent cylindrical shell (grey squares) are close to those of the actual cylindrical shell (green circles).

Table A4.1 includes the expressions used for parameter conversion for the design of ring-stiffened truncated cones in several rules [2].

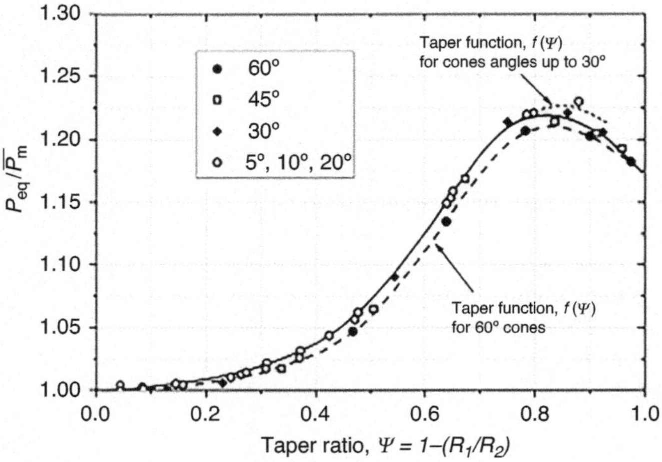

FIGURE A4.5 Buckling pressure of conical shells and their equivalent cylindrical shell under hydrostatic pressure [2].

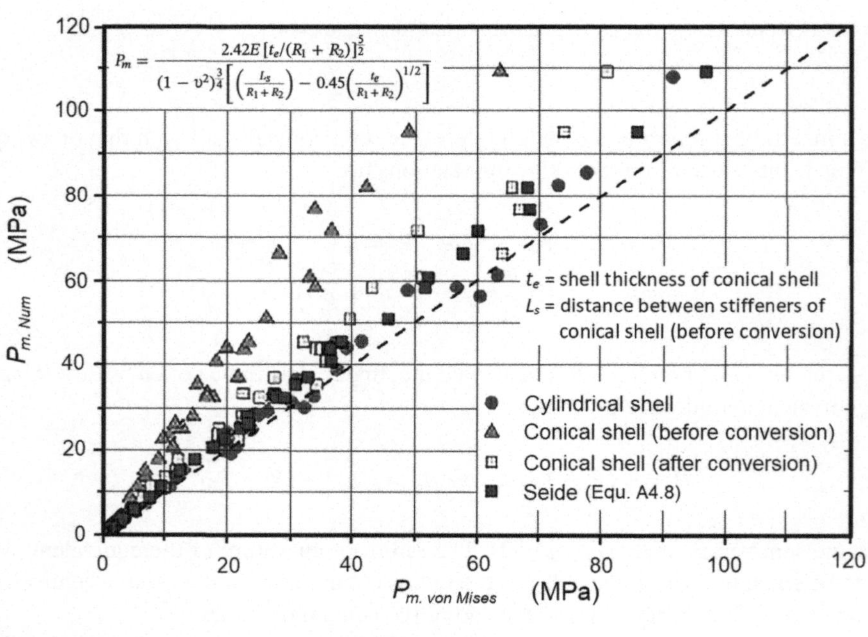

FIGURE A4.6 Validation of parameters for equivalent cylindrical shell [2].

TABLE A4.1
Parameter Conversion for Conical Shell

Converted Parameter	Before Conversion	Conversion			
		ABS	PD5500	DNV-GL	Equation (A4.2)
Shell thickness, t_e	T	$t\cos\alpha$	$t\cos\alpha$	$t\cos\alpha$	$t\cos\alpha$
Shell radius, R_e	R_2	$\frac{(R_1+R_2)}{2}$	$\frac{(R_1+R_2)}{2}\cos^2\alpha$	$\frac{(R_1+R_2)}{2}\cos\alpha$	$\frac{(R_1+R_2)}{2}\cos^2\alpha$
Length between stiffeners, L_e	L_S	L_S	$L_S\cos\alpha$	$L_S/\cos\alpha$	L_S
Overall length, L_e	L_c	$L_c/\cos\alpha$	$L_c\cos\alpha$	$L_c/\cos\alpha$	L_c
Young's Modulus	E	E	E	E	E

REFERENCES

1. Cho, S.-R., Muttaqie, T., Do, Q.-T., So, H.-Y., Sohn, J.-M. (2018). Ultimate strength formulation considering failure mode interactions of ring-stiffened cylinders subjected to hydrostatic pressure. *Ocean Eng.* Vol. 161, 242–256.
2. Cho, S.-R., Muttaqie, T., Do, Q.-T., Park S.-H., Kim S.-M., So, H.-Y., Sohn, J.-M. (2019). Experimental study on ultimate strength of steel-welded ring-stiffened conical shell under external hydrostatic pressure. *Mar. Struct.* Vol. 67, 102634.
3. Faulkner, D. (1991). Application of reliability theory in submarine design. *Adv. Mar. Struct.* Vol. 2, 565–585.
4. Caridis, P.A. (2024). *Local Strength of Ship Structures*. Boca Raton, FL: CRC Press, 774 p.
5. Seide, P. (1960). On the buckling of truncated conical shells under uniform hydrostatic pressure. In: Koiter, W.T. (Ed). *Proceedings of the I.U.T.A.M. Symposium of the Theory of Thin Elastic Shells*. Berlin: Springer, 368–388.

Appendix 5
DTMB Test Specimen Particulars for Spherical Shells

In this appendix, DTMB test specimen particulars for spherical shells discussed in Section 11.1.5 are given [1].

Both S and DSS series models were tested in a 5-inch diameter tank shown in Figure A5.1 (Tables A5.1 and A5.2).

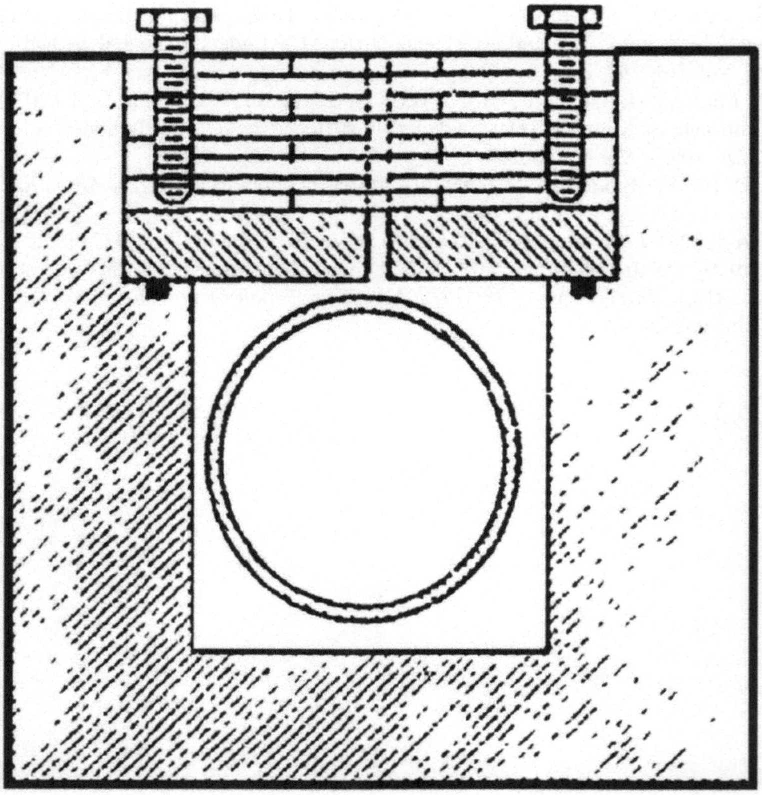

FIGURE A5.1 Test setup (S and DSS series models).

TABLE A5.1
Test Specimen Particulars for S Series Models

Model	Measured Wall Thickness (in)	Measured Variation in Thickness (in)	Cap Thickness (in)	Experimental Collapse Pressure (psi)
S-1	0.193	+0.004 −0.001	0.227	14,300
S-2	0.121	+0.0010 −0.0015	0.141	8,500
S-3	0.0498	+0.0007 −0.0013	0.056	3,000
S-4	0.0260	±0.0019	0.029	1,500

TABLE A5.2
Test Specimen Particulars for DSS Series Models

Model	Measured Wall Thickness (in)	Measured Variation in Thickness (in)	Experimental Collapse Pressure (psi)
DSS-1	0.160	+0.003 −0.002	11,100
DSS-2	0.165	±0.005	11,375
DSS-3	0.0705	+0.0035 −0.0015	4,720
DSS-4	0.125	+0.005 −0.002	8,400
DSS-5	0.0505	+0.0015 −0.0005	2,800
DSS-6	0.0471	+0.0004 −0.0012	2,925
DSS-7	0.0224	+0.0006 −0.0006	1,200
DSS-8	0.0248	+0.0007 −0.0010	1,290

REFERENCE

1. Krenzke, M.A., Kiernan, T.J. (1963). "Tests of stiffened and unstiffened machined spherical shells under external hydrostatic pressure". DTMB Report 1741, Washington, DC.

Appendix 6
Weight-to-Displacement Ratios (W/Δ) for Rescue and Search Deep-Diving Submersibles

The following diagrams include values of W/Δ for a variety of pressure hull forms and materials used for (a) a deep submergence vehicle and (b) a search vehicle, as specified in the DTMB study described in [1]. In these diagrams, Δ_8 is the assumed maximum initial departure from sphericity for an 8 ft diameter shell (Figures A6.1 and A6.2).

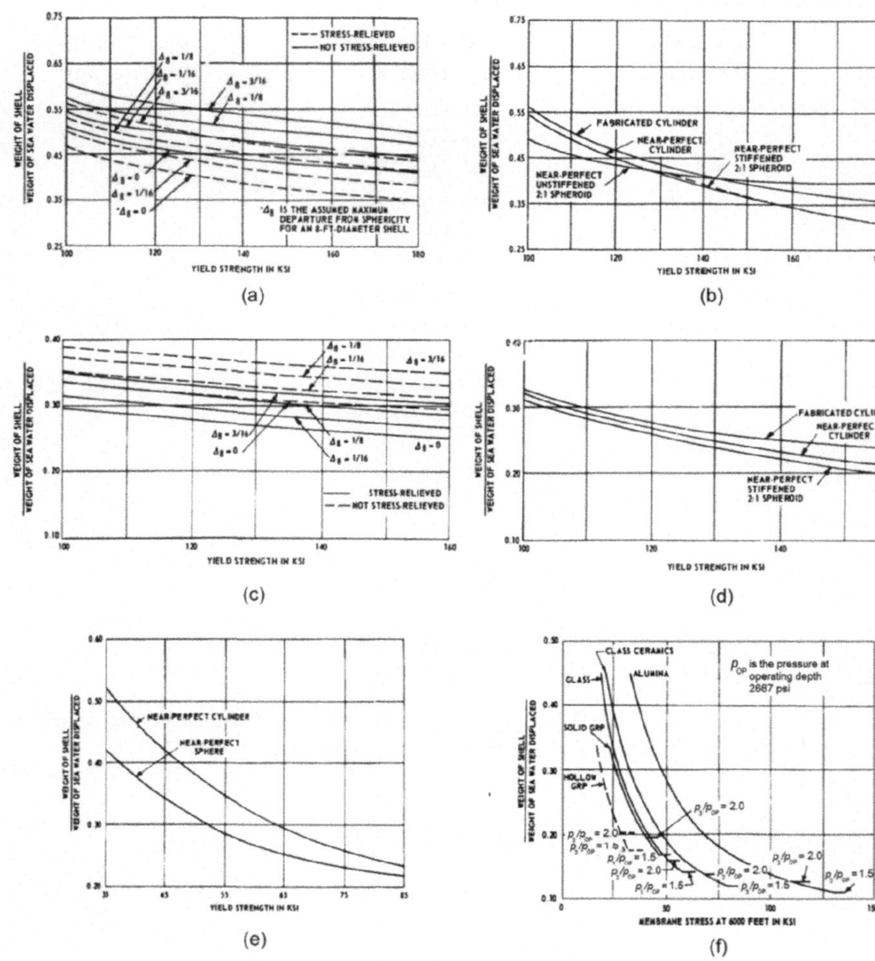

FIGURE A6.1 *W/Δ* ratios for the rescue vehicle. (a) High-strength steel spheres, (b) high-strength steel cylinders and spheroids, (c) titanium spheres, (d) titanium cylinders and spheroids, (e) aluminium alloy shells, and (f) near-perfect GRP, glass, ceramic spherical shells.

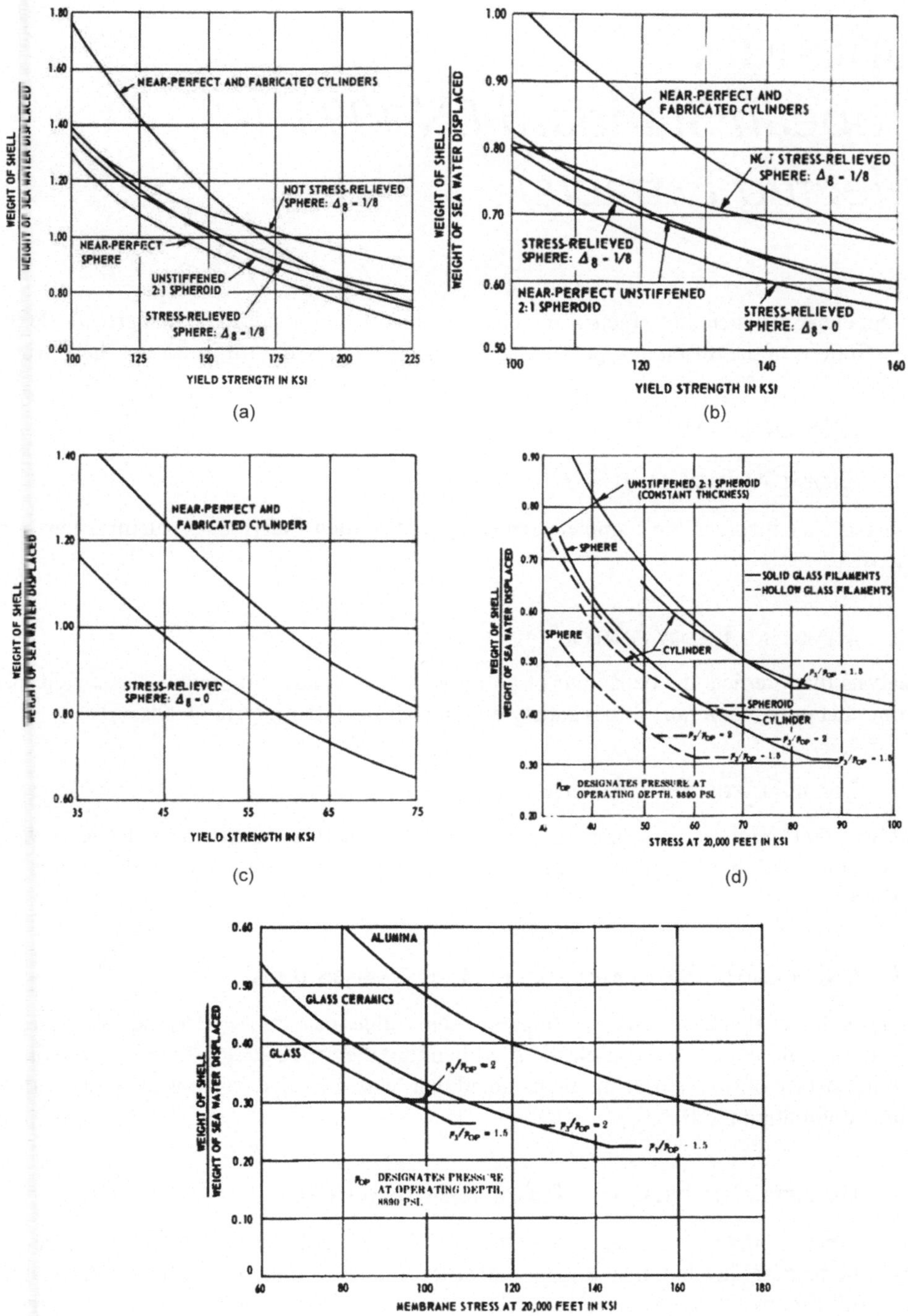

FIGURE A6.2 *W*/Δ ratios for the search vehicle. (a) High-strength steel, (b) titanium alloy, (c) aluminium alloy, (d) near-perfect GRP, and (e) near-perfect glass and ceramic spherical shells.

REFERENCE

1. Krenzke, M., Hom, K., Proffitt, J. (1965). "Potential hull structures for rescue and search vehicles of the deep submergence systems project". DTMB Report 1985, Washington, DC.

Appendix 7

European Standard EN 1993-1-6 (Selected Extracts)

In this appendix, we include extracts from the above-mentioned standard [1] that provide definitions and descriptions of the different types of analysis that can be used for thin-walled shell structures,

A7.1 TYPES OF ANALYSIS

A7.1.1 Global Analysis

An analysis that includes the complete structure, rather than individual structural parts treated separately.

A7.1.2 Membrane Theory Analysis

An analysis that predicts the behaviour of a thin-walled shell structure under distributed loads by assuming that only membrane forces satisfy equilibrium with the external loads.

A7.1.3 Linear Elastic Shell Analysis (LA)

An analysis that predicts the behaviour of a thin-walled shell structure on the basis of the small deflection linear elastic shell bending theory, related to the perfect geometry of the middle surface of the shell.

A7.1.4 Linear Elastic Bifurcation (Eigenvalue) Analysis (LBA)

An analysis that evaluates the linear bifurcation eigenvalue for a thin-walled shell structure on the basis of the small detlection linear elastic shell bending theory, related to the perfect geometry of the middle surface of the shell. It should be noted that, where an eigenvalue is mentioned, this does not relate to vibration modes.

A7.1.5 Geometrically Nonlinear Elastic Analysis (GNA)

An analysis based on the principles of shell bending theory applied to the perfect structure, using a linear elastic material law but including nonlinear large deflection theory for the displacements that accounts full for any change in geometry due to the actions on the shell. A bifurcation eigenvalue check is included at each load level.

A7.1.6 Materially Nonlinear Analysis (MNA)

An analysis based on shell bending theory applied to the perfect structure, using the assumption of small deflections, as in Section A7.1.3, but adopting a nonlinear elasto-plastic material law.

A7.1.7 Geometrically and Materially Nonlinear Analysis (GMNA)

An analysis based on shell bending theory applied to the perfect structure, using the assumptions of nonlinear large deflection theory for the displacements and a nonlinear elasto-plastic material law. A bifurcation eigenvalue check is included at each load level.

A7.1.8 Geometrically Nonlinear Elastic Analysis with Imperfections Included (GNIA)

An analysis with imperfections explicitly included, similar to a GNA analysis as defined in Section A7.1.5, but adopting a model for the geometry of the structure that includes the imperfect shape (i.e. the geometry of the middle surface includes unintended deviations from the ideal shape). The imperfection may also cover the effects of deviations in boundary conditions and/or the effects of residual stresses. A bifurcation eigenvalue check is included at each load level.

A7.1.9 Geometrically and Materially Nonlinear Analysis with Imperfections Included (GMNIA)

An analysis with imperfections explicitly included, based on the principles of shell bending theory applied to the imperfect structure (i.e. the geometry of the middle surface includes unintended deviations from the ideal shape), including nonlinear large deflection theory for the displacements that accounts full for any change in geometry due to the actions on the shell and a nonlinear elasto-plastic material law. The imperfections may also include imperfections in boundary conditions and residual stresses. A bifurcation eigenvalue check is included at each load level.

A7.2 TYPES OF ANALYSIS

A7.2.1 General

1. One or more of the following types of analysis should be used as detailed in Section A7.4, depending on the limit state and other considerations:
   - Global analysis, see Section A7.2.2;
   - Membrane theory analysis, see Section A7.2.3;
   - Linear elastic shell analysis, see Section A7.2.4;
   - Linear elastic bifurcation analysis, see Section A7.2.5;
   - Geometrically nonlinear elastic analysis, see Section A7.2.6;
   - Materially nonlinear analysis, see Section A7.2.7;
   - Geometrically and materially nonlinear analysis, see Section A7.2.8;
   - Geometrically nonlinear elastic analysis with imperfections included, see Section A7.2.9;
   - Geometrically and materially nonlinear analysis with imperfections included, see Section A7.2.10.

A7.2.2 Global Analysis

1. In a global analysis, simplified treatments may be used for certain parts of the structure.

A7.2.3 Membrane Theory Analysis

1. A membrane theory analysis should only be used provided that the following conditions are met: the boundary conditions are appropriate for transfer of the stresses in the shell into

support reactions without causing significant bending effects; the shell geometry varies smoothly in shape (without discontinuities); the loads have a smooth distribution (without locally concentrated or point loads).

2. A membrane theory analysis does not necessarily fulfil the compatibility of deformations at boundaries or between shell segments of different shape or between shell segments subjected to different loading. However, the resulting field of membrane forces satisfies the requirements of primary stresses (LS 1).

A7.2.4 Linear Elastic Shell Analysis (LA)

1. The linearity of the theory results from the assumptions of a linear elastic material law and the linear small deflection theory. Small deflection theory implies that the assumed geometry remains that of the undeformed structure.
2. An LA analysis satisfies compatibility in the deformations as well as equilibrium. The resulting field of membrane and bending stresses satisfy the requirements of primary plus secondary stresses (LS1–LS4)

A7.2.5 Linear Elastic Bifurcation Analysis (LBA)

1. The conditions of Section A7.2.4 concerning the material and geometric assumptions are met. However, this linear bifurcation analysis obtains the lowest eigenvalue at which the shell may buckle into a different deformation mode, assuming no change of geometry, no change in the direction of action of the loads, and no material degradation. Imperfections of all kinds are ignored. This analysis provides the elastic critical buckling resistance r_{Rcr} (LS3).

A7.2.6 Geometrically Nonlinear Elastic Analysis (GNA)

1. A GNA analysis satisfies both equilibrium and compatibility of the deflections under conditions in which the change in the geometry of the structure caused by loading is included. The resulting field of stresses matches the definition of primary plus secondary stresses (LS2 and LS4).
2. Where compression or shear stresses are predominant in some part of the shell, a GNA analysis delivers the elastic buckling load of the structure, including changes in geometry, that may be of assistance in checking the limit state LS3.
3. Where this analysis is used for a buckling load evaluation, the eigenvalues of the system must be checked to ensure that the numerical process does not fail to detect a bifurcation in the load path.

A7.2.7 Materially Nonlinear Analysis (MNA)

1. The result of an MNA analysis gives the plastic limit load, which can be interpreted as a load amplification factor $r_{R_{pl}}$ on the design value of the loads F_{Ed}. This analysis provides the plastic reference resistance ratio $r_{R_{pl}}$.
2. An MNA analysis may be used to verify limit states LS1 and LS3
3. An MNA analysis may be used to the plastic strain increment L1£ during one cycle of cyclic loading that may be used to verify limit state LS2. 16 BS EN 1993-1-6:2007 EN 1993-1-6: 2007 (E)

A7.2.8 Geometrically and Materially Nonlinear Analysis (GMNA)

1. The result of a GMNA analysis, analogously to A7.2.7, gives the geometrically nonlinear plastic limit load of the perfect structure and the plastic strain increment that may be used for checking the limit states LS1 and LS2.
2. Where compression or shear stresses are predominant in some part of the shell, a GMN A analysis the elasto-plastic buckling load of the perfect structure, that may be of assistance in checking the limit state LS3, see 8.7.
3. Where this analysis is used for a buckling load evaluation, the eigenvalues of the system should be checked to ensure that the numerical process does not fail to detect a bifurcation in the load path.

A7.2.9 Geometrically Nonlinear Elastic Analysis with Imperfections Included (GNIA)

1. A GNIA analysis is used in cases where compression or shear stresses dominate in the shell. It delivers elastic buckling loads of the imperfect structure, that may be of assistance in checking the limit state LS3.
2. Where this analysis is used for a buckling load evaluation (LS3), the eigenvalues of the system should be checked to ensure that the numerical process does not fail to detect a bifurcation in the load path. Care must be taken to ensure that the local stresses do not exceed values at which material nonlinearity may affect the behaviour.

A7.2.10 Geometrically and Materially Nonlinear Analysis with Imperfections Included (GMNIA)

1. A GMNIA analysis is used in cases where compression or shear stresses are dominant in the shell. It delivers elasto-plastic buckling loads for the real imperfect structure, that may be used for checking the limit state LS3.
2. Where this analysis is used for a buckling load evaluation, the eigenvalues of the system should be checked to ensure that the numerical process does not fail to detect a bifurcation in the load path.
3. Where this analysis is used for a buckling load evaluation, an additional GMNA analysis of the perfect shell should always be conducted to ensure that the imperfection sensitivity of the structural system is identified.

A7.3 SHELL BOUNDARY CONDITIONS

1. The boundary conditions assumed in the design calculation should be chosen in such a way as to ensure that they achieve a realistic or conservative model of the real construction. Special attention should be given not only to the constraint of displacements normal to the shell wall (deflections) but also to the constraint of the displacements in the plane of the shell wall (meridional and circumferential) because of the significant effect these have on shell strength and buckling resistance.
2. In shell buckling (eigenvalue) calculations (limit state LS3), the definition of the boundary conditions should refer to the incremental displacements during the buckling process and not to total displacements induced by the applied actions before buckling.
3. The boundary conditions at a continuously supported lower edge of a shell should take into account whether local uplifting of the shell is prevented or not.
4. The shell edge rotation β_φ should be particularly considered in short shells and in the calculation of secondary stresses in longer shells (according to the limit states LS2 and LS4).
5. The boundary conditions set out in 5.2.2 should be used in computer analyses and in selecting expressions from Annexes A to D of EN 1993-1-6.

A7.4 ULTIMATE LIMIT STATES IN STEEL SHELLS

A7.4.1 Ultimate Limit States to Be Considered

A7.4.1.1 LS1: Plastic Limit

1. The limit state of the plastic limit should be taken as the condition in which the capacity of the structure to resist the actions on it is exhausted by yielding of the material. The resistance offered by the structure at the plastic limit state may be derived as the plastic collapse load obtained from a mechanism based on small displacement theory.
2. The limit state of tensile rupture should be taken as the condition in which the shell wall experiences gross section tensile failure, leading to separation of the two parts of the shell.
3. In the absence of fastener holes, verification at the limit state of tensile rupture may be assumed to be covered by the check for the plastic limit state. However, where holes for fasteners occur, a supplementary check in accordance with 6.2 of EN 1993-1-1 should be carried out.
4. In verifying the plastic limit state, plastic or partially plastic behaviour of the structure may be assumed (i.e. elastic compatibility considerations may be neglected),

   NOTE: The basic characteristic of this limit state is that the load or actions sustained cannot be increased without exploiting a significant change in the geometry of the structure or strain-hardening of the material.
5. All relevant load combinations should be accounted for when checking LS1.
6. One or more of the following methods of analysis (see the design stresses and stress resultants when checking LS1:

   - membrane theory;
   - expressions in Annexes A and B of EN 1993-1-6;
   - linear elastic analysis (LA);
   - materially nonlinear analysis (MNA);
   - geometrically and materially nonlinear analysis (GMNA).

A7.4.1.2 LS2: Cyclic Plasticity

1. The limit state of cyclic plasticity should be taken as the condition in which repeated cycles of loading and unloading produce yielding in tension and in compression at the same point, thus causing plastic work to be repeatedly done on the structure, eventually leading to local cracking by exhaustion of the energy absorption capacity of the material.

   NOTE: The stresses that are associated with this limit state develop under a combination of all actions and the compatibility conditions for the structure.
2. All variable actions (such as imposed loads and temperature variations) that can lead to yielding, and which might be applied with more than three cycles in the life of the structure, should be accounted for when checking LS2.
3. In the verification of this limit state, compatibility of the deformations under elastic or elasticplastic conditions should be considered.
4. One or more of the following methods of analysis (see should be used for the calculation of the design stresses and stress resultants when checking LS2:
   expressions in Annex C of EN 1993-1-6, using stress concentration factors;
   elastic analysis (LA or GNA), using stress concentration factors.
   MNA or GMNA to determine the plastic strain range.
5. Low cycle fatigue failure may be assumed to be prevented if the procedures set out in this standard are adopted.

A7.4.1.3 LS3: Buckling

1. The limit state of buckling should be taken as the condition in which all or part of the structure suddenly develops large displacements normal to the shell surface, caused by loss of stability under compressive membrane or shear membrane stresses in the shell wall, leading to inability to sustain any increase in the stress resultants, possibly causing total collapse of the structure.
2. One or more of the following methods of analysis (see Section A7.3) should be used for the calculation of the design stresses and stress resultants when checking LS3:
   - membrane theory for axisymmetric conditions only (for exceptions, see relevant application parts of EN 1993)
   - expressions in Annex A of EN 1993-1-6; linear elastic analysis (LA), which is a minimum requirement for stress analysis under general loading conditions (unless the load case is given in Annex A of EN 1993-1-6);
   - linear elastic bifurcation analysis (LBA), which is required for shells under general loading conditions if the critical buckling resistance is to be used;
   - materially nonlinear analysis (MNA), which is required for shells under general loading conditions if the reference plastic resistance is to be used;
   - GMNIA, coupled with MNA, LBA, and GMNA, using appropriate imperfections and calculated calibration factors.
3. All relevant load combinations causing compressive membrane or shear membrane stresses in the shell should be accounted for when checking LS3.

Because the strength under limit state LS3 depends strongly on the quality of construction, the strength assessment should take account of the associated requirements for execution tolerances.

NOTE: For this purpose, three classes of geometrical tolerances, termed "fabrication quality classes" are given in Section 8 of EN 1993-1-6.

A7.4.1.4 LS4: Fatigue

1. The limit state of fatigue should be taken as the condition in which repeated cycles of increasing and decreasing stress lead to the development of a fatigue crack.
2. The following methods of analysis (see Section A7.3) should be used for the calculation of the design stresses and stress resultants when checking LS4:
   - expressions in Annex C of EN 1993-1-6, using stress concentration factors;
   - elastic analysis (LA or GNA), using stress concentration factors.
3. All variable actions that will be applied with more than N_f cycles in the design life time of the structure according to the relevant action spectrum in EN 1991 in accordance with the appropriate application part of EN 1993-3 or EN 1993-4, should be accounted for when checking LS4.

NOTE: The National Annex may choose the value of N_f. The value $N_f = 10{,}000$ is recommended (Table A7.1).

A7.5 TYPES OF ANALYSIS

1. The design should be based on one or more of the types of analysis given in Table A7.2. Reference should be made to Section A7.3 for the conditions governing the use of each type of analysis.

TABLE A7.1
Boundary Conditions for Shells

Boundary Condition Code	Simple Term	Description	Normal DisplacementS	Meridional Displacements	Meridional Rotation
BC1r	Clamped	Radially restrained Meridionally restrained Rotation restrained	$w=0$	$u=0$	$\beta_\varphi=0$
BC1f		Radially restrained Meridionally restrained Rotation free	$w=0$	$u=0$	$\beta_\varphi\neq0$
BC2r		Radially restrained Meridionally free Rotation restrained	$w=0$	$u\neq0$	$\beta_\varphi=0$
BC2f	Pinned	Radially restrained Meridionally free Rotation free	$w=0$	$u\neq0$	$\beta_\varphi\neq0$
BC3	Free edge	Radially free Meridionally free Rotation free	$w\neq0$	$u\neq0$	$\beta_\varphi\neq0$

Note: The circumferential displacement v is closely linked to the displacement w normal to the surface, so separate boundary conditions are not identified for these two parameters (see (4)) but the values in column 4 should be adopted for displacement v.

TABLE A7.2
Types of Shell Analysis

Type of Analysis	Shell Theory	Material Law	Shell Geometry
Membrane theory of shells	Membrane equilibrium	Not applicable	Perfect
Linear elastic shell analysis (LA)	Linear bending	linear	Perfect
Linear elastic bifurcation analysis (LBA)	Linear bending and stretching	Linear	Perfect
Geometrically non-linear elastic analysis (GNA)	Non-linear	Linear	Perfect
Materially non-linear elastic analysis (MNA)	Linear	Non-linear	Perfect
Geometrically and materially non-linear analysis (GMNA)	Non-linear	Non-linear	Perfect
Geometrically non-linear elastic analysis with imperfections (GNIA)	Non-linear	Linear	Imperfect
Geometrically and materially non-linear analysis with imperfections (GMNIA)	Non-linear	Non-linear	Imperfect

REFERENCE

1. European Standard. (2007). "EN 1993-1-6 eurocode 3 – design of steel structures – part 1-6: Strength and stability of shell structures". The European Union Per Regulation 305/2011, Directive 98/34/EC, Directive 2004/18/EC.

Appendix 8
Design of Spherical Shells According to ASME PHVO-1-2007

In this appendix, extracts from the ASME rules for pressure vessels for human occupancy, [1], are included verbatim, for ease of reference in the design of an acrylic spherical deep-diving submersible discussed in Section 12.3.5. This extract details the design procedure for determining the minimum required thickness for unstiffened spheres and formed heads.

A8.1 MINIMUM REQUIRED THICKNESS FOR UNSTIFFENED SPHERES AND FORMED HEADS

a. *Limitations*: The allowable pressure for spheres and spherical segments is derived using the iterative procedure following. This procedure applies to spheres and hemispherical formed heads directly.

   An adjustment is made for 2:1 ellipsoidal heads. These rules do not apply to other shaped ellipsoidal or to torispherical heads. For PVHOs not conforming to these and the following limitations, the external pressure design shall be as required by the specified Division of Section VIII of the Code.

   1. The maximum outside radius, R_o shall not exceed 60 in. (1,500 mm).
   2. The maximum shell thickness, including corrosion allowance, shall not exceed 2 in. (50 mm).
   3. The minimum shell thickness, excluding corrosion allowance shall not be less than 3/8 in. (10 mm).

b. *2:1 Ellipsoidal Heads*: For 2:1 ellipsoidal heads, use the procedure specified in para. (c) below using $R_o = 0.9D_0$

c. *Minimum Thickness*: The minimum required thickness for the spherical shell or formed head exclusive of corrosion allowance shall be determined by the following procedure:

   Step 1: Calculate the value of C from the following two equations:

   $C =$ the larger of C_1 or C_2

$$C_1 = \frac{0.75PT}{F_y}$$

$$C_2 = \sqrt{\frac{1.79PT}{E}}$$

   Step 2: Enter the left ordinate of Figure A8.1 with the value of C calculated in step 1. Move horizontally to an intersection with the solid curve. Extrapolation beyond the upper or lower limits of the curve is prohibited. When values of C fall outside the limits of Figure A8.1, design shall follow rules of Section VIII, UG-2S(d) for Division 1 or AD-320 for Division 2.

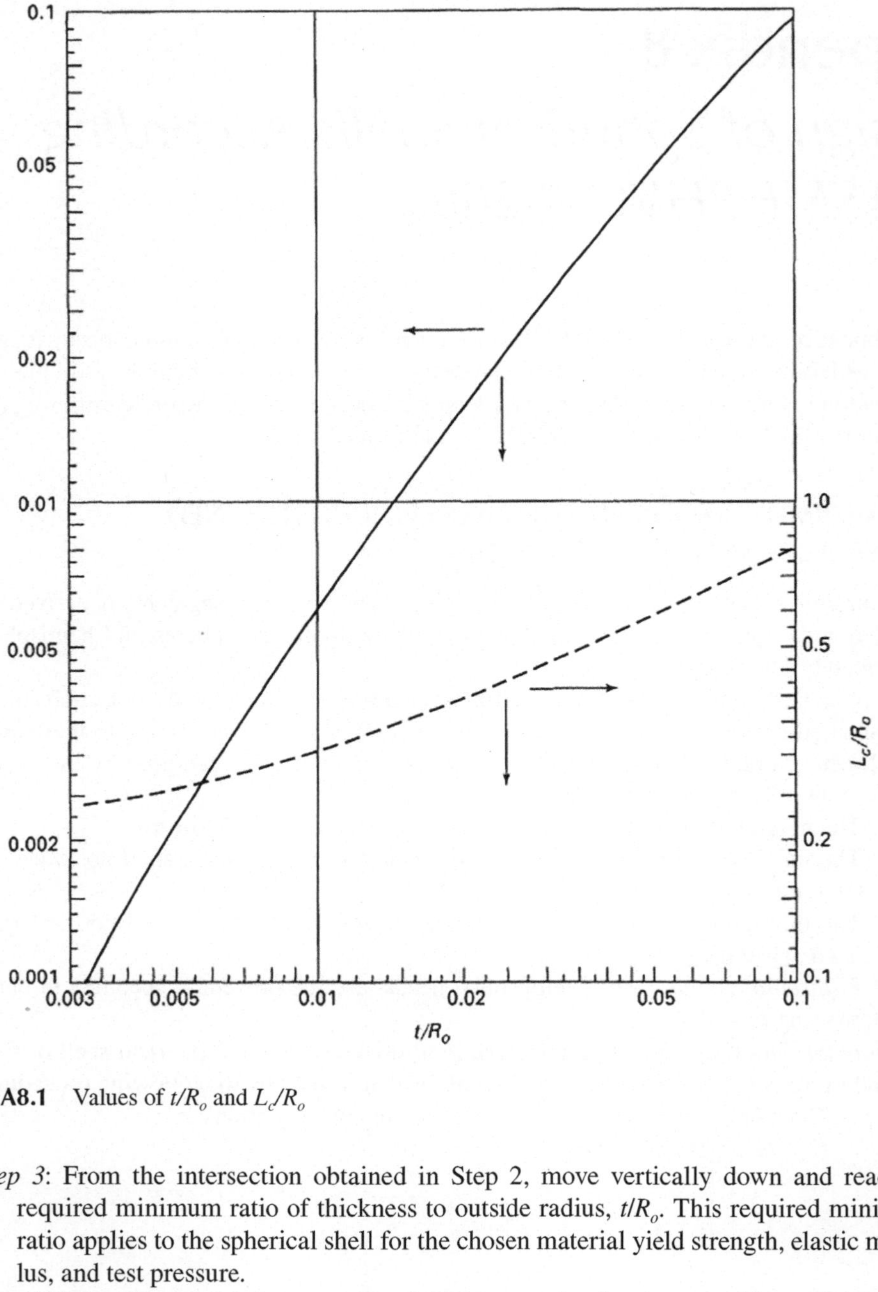

FIGURE A8.1 Values of t/R_o and L_c/R_o

Step 3: From the intersection obtained in Step 2, move vertically down and read the required minimum ratio of thickness to outside radius, t/R_o. This required minimum ratio applies to the spherical shell for the chosen material yield strength, elastic modulus, and test pressure.

Step 4: Determine the minimum required thickness, t, for the given outside radius, R_o. The value of t shall neither be less than 3/8 in. (10 mm) nor greater than 2 in. (50 mm). If the maximum thickness of the spherical shell including corrosion allowance exceeds 2 in. (50 mm), the rules of Section VIII, UG-2S(d) (Division 1) or Article D-3 (Division 2) shall apply.

d. Tolerances for Spherical Shells and Spherical Segments Subjected to External Pressure
   1. *Out-of-Roundness*: The difference between the maximum and minimum inside diameters at any cross section shall not exceed 1% of the nominal inside diameter at the cross section under consideration. The diameters may be measured on the inside or outside of the sphere. If measured on the outside, the diameters shall be corrected

for the plate thickness at the cross section under consideration. When the cross section passes through an opening, the permissible difference in the inside diameters just given may be increased by 2% of the inside diameter of the opening.

2. *Local Shell Tolerances*: The maximum plus or minus deviations from true spherical form, measured radially on the outside or inside of the vessel, shall not exceed 0.5% of the nominal outside radius of the spherical shell and shall not be abrupt. Measurements shall be made from a segmental template having the design inside or outside radius (depending where the measurements are taken) and a chord length, L_c, equal to the arc length determined as follows:

Using the required minimum ratio of thickness to the outside radius, t/R_o obtained in Step 3, move vertically upward on Figure A8.1 to the intersection of the dashed line. Move horizontally to the right from the dashed line and determine the ratio of critical arc length to outside radius, L_c/R_o. The chord length, L_c, is obtained by multiplying this ratio by the outside radius, R_o.

NOMENCLATURE

Do outside diameter of cylinder, in.
E modulus of elasticity of material at design temperature, determined from the applicable material chart in Subpart 2 of Section II, Part D, ksi.
Fy yield strength of material at design metal temperature from applicable table in Subpart 1 of Section II, Part D, ksi.
Lc chord length of template used to measure deviation from nominal circularity, in.
P applied external pressure, psi
PT external test pressure, equal to 1.25*P*, psi.
Ro radius to outside of shell, in.
t thickness of shell, less corrosion allowance, in.

REFERENCE

1. ASME. (2007). "Safety standard for pressure vessels for human occupancy." ASME PVHO-1-2007.

Appendix 9
Abbreviations and Acronyms

ABS	American Bureau of Shipping
AC	alternating current
ADS	atmospheric diving suit
API	American Petroleum Institute
ARE	Admiralty Research Establishment (UK)
ARV	autonomous and remotely operated underwater vehicle
ASAS	Atkins Stress Analysis System
ASME	American Society of Mechanical Engineers
AUV	autonomous underwater vehicle
BCC	body-centred cubic
BCD	body-centred tetragonal
BOSOR5	Buckling of Shells of Revolution
BS	British Standard
CFRP	carbon-fibre-reinforced polymer (plastic)
CG	centre of gravity
CLT	classical laminate theory
COV	coefficient of variation
DC	direct current
DCD	collapse depth
DDD	deep diving depth
DFP	Davidon–Fletcher–Powell method
DGEBA	Bisphenol A (*Di*Glycidyl *E*ther of *B*isphenol *A*)
DNVGL	Det Norske Veritas-Germanischer Lloyd
DOF	degree of freedom
DRDC	Defence Research and Development Canada
DSRV	deep submergence rescue vehicle
DSV	deep submergence vessel
DTMB	David Taylor Model Basin (now David Taylor Naval Ship Research and Development Center – DTNSRDC)
DTS	deep-towed seismic system
ECCS	European Convention on Constructional Steelwork
EEZ	Exclusive Economic Zone
EIPF	extended interior penalty function method
EN	EuroNorm
EN 13445-3	European Norm
FCAW	flux-core arc welding
FD	finite difference
FEA	finite element analysis
FMI	failure mode interaction

FRP	fibre-reinforced polymer (plastic)
FSDT	first-order shear deformation theory
GMNA	geometric and material nonlinear analysis
GRP	glass-reinforced polymer (plastic)
HAZ	heat-affected zone
HOV	human operated (underwater) vehicle
HRC	Hardness Rockwell C
HSLA	high strength low alloy (steel)
HTS	high tensile steel
HWV	heavyweight vehicle
HY	high yield (steel)
HY80	High Yield 80
ISO	International Standards Organisation
KDF	knockdown factor
LARS	launch and recovery system
LEFM	linear elastic fracture mechanics
LRSM	locally reduced stiffness method
LS	Lunchick-Short solution
LSD	limit state design
LWV	lightweight vehicle
MAESTRO	Method for Analysis Evaluation and Structural Optimization
MMC	metal–matrix composite
MUV	manned underwater vehicle
NAVSEA	Naval Sea Systems Command
NEMO	*N*aval *E*xperimental *M*anned *O*bservatory
NLFEA	nonlinear finite element analysis
ODE	ordinary differential equation
OOC	out-of-circularity
PD5500	Published Document 5500
PHP	pressure hull penetrator
PMMA	polymethyl methacrylate (acrylic)
PS	Pulos–Salerno solution
PSF	partial safety factor
PTFE	polytetrafluoroethylene
PVC	polyvinyl chloride
QIN	Quality 1 Nickel-based steel
RMS	root-mean-square
ROUV	remotely operated underwater vehicle
ROV	remotely operated (underwater) vehicle
RSM	reduced stiffness method
SAR	search and rescue
SAW	submerged arc welding
SDF	submarine design formula
SLI	single load indentation
SMD	simultaneous mode design
SOR	shell of revolution
SPLA	single perturbation load analysis

SSE	Submerged signal ejector
SSN	Ship Submersible Nuclear (Nuclear-powered attack submarine)
SUMT	sequential unconstrained minimisation technique
TIG	tungsten inert gas
TMS	tethered management system
UHTS	ultrahigh tensile steel
UK	United Kingdom
US	United States
USN	United States Navy
UUV	unmanned underwater vehicle
UV	underwater vehicle
V&V	validation and verification
VOC	volatile organic compound
WMPLA	worst multiple perturbation load approach
XLUUV	extra large unmanned underwater vehicle

Index

Note: *Italic* page numbers refer to figures and page numbers followed by "n" refer to end notes.